Introductory Algebra

An Applied Approach

Introductory Algebra

An Applied Approach

THIRD EDITION

Richard N. Aufmann

Palomar College, California

Vernon C. Barker

Palomar College, California

HOUGHTON MIFFLIN COMPANY   Boston

Dallas   Geneva, Illinois
Palo Alto   Princeton, New Jersey

Chapter Opener Art designed and rendered by Daniel P. Derdula.

Most Interest Grabber Art designed and illustrated by Daniel P. Derdula, with airbrushing by Linda Phinney on Chapters 6 and 11, and calligraphy by Susan Fong on Chapter 8. Linda Phinney is credited for illustrating Chapter 7, and Networkgraphics is credited for illustrating Chapter 9.

All Interior Math Figures rendered by Networkgraphics (135 Fell Court, Hauppauge, New York 11788).

Cover design by Slide Graphics of New England, Inc.

Interior design by George McLean.

Printed in the U.S.A.

ISBN Numbers:
Text: 0-395-43191-3
Instructor's Annotated Edition: 0-395-57067-0
Solutions Manual: 0-395-57068-9
Instructor's Resource Manual/Testing Program: 0-395-57072-7
Test Bank for the Computerized Test Generator: 0-395-57069-7
Transparencies: 0-395-57070-0
Videos: 0-395-53368-6

BCDEFGHIJ-VH-987654321

Contents

Preface

The third edition of *Introductory Algebra: An Applied Approach* provides mathematically sound and comprehensive coverage of the topics considered essential in an introductory algebra course. Our strategy in preparing this revision has been to build on the successful features of the second edition, features designed to enhance the student's mastery of math skills. In addition, we have expanded the ancillary package for both the instructor and the student by adding transparencies and videos.

Features

The Interactive Approach

Instructors have long recognized the need for a text that requires the student to use a skill as it is being taught. *Introductory Algebra: An Applied Approach* uses an interactive technique that meets this need. Each section is divided into objectives, and every objective contains one or more sets of matched-pair examples. The first example in each set is worked out; the second example is not. By solving this second problem, the student interacts with the text. The complete worked-out solutions to these examples are provided in an appendix at the end of the book, so the student can obtain immediate feedback on and reinforcement of the skill being learned.

Emphasis on Problem-Solving Strategies

Introductory Algebra: An Applied Approach features a carefully developed approach to problem solving that emphasizes developing strategies to solve problems. For each type of word problem contained in the text, the student is prompted to use a "strategy step" before performing the actual manipulation of numbers and variables. By developing problem-solving strategies, the student will know better how to analyze and solve those word problems encountered in an introductory algebra course.

Applications

The traditional approach to teaching or reviewing algebra covers only the straightforward manipulation of numbers and variables and thereby fails to teach students the practical value of algebra. By contrast, *Introductory Algebra: An Applied Approach* emphasizes applications. Wherever appropriate, the last objective in each section presents applications that require the student to use the skills covered in that section to solve practical problems. Also, all of Chapter 4, *"Solving Equations: Applications,"* and portions of several other chapters are devoted entirely to applications. This carefully integrated applied approach generates awareness on the student's part of the value of algebra as a real-life tool.

Complete Integrated Learning System Organized by Objectives

Each chapter begins with a list of the learning objectives included within that chapter. Each of the objectives is then restated in the chapter to remind the student of the current topic of discussion. The same objectives that organize the text organize each ancillary. The Solutions Manual, Computerized Test Generator, Computer Tutor™, Videos, Transparencies, Test Bank, and the Printed Testing Program have all been prepared so that both the student and instructor can easily connect all of the different aids.

Exercises

There are more than 6000 exercises in the text, grouped in the following categories:

- **End-of-section exercise sets,** which are keyed to the corresponding learning objectives, provide ample practice and review of each skill.
- **Chapter review exercises,** which appear at the end of each chapter, help the student integrate all of the skills presented in the chapter.
- **Chapter tests,** which appear at the end of each chapter, are typical one-hour exams that the student can use to prepare for an in-class test.
- **Cumulative review exercises,** which appear at the end of each chapter (beginning with Chapter 2), help the student retain math skills learned in earlier chapters.
- **Final exam,** which follows the last chapter, can be used as a review item or practice final.
- **Calculator exercises,** which appear in the end-of-section exercises, are included throughout the text. These exercises provide the student with the opportunity to practice using a hand-held calculator and are identified by a special color box printed over the exercise number for each calculator exercise.

Calculator and Computer Enrichment Topics

Each chapter also contains optional calculator or computer enrichment topics. Calculator topics provide the student with valuable key-stroking instructions and practice in using a hand-held calculator. Computer topics correspond directly to the programs found on the Math ACE (Additional Computer Exercises) Disk. These topics range from solving first-degree equations to graphing linear equations in two variables.

New To This Edition

Topical Coverage

In Chapter 5, negative exponents are introduced before division of monomials. With this change, a single rule for division of monomials can be stated.

The coverage of factoring has been expanded to include the ac method of factoring as an alternative to factoring by using trial factors.

In Chapter 7, a wider variety of exercises with complex fractions have been added.

The introduction to *Systems of Equations* has been rewritten to emphasize the different types of systems of equations.

The section on radical equations has been enhanced and now includes exercises that require squaring twice to find the solution of an equation.

Keeping with our commitment to applications, over two hundred word problems have been rewritten to reflect contemporary situations.

Chapter Review and Chapter Test

There is now a Chapter Review and a Chapter Test at the end of each chapter. Problems in the Chapter Review are grouped according to the section heads for that chapter. Problems in the Chapter Test are not divided into sections or objectives. Instead the objective references are given in the Answer Section at the back of the book so the student will know which objective to restudy if necessary.

New Testing Program

Both the Computerized Testing program and the Printed Testing Program have been completely rewritten to provide instructors with the option of creating countless new tests.

Supplements For The Student

Two computerized study aids, the Computer Tutor™ and the Math ACE (Additional Computer Exercises) Disk have been carefully designed for the student. The Computer Tutor™ has been expanded to include nine "you-try-it" examples with specific help screens for each lesson.

The COMPUTER TUTOR™

The Computer Tutor™ is an interactive instructional microcomputer program for student use. Each learning objective in the text is supported by a lesson on the Computer Tutor™. As a reminder of this, a small computer icon appears to the right of each objective title in the text. Lessons on the tutor provide additional instruction and practice and can be used in several ways: (1) to cover material the student missed because of absence from class; (2) to repeat instruction on the skill or concept that the student has not yet mastered; or (3) to review material in preparation for examinations. This tutorial program is available for the IBM PC and compatible computers.

Math ACE (Additional Computer Exercises) Disk

The Math ACE Disk contains a number of computational and drill-and-practice programs that correspond to selected Calculator and Computer Enrichment Topics in the text. These programs available for the IBM PC and compatible computers.

Supplements For The Instructor

Introductory Algebra: An Applied Approach has an unusually complete set of teaching aids for the instructor.

Instructor's Annotated Edition

The Instructor's Annotated Edition is an exact replica of the student text except that the answers to all of the exercises are printed in color next to the problems.

Solutions Manual

The Solutions Manual contains worked-out solutions for all end-of-section exercise sets, chapter reviews, chapter tests, cumulative reviews, and the final exam.

Instructor's Resource Manual/Testing Program

The Instructor's Resource Manual/Testing Program contains the printed testing program, which is the first of three sources of testing material available to users of *Introductory Algebra: An Applied Approach*. Eight printed tests (in two formats—free response and multiple choice) are provided for each chapter, as are cumulative and final exams. In addition, the Instructor's Manual includes the documentation for all the software ancillaries (ACE, the Computer Tutor, and the Instructor's Computerized Test Generator) as well as suggested course sequences and class assignments.

Instructor's Computerized Test Generator

The Instructor's Computerized Test Generator is the second source of testing material for use with *Introductory Algebra: An Applied Approach*. The database contains over 1800 new test items. These questions are unique to the test generator and do not repeat items provided in the Instructor's Resource Manual/ Testing Program. Organized according to the keyed objectives in the text, the Test Generator is designed to produce an unlimited number of tests for each chapter of the text, including cumulative tests and final exams. It is available for the IBM PC or compatible computers with editing capabilities for all nongraphic questions.

Printed Test Bank

The Printed Test Bank, the third component of the testing materials, is a printout of all items in the Instructor's Computerized Test Generator. Instructors using the Test Generator can use the test bank to select specific items from the database. Instructors who do not have access to a computer can use the test bank to select items to be included on a test being prepared by hand.

Videotapes

Approximately 33 half-hour videotape lessons accompany *Introductory Algebra: An Applied Approach*. These lessons follow the format and style of the text and are closely tied to specific sections of the text.

Transparencies

Approximately 200 transparencies accompany *Introductory Algebra: An Applied Approach*. These transparencies contain the complete solution to every "you-try-it" example in the text.

Acknowledgements

The authors would like to thank the people who have reviewed this manuscript and provided many valuable suggestions:

Geoffrey Akst
Borough of Manhattan Community College, NY

Betty Jo Baker
Lansing Community College, MI

Judith Brower
North Idaho College, ID

Mary Cabral
Middlesex Community College, MA

Carmy Carranga
Indiana University of Pennsylvania, PA

Patricia Confort
Roger Williams College, RI

Michael Contino
California State University, CA

Bob C. Denton
Orange Coast College, CA

Sharon Edgmon
Bakersfield College, CA

Joel Greenstein
New York Technical College, NY

Bonnie-Lou Guertin
Mt. Wachusett Community College, MA

Lynn Hartsell
Dona Ana Branch Community College, NM

Ida E. Hendricks
Shepherd College, WV

Diana L. Hestwood
Minneapolis Community College, MN

Pamela Hunt
Paris Junior College, TX

Arlene Jesky
Rose State College, OK

Sue Korsak
New Mexico State University, NM

Randy Leifson
Pierce College, WA

Virginia M. Licata
Camden County College, NJ

Judy Liles
North Harris County College–South Campus, TX

David Longshore
Victor Valley College, CA

Margaret Luciano
Broome Community College, NY

Carl C. Maneri
Wright State University, OH

Rudy Maglio
DePaul University, IL

Patricia McCann
Franklin University, OH

Dale Miller
Highland Park Community College, MI

Judy Miller
Delta College, MI

Ellen Milosheff
Triton College, IL

Scott L. Mortensen
Dixie College, UT

Michelle Mosman
Des Moines Area Community College, IA

Linda Murphy
Northern Essex Community College, MA

Wendell Neal
Houston Community College, TX

Doris Nice
University of Wisconsin–Parkside, WI

Kent Pearce
Texas Tech University, TX

Sue Porter
Davenport College, MI

Jack Rotman
Lansing Community College, MI

Karen Schwitters
Seminole Community College, FL

Dorothy Smith
Del Mar College, TX

James T. Sullivan
Massachusetts Bay Community College, MA

Lana Taylor
Siena Heights College, MI

William N. Thomas, Jr.
University of Toledo, OH

Dana Mignogna Thompson
Mount Aloysius Junior College, PA

Paul Treuer
University of Minnesota–Duluth, MN

Thomas Wentland
Columbus College, GA

William T. Wheeler
Abraham Baldwin College, GA

Harvey Wilensky
San Diego Miramar College, CA

Professor Warren Wise
Blue Ridge Community College, VA

Wayne Wolfe
Orange Coast College, CA

Kenneth Word
Central Texas College, TX

Justane Valudez-Ortiz
San Jose College, CA

To the Student

Many students feel that they will never understand math while others appear to do very well with little effort. Often times what makes the difference is that successful students take an active role in the learning process.

Learning mathematics requires your *active* participation. Although doing homework is one way you can actively participate, it is not the only way. First, you must attend class regularly and become an active participant. Secondly, you must become actively involved with the textbook.

Introductory Algebra: An Applied Approach was written and designed with you in mind as a participant. Here are some suggestions on how to use the features of this textbook.

There are 12 chapters in this text. Each chapter is divided into sections and each section is subdivided into learning objectives. Each learning objective is labeled with a letter from A–E.

First, read each objective statement carefully so you will understand the learning goal that is being presented. Next, read the objective material carefully, being sure to note each bold word. These words indicate important concepts that you should familiarize yourself with. Study each in-text example carefully, noting the techniques and strategies used to solve the example.

You will then come to the key learning feature of this text, the *boxed examples*. These examples have been designed to aid you in a very specific way. Notice that in each example box, the example on the left is completely worked out and the example on the right is not. The reason for this is that *you* are expected to work the right-hand example (in the space provided) in order to immediately test your understanding of the material you have just studied.

You should study the worked-out example carefully by working through each step presented. This allows you to focus on each step and reinforces the technique for solving that type of problem. You can then use the worked-out example as a model for solving similar problems.

Then you should solve the right-hand example using the problem solving techniques that you have just studied. When you have completed your solution, check your work by turning to the page in the appendix where the complete solution can be found. The page number on which the solution appears is printed at the bottom of the example box in the right-hand corner. By checking your solution, you will know immediately whether or not you fully understand the skill just studied.

When you have completed studying an objective, do all of the exercises in the exercise set that correspond with that objective. The exercises will be labeled with the same letter as the objective. Algebra is a subject that needs to be learned in small sections and practiced continually in order to be mastered. Doing all of the exercises in each exercise set will help you master the problem solving techniques necessary for success.

Once you have completed the exercises to an objective, you should check your answers to the odd-numbered exercises with those found in the back of the book.

After completing a chapter, read the Chapter Summary. This summary highlights the important topics covered in the chapter. Following the Chapter Summary are Chapter Review Exercises, a Chapter Test, and a Cumulative Review (beginning with Chapter 2). Doing the review exercises is an important way of testing your understanding of the chapter. The answer to each review exercise is in an appendix at the back of the book. Each answer is followed by a reference that tells which objective that exercise was taken from. For example, (4.2B) means Section 4.2, Objective B. After checking your answers, restudy any objective that you missed. It may be very helpful to retry some of the exercises for that objective to reinforce your problem solving techniques.

The Chapter Test should be used to prepare for an exam. We suggest that you try the Chapter Test a few days before your actual exam. Take the test in a quiet place and try to complete the test in the same amount of time you will be allowed for your exam. When taking the Chapter Test, practice the strategies of successful test takers: 1) scan the entire test to get a feel for the questions; 2) read the directions carefully; 3) work the problems that are easiest for you first; and perhaps most importantly, 4) try to stay calm.

When you have completed the Chapter Test, check your answers. If you missed a question, review the material in that objective and rework some of the exercises from that objective. This will strengthen your ability to perform the skills in that objective.

The Cumulative Review allows you to refresh the skills you have learned in previous chapters. This is very important in mathematics. By consistently reviewing previous materials, you will retain the previous skills as you build new ones.

Remember, to be successful, attend class regularly; read the textbook carefully; actively participate in class; work with your textbook using the boxed examples for immediate feedback and reinforcement of each skill; do all the homework assignments; review constantly; and work carefully.

Introductory Algebra

An Applied Approach

1

Real Numbers

OBJECTIVES

▶ To use the inequality symbols < and > with integers
▶ To use opposites and absolute value
▶ To add integers
▶ To subtract integers
▶ To multiply integers
▶ To divide integers
▶ To solve application problems
▶ To write a rational number as a decimal
▶ To add or subtract rational numbers
▶ To multiply or divide rational numbers
▶ To evaluate exponential expressions
▶ To use the Order of Operations Agreement to simplify expressions

Early Egyptian Fractions

One of the earliest written documents of mathematics is the Rhind Papyrus. This tablet was found in Egypt in 1858, but it is estimated that the writings date back to 1650 B.C.

The Rhind Papyrus contains over 80 problems. A study of these problems has enabled mathematicians and scientists to understand some of the methods by which the early Egyptians used mathematics.

Evidence gained from the Papyrus shows that the Egyptian method of calculating with fractions was much different from the methods used today. All fractions were represented in terms of what are called "unit fractions." A unit fraction is a fraction in which the numerator is one. This fraction was symbolized (using modern numbers) with a bar over the number.

For example, $\overline{3} = \dfrac{1}{3}$; $\qquad \overline{15} = \dfrac{1}{15}$

The early Egyptians also tended to deal with powers of two (2, 4, 8, 16, . . .). As a result, representing fractions with a two in the numerator in terms of unit fractions was an important matter. The Rhind Papyrus has a table giving the equivalent unit fractions for all odd denominators from 5 to 101 and 2 as the numerator. Some of these are listed below.

$$\frac{2}{5} = \overline{3}\ \overline{15} \qquad \left(\frac{2}{5} = \frac{1}{3} + \frac{1}{15}\right)$$

$$\frac{2}{7} = \overline{4}\ \overline{28}$$

$$\frac{2}{11} = \overline{6}\ \overline{66}$$

$$\frac{2}{19} = \overline{12}\ \overline{76}\ \overline{114}$$

SECTION 1.1   # Introduction to Integers

Objective A   **To use the inequality symbols < and > with integers**

The **natural numbers** are 1, 2, 3, 4, 5, 6, 7, 8, ...

The three dots mean that the list continues on and on and that there is no largest natural number.

The **integers** are . . . −4, −3, −2, −1, 0, 1, 2, 3, 4, ...

Each integer can be shown on a number line. The **graph** of an integer is shown by placing a heavy dot on the number line directly above the number. The graph of −4 is shown below.

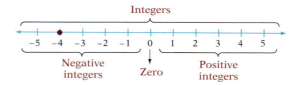

The integers to the left of zero on the number line are called **negative integers.** The integers to the right of zero are called **positive integers** or natural numbers. Zero is neither a positive nor a negative number.

Just as the word "it" is used in language to stand for an object, a letter of the alphabet can be used in mathematics to stand for a number. Such a letter is called a **variable.**

A number line can be used to visualize the relative order of two integers. If a and b are two integers, and a is to the left of b on the number line, then a is **less than** b ($a < b$). If a is to the right of b on the number line, then a is **greater than** b ($a > b$).

Negative 4 is less than negative 1.

$-4 < -1$

5 is greater than 0.

$5 > 0$

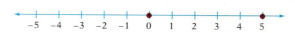

Example 1   Place the correct symbol, < or >, between the two numbers.
a. −17   6   b. −30   −3

Solution   a. $-17 < 6$ b. $-30 < -3$

Example 2   Place the correct symbol, < or >, between the two numbers.
a. 5   −13   b. −8   −22

Your solution   a. $5 > -13$ b. $-8 > -22$

Solution on p. A5

Objective B

To use opposites and absolute value

The numbers 5 and -5 are the same distance from zero but on opposite sides of zero. The numbers 5 and -5 are called **opposites** or **additive inverses** of each other.

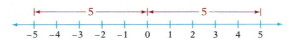

The opposite or additive inverse of 5 is -5. The opposite or additive inverse of -5 is 5.

The **absolute value** of a number is its distance from zero on the number line. Therefore, the absolute value of a number is a positive number or zero. The symbol for absolute value is | |.

The distance from 0 to 3 is 3.
Therefore, the absolute value of 3 is 3.

$|3| = 3$

The distance from 0 to -3 is 3.
Therefore, the absolute value of -3 is 3.

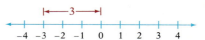

$|-3| = 3$

The absolute value of a positive number is the number itself. The absolute value of a negative number is the opposite of the negative number. The absolute value of zero is zero.

Evaluate $-|-7|$.

The absolute value sign does not affect the negative sign in front of the absolute value sign.

$-|-7| = -7$

Example 3   Find the opposite number.
a. 6   b. -51

Solution   a. -6 b. 51

Example 4   Find the opposite number.
a. -9 b. 62

Your solution   a. 9   b. -62

Example 5   Evaluate $|-4|$ and $-|-10|$.

Solution $|-4| = 4$

$-|-10| = -10$

Example 6   Evaluate $|-5|$ and $-|-9|$.

Your solution   5
-9

Solutions on p. A5

1.1 EXERCISES

▶ **Objective A**

Place the correct symbol, < or >, between the two numbers.

1. $3 < 5$ **2.** $7 > 4$ **3.** $-2 > -5$ **4.** $-6 < -1$ **5.** $-16 < 1$

6. $-2 < 13$ **7.** $3 > -7$ **8.** $5 > -6$ **9.** $-11 < -8$ **10.** $-4 > -10$

11. $-1 > -6$ **12.** $-9 < -4$ **13.** $0 > -3$ **14.** $8 > 0$ **15.** $6 > -8$

16. $8 > -6$ **17.** $-14 < 16$ **18.** $-12 < 1$ **19.** $35 > 28$ **20.** $42 > 19$

21. $-42 < 27$ **22.** $-36 < 49$ **23.** $21 > -34$ **24.** $53 > -46$ **25.** $-27 > -39$

26. $-51 < -20$ **27.** $-87 < 63$ **28.** $-75 < 92$ **29.** $68 > -79$ **30.** $95 > -71$

31. $-62 > -84$ **32.** $-91 < -70$ **33.** $94 > 83$ **34.** $76 < 81$ **35.** $59 > -67$

36. $48 > -66$ **37.** $-93 < -55$ **38.** $-64 > -86$ **39.** $-88 < 57$ **40.** $-58 < 82$

41. $0 < 129$ **42.** $-136 < 0$ **43.** $-131 < 101$ **44.** $127 > -150$ **45.** $-194 < -180$

▶ **Objective B**

Find the opposite number.

46. 4
        −4

47. 16
        −16

48. −2
        2

49. −3
        3

50. 22
        −22

51. 45
        −45

52. −31
        31

53. −88
        88

Evaluate.

54. $|2|$
        2

55. $|-2|$
        2

56. $|-6|$
        6

57. $|6|$
        6

58. $|8|$
        8

59. $|5|$
        5

60. $|-9|$
        9

61. $|-1|$
        1

62. $-|-1|$
        −1

63. $-|-8|$
        −8

64. $-|-5|$
        −5

65. $-|0|$
        0

66. $|16|$
        16

67. $|19|$
        19

68. $|-12|$
        12

69. $|-22|$
        22

70. $-|29|$
        −29

71. $-|20|$
        −20

72. $-|-14|$
        −14

73. $-|-18|$
        −18

74. $|-15|$
        15

75. $|-23|$
        23

76. $-|33|$
        −33

77. $-|27|$
        −27

78. $-|-36|$
        −36

79. $-|-41|$
        −41

80. $|32|$
        32

81. $|25|$
        25

82. $|-38|$
        38

83. $|-30|$
        30

84. $-|37|$
        −37

85. $-|34|$
        −34

86. $-|-42|$
        −42

87. $-|-45|$
        −45

88. $|44|$
        44

89. $|36|$
        36

90. $|-74|$
        74

91. $|-61|$
        61

92. $-|88|$
        −88

93. $-|52|$
        −52

94. $-|-81|$
        −81

95. $-|-93|$
        −93

96. $|-107|$
        107

97. $|-119|$
        119

Addition and Subtraction of Integers

Objective A

To add integers

A number can be represented anywhere along the number line by an arrow. A positive number is represented by an arrow pointing to the right, and a negative number is represented by an arrow pointing to the left. The size of a number is represented by the length of the arrow.

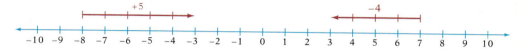

Addition of integers can be shown on the number line. To add integers, find the point on the number line corresponding to the first addend. At that point draw an arrow representing the second addend. The sum is the number directly below the tip of the arrow.

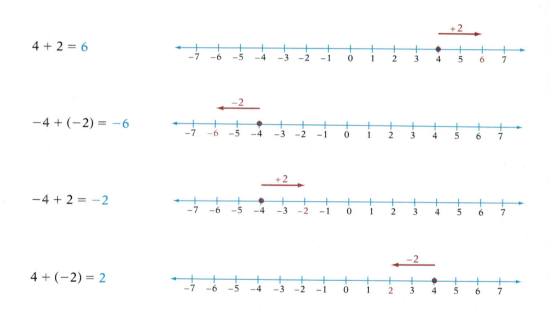

$4 + 2 = 6$

$-4 + (-2) = -6$

$-4 + 2 = -2$

$4 + (-2) = 2$

The pattern for addition shown on the number lines above is summarized in the following rules for adding integers.

Addition of Integers with the Same Sign

To add numbers with the same sign, add the absolute values of the numbers. Then attach the sign of the addends.

$$2 + 8 = 10 \qquad\qquad -2 + (-8) = -10$$

Addition of Integers with Different Signs

> To add numbers with different signs, subtract the absolute value of the smaller number from the absolute value of the larger number. Then attach the sign of the number with the greater absolute value.

$$-2 + 8 = 6 \qquad\qquad 2 + (-8) = -6$$

Add: $162 + (-247)$

Because the signs are different, subtract the absolute value of the smaller number from the absolute value of the larger number. Then attach the sign of the number with the greater absolute value.

$162 + (-247)$
-85

Add: $-14 + (-47)$

Because the signs are the same, add the absolute values of the numbers. Then attach the sign of the addends.

$-14 + (-47)$
-61

Add: $-4 + (-6) + (-8) + 9$

To add more than two numbers, add the first two numbers. Then add the sum to the third number. Continue until all the numbers have been added.

$$-4 + (-6) + (-8) + 9$$
$$-10 \qquad + (-8) + 9$$
$$-18 \qquad\quad + 9$$
$$-9$$

Example 1   Add: $-162 + 98$

Solution $-162 + 98$
-64

Example 2   Add: $-154 + (-37)$

Your solution -191

Example 3   Add: $42 + (-12) + (-30)$

Solution $42 + (-12) + (-30)$
$30 + (-30)$
0

Example 4   Add: $-36 + 17 + (-21)$

Your solution -40

Example 5   Add: $-2 + (-7) + 4 + (-6)$

Solution $-2 + (-7) + 4 + (-6)$
$-9 + 4 + (-6)$
$-5 + (-6)$
-11

Example 6   Add: $-5 + (-2) + 9 + (-3)$

Your solution -1

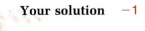

Solutions on p. A5

Objective B   **To subtract integers**

Look at the two expressions below and note that each expression equals the same number.

$$8 - 3 = 5 \qquad \text{8 minus 3 is 5}$$
$$8 + (-3) = 5 \qquad \text{8 plus the opposite of 3 is 5}$$

This example suggests that to subtract two numbers, add the opposite of the second number to the first number.

first number	−	second number	=	first number	+	the opposite of the second number	
40	−	60	=	40	+	(−60)	= −20
−40	−	60	=	−40	+	(−60)	= −100
−40	−	(−60)	=	−40	+	60	= 20
40	−	(−60)	=	40	+	60	= 100

Here are two more examples.

Subtract: $-17 - (-20)$

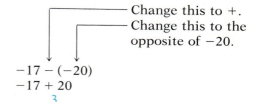

Change the subtraction symbol to addition, and change the sign of the second number. Then add.

$$-17 - (-20)$$
$$-17 + 20$$
$$3$$

Subtract: $9 - 16$

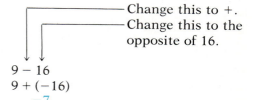

Change the subtraction symbol to addition, and change the sign of the second number. Then add.

$$9 - 16$$
$$9 + (-16)$$
$$-7$$

When subtraction occurs several times in a problem, rewrite each subtraction as addition of the opposite. Then add.

Subtract: $-12 - 4 - (-15)$

Rewrite each subtraction as addition of the opposite. Add.

$$-12 - 4 - (-15)$$
$$-12 + (-4) + 15$$
$$-16 \qquad + 15$$
$$-1$$

Example 7   Subtract: $-12 - 8$

Solution $-12 - 8$
$-12 + (-8)$
-20

Example 8   Subtract: $-8 - 14$

Your solution -22

Example 9   Subtract: $8 - 6 - (-20)$

Solution $8 - 6 - (-20)$
$8 + (-6) + 20$
$2 + 20$
22

Example 10   Subtract: $3 - (-4) - 15$

Your solution -8

Example 11   Subtract:
$-8 - 30 - (-12) - 7 - (-14)$

Solution $-8 - 30 - (-12) - 7 - (-14)$
$-8 + (-30) + 12 + (-7) + 14$
$-38 + 12 + (-7) + 14$
$-26 + (-7) + 14$
$-33 + 14$
-19

Example 12   Subtract:
$4 - (-3) - 12 - (-7) - 20$

Your solution -18

Example 13   Subtract:
$12 - 12 - (-3) - 5 - 7$

Solution $12 - 12 - (-3) - 5 - 7$
$12 + (-12) + 3 + (-5) + (-7)$
$0 + 3 + (-5) + (-7)$
$3 + (-5) + (-7)$
$-2 + (-7)$
-9

Example 14   Subtract:
$17 - 10 - 2 - (-6) - 9$

Your solution 2

Solutions on p. A5

| 1.2 | **EXERCISES** |

▶ **Objective A**

Add.

1. $3 + (-5)$
-2

2. $-4 + 2$
-2

3. $8 + 12$
20

4. $16 + 23$
39

5. $-3 + (-8)$
-11

6. $-12 + (-1)$
-13

7. $-4 + (-5)$
-9

8. $-12 + (-12)$
-24

9. $6 + (-9)$
-3

10. $4 + (-9)$
-5

11. $-6 + 7$
1

12. $-12 + 6$
-6

13. $2 + (-3) + (-4)$
-5

14. $7 + (-2) + (-8)$
-3

15. $-3 + (-12) + (-15)$
-30

16. $9 + (-6) + (-16)$
-13

17. $-17 + (-3) + 29$
9

18. $13 + 62 + (-38)$
37

19. $-3 + (-8) + 12$
1

20. $-27 + (-42) + (-18)$
-87

21. $13 + (-22) + 4 + (-5)$
-10

22. $-14 + (-3) + 7 + (-6)$
-16

23. $-22 + 10 + 2 + (-18)$
-28

24. $-6 + (-8) + 13 + (-4)$
-5

25. $-16 + (-17) + (-18) + 10$
-41

26. $-25 + (-31) + 24 + 19$
-13

27. $-126 + (-247) + (-358) + 339$
-392

28. $-651 + (-239) + 524 + 487$
121

▶ **Objective B**

Subtract.

29. $16 - 8$
8

30. $12 - 3$
9

31. $7 - 14$
-7

32. $6 - 9$
-3

33. $-7 - 2$
-9

34. $-9 - 4$
-13

35. $7 - (-2)$
9

36. $3 - (-4)$
7

37. $-6 - (-3)$
-3

38. $-4 - (-2)$
-2

39. $6 - (-12)$
18

40. $-12 - 16$
-28

41. $-4 - 3 - 2$
-9

42. $4 - 5 - 12$
-13

43. $12 - (-7) - 8$
11

44. $-12 - (-3) - (-15)$
6

45. $4 - 12 - (-8)$
0

46. $13 - 7 - 15$
-9

47. $-6 - (-8) - (-9)$
11

48. $7 - 8 - (-1)$
0

49. $-30 - (-65) - 29 - 4$
2

50. $42 - (-82) - 65 - 7$
52

51. $-16 - 47 - 63 - 12$
-138

52. $42 - (-30) - 65 - (-11)$
18

53. $-47 - (-67) - 13 - 15$
-8

54. $-18 - 49 - (-84) - 27$
-10

55. $167 - 432 - (-287) - 359$
-337

56. $-521 - (-350) - 164 - (-299)$
-36

SECTION 1.3   Multiplication and Division of Integers

Objective A

To multiply integers

Multiplication is the repeated addition of the same number. For example,

$$5 \text{ is added 3 times}$$
$$3 \times 5 = 5 \ + \ 5 \ + \ 5 = 15$$

To indicate multiplication, several different symbols are used: a ×, a raised dot, and parentheses.

$$3 \times 5 = 15$$
$$3 \cdot 5 = 15$$
$$(3)(5) = 15$$

In the last example, note that when parentheses surround two expressions and there is no arithmetic operation symbol, you are to assume the operation is multiplication.

Now consider the product $3 \times (-5)$.

$$(-5) \text{ is added 3 times}$$
$$3 \times (-5) = (-5) \ + \ (-5) \ + \ (-5) = -15$$

This example suggests that the product of a positive number and a negative number is negative. Here are a few more examples.

$$(-4) \cdot 6 = -24$$
$$(7)(-8) = -56$$
$$(-12) \times 21 = -252$$

To find the product of two negative numbers, look at the following pattern. When -5 is multiplied by a sequence of decreasing integers, each product is 5 greater than the one before it.

$$-5 \times 3 = -15$$
$$-5 \times 2 = -10$$
$$-5 \times 1 = -5$$
$$-5 \times 0 = 0$$

The pattern developed can be continued so that -5 is multiplied by a sequence of negative numbers. The resulting products must be positive in order to maintain the pattern of increasing by 5.

$$-5 \times (-1) = 5$$
$$-5 \times (-2) = 10$$
$$-5 \times (-3) = 15$$
$$-5 \times (-4) = 20$$

This illustrates that the product of two negative numbers is positive.

The pattern for multiplication shown above is summarized in the following rules for multiplying integers.

Multiplication of Integers with the Same Sign

To multiply two numbers with the same sign, multiply the absolute values of the factors. The product is positive.

$$4 \cdot 8 = 32 \qquad\qquad (-4)(-8) = 32$$

Multiplication of Integers with Different Signs

To multiply two numbers with different signs, multiply the absolute values of the factors. The product is negative.

$$-4 \cdot 8 = -32 \qquad\qquad 4(-8) = -32$$

Multiply: $2(-3)(-5)(-7)$

To multiply more than two numbers, multiply the first two numbers. Then multiply the product by the third number. Continue until all the numbers have been multiplied.

$2(-3)(-5)(-7)$

$-6 \cdot (-5)(-7)$

$30 \cdot (-7)$

-210

Example 1   Multiply: $(-2) \cdot 3 \cdot 6$

Solution $(-2) \cdot 3 \cdot 6$
$-6 \cdot 6$
-36

Example 2   Multiply: $(-3) \cdot 4 \cdot (-5)$

Your solution   60

Example 3   Multiply: $-42 \cdot 62$

Solution $-42 \cdot 62$
-2604

Example 4   Multiply: $-38 \cdot 51$

Your solution -1938

Example 5   Multiply: $-2 \cdot (-10) \cdot 7 \cdot 12$

Solution $-2 \cdot (-10) \cdot 7 \cdot 12$
$20 \cdot 7 \cdot 12$
$140 \cdot 12$
1680

Example 6   Multiply: $-6 \cdot 8 \cdot (-11) \cdot 3$

Your solution   1584

Example 7   Multiply: $-5(-4)(6)(-3)$

Solution $-5(-4)(6)(-3)$
$20(6)(-3)$
$120(-3)$
-360

Example 8   Multiply: $-7(-8)(9)(-2)$

Your solution -1008

Solutions on p. A5

| Objective B | **To divide integers** | |

For every division problem there is a related multiplication problem.

$$\frac{8}{2} = 4 \qquad \text{because} \qquad 2 \cdot 4 = 8$$

Division   Related multiplication

This fact and the rules for multiplying integers can be used to illustrate the rules for dividing integers.

Note in the following examples that the quotient of two numbers with the same sign is positive.

$$\frac{12}{3} = 4 \text{ because } 4 \cdot 3 = 12 \qquad \frac{-12}{-3} = 4 \text{ because } 4 \cdot (-3) = -12$$

The next two examples illustrate that the quotient of two numbers with different signs is negative.

$$\frac{12}{-3} = -4 \text{ because } (-4)(-3) = 12 \qquad \frac{-12}{3} = -4 \text{ because } (-4) \cdot 3 = -12$$

Division of Integers with the Same Sign

The quotient of two numbers with the same sign is positive.

Division of Integers with Different Signs

The quotient of two numbers with different signs is negative.

Note that $\frac{12}{-3}$, $\frac{-12}{3}$, and $-\frac{12}{3}$ are all equal to -4.

If a and b are two integers, and $b \neq 0$, then $\frac{a}{-b} = \frac{-a}{b} = -\frac{a}{b}$.

Properties of Zero and One in Division

Zero divided by any number other than zero is zero.

$\frac{0}{a} = 0$ because $0 \cdot a = 0$

Division by zero is not defined.

$\frac{4}{0} = ?$ $? \times 0 = 4$
There is no number whose product with zero is 4.

Any number other than zero divided by itself is 1.

$\frac{a}{a} = 1$ because $1 \cdot a = a$

Example 9   Divide: $(-120) \div (-8)$

Solution $(-120) \div (-8)$
15

Example 10   Divide: $(-135) \div (-9)$

Your solution   15

Example 11   Divide: $-81 \div 3$

Solution $-81 \div 3$
-27

Example 12   Divide: $-72 \div 4$

Your solution -18

Example 13   Divide: $95 \div (-5)$

Solution $95 \div (-5)$
-19

Example 14   Divide: $84 \div (-6)$

Your solution -14

Solutions on p. A5

Objective C   To solve application problems

To solve an application problem, first read the problem carefully. The *Strategy* involves identifying the quantity to be found and planning the steps that are necessary to find that quantity. The *Solution* involves performing each operation stated in the Strategy and writing the answer.

Example 15

The average temperature on the sunlit side of the moon is approximately 215°F. On the dark side, it is approximately −250°F. Find the difference between these average temperatures.

Strategy

To find the difference, subtract the average temperature on the dark side of the moon (−250°) from the average temperature on the sunlit side (215°).

Solution

$215 - (-250)$
$215 + 250$
465

The difference is 465°F.

Example 16

The average temperature throughout the earth's stratosphere is −70°F. The average temperature on the earth's surface is 57°F. Find the difference between these average temperatures.

Your strategy

Your solution

127°F

Example 17

The daily high temperatures (in Celsius) during one week were recorded as follows: −9°, 3°, 0°, −8°, 2°, 1°, 4°. Find the average daily high temperature for the week.

Strategy

To find the average daily high temperature:
- Add the seven temperature readings.
- Divide by 7.

Solution

$-9 + 3 + 0 + (-8) + 2 + 1 + 4$
$-6 + 0 + (-8) + 2 + 1 + 4$
$-6 + (-8) + 2 + 1 + 4$
$-14 + 2 + 1 + 4$
$-12 + 1 + 4$
$-11 + 4$
-7

$-7 \div 7 = -1$

The average daily high temperature was −1°C.

Example 18

The daily low temperatures (in Celsius) during one week were recorded as follows: −6°, −7°, 1°, 0°, −5°, −10°, −1°. Find the average daily low temperature for the week.

Your strategy

Your solution

−4°C

Solutions on p. A6

1.3   EXERCISES

▶ **Objective A**

Multiply.

1. 14×3
42

2. 62×9
558

3. $-4 \cdot 6$
-24

4. $-7 \cdot 3$
-21

5. $-2 \cdot (-3)$
6

6. $-5 \cdot (-1)$
5

7. $(9)(2)$
18

8. $(3)(8)$
24

9. $5(-4)$
-20

10. $4(-7)$
-28

11. $-8(2)$
-16

12. $-9(3)$
-27

13. $(-5)(-5)$
25

14. $(-3)(-6)$
18

15. $(-7)(0)$
0

16. -32×4
-128

17. -24×3
-72

18. $19 \cdot (-7)$
-133

19. $6(-17)$
-102

20. $-8(-26)$
208

21. $-4(-35)$
140

22. $-5 \cdot (23)$
-115

23. $-6 \cdot (38)$
-228

24. $9(-27)$
-243

25. $8(-40)$
-320

26. $-7(-34)$
238

27. $-4(39)$
-156

28. $4 \cdot (-8) \cdot 3$
-96

29. $5 \times 7 \times (-2)$
-70

30. $8 \cdot (-6) \cdot (-1)$
48

31. $(-9)(-9)(2)$
162

32. $-8(-7)(-4)$
-224

33. $-5(8)(-3)$
120

34. $(-6)(5)(7)$
-210

35. $-1(4)(-9)$
36

36. $6(-3)(-2)$
36

37. $4(-4) \cdot 6(-2)$
192

38. $-5 \cdot 9(-7) \cdot 3$
945

39. $-9(4) \cdot 3(1)$
-108

40. $8(8)(-5)(-4)$
1280

41. $(-6) \cdot 7 \cdot (-10)(-5)$
-2100

42. $-9(-6)(11)(-2)$
-1188

43. $-6(-5)(12)(0)$
0

44. $7(9) \cdot 10 \cdot (-1)$
-630

45. $-19(28)(-43)(-11)$
$-251,636$

46. $-65(13)(-47)(-92)$
$-3,653,780$

▶ **Objective B**

Divide.

47. $12 \div (-6)$
-2

48. $18 \div (-3)$
-6

49. $(-72) \div (-9)$
8

50. $(-64) \div (-8)$
8

51. $0 \div (-6)$
0

52. $-49 \div 7$
-7

53. $45 \div (-5)$
-9

54. $-24 \div 4$
-6

55. $-36 \div 4$
-9

56. $-56 \div 7$
-8

57. $-81 \div (-9)$
9

58. $-40 \div (-5)$
8

59. $72 \div (-3)$
-24

60. $44 \div (-4)$
-11

61. $-60 \div 5$
-12

62. $-66 \div 6$
-11

63. $-93 \div (-3)$
31

64. $-98 \div (-7)$
14

65. $(-85) \div (-5)$
17

66. $(-60) \div (-4)$
15

67. $120 \div 8$
15

68. $144 \div 9$
16

69. $78 \div (-6)$
-13

70. $84 \div (-7)$
-12

71. $-72 \div 4$
-18

72. $-80 \div 5$
-16

73. $-114 \div (-6)$
19

74. $-91 \div (-7)$
13

75. $-104 \div (-8)$
13

76. $-126 \div (-9)$
14

77. $57 \div (-3)$
-19

78. $162 \div (-9)$
-18

79. $-136 \div (-8)$
17

80. $-128 \div 4$
-32

81. $-130 \div (-5)$
26

82. $(-280) \div 8$
-35

83. $(-92) \div (-4)$
23

84. $-196 \div (-7)$
28

85. $-150 \div (-6)$
25

86. $(-261) \div 9$
-29

Divide.

87. $204 \div (-6)$
    -34

88. $165 \div (-5)$
    -33

89. $-132 \div (-12)$
    11

90. $-156 \div (-13)$
    12

91. $-182 \div 14$
    -13

92. $-144 \div 12$
    -12

93. $143 \div 11$
    13

94. $168 \div 14$
    12

95. $-180 \div (-15)$
    12

96. $-169 \div (-13)$
    13

97. $154 \div (-11)$
    -14

98. $274,883 \div 367$
    749

99. $398,750 \div 1375$
    290

100. $841,662 \div 2461$
    342

▶ **Objective C**   *Application Problems*

The elevation, or height of places on earth is measured in relation to sea level, or the average level of the ocean's surface. The following table shows height above sea level as a positive number, depth below sea level as a negative number.

Place	*Elevation (in feet)*
Mt. Everest	29,028
Mt. Aconcagua	23,035
Mt. McKinley	20,320
Mt. Kilimanjaro	19,340
Salinas Grandes	−131
Death Valley	−282
Qattara Depression	−436
Dead Sea	−1286

101. Use the table to find the difference in elevation between Mt. McKinley and Death Valley (the highest and lowest points in North America).
    20,602 ft

102. Use the table to find the difference in elevation between Mt. Kilimanjaro and the Qattara Depression (the highest and lowest points in Africa).
    19,776 ft

103. Use the table to find the difference in elevation between Mt. Everest and the Dead Sea (the highest and lowest points in Asia).
    30,314 ft

104. Use the table to find the difference in elevation between Mt. Aconcagua and Salinas Grandes (the highest and lowest points in South America).
23,166 ft

105. The daily low temperatures (in Celsius) during one week were recorded as follows: 5°, −4°, 9°, 0°, −11°, −13°, −7°. Find the average daily low temperature for the week.
−3°C

106. The daily high temperatures (in Fahrenheit) during one week were recorded as follows: −7°, −10°, 4°, 6°, −2°, −8°, −4°. Find the average daily high temperature for the week.
−3° F

107. During one day of a fall month, the temperature in Palm Springs, California, was 82° F. On the same day, the temperature in Anchorage, Alaska, was −17° F. Find the difference between the temperature in Palm Springs and that in Anchorage.
99° F

108. In one year, the coldest temperature in the United States was −68° F and the warmest temperature was 127° F. Find the difference between the coldest and the warmest temperature that year.
−195° F

109. Accountants frequently indicate a negative number by placing parentheses around the number. For example, ($1400) indicates a loss of $1400. At the beginning of June, a compact disc manufacturer had a balance of ($36,291). At the beginning of July, the manufacturer had a balance of $28,774. Find the difference between the balance in July and the balance in June.
$65,065

110. Accountants frequently indicate a negative number by placing parentheses around the number. For example, ($1400) indicates a loss of $1400. The beginning balance in January for a car stereo store was ($12,540). The ending balance in January was ($15,005). Find the difference between the ending and the beginning balance for January.
($2465)

SECTION 1.4 Rational Numbers

Objective A   To write a rational number as a decimal

A **rational number** is the quotient of two integers. A rational number written in this way is commonly called a fraction. Here are some examples of rational numbers

$$\frac{2}{3}, \frac{-4}{9}, \frac{18}{-5}, \frac{4}{1}$$

Rational Numbers

> A rational number is a number that can be written in the form $\frac{a}{b}$, where a and b are integers and b is not zero.

Because an integer can be written as the quotient of the integer and 1, every integer is a rational number.

$$5 = \frac{5}{1} \qquad -3 = \frac{-3}{1}$$

A number written in **decimal notation** is also a rational number.

three-tenths $0.3 = \frac{3}{10}$

thirty-five hundredths $0.35 = \frac{35}{100}$

negative four-tenths $-0.4 = -\frac{4}{10}$

A rational number written as a fraction can be written in decimal notation.

Write $\frac{5}{8}$ as a decimal.

The fraction bar can be read "÷".

$$\frac{5}{8} = 5 \div 8$$

```
  0.625   ← This is called
8)5.000       a terminating
 -4 8         decimal.
   20
  -16
   40
  -40
    0   ← The remainder
            is zero.
```

$$\frac{5}{8} = 0.625$$

Write $\frac{4}{11}$ as a decimal.

```
   0.3636. . .  ← This is called
11)4.0000          a repeating
  -3 3             decimal.
    70
   -66
    40
   -33
    70
   -66
     4   ← The remainder
             is never zero.
```

$$\frac{4}{11} = 0.\overline{36}$$ The bar over the digits 3 and 6 is used to show that these digits repeat.

Every rational number can be written as a terminating or a repeating decimal.

Some numbers—for example, $\sqrt{2}$ and π, have decimal representations that never terminate or repeat. These numbers are called **irrational numbers.**

$$\sqrt{2} = 1.414213562\ldots \qquad \pi = 3.141592654\ldots$$

The rational numbers and the irrational numbers taken together are called the **real numbers.**

Example 1   Write $\frac{3}{20}$ as a decimal.

Solution
$$
\begin{array}{r}
0.15 \\
20\overline{)3.00} \\
-2\,0 \\
\hline
1\,00 \\
-1\,00 \\
\hline
0
\end{array}
$$

$\dfrac{3}{20} = 0.15$

Example 2   Write $\frac{4}{25}$ as a decimal.

Your solution   0.16

Example 3   Write $\frac{3}{22}$ as a decimal. Place a bar over the repeating digits of the decimal.

Solution
$$
\begin{array}{r}
0.13636 \\
22\overline{)3.00000} \\
-2\,2 \\
\hline
80 \\
-66 \\
\hline
140 \\
-132 \\
\hline
80 \\
-66 \\
\hline
140 \\
-132 \\
\hline
8
\end{array}
$$

$\dfrac{3}{22} = 0.1\overline{36}$

Example 4   Write $\frac{4}{9}$ as a decimal. Place a bar over the repeating digits of the decimal.

Your solution $0.\overline{4}$

Solution on p. A6

| **Objective B** | **To add or subtract rational numbers** |

To add or subtract fractions, first rewrite the fractions as equivalent fractions with a common denominator, using the least common multiple (LCM) of the denominators as the common denominator. Then add the numerators and place the sum over the common denominator. Write the answer in simplest form.

Simplify $-\frac{5}{6} + \frac{3}{10}$.

| Find the LCM of the denominators 6 and 10. The LCM of denominators is sometimes called the **least common denominator** (LCD). | Rewrite the fractions as equivalent fractions, using the LCM of the denominators as the common denominator. | Add the numerators, and place the sum over the common denominator. | Write the answer in simplest form. |

The LCM = 30.

$$-\frac{5}{6} + \frac{3}{10} = -\frac{25}{30} + \frac{9}{30} = \frac{-25 + 9}{30} = \frac{-16}{30} = -\frac{8}{15}$$

Simplify $-\frac{5}{9} - \frac{7}{12}$.

First find the LCM. The LCM = 36. Rewrite the fraction as equivalent fractions, using the LCM of the denominators as the common denominator.

$$-\frac{5}{9} - \frac{7}{12} = -\frac{20}{36} - \frac{21}{36} = \frac{-20}{36} + \frac{-21}{36} = \frac{-20 + (-21)}{36} = \frac{-41}{36} = -\frac{41}{36}$$

To add or subtract decimals, write the numbers so that the decimal points are in a vertical line. Then proceed as in the addition or subtraction of integers. Write the decimal point in the answer directly below the decimal points in the problem.

Simplify $14.02 + 137.6 + 9.852$.

Write the decimals so that the decimal points are in a vertical line.

Write the decimal point in the sum directly below the decimal points in the problem.

$$\begin{array}{r} 14.02 \\ 137.6 \\ +\ \ 9.852 \\ \hline 161.472 \end{array}$$

Simplify $-114.039 + 84.76$.

Because the signs are different, subtract the absolute value of the smaller number from the absolute value of the larger number.

$$\begin{array}{r} \overset{\scriptstyle 10\ \ 13\ \ \ \ 9}{\cancel{1}\ \cancel{1}\ 4\,.\,\cancel{0}\ \cancel{3}\ 9} \\ -\ \ \ 8\ 4\,.\,7\ 6 \\ \hline 2\ 9\,.\,2\ 7\ 9 \end{array}$$

Attach the sign of the number with the greater absolute value.

$-114.039 + 84.76 = -29.279$

Example 5 Simplify $\frac{5}{16} - \frac{7}{40}$.

Solution The LCM of 16 and 40 is 80.

$$\frac{5}{16} - \frac{7}{40} = \frac{25}{80} - \frac{14}{80}$$
$$= \frac{25}{80} + \frac{-14}{80}$$
$$= \frac{25 + (-14)}{80}$$
$$= \frac{11}{80}$$

Example 6 Simplify $\frac{5}{9} - \frac{11}{12}$.

Your solution $-\frac{13}{36}$

Example 7 Simplify $-\frac{3}{4} + \frac{1}{6} - \frac{5}{8}$.

Solution The LCM of 4, 6, and 8 is 24.

$$-\frac{3}{4} + \frac{1}{6} - \frac{5}{8}$$
$$= -\frac{18}{24} + \frac{4}{24} - \frac{15}{24}$$
$$= \frac{-18}{24} + \frac{4}{24} + \frac{-15}{24}$$
$$= \frac{-18 + 4 + (-15)}{24}$$
$$= \frac{-29}{24} = -\frac{29}{24}$$

Example 8 Simplify $-\frac{7}{8} - \frac{5}{6} + \frac{1}{2}$.

Your solution $-\frac{29}{24}$

Example 9 Simplify 4.027 + 19.66 + 3.09.

Solution

$$\begin{array}{r} 4.027 \\ 19.66 \\ +\ 3.09 \\ \hline 26.777 \end{array}$$

Example 10 Simplify 3.907 + 4.9 + 6.63.

Your solution 15.437

Example 11 Simplify 42.987 − 98.61.

Solution Find the difference between the absolute values.

$$\begin{array}{r} 98.61 \\ -42.987 \\ \hline 55.623 \end{array}$$

42.987 − 98.61 = −55.623

Example 12 Simplify 16.127 − 67.91.

Your solution −51.783

Example 13 Simplify 1.02 + (−3.6) + 9.24.

Solution 1.02 + (−3.6) + 9.24
−2.58 + 9.24
6.66

Example 14 Simplify 2.7 + (−9.44) + 6.2.

Your solution −0.54

Solution on p. A6

Objective C   To multiply or divide rational numbers

The product of two fractions is the product of the numerators divided by the product of the denominators.

Simplify $\frac{3}{8} \times \frac{12}{17}$.

Multiply the numerators.
Multiply the denominators.

$$\frac{3}{8} \times \frac{12}{17} = \frac{3 \cdot 12}{8 \cdot 17}$$

Write the prime factorization of each factor. Divide by the common factors.

$$= \frac{3 \cdot \overset{1}{\cancel{2}} \cdot \overset{1}{\cancel{2}} \cdot 3}{2 \cdot \underset{1}{\cancel{2}} \cdot \underset{1}{\cancel{2}} \cdot 17}$$

Multiply the factors remaining in the numerator and denominator.

$$= \frac{9}{34}$$

To divide fractions, invert the divisor. Then proceed as in the multiplication of fractions.

Simplify $\frac{3}{10} \div \left(-\frac{18}{25}\right)$.

The signs are different. The quotient is negative.

$$\frac{3}{10} \div \left(-\frac{18}{25}\right) = -\left(\frac{3}{10} \div \frac{18}{25}\right) = -\left(\frac{3}{10} \times \frac{25}{18}\right) = -\left(\frac{3 \cdot 25}{10 \cdot 18}\right) = -\left(\frac{\overset{1}{\cancel{3}} \cdot \overset{1}{\cancel{5}} \cdot 5}{2 \cdot \cancel{5} \cdot 2 \cdot \cancel{3} \cdot 3}\right) = -\frac{5}{12}$$

To multiply decimals, multiply as with integers. Write the decimal point in the product so that the number of decimal places in the product equals the sum of the decimal places in the factors.

Simplify -6.89×0.00035.

The signs are different.
Multiply the absolute values.

$$\begin{array}{rl} 6.89 & \text{2 decimal places} \\ \times 0.00035 & \text{5 decimal places} \\ \hline 3445 & \\ 2067 & \\ \hline 0.0024115 & \text{7 decimal places} \end{array}$$

The product is negative. $-6.89 \times 0.00035 = -0.0024115$

To divide decimals, move the decimal point in the divisor far enough to make the divisor a whole number. Move the decimal point in the dividend the same number of places to the right. Place the decimal point in the quotient directly over the decimal point in the dividend. Then divide as with whole numbers.

Simplify $1.32 \div 0.27$. Round to the nearest tenth.

$$0.27\overline{)1.32}$$

Move the decimal point 2 places to the right in the divisor and then in the dividend. Place the decimal point in the quotient.

$$\begin{array}{r} 4.88 \approx 4.9 \\ 27\overline{)132.00} \\ -108 \\ \hline 240 \\ -216 \\ \hline 240 \\ -216 \\ \hline 24 \end{array}$$

The symbol $\approx$ is used to indicate that the quotient is an approximate value that has been rounded off.

Example 15 Simplify $\frac{2}{3} \times \left(-\frac{9}{10}\right)$.

Solution The product is negative.

$$\frac{2}{3} \times \left(-\frac{9}{10}\right) = -\left(\frac{2}{3} \times \frac{9}{10}\right) =$$

$$-\frac{2 \cdot 9}{3 \cdot 10} = -\frac{\overset{1}{\cancel{2}} \cdot \overset{1}{\cancel{3}} \cdot 3}{\underset{1}{\cancel{3}} \cdot \underset{1}{\cancel{2}} \cdot 5} = -\frac{3}{5}$$

Example 16 Simplify $-\frac{7}{12} \times \frac{9}{14}$.

Your solution $-\frac{3}{8}$

Example 17 Simplify $-\frac{5}{8} \div \left(-\frac{5}{40}\right)$.

Solution The quotient is positive.

$$-\frac{5}{8} \div \left(-\frac{5}{40}\right) = \frac{5}{8} \div \frac{5}{40} =$$

$$\frac{5}{8} \times \frac{40}{5} = \frac{5 \cdot 40}{8 \cdot 5} =$$

$$\frac{\overset{1}{\cancel{5}} \cdot \overset{1}{\cancel{2}} \cdot \overset{1}{\cancel{2}} \cdot \overset{1}{\cancel{2}} \cdot 5}{\underset{1}{\cancel{2}} \cdot \underset{1}{\cancel{2}} \cdot \underset{1}{\cancel{2}} \cdot \underset{1}{\cancel{5}}} = \frac{5}{1} = 5$$

Example 18 Simplify $-\frac{3}{8} \div \left(-\frac{5}{12}\right)$.

Your solution $\frac{9}{10}$

Example 19 Simplify -4.29×8.2.

Solution The product is negative.

$$\begin{array}{r} 4.29 \\ \times 8.2 \\ \hline 858 \\ 3432 \\ \hline 35.178 \end{array}$$

$$-4.29 \times 8.2 = -35.178$$

Example 20 Simplify -5.44×3.8.

Your solution -20.672

Example 21 Simplify $-3.2 \times (-0.4) \times 6.9$.

Solution $-3.2 \times (-0.4) \times 6.9$
1.28×6.9
8.832

Example 22 Simplify $3.44 \times (-1.7) \times 0.6$.

Your solution -3.5088

Example 23 Simplify $-0.0792 \div (-0.42)$. Round to the nearest hundredth.

Solution

$$\begin{array}{r} 0.188 \approx 0.19 \\ 0.42\overline{)0.07\,920} \\ \underline{-4\,2} \\ 3\,72 \\ \underline{-3\,36} \\ 360 \\ \underline{-336} \\ 24 \end{array}$$

$$-0.0792 \div (-0.42) \approx 0.19$$

Example 24 Simplify $-0.394 \div 1.7$. Round to the nearest hundredth.

Your solution -0.23

Solutions on pp. A6–A7

1.4   EXERCISES

▶ **Objective A**

Write as a decimal. Place a bar over the repeating digits of a repeating decimal.

1. $\dfrac{1}{3}$ **2.** $\dfrac{2}{3}$ **3.** $\dfrac{1}{4}$ **4.** $\dfrac{3}{4}$ **5.** $\dfrac{2}{5}$

$0.\overline{3}$ $0.\overline{6}$ 0.25 0.75 0.4

6. $\dfrac{4}{5}$ **7.** $\dfrac{1}{6}$ **8.** $\dfrac{5}{6}$ **9.** $\dfrac{1}{8}$ **10.** $\dfrac{7}{8}$

0.8 $0.1\overline{6}$ $0.8\overline{3}$ 0.125 0.875

11. $\dfrac{2}{9}$ **12.** $\dfrac{8}{9}$ **13.** $\dfrac{5}{11}$ **14.** $\dfrac{10}{11}$ **15.** $\dfrac{7}{12}$

$0.\overline{2}$ $0.\overline{8}$ $0.\overline{45}$ $0.\overline{90}$ $0.583\overline{3}$

16. $\dfrac{11}{12}$ **17.** $\dfrac{4}{15}$ **18.** $\dfrac{8}{15}$ **19.** $\dfrac{9}{16}$ **20.** $\dfrac{15}{16}$

$0.91\overline{6}$ $0.2\overline{6}$ $0.5\overline{3}$ 0.5625 0.9375

21. $\dfrac{7}{18}$ **22.** $\dfrac{17}{18}$ **23.** $\dfrac{1}{20}$ **24.** $\dfrac{13}{20}$ **25.** $\dfrac{6}{25}$

$0.3\overline{8}$ $0.9\overline{4}$ 0.05 0.65 0.24

26. $\dfrac{14}{25}$ **27.** $\dfrac{7}{30}$ **28.** $\dfrac{19}{30}$ **29.** $\dfrac{9}{40}$ **30.** $\dfrac{21}{40}$

0.56 $0.2\overline{3}$ $0.6\overline{3}$ 0.225 0.525

31. $\dfrac{13}{36}$ **32.** $\dfrac{29}{36}$ **33.** $\dfrac{15}{22}$ **34.** $\dfrac{19}{22}$ **35.** $\dfrac{11}{24}$

$0.36\overline{1}$ $0.80\overline{5}$ $0.68\overline{1}$ $0.8\overline{63}$ $0.4583\overline{3}$

36. $\dfrac{19}{24}$ **37.** $\dfrac{5}{33}$ **38.** $\dfrac{25}{33}$ **39.** $\dfrac{3}{37}$ **40.** $\dfrac{14}{37}$

$0.791\overline{6}$ $0.\overline{15}$ $0.\overline{75}$ $0.\overline{081}$ $0.\overline{378}$

▶ **Objective B**

Simplify.

41. $\dfrac{2}{3} + \dfrac{5}{12}$

$\dfrac{13}{12}$

42. $\dfrac{1}{2} + \dfrac{3}{8}$

$\dfrac{7}{8}$

43. $\dfrac{5}{8} - \dfrac{5}{6}$

$-\dfrac{5}{24}$

44. $\dfrac{1}{9} - \dfrac{5}{27}$

$-\dfrac{2}{27}$

45. $-\dfrac{5}{12} - \dfrac{3}{8}$

$-\dfrac{19}{24}$

46. $-\dfrac{5}{6} - \dfrac{5}{9}$

$-\dfrac{25}{18}$

47. $-\dfrac{6}{13} + \dfrac{17}{26}$

$\dfrac{5}{26}$

48. $-\dfrac{7}{12} + \dfrac{5}{8}$

$\dfrac{1}{24}$

49. $-\dfrac{5}{8} - \left(-\dfrac{11}{12}\right)$

$\dfrac{7}{24}$

50. $\dfrac{1}{3} + \dfrac{5}{6} - \dfrac{2}{9}$

$\dfrac{17}{18}$

51. $\dfrac{1}{2} - \dfrac{2}{3} + \dfrac{1}{6}$

0

52. $-\dfrac{3}{8} - \dfrac{5}{12} - \dfrac{3}{16}$

$-\dfrac{47}{48}$

53. $-\dfrac{5}{16} + \dfrac{3}{4} - \dfrac{7}{8}$

$-\dfrac{7}{16}$

54. $\dfrac{1}{2} - \dfrac{3}{8} - \left(-\dfrac{1}{4}\right)$

$\dfrac{3}{8}$

55. $\dfrac{3}{4} - \left(-\dfrac{7}{12}\right) - \dfrac{7}{8}$

$\dfrac{11}{24}$

56. $\dfrac{1}{3} - \dfrac{1}{4} - \dfrac{1}{5}$

$-\dfrac{7}{60}$

57. $\dfrac{2}{3} - \dfrac{1}{2} + \dfrac{5}{6}$

1

58. $\dfrac{5}{16} + \dfrac{1}{8} - \dfrac{1}{2}$

$-\dfrac{1}{16}$

59. $\dfrac{5}{8} - \left(-\dfrac{5}{12}\right) + \dfrac{1}{3}$

$\dfrac{11}{8}$

60. $\dfrac{1}{8} - \dfrac{11}{12} + \dfrac{1}{2}$

$-\dfrac{7}{24}$

61. $-\dfrac{7}{9} + \dfrac{14}{15} + \dfrac{8}{21}$

$\dfrac{169}{315}$

62. $1.09 + 6.2$

7.29

63. $-32.1 - 6.7$

-38.8

64. $5.13 - 8.179$

-3.049

65. $-13.092 + 6.9$

-6.192

66. $2.54 - 3.6$

-1.06

67. $5.43 + 7.925$

13.355

68. $-16.92 - 6.925$

-23.845

69. $-3.87 + 8.546$

4.676

70. $6.9027 - 17.692$

-10.7893

Simplify.

71. $2.09 - 6.72 - 5.4$

-10.03

72. $16.4 + 3.09 - 7.93$

11.56

73. $-18.39 + 4.9 - 23.7$

-37.19

74. $19 - (-3.72) - 82.75$

-60.03

75. $-3.07 - (-2.97) - 17.4$

-17.5

76. $-3.09 - 4.6 - 27.3$

-34.99

77. $317.09 - 46.902 + 583.0714$

853.2594

78. $71.0235 - 86.0974 + 254.309$

239.2351

▶ **Objective C**

Simplify.

79. $\dfrac{1}{2} \times \left(-\dfrac{3}{4}\right)$

$-\dfrac{3}{8}$

80. $-\dfrac{2}{9} \times \left(-\dfrac{3}{14}\right)$

$\dfrac{1}{21}$

81. $\left(-\dfrac{3}{8}\right)\left(-\dfrac{4}{15}\right)$

$\dfrac{1}{10}$

82. $\left(-\dfrac{3}{4}\right)\left(-\dfrac{8}{27}\right)$

$\dfrac{2}{9}$

83. $-\dfrac{1}{2} \times \dfrac{8}{9}$

$-\dfrac{4}{9}$

84. $\dfrac{5}{12} \times \left(-\dfrac{8}{15}\right)$

$-\dfrac{2}{9}$

85. $\dfrac{5}{8} \times \left(-\dfrac{7}{12}\right) \times \dfrac{16}{25}$

$-\dfrac{7}{30}$

86. $\left(\dfrac{1}{2}\right)\left(-\dfrac{3}{4}\right)\left(-\dfrac{5}{8}\right)$

$\dfrac{15}{64}$

87. $\left(\dfrac{5}{12}\right)\left(-\dfrac{8}{15}\right)\left(-\dfrac{1}{3}\right)$

$\dfrac{2}{27}$

88. $\dfrac{3}{8} \div \dfrac{1}{4}$

$\dfrac{3}{2}$

89. $\dfrac{5}{6} \div \left(-\dfrac{3}{4}\right)$

$-\dfrac{10}{9}$

90. $-\dfrac{5}{12} \div \dfrac{15}{32}$

$-\dfrac{8}{9}$

91. $-\dfrac{7}{8} \div \dfrac{4}{21}$

$-\dfrac{147}{32}$

92. $\dfrac{7}{10} \div \dfrac{2}{5}$

$\dfrac{7}{4}$

93. $-\dfrac{15}{64} \div \left(-\dfrac{3}{40}\right)$

$\dfrac{25}{8}$

Simplify.

94. $\dfrac{1}{8} \div \left(-\dfrac{5}{12}\right)$ **95.** $-\dfrac{4}{9} \div \left(-\dfrac{2}{3}\right)$ **96.** $-\dfrac{6}{11} \div \dfrac{4}{9}$ **97.** 1.2×3.47

$-\dfrac{3}{10}$ $\dfrac{2}{3}$ $-\dfrac{27}{22}$ 4.164

98. -0.8×6.2 **99.** $(-1.89)(-2.3)$ **100.** $(6.9)(-4.2)$

-4.96 4.347 -28.98

101. $1.06 \times (-3.8)$ **102.** $-2.7 \times (-3.5)$ **103.** $1.2 \times (-0.5) \times 3.7$

-4.028 9.45 -2.22

104. $-2.4 \times 6.1 \times 0.9$ **105.** $(-0.8)(3.006)(-5.1)$ **106.** $(-3.4)(-0.08)(1.06)$

-13.176 12.26448 0.28832

Simplify. Round to the nearest hundredth.

107. $-24.7 \div 0.09$ **108.** $-1.27 \div (-1.7)$ **109.** $9.07 \div (-3.5)$ **110.** $0.0976 \div 0.042$

-274.44 0.75 -2.59 2.32

111. $-6.904 \div 1.35$ **112.** $-7.894 \div (-2.06)$ **113.** $-354.2086 \div 0.1719$

-5.11 3.83 -2060.55

114. $-2658.3109 \div (-0.0473)$ **115.** $(-3.92)(-27.1)(45.008)$ **116.** $-0.461 \times 0.087 \times (-9.675)$

$56{,}201.08$ 4781.29 0.39

Content and Format © 1991 HMCo.

SECTION 1.5

Exponents and the Order of Operations Agreement

Objective A

To evaluate exponential expressions

Repeated multiplication of the same factor can be written using an exponent.

$$2 \cdot 2 \cdot 2 \cdot 2 \cdot 2 = 2^5 \leftarrow \textbf{exponent}$$
$$\underset{\textstyle\raisebox{0pt}{\textbf{base}}}{\uparrow}$$

$$a \cdot a \cdot a \cdot a = a^4 \leftarrow \textbf{exponent}$$
$$\underset{\textstyle\raisebox{0pt}{\textbf{base}}}{\uparrow}$$

The **exponent** indicates how many times the factor, called the **base,** occurs in the multiplication. The multiplication $2 \cdot 2 \cdot 2 \cdot 2 \cdot 2$ is in **factored form.** The exponential expression 2^5 is in **exponential form.**

2^1 is read "the first power of two" or just "two."   Usually the exponent 1 is not written.

2^2 is read "the second power of two" or "two squared."

2^3 is read "the third power of two" or "two cubed."

2^4 is read "the fourth power of two."

2^5 is read "the fifth power of two."

a^5 is read "the fifth power of a."

To evaluate an exponential expression, write each factor as many times as indicated by the exponent. Then multiply.

$$3^5 = 3 \cdot 3 \cdot 3 \cdot 3 \cdot 3 = 243$$

$$2^3 \cdot 3^2 = (2 \cdot 2 \cdot 2) \cdot (3 \cdot 3) = 8 \cdot 9 = 72$$

Evaluate $(-4)^2$ and -4^2.

The -4 is squared only when the negative sign is *inside* parentheses.

$$(-4)^2 = (-4)(-4) = 16$$

$$-4^2 = -(4 \times 4) = -16$$

Evaluate $(-2)^4$ and $(-2)^5$.

The product of an even number of negative factors is positive.

$$(-2)^4 = (-2)(-2)(-2)(-2) = 4(-2)(-2)$$
$$= -8(-2) = 16$$

The product of an odd number of negative factors is negative.

$$(-2)^5 = (-2)(-2)(-2)(-2)(-2) = 4(-2)(-2)(-2)$$
$$= -8(-2)(-2) = 16(-2) = -32$$

Example 1   Evaluate -5^3.

Solution $-5^3 = -(5 \cdot 5 \cdot 5) = -125$

Example 2   Evaluate -6^3.

Your solution -216

Example 3   Evaluate $(-4)^4$.

Solution $(-4)^4 = (-4)(-4)(-4)(-4)$
   $= 256$

Example 4   Evaluate $(-3)^4$.

Your solution 81

Example 5   Evaluate $(-3)^2 \cdot 2^3$

Solution $(-3)^2 \cdot 2^3 = (-3)(-3) \cdot (2)(2)(2)$
   $= 9 \cdot 8 = 72$

Example 6   Evaluate $(3^3)(-2)^3$.

Your solution -216

Example 7   Evaluate $\left(-\frac{2}{3}\right)^3$.

Solution $\left(-\frac{2}{3}\right)^3 = \left(-\frac{2}{3}\right)\left(-\frac{2}{3}\right)\left(-\frac{2}{3}\right)$
   $= -\frac{2 \cdot 2 \cdot 2}{3 \cdot 3 \cdot 3} = -\frac{8}{27}$

Example 8   Evaluate $\left(-\frac{2}{5}\right)^2$.

Your solution $\frac{4}{25}$

Example 9   Evaluate $-4(0.7)^2$.

Solution $-4(0.7)^2 = -4(0.7)(0.7)$
   $= -2.8(0.7) = -1.96$

Example 10   Evaluate $-3(0.3)^3$.

Your solution -0.081

Solutions on p. A7

Objective B

To use the Order of Operations Agreement to simplify expressions

Evaluate $2 + 3 \cdot 5$.

There are two arithmetic operations, addition and multiplication, in this expression. The operations could be performed in different orders.

Add first. $\underline{2 + 3} \cdot 5$

Then multiply. $\underline{5 \quad \cdot 5}$

   25

Multiply first. $2 + \underline{3 \cdot 5}$

Then add. $\underline{2 + \quad 15}$

   17

In order to prevent more than one answer for a numerical expression, an Order of Operations Agreement has been established.

The Order of Operations Agreement

Step 1   Perform operations inside grouping symbols. Grouping symbols include parentheses ( ), brackets [ ], and the fraction bar.

Step 2   Simplify exponential expressions.

Step 3   Do multiplication and division as they occur from left to right.

Step 4   Do addition and subtraction as they occur from left to right.

Evaluate $12 - 24(8 - 5) \div 2^2$.

1. Perform operations inside grouping symbols.

2. Simplify exponential expressions.

3. Do multiplication and division as they occur from left to right.

4. Do addition and subtraction as they occur from left to right.

$12 - 24\underbrace{(8 - 5)} \div 2^2$

$12 - 24(3) \div \underbrace{2^2}$

$12 - \underbrace{24(3)} \div 4$

$12 - \underbrace{72 \div 4}$

$12 - \quad 18$

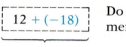
$\boxed{12 + (-18)}$ Do this step
  -6 mentally.

One or more of the above steps may not be needed to evaluate an expression. In that case, proceed to the next step in the Order of Operations Agreement.

Evaluate $\frac{4 + 8}{2 + 1} - (3 - 1) + 2$.

1. Perform operations inside grouping symbols.

3. Do multiplication and division as they occur from left to right.

4. Do addition and subtraction as they occur from left to right.

$\underbrace{\frac{4 + 8}{2 + 1}} - \underbrace{(3 - 1)} + 2$

$\underbrace{\frac{12}{3}} - \quad 2 \quad + 2$

$4 - 2 + 2$

$\boxed{4 + (-2) + 2}$ Do this step
                         mentally.

$\underbrace{2 \qquad + 2}$

$\qquad 4$

When an expression has grouping symbols inside grouping symbols, perform the operations inside the inner grouping symbols first.

Evaluate $6 \div [4 - (6 - 8)] + 2^2$.

1. Perform operations inside grouping symbols.

$6 \div [4 - \underbrace{(6 - 8)}] + 2^2$

$6 \div [4 - (-2)] + 2^2$

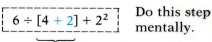
$\boxed{6 \div [4 + 2] + 2^2}$ Do this step
                                mentally.

2. Simplify exponential expressions.

$6 \div \quad 6 \quad + 2^2$

3. Do multiplication and division as they occur from left to right.

$\underbrace{6 \div \quad 6} \quad + 4$

4. Do addition and subtraction as they occur from left to right.

$\underbrace{1 \qquad + 4}$

$\qquad 5$

Example 11   Evaluate
$4 - 3[4 - 2(6 - 3)] \div 2.$

Solution $4 - 3[4 - 2(6 - 3)] \div 2$
$4 - 3[4 - 2 \cdot 3] \div 2$
$4 - 3[4 - 6] \div 2$
$4 - 3[-2] \div 2$
$4 + 6 \div 2$
$4 + 3$
7

Example 12   Evaluate
$18 - 5[8 - 2(2 - 5)] \div 10.$

Your solution   11

Example 13   Evaluate
$27 \div (5 - 2)^2 + (-3)^2 \cdot 4.$

Solution $27 \div (5 - 2)^2 + (-3)^2 \cdot 4$
$27 \div 3^2 + (-3)^2 \cdot 4$
$27 \div 9 + 9 \cdot 4$
$3 + 9 \cdot 4$
$3 + 36$
39

Example 14   Evaluate
$36 \div (8 - 5)^2 - (-3)^2 \cdot 2.$

Your solution -14

Example 15   Evaluate
$(1.75 - 1.3)^2 \div 0.025 + 6.1.$

Solution $(1.75 - 1.3)^2 \div 0.025 + 6.1$
$(0.45)^2 \div 0.025 + 6.1$
$0.2025 \div 0.025 + 6.1$
$8.1 + 6.1$
14.2

Example 16   Evaluate
$(6.97 - 4.72)^2 \times 4.5 \div 0.05.$

Your solution   455.625

Example 17   Evaluate $\frac{5}{8} - \left(\frac{2}{5} - \frac{1}{2}\right) \div \left(\frac{2}{3}\right)^2.$

Solution $\frac{5}{8} - \left(\frac{2}{5} - \frac{1}{2}\right) \div \left(\frac{2}{3}\right)^2$

$\frac{5}{8} - \left(-\frac{1}{10}\right) \div \left(\frac{2}{3}\right)^2$

$\frac{5}{8} + \frac{1}{10} \div \frac{4}{9}$

$\frac{5}{8} + \frac{1}{10} \cdot \frac{9}{4}$

$\frac{5}{8} + \frac{9}{40}$

$\frac{25}{40} + \frac{9}{40}$

$\frac{34}{40} = \frac{17}{20}$

Example 18   Evaluate $\frac{5}{8} \div \left(\frac{1}{3} - \frac{3}{4}\right) + \frac{7}{12}.$

Your solution $-\frac{11}{12}$

Solutions on p. A7

Content and Format © 1991 HMCo.

1.5 EXERCISES

▶ **Objective A**

Evaluate.

1. 6^2
36

2. 7^4
2401

3. -7^2
-49

4. -4^3
-64

5. $(-3)^2$
9

6. $(-2)^3$
-8

7. $(-3)^4$
81

8. $(-5)^3$
-125

9. $\left(\dfrac{1}{2}\right)^2$
$\dfrac{1}{4}$

10. $\left(-\dfrac{3}{4}\right)^3$
$-\dfrac{27}{64}$

11. $(0.3)^2$
0.09

12. $(1.5)^3$
3.375

13. $\left(\dfrac{2}{3}\right)^2 \cdot 3^3$
12

14. $\left(-\dfrac{1}{2}\right)^3 \cdot 8$
-1

15. $(0.3)^3 \cdot 2^3$
0.216

16. $(0.5)^2 \cdot 3^3$
6.75

17. $(-3) \cdot 2^2$
-12

18. $(-5) \cdot 3^4$
-405

19. $(-2) \cdot (-2)^3$
16

20. $(-2) \cdot (-2)^2$
-8

21. $2^3 \cdot 3^3 \cdot (-4)$
-864

22. $(-3)^3 \cdot 5^2 \cdot 10$
-6750

23. $(-7) \cdot 4^2 \cdot 3^2$
-1008

24. $(-2) \cdot 2^3 \cdot (-3)^2$
-144

25. $\left(\dfrac{2}{3}\right)^2 \cdot \dfrac{1}{4} \cdot 3^3$
3

26. $\left(\dfrac{3}{4}\right)^2 \cdot (-4) \cdot 2^3$
-18

27. $8^2 \cdot (-3)^5 \cdot 5$
$-77,760$

▶ **Objective B**

Evaluate by using the Order of Operations Agreement.

28. $4 - 8 \div 2$
0

29. $2^2 \cdot 3 - 3$
9

30. $2(3 - 4) - (-3)^2$
-11

31. $16 - 32 \div 2^3$
12

32. $24 - 18 \div 3 + 2$
20

33. $8 - (-3)^2 - (-2)$
1

Evaluate by using the Order of Operations Agreement.

34. $16 + 15 \div (-5) - 2$
11

35. $14 - 2^2 - (4 - 7)$
13

36. $3 - 2[8 - (3 - 2)]$
−11

37. $-2^2 + 4[16 \div (3 - 5)]$
−36

38. $6 + \dfrac{16 - 4}{2^2 + 2} - 2$
6

39. $24 \div \dfrac{3^2}{8 - 5} - (-5)$
13

40. $96 \div 2[12 + (6 - 2)] - 3^3$
741

41. $4 \cdot [16 - (7 - 1)] \div 10$
4

42. $16 \div 2 - 4^2 - (-3)^2$
−17

43. $18 \div (9 - 2^3) + (-3)$
15

44. $16 - 3(8 - 3)^2 \div 5$
1

45. $4(-8) \div [2(7 - 3)^2]$
−1

46. $\dfrac{(-10) + (-2)}{6^2 - 30} \div (2 - 4)$
1

47. $16 - 4 \cdot \dfrac{3^3 - 7}{2^3 + 2} - (-2)^2$
4

48. $(0.2)^2 \cdot (-0.5) + 1.72$
1.7

49. $0.3(1.7 - 4.8) + (1.2)^2$
0.51

50. $(1.8)^2 - 2.52 \div (1.8)$
1.84

51. $(1.65 - 1.05)^2 \div 0.4 + 0.8$
1.7

52. $\left(\dfrac{1}{2}\right)^2 - \left(\dfrac{3}{4} - \dfrac{1}{2}\right)$
0

53. $\dfrac{3}{8} \div \left(\dfrac{5}{6} + \dfrac{2}{3}\right)$
$\dfrac{1}{4}$

54. $\left(\dfrac{5}{12} - \dfrac{9}{16}\right) \cdot \dfrac{3}{7}$
$-\dfrac{1}{16}$

55. $\left(\dfrac{3}{4}\right)^2 - \left(\dfrac{1}{2}\right)^3 \div \dfrac{3}{5}$
$\dfrac{17}{48}$

Calculators and Computers

Extended Precision on an Electronic Calculator

Consider the decimal equivalents of the following fractions:

$$\frac{7}{33} = 0.\overline{21} \qquad \text{and} \qquad \frac{15}{37} = 0.\overline{405}$$

These decimal equivalents were calculated using an electronic calculator that displays 7 decimal places. Some fractions, however, do not have decimal equivalents that repeat until well after 7 places.

Examples of fractions with repeating cycles that are longer than 7 places include

$$\frac{4}{17} = 0.\overline{2352941176470588} \qquad \text{and} \qquad \frac{9}{23} = 0.\overline{3913043478260869565217}$$

A calculator that displays 7 decimal places was used to determine each of the above decimal equivalents. The procedure for these calculations is illustrated below.

Find the repeating decimal expression for $\frac{8}{17}$.

1. Divide the numerator by the denominator. $\qquad \frac{8}{17} = 0.4705882$

    The decimal approximation is the first four digits.

    $$\frac{8}{17} = 0.4705$$

2. Take the last three digits (as a decimal) and form the product with the denominator.

    $$0.882 \times 17 = 14.994 \approx 15$$

    Round this to the nearest integer and divide by the denominator.

    $$\frac{15}{17} = 0.8823529$$

    The new decimal approximation is the approximation from Step 1 and the first four digits from Step 2.

    $$\frac{8}{17} = 0.47058823$$

3. Repeat Step 2. Continue to repeat Step 2 until the decimal representation repeats.

    $$0.529 \times 17 = 8.993 \approx 9 \quad 0.117 \times 17 = 1.989 \approx 2 \quad 0.470 \times 17 = 7.99 \approx 8$$

    $$\frac{9}{17} \approx 0.5294117 \qquad\qquad \frac{2}{17} \approx 0.1176470 \qquad\qquad \frac{8}{17} \approx 0.4705882$$

    $$\frac{8}{17} \approx 0.470588235294 \qquad \frac{8}{17} \approx 0.4705882352941176$$

    The decimal begins to repeat with the digits 4705.

    The decimal equivalent of $\frac{8}{17}$ is $0.\overline{4705882352941176}$.

Exercises: Find the decimal equivalents of each of the following:

1. $\frac{3}{17}$ **2.** $\frac{10}{19}$ **3.** $\frac{10}{23}$

1. $0.\overline{1764705882352941}$ 2. $0.\overline{526315789473684210}$ 3. $0.\overline{4347826086956521739130}$

Chapter Summary

Key Words   The *natural numbers* are 1, 2, 3, 4, 5, 6, 7 . . . .

The *integers* are . . . , −4, −3, −2, −1, 0, 1, 2, 3, 4 . . . .

The *absolute value* of a number is its distance from zero on a number line.

A *rational number* is a number of the form $\frac{a}{b}$, where a and b are integers and b is not equal to zero. A rational number written in this form is commonly called a *fraction*.

An *irrational number* is a number that has a decimal representation that never terminates or repeats.

An expression of the form a^n is in *exponential form*, where a is the base and n is the exponent.

Essential Rules

Addition of Integers with the Same Sign	To add two numbers with the same sign, add the absolute values of the numbers. Then attach the sign of the addends.
Addition of Integers with Different Signs	To add two numbers with different signs, subtract the absolute value of the smaller number from the absolute value of the larger number. Then attach the sign of the number with the greater absolute value.
Subtraction of Integers	To subtract two integers, add the opposite of the second integer to the first integer.
Multiplication of Integers with the Same Sign	To multiply two numbers with the same sign, multiply the absolute values of the numbers. The product is positive.
Multiplication of Integers with Different Signs	To multiply two numbers with different signs, multiply the absolute values of the numbers. The product is negative.
Division of Integers with the Same Sign	The quotient of two numbers with the same sign is positive.
Division of Integers with Different Signs	The quotient of two numbers with different signs is negative.

Order of Operations Agreement

Step 1   Perform operations inside grouping symbols.
Step 2   Simplify exponential expressions.
Step 3   Do multiplication and division as they occur from left to right.
Step 4   Do addition and subtraction as they occur from left to right.

Chapter Review

SECTION 1.1

1. Place the correct symbol, < or >, between the two numbers.
14 < 29

2. Place the correct symbol, < or >, between the two numbers.
−4 > −6

3. Find the opposite of −17.
17

4. Find the opposite of 6.
−6

5. Evaluate |16|.
16

6. Evaluate −|7|.
−7

SECTION 1.2

7. Add: −13 + 7
−6

8. Add: −3 + (−12) + 6 + (−4)
−13

9. Subtract: 9 − 13
−4

10. Subtract: 16 − (−3) − 18
1

SECTION 1.3

11. Multiply: (−6)(7)
−42

12. Multiply: −9(−9)
81

13. Divide: −32 ÷ (−4)
8

14. Divide: −100 ÷ 5
−20

15. The temperature at which mercury boils, that is, its boiling point, is 357° C. The temperature at which mercury freezes, or its freezing point, is −39° C. Find the difference between the boiling point and the freezing point of mercury.
396° C

16. The temperature on the surface of the planet Venus is 480° C. The temperature on the planet Pluto is −234° C. Find the difference between the temperature on Venus and the temperature on Pluto.
714° C

SECTION 1.4

17. Write $\frac{7}{25}$ as a decimal.
0.28

18. Write $\frac{2}{15}$ as a decimal. Place a bar over the repeating digits of the decimal.
$0.1\bar{3}$

19. Simplify $\frac{1}{3} - \frac{1}{6} + \frac{5}{12}$.
$\frac{7}{12}$

20. Simplify $5.17 - 6.238$.
−1.068

21. Simplify $-\frac{18}{35} \div \frac{27}{28}$.
$-\frac{8}{15}$

22. Simplify $4.32 \cdot (-1.07)$.
−4.6224

SECTION 1.5

23. Evaluate $(-2)^3$.
−8

24. Evaluate -5^2.
−25

25. Evaluate $\left(-\frac{2}{3}\right)^4$.
$\frac{16}{81}$

26. Evaluate $(-5)(3)^2$.
−45

27. Evaluate $5 - 2^2 + 9$.
10

28. Evaluate $15 \cdot (6 - 4)^2$.
60

29. Evaluate $36 \div 9 - 5$.
−1

30. Evaluate $-3^2 + 4[18 + (12 - 20)]$.
31

31. Evaluate $(18.25 - 14.75)^2 \div 0.7 + 0.5$.
18

32. Evaluate $\frac{5^2 + 11}{2^2 + 5} \div (2^3 - 2^2)$.
1

Chapter Test

1. Place the correct symbol, < or >, between the two numbers.

   $-2 \quad -40$

   $>$ [1.1A]

2. Place the correct symbol, < or >, between the two numbers.

   $-1 \quad 0$

   $<$ [1.1A]

3. Find the opposite of -4.

   4 [1.1B]

4. Evaluate $-|-4|$.

   -4 [1.1B]

5. Add: $13 + (-16)$

   -3 [1.2A]

6. Add: $-22 + 14 + (-8)$

   -16 [1.2A]

7. Subtract: $16 - 30$

   -14 [1.2B]

8. Subtract: $16 - (-30) - 42$

   4 [1.2B]

9. Multiply: -4×12

   -48 [1.3A]

10. Multiply: $-5 \times (-6) \times 3$

    90 [1.3A]

11. Divide: $-72 \div 8$

    -9 [1.3B]

12. Divide: $-561 \div (-33)$

    17 [1.3B]

13. Find the temperature after a rise of 11°C from -4°C.

    7°C [1.3C]

14. The daily low temperature (in Fahrenheit) readings for a three-day period were as follows: -7°, 9°, -8°. Find the average low temperature for the three-day period.

    -2°F [1.3C]

15. Write $\frac{17}{20}$ as a decimal.

0.85   [1.4A]

16. Write $\frac{7}{9}$ as a decimal. Place a bar over the repeating digits of the decimal.

$0.\overline{7}$ [1.4A]

17. Simplify $-\frac{2}{5} + \frac{7}{15}$.

$\frac{1}{15}$ [1.4B]

18. Simplify $6.039 - 12.92$.

-6.881 [1.4B]

19. Simplify $\frac{5}{12} \div \left(-\frac{5}{6}\right)$.

$-\frac{1}{2}$ [1.4C]

20. Simplify $6.02 \times (-0.89)$.

-5.3578 [1.4C]

21. Evaluate $(-3^3) \cdot 2^2$.

-108 [1.5A]

22. Evaluate $\frac{3}{4} \cdot (4)^2$.

12   [1.5A]

23. Evaluate $3^2 - 4 + 20 \div 5$.

9   [1.5B]

24. Evaluate $\frac{-10 + 2}{2 + (-4)} \div 2 + 6$.

8   [1.5B]

25. Evaluate $16 \div 2[8 - 3(4 - 2)] + 1$

17   [1.5B]

2

Variable Expressions

OBJECTIVES

► To evaluate a variable expression
► To simplify a variable expression using the Properties of Addition
► To simplify a variable expression using the Properties of Multiplication
► To simplify a variable expression using the Distributive Property
► To simplify general variable expressions
► To translate a verbal expression into a variable expression given the variable
► To translate a verbal expression into a variable expression and then simplify
► To translate application problems

History of Variables

Prior to the 16th century, unknown quantities were represented by words. In Latin, the language in which most scholarly works were written, the word *res,* meaning "thing," was used. In Germany, the word *zahl,* meaning "number," was used. In Italy, the word *cosa,* also meaning "thing," was used.

Then in 1637 René Descartes, a French mathematician, began using the letters x, y, and z to represent variables. It is interesting to note, upon examining Descartes's work, that toward the end of the book the letters y and z were no longer used and x became the choice for a variable.

One explanation of why the letters y and z appeared less frequently has to do with the nature of printing presses during Descartes's time. A printer had a large tray that contained all the letters of the alphabet. There were many copies of each letter, especially those letters that are used frequently. For example, there were more e's than q's. Since the letters y and z do not occur frequently in French, a printer would have few of these letters on hand. Consequently when Descartes started using these letters as variables, it quickly depleted the printer's supply and x's had to be used instead.

Today, x is used by most nations as the standard letter for a single unknown. In fact, x-rays were so named because the scientists who discovered them did not know what they were and thus labeled them the "unknown rays" or x-rays.

SECTION 2.1   # Evaluating Variable Expressions

| Objective A |

To evaluate a variable expression

Often we discuss a quantity without knowing its exact value—for example, the price of gold next month, the cost of a new automobile next year, or the tuition cost for next semester. In algebra, a letter of the alphabet is used to stand for a quantity that is unknown or that can change, or *vary*. The letter is called a **variable.** An expression that contains one or more variables is called a **variable expression.**

A variable expression is shown at the right. The expression can be re-written by writing subtraction as the addition of the opposite.

$$3x^2 - 5y + 2xy - x - 7$$

$$3x^2 + (-5y) + 2xy + (-x) + (-7)$$

Note that the expression has 5 addends. The **terms** of a variable expression are the addends of the expression. The expression has 5 terms.

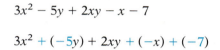

The terms $3x^2$, $-5y$, $2xy$, and $-x$ are **variable terms.**

The term -7 is a **constant term,** or simply a **constant.**

Each variable term is composed of a **numerical coefficient** and a **variable part** (the variable or variables and their exponents).

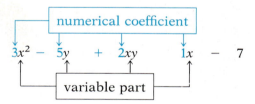

When the numerical coefficient is 1 or -1, the 1 is usually not written ($x = 1x$ and $-x = -1x$).

Variable expressions occur naturally in science. In a physics lab, a student may discover that a weight of one pound will stretch a spring $\frac{1}{2}$ inch. Two pounds will stretch the spring 1 inch. By experimenting, the student can discover that the distance the spring will stretch is found by multiplying the weight by $\frac{1}{2}$. By letting W represent the weight attached to the spring, the student can represent the distance the spring stretches by the variable expression $\frac{1}{2}W$.

With a weight of W pounds, the spring will stretch $\frac{1}{2} \cdot W = \frac{1}{2}W$ inches.

With a weight of 10 pounds, the spring will stretch $\frac{1}{2} \cdot 10 = 5$ inches. Note that 10 is called the **value** of the variable W.

With a weight of 3 pounds, the spring will stretch $\frac{1}{2} \cdot 3 = 1\frac{1}{2}$ inches.

Replacing each variable by its value and then simplifying the resulting numerical expression is called **evaluating the variable expression.**

Evaluate $ab - b^2$ when $a = 2$ and $b = -3$.

Replace each variable in the expression by its value.	$ab - b^2$ $2(-3) - (-3)^2$
Use the Order of Operations Agreement to simplify the resulting numerical expression.	$-6 - 9$ -15

Example 1   Name the variable terms of the expression $2a^2 - 5a + 7$.

Solution $2a^2$ and $-5a$

Example 2   Name the constant term of the expression $6n^2 + 3n - 4$.

Your solution -4

Example 3   Evaluate $x^2 - 3xy$ when $x = 3$ and $y = -4$.

Solution
$x^2 - 3xy$
$3^2 - 3(3)(-4)$
$9 - 3(3)(-4)$
$9 - 9(-4)$
$9 - (-36)$
45

Example 4   Evaluate $2xy + y^2$ when $x = -4$ and $y = 2$.

Your solution -12

Example 5   Evaluate $\dfrac{a^2 - b^2}{a - b}$ when $a = 3$ and $b = -4$.

Solution
$$\dfrac{a^2 - b^2}{a - b}$$
$$\dfrac{3^2 - (-4)^2}{3 - (-4)}$$
$$\dfrac{9 - 16}{3 - (-4)}$$
$$\dfrac{-7}{7} = -1$$

Example 6   Evaluate $\dfrac{a^2 + b^2}{a + b}$ when $a = 5$ and $b = -3$.

Your solution 17

Example 7   Evaluate $x^2 - 3(x - y) - z^2$ when $x = 2$, $y = -1$, and $z = 3$.

Solution
$x^2 - 3(x - y) - z^2$
$2^2 - 3[2 - (-1)] - 3^2$
$2^2 - 3(3) - 3^2$
$4 - 3(3) - 9$
$4 - 9 - 9$
$-5 - 9$
-14

Example 8   Evaluate $x^3 - 2(x + y) + z^2$ when $x = 2$, $y = -4$, and $z = -3$.

Your solution 21

Solutions on p. A8

2.1   EXERCISES

▶ **Objective A**

Name the terms of the variable expression. Then underline the constant term.

1. $2x^2 + 5x - 8$
$2x^2$, $5x$, $\underline{-8}$

2. $-3n^2 - 4n + 7$
$-3n^2$, $-4n$, $\underline{7}$

3. $6 - a^4$
$\underline{6}$, $-a^4$

Name the variable terms of the expression. Then underline the variable part of each term.

4. $9b^2 - 4ab + a^2$
$9\underline{b^2}$, $-4\underline{ab}$, $\underline{a^2}$

5. $7x^2y + 6xy^2 + 10$
$7\underline{x^2y}$, $6\underline{xy^2}$

6. $5 - 8n - 3n^2$
$-8\underline{n}$, $-3\underline{n^2}$

Name the coefficients of the variable terms.

7. $x^2 - 9x + 2$
1, -9

8. $12a^2 - 8ab - b^2$
12, -8, -1

9. $n^3 - 4n^2 - n + 9$
1, -4, -1

Evaluate the variable expression when $a = 2$, $b = 3$, and $c = -4$.

10. $3a + 2b$
12

11. $a - 2c$
10

12. $-a^2$
-4

13. $2c^2$
32

14. $-3a + 4b$
6

15. $3b - 3c$
21

16. $b^2 - 3$
6

17. $-3c + 4$
16

18. $16 \div (2c)$
-2

19. $6b \div (-a)$
-9

20. $bc \div (2a)$
-3

21. $-2ab \div c$
3

22. $a^2 - b^2$
-5

23. $b^2 - c^2$
-7

24. $(a + b)^2$
25

25. $a^2 + b^2$
13

26. $2a - (c + a)^2$
0

27. $(b - a)^2 + 4c$
-15

28. $b^2 - \dfrac{ac}{8}$
10

29. $\dfrac{5ab}{6} - 3cb$
41

30. $(b - 2a)^2 + bc$
-11

Evaluate the variable expression when $a = -2$, $b = 4$, $c = -1$, and $d = 3$.

31. $\dfrac{b + c}{d}$
1

32. $\dfrac{d - b}{c}$
1

33. $\dfrac{2d + b}{-a}$
5

34. $\dfrac{b + 2d}{b}$
$\dfrac{5}{2}$

35. $\dfrac{b - d}{c - a}$
1

36. $\dfrac{2c - d}{-ad}$
$-\dfrac{5}{6}$

37. $(b + d)^2 - 4a$
57

38. $(d - a)^2 - 3c$
28

39. $(d - a)^2 \div 5$
5

40. $(b - c)^2 \div 5$
5

41. $b^2 - 2b + 4$
12

42. $a^2 - 5a - 6$
8

43. $\dfrac{bd}{a} \div c$
6

44. $\dfrac{2ac}{b} \div (-c)$
1

45. $2(b + c) - 2a$
10

46. $3(b - a) - bc$
22

47. $\dfrac{b - 2a}{bc^2 - d}$
8

48. $\dfrac{b^2 - a}{ad + 3c}$
-2

49. $\dfrac{1}{3}d^2 - \dfrac{3}{8}b^2$
-3

50. $\dfrac{5}{8}a^4 - c^2$
9

51. $\dfrac{-4bc}{2a - b}$
-2

52. $\dfrac{abc}{b - d}$
8

53. $a^3 - 3a^2 + a$
-22

54. $d^3 - 3d - 9$
9

55. $-\dfrac{3}{4}b + \dfrac{1}{2}(ac + bd)$
4

56. $-\dfrac{2}{3}d - \dfrac{1}{5}(bd - ac)$
-4

57. $(b - a)^2 - (d - c)^2$
20

58. $(b + c)^2 + (a + d)^2$
10

59. $4ac + (2a)^2$
24

60. $3dc - (4c)^2$
-25

Evaluate the variable expression when $a = 2.7$, $b = -1.6$, and $c = -0.8$.

61. $c^2 - ab$
4.96

62. $(a + b)^2 - c$
2.01

63. $\dfrac{b^3}{c} - 4a$
-5.68

<table>
<tr><td>**SECTION 2.2**</td><td></td></tr>
</table>

Simplifying Variable Expressions

Objective A	To simplify a variable expression using the Properties of Addition

Like terms of a variable expression are the terms with the same variable part. (Because $x^2 = x \cdot x$, x^2 and x are not like terms.)

$$3x \;+\; 4 \;-\; 7x \;+\; 9 \;-\; x^2$$

like terms

like terms

Constant terms are like terms. 4 and 9 are like terms.

To simplify a variable expression, use the Distributive Property to combine like terms by adding the numerical coefficients. The variable part remains unchanged.

Distributive Property

> If a, b, and c are real numbers, then $a(b + c) = ab + ac$.

The Distributive Property can also be written as $ba + ca = (b + c)a$. This form is used to simplify a variable expression.

Simplify $2x + 3x$.

Use the Distributive Property to add the numerical coefficients of the like variable terms.

$$2x + 3x$$
$$(2 + 3)x$$
$$5x$$

Simplify $5x - 11x$.

Add the numerical coefficients of the like variable terms.

$$5x - 11x$$
$$[5 + (-11)]x$$
$$-6x$$

In simplifying more complicated expressions, the following Properties of Addition are used.

The Associative Property of Addition

> If a, b, and c are real numbers, then $(a + b) + c = a + (b + c)$.

When adding three or more terms, the terms can be grouped in any order. The sum is the same. For example,

$$(3x + 5x) + 9x = 3x + (5x + 9x)$$
$$8x + 9x = 3x + 14x$$
$$17x = 17x$$

The Commutative Property of Addition

If a and b are real numbers, then $a + b = b + a$.

When adding two like terms, the terms can be added in either order. The sum is the same. For example,

$$2x + (-4x) = -4x + 2x$$
$$-2x = -2x$$

The Addition Property of Zero

If a is a real number, then $a + 0 = 0 + a = a$.

The sum of a term and zero is the term. For example,

$$5x + 0 = 0 + 5x = 5x.$$

The Inverse Property of Addition

If a is a real number, then $a + (-a) = (-a) + a = 0$.

The sum of a term and its opposite is zero. The opposite of a number is called its **additive inverse.**

$$7x + (-7x) = -7x + 7x = 0$$

Simplify $8x + 3y - 8x$.

Use the Commutative and Associative Properties of Addition to rearrange and group like terms.
Combine like terms.

$8x + 3y - 8x$

$3y + (8x - 8x)$ Do these steps
$3y + 0$ mentally.

$3y$

Simplify $4x^2 + 5x - 6x^2 - 2x$.

Use the Commutative and Associative Properties of Addition to rearrange and group like terms.

$4x^2 + 5x - 6x^2 - 2x$

$(4x^2 - 6x^2) + (5x - 2x)$ Do these steps mentally.

Combine like terms.

$-2x^2 + 3x$

Example 1   Simplify $3x + 4y - 10x + 7y$.

Solution $3x + 4y - 10x + 7y$
$-7x + 11y$

Example 2   Simplify $3a - 2b - 5a + 6b$.

Your solution $-2a + 4b$

Example 3   Simplify $x^2 - 7 + 4x^2 - 16$.

Solution $x^2 - 7 + 4x^2 - 16$
$5x^2 - 23$

Example 4   Simplify
$-3y^2 + 7 + 8y^2 - 14$.

Your solution $5y^2 - 7$

Solutions on p. A8

Objective B	**To simplify a variable expression using the Properties of Multiplication**

In simplifying variable expressions, the following Properties of Multiplication are used.

The Associative Property of Multiplication

If a, b, and c are real numbers, then $(a \cdot b) \cdot c = a \cdot (b \cdot c)$.

When multiplying three or more factors, the factors can be grouped in any order. The product is the same. $2(3x) = (2 \cdot 3)x = 6x$

The Commutative Property of Multiplication

If a and b are real numbers, then $a \cdot b = b \cdot a$.

When multiplying two factors, the factors can be multiplied in either order. The product is the same. $(2x) \cdot 3 = 3 \cdot (2x) = 6x$

The Multiplication Property of One

If a is a real number, then $a \cdot 1 = 1 \cdot a = a$.

The product of a term and one is the term. $(8x)(1) = (1)(8x) = 8x$

The Inverse Property of Multiplication

If a is a real number, and a is not equal to zero, then $a \cdot \frac{1}{a} = \frac{1}{a} \cdot a = 1$.

$\frac{1}{a}$ is called the **reciprocal** of a. $\frac{1}{a}$ is also called the **multiplicative inverse** of a.

The product of a number and its reciprocal is one. $7 \cdot \frac{1}{7} = \frac{1}{7} \cdot 7 = 1$

The multiplication properties just discussed are used to simplify variable expressions.

Simplify $2(-x)$.

Use the Associative Property of Multiplication to group factors.

$2(-x)$

$\boxed{\begin{array}{l} 2(-1 \cdot x) \\ [2 \cdot (-1)]x \end{array}}$ Do these steps mentally.

$-2x$

Simplify $\frac{3}{2}\left(\frac{2x}{3}\right)$.

Note that $\frac{2x}{3} = \frac{2}{3} \cdot \frac{x}{1} = \frac{2}{3}x$.

Use the Associative Property of Multiplication to group factors. Use the Inverse Property of Multiplication and the Multiplication Property of One.

$\frac{3}{2}\left(\frac{2x}{3}\right)$

$\frac{3}{2}\left(\frac{2}{3}x\right)$

$\left(\frac{3}{2} \cdot \frac{2}{3}\right)x$ Do these steps

$1x$ mentally.

x

Simplify $(16x)2$.

Use the Commutative and Associative Properties of Multiplication to rearrange and group factors.

$(16x)2$

$2(16x)$ Do these steps
$(2 \cdot 16)x$ mentally.

$32x$

Example 5   Simplify $-2(3x^2)$.

Solution $-2(3x^2)$
$-6x^2$

Example 6   Simplify $-5(4y^2)$.

Your solution $-20y^2$

Example 7   Simplify $-5(-10x)$.

Solution $-5(-10x)$
$50x$

Example 8   Simplify $-7(-2a)$.

Your solution $14a$

Example 9   Simplify $(6x)(-4)$.

Solution $(6x)(-4)$
$-24x$

Example 10   Simplify $(-5x)(-2)$.

Your solution $10x$

Solutions on p. A8

Objective C   **To simplify a variable expression using the Distributive Property**

Recall that the Distributive Property states that if a, b, and c are real numbers, then

$$a(b + c) = ab + ac$$

The Distributive Property is used to remove parentheses from a variable expression.

Simplify $3(2x + 7)$.

$3(2x + 7)$

Use the Distributive Property.

$3(2x) + 3(7)$ Do this step
    mentally.

$6x + 21$

Content and Format © 1991 HMCo.

Simplify $-5(4x + 6)$.

$$-5(4x + 6)$$

Use the Distributive Property.

$-5(4x) + (-5) \cdot 6$ Do this step mentally.

$$-20x - 30$$

Simplify $-(2x - 4)$.

$$-(2x - 4)$$

Use the Distributive Property to remove parentheses from the variable expression.

$-1(2x - 4)$
$-1(2x) - (-1)(4)$ Do these steps mentally.

$$-2x + 4$$

Note: When a negative sign immediately precedes the parentheses, the sign of each term inside the parentheses is changed.

Simplify $-4(3y + 8)$.

$$-4(3y + 8)$$

Use the Distributive Property.

$-4(3y) + (-4)(8)$ Do this step mentally.

$$-12y - 32$$

An extension of the Distributive Property is used when an expression contains more than two terms.

Simplify $3(4x - 2y - z)$.

$$3(4x - 2y - z)$$

Use the Distributive Property.

$3(4x) - 3(2y) - 3(z)$ Do this step mentally.

$$12x - 6y - 3z$$

Example 11 Simplify $7(4 + 2x)$.

Solution $7(4 + 2x)$
$28 + 14x$

Example 12 Simplify $5(3 + 7b)$.

Your solution $15 + 35b$

Example 13 Simplify $(2x - 6)2$

Solution $(2x - 6)2$
$4x - 12$

Example 14 Simplify $(3a - 1)5$.

Your solution $15a - 5$

Example 15 Simplify $-3(-5a + 7b)$.

Solution $-3(-5a + 7b)$
$15a - 21b$

Example 16 Simplify $-8(-2a + 7b)$.

Your solution $16a - 56b$

Solutions on p. A8

Example 17 Simplify $-(3a - 2)$.	**Example 18** Simplify $-(5x - 12)$.
Solution $-(3a - 2)$ $-3a + 2$	**Your solution** $-5x + 12$
Example 19 Simplify $-2(x^2 + 5x - 4)$.	**Example 20** Simplify $3(-a^2 - 6a + 7)$.
Solution $-2(x^2 + 5x - 4)$ $-2x^2 - 10x + 8$	**Your solution** $-3a^2 - 18a + 21$

Solutions on p. A8

Objective D   **To simplify general variable expressions**

When simplifying variable expressions, use the Distributive Property to remove parentheses and brackets used as grouping symbols.

Simplify $4(x - y) - 2(-3x + 6y)$.

Use the Distributive Property to remove parentheses. Combine like terms.	$4(x - y) - 2(-3x + 6y)$ $4x - 4y + 6x - 12y$ $10x - 16y$

Example 21 Simplify $2x - 3(2x - 7y)$.	**Example 22** Simplify $3y - 2(y - 7x)$.
Solution $2x - 3(2x - 7y)$ $2x - 6x + 21y$ $-4x + 21y$	**Your solution** $y + 14x$
Example 23 Simplify $7(x - 2y) - 3(-x - 2y)$.	**Example 24** Simplify $-2(x - 2y) + 4(x - 3y)$.
Solution $7(x - 2y) - 3(-x - 2y)$ $7x - 14y + 3x + 6y$ $10x - 8y$	**Your solution** $2x - 8y$
Example 25 Simplify $-2(-3x + 7y) - 14x$.	**Example 26** Simplify $-5(-2y - 3x) + 4y$.
Solution $-2(-3x + 7y) - 14x$ $6x - 14y - 14x$ $-8x - 14y$	**Your solution** $15x + 14y$
Example 27 Simplify $2x - 3[2x - 3(x + 7)]$.	**Example 28** Simplify $3y - 2[x - 4(2 - 3y)]$.
Solution $2x - 3[2x - 3(x + 7)]$ $2x - 3[2x - 3x - 21]$ $2x - 3[-x - 21]$ $2x + 3x + 63$ $5x + 63$	**Your solution** $-2x - 21y + 16$

Solutions on p. A8

| 2.2 | **EXERCISES** |

▶ Objective A

Simplify.

1. $6x + 8x$
$14x$

2. $12x + 13x$
$25x$

3. $9a - 4a$
$5a$

4. $12a - 3a$
$9a$

5. $4y + (-10y)$
$-6y$

6. $8y + (-6y)$
$2y$

7. $-3b - 7$
$-3b - 7$

8. $-12y - 3$
$-12y - 3$

9. $-12a + 17a$
$5a$

10. $-3a + 12a$
$9a$

11. $5ab - 7ab$
$-2ab$

12. $9ab - 3ab$
$6ab$

13. $-12xy + 17xy$
$5xy$

14. $-15xy + 3xy$
$-12xy$

15. $-3ab + 3ab$
0

16. $-7ab + 7ab$
0

17. $-\dfrac{1}{2}x - \dfrac{1}{3}x$
$-\dfrac{5}{6}x$

18. $-\dfrac{2}{5}y + \dfrac{3}{10}y$
$-\dfrac{1}{10}y$

19. $\dfrac{3}{8}x^2 - \dfrac{5}{12}x^2$
$-\dfrac{1}{24}x^2$

20. $\dfrac{2}{3}y^2 - \dfrac{4}{9}y^2$
$\dfrac{2}{9}y^2$

21. $3x + 5x + 3x$
$11x$

22. $8x + 5x + 7x$
$20x$

23. $5a - 3a + 5a$
$7a$

24. $10a - 17a + 3a$
$-4a$

25. $-5x^2 - 12x^2 + 3x^2$
$-14x^2$

26. $-y^2 - 8y^2 + 7y^2$
$-2y^2$

27. $7x + (-8x) + 3y$
$-x + 3y$

28. $8y + (-10x) + 8x$
$-2x + 8y$

29. $7x - 3y + 10x$
$17x - 3y$

30. $8y + 8x - 8y$
$8x$

31. $3a + (-7b) - 5a + b$
$-2a - 6b$

32. $-5b + 7a - 7b + 12a$
$19a - 12b$

33. $3x + (-8y) - 10x + 4x$
$-3x - 8y$

34. $3y + (-12x) - 7y + 2y$
$-12x - 2y$

35. $x^2 - 7x + (-5x^2) + 5x$
$-4x^2 - 2x$

36. $3x^2 + 5x - 10x^2 - 10x$
$-7x^2 - 5x$

▶ Objective B

Simplify.

37. $4(3x)$
12x

38. $12(5x)$
60x

39. $-3(7a)$
−21a

40. $-2(5a)$
−10a

41. $-2(-3y)$
6y

42. $-5(-6y)$
30y

43. $(4x)2$
8x

44. $(6x)12$
72x

45. $(3a)(-2)$
−6a

46. $(7a)(-4)$
−28a

47. $(-3b)(-4)$
12b

48. $(-12b)(-9)$
108b

49. $-5(3x^2)$
−15x²

50. $-8(7x^2)$
−56x²

51. $\frac{1}{3}(3x^2)$
x²

52. $\frac{1}{6}(6x^2)$
x²

53. $\frac{1}{5}(5a)$
a

54. $\frac{1}{8}(8x)$
x

55. $-\frac{1}{2}(-2x)$
x

56. $-\frac{1}{4}(-4a)$
a

57. $-\frac{1}{7}(-7n)$
n

58. $-\frac{1}{9}(-9b)$
b

59. $(3x)\left(\frac{1}{3}\right)$
x

60. $(12x)\left(\frac{1}{12}\right)$
x

61. $(-6y)\left(-\frac{1}{6}\right)$
y

62. $(-10n)\left(-\frac{1}{10}\right)$
n

63. $\frac{1}{3}(9x)$
3x

64. $\frac{1}{7}(14x)$
2x

65. $-\frac{1}{5}(10x)$
−2x

66. $-\frac{1}{8}(16x)$
−2x

67. $-\frac{2}{3}(12a^2)$
−8a²

68. $-\frac{5}{8}(24a^2)$
−15a²

69. $-\frac{1}{2}(-16y)$
8y

70. $-\frac{3}{4}(-8y)$
6y

71. $(16y)\left(\frac{1}{4}\right)$
4y

72. $(33y)\left(\frac{1}{11}\right)$
3y

73. $(-6x)\left(\frac{1}{3}\right)$
−2x

74. $(-10x)\left(\frac{1}{5}\right)$
−2x

75. $(-8a)\left(-\frac{3}{4}\right)$
6a

▶ **Objective C**

Simplify.

76. $-(x + 2)$
$-x - 2$

77. $-(x + 7)$
$-x - 7$

78. $2(4x - 3)$
$8x - 6$

79. $5(2x - 7)$
$10x - 35$

80. $-2(a + 7)$
$-2a - 14$

81. $-5(a + 16)$
$-5a - 80$

82. $-3(2y - 8)$
$-6y + 24$

83. $-5(3y - 7)$
$-15y + 35$

84. $(5 - 3b)7$
$35 - 21b$

85. $(10 - 7b)2$
$20 - 14b$

86. $-3(3 - 5x)$
$-9 + 15x$

87. $-5(7 - 10x)$
$-35 + 50x$

88. $3(5x^2 + 2x)$
$15x^2 + 6x$

89. $6(3x^2 + 2x)$
$18x^2 + 12x$

90. $-2(-y + 9)$
$2y - 18$

91. $-5(-2x + 7)$
$10x - 35$

92. $(-3x - 6)5$
$-15x - 30$

93. $(-2x + 7)7$
$-14x + 49$

94. $2(-3x^2 - 14)$
$-6x^2 - 28$

95. $5(-6x^2 - 3)$
$-30x^2 - 15$

96. $-3(2y^2 - 7)$
$-6y^2 + 21$

97. $-8(3y^2 - 12)$
$-24y^2 + 96$

98. $3(x^2 - y^2)$
$3x^2 - 3y^2$

99. $5(x^2 + y^2)$
$5x^2 + 5y^2$

100. $-2(x^2 - 3y^2)$
$-2x^2 + 6y^2$

101. $-4(x^2 - 5y^2)$
$-4x^2 + 20y^2$

102. $-(6a^2 - 7b^2)$
$-6a^2 + 7b^2$

103. $3(x^2 + 2x - 6)$
$3x^2 + 6x - 18$

104. $4(x^2 - 3x + 5)$
$4x^2 - 12x + 20$

105. $-2(y^2 - 2y + 4)$
$-2y^2 + 4y - 8$

106. $-3(y^2 - 3y - 7)$
$-3y^2 + 9y + 21$

107. $2(-a^2 - 2a + 3)$
$-2a^2 - 4a + 6$

108. $4(-3a^2 - 5a + 7)$
$-12a^2 - 20a + 28$

109. $-5(-2x^2 - 3x + 7)$
$10x^2 + 15x - 35$

110. $-3(-4x^2 + 3x - 4)$
$12x^2 - 9x + 12$

111. $3(2x^2 + xy - 3y^2)$
$6x^2 + 3xy - 9y^2$

112. $5(2x^2 - 4xy - y^2)$
$10x^2 - 20xy - 5y^2$

113. $-(3a^2 + 5a - 4)$
$-3a^2 - 5a + 4$

114. $-(8b^2 - 6b + 9)$
$-8b^2 + 6b - 9$

▶ **Objective D**

Simplify.

115. $4x - 2(3x + 8)$
$-2x - 16$

116. $6a - (5a + 7)$
$a - 7$

117. $9 - 3(4y + 6)$
$-12y - 9$

118. $10 - (11x - 3)$
$-11x + 13$

119. $5n - (7 - 2n)$
$7n - 7$

120. $8 - (12 + 4y)$
$-4y - 4$

121. $3(x + 2) - 5(x - 7)$
$-2x + 41$

122. $2(x - 4) - 4(x + 2)$
$-2x - 16$

123. $12(y - 2) + 3(7 - 3y)$
$3y - 3$

124. $6(2y - 7) - 3(3 - 2y)$
$18y - 51$

125. $3(a - b) - 4(a + b)$
$-a - 7b$

126. $2(a + 2b) - (a - 3b)$
$a + 7b$

127. $4[x - 2(x - 3)]$
$-4x + 24$

128. $2[x + 2(x + 7)]$
$6x + 28$

129. $-2[3x + 2(4 - x)]$
$-2x - 16$

130. $-5[2x + 3(5 - x)]$
$5x - 75$

131. $-3[2x - (x + 7)]$
$-3x + 21$

132. $-2[3x - (5x - 2)]$
$4x - 4$

133. $2x - 3[x - 2(4 - x)]$
$-7x + 24$

134. $-7x + 3[x - 7(3 - 2x)]$
$38x - 63$

135. $-5x - 2[2x - 4(x + 7)] - 6$
$-x + 50$

136. $4a - 2[2b - (b - 2a)] + 3b$
b

137. $2x + 3(x - 2y) + 5(3x - 7y)$
$20x - 41y$

138. $5y - 2(y - 3x) + 2(7x - y)$
$20x + y$

SECTION 2.3

Translating Verbal Expressions into Variable Expressions

Objective A

To translate a verbal expression into a variable expression, given the variable

One of the major skills required in applied mathematics is to translate a verbal expression into a variable expression. This requires recognizing the verbal phrases that translate into mathematical operations. A partial list of the verbal phrases used to indicate the different mathematical operations is given below.

Addition	added to	6 added to y	$y + 6$
	more than	8 more than x	$x + 8$
	the sum of	the sum of x and z	$x + z$
	increased by	t increased by 9	$t + 9$
	the total of	the total of 5 and y	$5 + y$
Subtraction	minus	x minus 2	$x - 2$
	less than	7 less than t	$t - 7$
	decreased by	m decreased by 3	$m - 3$
	the difference between	the difference between y and 4	$y - 4$
Multiplication	times	10 times t	$10t$
	of	one half of x	$\frac{1}{2}x$
	the product of	the product of y and z	yz
	multiplied by	y multiplied by 11	$11y$
Division	divided by	x divided by 12	$\frac{x}{12}$
	the quotient of	the quotient of y and z	$\frac{y}{z}$
	the ratio of	the ratio of t to 9	$\frac{t}{9}$
Power	the square of	the square of x	x^2
	the cube of	the cube of a	a^3

Translate "14 less than the cube of x" into a variable expression.

Identify the words that indicate the mathematical operations.   14 <u>less than</u> the <u>cube</u> of x

Use the identified operations to write the variable expression. $x^3 - 14$

Translate "the difference between the square of x and the sum of y and z" into a variable expression.

Identify words that indicate the mathematical operations.	the <u>difference between</u> the <u>square of</u> x and the <u>sum</u> of y and z
Use the identified operations to write the variable expression.	$x^2 - (y + z)$

Example 1

Translate "the total of 3 times n and n" into a variable expression.

Solution

the <u>total of</u> 3 <u>times</u> n and n

$3n + n$

Example 2

Translate "the difference between twice n and one-third of n" into a variable expression.

Your solution

$2n - \dfrac{1}{3}n$

Example 3

Translate "m decreased by the sum of n and 12" into a variable expression.

Solution

m <u>decreased by</u> the <u>sum</u> of n and 12

$m - (n + 12)$

Example 4

Translate "the quotient of 7 less than b and 15" into a variable expression.

Your solution

$\dfrac{b - 7}{15}$

Solutions on p. A9

Objective B

To translate a verbal expression into a variable expression and then simplify

In most applications that involve translating phrases into variable expressions, the variable to be used is not given. To translate these phrases, a variable must be assigned to an unknown quantity before the variable expression can be written.

Translate "the sum of two consecutive integers" into a variable expression.

Assign a variable to one of the unknown quantities.	the first integer: n
Use the assigned variable to write an expression for any other unknown quantity.	the next consecutive integer: $n + 1$
Use the assigned variable to write the variable expression.	$n + (n + 1)$

Translate "the quotient of twice a number and the difference between the number and twelve" into a variable expression.

Assign a variable to one of the unknown quantities.

the unknown number: n

Use the assigned variable to write an expression for any other unknown quantity.

twice a number: $2n$
the difference between the number and twelve: $n - 12$

Use the assigned variable to write the variable expression.

$$\frac{2n}{n - 12}$$

Example 5

Translate "a number added to the product of four and the square of the number" into a variable expression.

Solution

the unknown number: n
the square of the number: n^2
the product of four and the square of the number: $4n^2$
$n + 4n^2$

Example 6

Translate "negative 4 multiplied by the total of ten and the cube of a number" into a variable expression.

Your solution

$-4(10 + n^3)$

Example 7

Translate "the sum of the squares of two and a number" into a variable expression. Then simplify.

Solution

the unknown number: x
$2^2 + x^2$
$4 + x^2$

Example 8

Translate "the sum of three consecutive integers" into a variable expression. Then simplify.

Your solution

$x + (x + 1) + (x + 2); 3x + 3$

Example 9

Translate "four times the sum of half of a number and fourteen" into a variable expression. Then simplify.

Solution

the unknown number: n
half of the number: $\frac{1}{2}n$
the sum of half of the number and fourteen: $\frac{1}{2}n + 14$

$4\left(\frac{1}{2}n + 14\right)$

$2n + 56$

Example 10

Translate "five times the difference between a number and sixty" into a variable expression. Then simplify.

Your solution

$5(x - 60); 5x - 300$

Solutions on p. A9

Objective C   To translate application problems

Many of the applications of mathematics require that you identify the unknown quantity, assign a variable to that quantity, and then attempt to express the other unknown quantities in terms of the variable.

A deluxe car stereo costs $10 more than twice the cost of the standard car stereo. Express the cost of the deluxe car stereo in terms of the cost of the standard car stereo.

Assign a variable to the cost of the standard car stereo.

the cost of the standard car stereo: C

Express the cost of the deluxe car stereo in terms of C.

the cost of the deluxe car stereo is $10 more than twice the standard car stereo: $2C + 10$

Example 11

The length of a swimming pool is 4 feet less than two times the width. Express the length of the pool in terms of the width.

Solution

the width of the pool: w
the length is 4 feet less than two times the width: $2w - 4$

Example 12

The speed of a new laptop computer is twice the speed of an older model. Express the speed of the new model in terms of the speed of the old model.

Your solution

the speed of the older model: s
the new laptop operates at twice the speed of the older model: $2s$

Example 13

A banker divided $5000 between two accounts, one paying 10% annual interest and the second paying 8% annual interest. Express the amount invested in the 10% account in terms of the amount invested in the 8% account.

Solution

the amount invested at 8%: x
the amount invested at 10%: $5000 - x$

Example 14

A guitar string 6 feet long was cut into two pieces. Express the length of the shorter piece in terms of the length of the longer piece.

Your solution

the length of the longer piece: y
the length of the shorter piece: $6 - y$

Solutions on p. A9

Content and Format © 1991 HMCo.

2.3 | **EXERCISES**

▶ **Objective A**

Translate into a variable expression.

1. the sum of 8 and y

$8 + y$

2. a less than 16

$16 - a$

3. t increased by 10

$t + 10$

4. p decreased by 7

$p - 7$

5. z added to 14

$z + 14$

6. q multiplied by 13

$13q$

7. 20 less than the square of x

$x^2 - 20$

8. 6 times the difference between m and seven

$6(m - 7)$

9. the sum of three-fourths of n and 12

$\frac{3}{4}n + 12$

10. b decreased by the product of 2 and b

$b - 2b$

11. 8 increased by the quotient of n and 4

$8 + \frac{n}{4}$

12. the product of -8 and y

$-8y$

13. the product of 3 and the total of y and 7

$3(y + 7)$

14. 8 divided by the difference between x and 6

$\frac{8}{x - 6}$

15. the product of t and the sum of t and 16

$t(t + 16)$

16. the quotient of 6 less than n and twice n

$\frac{n - 6}{2n}$

17. 15 more than one half of the square of x

$\frac{1}{2}x^2 + 15$

18. 19 less than the product of n and -2

$-2n - 19$

19. the total of 5 times the cube of n and the square of n

$5n^3 + n^2$

20. the ratio of 9 more than m to m

$\frac{m + 9}{m}$

21. r decreased by the quotient of r and 3

$r - \frac{r}{3}$

22. four-fifths of the sum of w and 10

$\frac{4}{5}(w + 10)$

23. the difference between the square of x and the total of x and 17

$x^2 - (x + 17)$

24. s increased by the quotient of 4 and s

$s + \frac{4}{s}$

25. the product of 9 and the total of z and 4

$9(z + 4)$

26. n increased by the difference between 10 times n and 9

$n + (10n - 9)$

▶ **Objective B**

Translate into a variable expression. Then simplify.

27. twelve minus a number

$12 - x$

28. a number divided by eighteen

$\dfrac{x}{18}$

29. two-thirds of a number

$\dfrac{2}{3}x$

30. twenty more than a number

$x + 20$

31. the quotient of twice a number and nine

$\dfrac{2x}{9}$

32. ten times the difference between a number and fifty

$10(x - 50); 10x - 500$

33. eight less than the product of eleven and a number

$11x - 8$

34. the sum of five-eighths of a number and six

$\dfrac{5}{8}x + 6$

35. nine less than the total of a number and two

$(x + 2) - 9; x - 7$

36. the difference between a number and three more than the number

$x - (x + 3); -3$

37. the quotient of seven and the total of five and a number

$\dfrac{7}{5 + x}$

38. four times the sum of a number and nineteen

$4(x + 19); 4x + 76$

39. five increased by one half of the sum of a number and three

$5 + \dfrac{1}{2}(x + 3); \dfrac{1}{2}x + \dfrac{13}{2}$

40. the quotient of fifteen and the sum of a number and twelve

$\dfrac{15}{x + 12}$

41. a number added to the difference between twice the number and four

$x + (2x - 4); 3x - 4$

42. the product of two-thirds and the sum of a number and seven

$\dfrac{2}{3}(x + 7); \dfrac{2}{3}x + \dfrac{14}{3}$

43. the product of five less than a number and seven

$(x - 5)7; 7x - 35$

44. the difference between forty and the quotient of a number and twenty.

$40 - \dfrac{x}{20}$

45. the quotient of five more than twice a number and the number

$\dfrac{2x + 5}{x}$

46. the sum of the square of a number and twice the number

$x^2 + 2x$

47. a number decreased by the difference between three times the number and eight

$x - (3x - 8); -2x + 8$

48. the sum of eight more than a number and one-third of the number

$(x + 8) + \dfrac{1}{3}x; \dfrac{4}{3}x + 8$

Translate into a variable expression. Then simplify.

49. a number added to the product of three and the number

    $x + 3x$; $4x$

50. a number increased by the total of the number and nine

    $x + (x + 9)$; $2x + 9$

51. five more than the sum of a number and six

    $(x + 6) + 5$; $x + 11$

52. a number decreased by the difference between eight and the number

    $x - (8 - x)$; $2x - 8$

53. a number minus the sum of the number and ten

    $x - (x + 10)$; -10

54. the difference between one-third of a number and five-eighths of the number

    $\frac{1}{3}x - \frac{5}{8}x$; $-\frac{7}{24}x$

55. the sum of one-sixth of a number and four-ninths of the number

    $\frac{1}{6}x + \frac{4}{9}x$; $\frac{11}{18}x$

56. two more than the total of a number and five

    $(x + 5) + 2$; $x + 7$

57. the sum of a number divided by three and the number

    $\frac{x}{3} + x$; $\frac{4x}{3}$

58. twice the sum of six times a number and seven

    $2(6x + 7)$; $12x + 14$

59. fourteen multiplied by one-seventh of a number

    $14\left(\frac{1}{7}x\right)$; $2x$

60. four times the product of six and a number

    $4(6x)$; $24x$

61. the difference between ten times a number and twice the number

    $10x - 2x$; $8x$

62. the total of twelve times a number and twice the number

    $12x + 2x$; $14x$

63. sixteen more than the difference between a number and six

    $(x - 6) + 16$; $x + 10$

64. a number plus the product of the number and nineteen

    $x + 19x$; $20x$

65. a number subtracted from the product of the number and four

    $4x - x$; $3x$

66. eight times the sum of the square of a number and three

    $8(x^2 + 3)$; $8x^2 + 24$

67. the difference between fifteen times a number and the product of the number and five

    $15x - 5x$; $10x$

68. two-thirds of the sum of nine times a number and three

    $\frac{2}{3}(9x + 3)$; $6x + 2$

▶ **Objective C**   *Application Problems*

69. A propeller-driven plane flies at a rate that is half that of a jet plane. Express the rate of the propeller plane in terms of the rate of the jet plane.
rate of jet plane: j; rate of propeller plane: $\frac{1}{2}j$

70. The length of a football field is 30 yards more than the width. Express the length of the field in terms of the width.
width of football field: w; length of football field: $w + 30$

71. The diameter of a basketball is approximately 4 times the diameter of a baseball. Express the diameter of the baseball in terms of the diameter of the basketball.
diameter of basketball: x; diameter of baseball: $\frac{1}{4}x$

72. An 18-carat gold chain has 6 grams less than twice the amount of gold in a 12-carat gold chain. Express the gold in the 18-carat gold chain in terms of the amount of gold in the 12-carat gold chain.
amount of gold in 12-carat gold chain: y;
amount of gold in 18-carat gold chain: $2y - 6$

73. The highest percent income tax bracket is 3 percent more than twice the lowest percent bracket. Express the highest income tax bracket in terms of the lowest bracket.
lowest tax bracket: t; highest tax bracket: $2t + 3$

74. A batter can swing a bat at a rate that is two-thirds the rate at which a pitcher can throw the ball. Express the rate of the ball in terms of the rate of the bat.
rate of bat: x; rate of ball: $\frac{3}{2}x$

75. Twenty gallons of crude oil were poured into two containers of different size. Express the amount of crude oil poured into the smaller container in terms of the amount poured into the larger container.
amount in larger container: L; amount in smaller container: $20 - L$

76. A rope was cut into two pieces in such a way that the length of the longer piece was 2 feet less than 3 times the length of the shorter piece. Express the length of the longer piece in terms of the shorter piece.
shorter piece: s; longer piece: $3s - 2$

77. A new design model of a laser printer can print pages at a rate that is seven pages more than one-half the speed of the older model. Express the speed of the newer model in terms of the older model.
speed of older model: x; speed of newer model: $\frac{1}{2}x + 7$

78. Two cars are traveling in opposite directions and at different rates. Two hours later the cars are 200 miles apart. Express the distance traveled by the slower car in terms of the distance traveled by the faster car.
distance traveled by faster car: x; distance traveled by slower car: $200 - x$

Calculators and Computers

Evaluating Variable Expressions

Evaluating variable expressions with your calculator will at times require the use of the $+/-$ key, the x^2 key, and the parentheses keys.

The $+/-$ key changes the sign of the number currently in the display. The parentheses keys are used as grouping symbols, and the x^2 key is used to square the number in the display. For the examples below, $a = 2$, $b = -3$, and $c = 4$.

Evaluate $a + bc$.

Replace a, b, and c by their values. $\qquad$ $2 + (-3)(4)$

Use the Order of Operations Agreement (multiply before adding) and enter the following on your calculator:

$$3 \boxed{+/-} \boxed{\times} 4 \boxed{+} 2 \boxed{=}$$

The answer in the display should be -10.

Evaluate $ac^2 - b$.

Replace a, b, and c by their values. $\qquad$ $(2)(4)^2 - (-3)$

Use the Order of Operations Agreement (exponents, multiply, subtract) and enter the following on your calculator:

$$4 \boxed{x^2} \boxed{\times} 2 \boxed{-} 3 \boxed{+/-} \boxed{=}$$

The answer in the display should be 35.

Evaluate $5a^2 - 6bc$.

This example is a little more complicated and requires the use of the parentheses keys. Because of the Order of Operations Agreement, the expression $5a^2 - 6bc$ is evaluated as though there were parentheses around each of the terms: $(5a^2) - (6bc)$. (Do all multiplications before subtraction.)

Replace a, b, and c by their values. $\qquad$ $5(2)^2 - 6(-3)(4)$

Use the Order of Operations Agreement and enter the following on your calculator:

$$\boxed{(} 2 \boxed{x^2} \boxed{\times} 5 \boxed{)} \boxed{-} \boxed{(} 6 \boxed{\times} 3 \boxed{+/-} \boxed{\times} 4 \boxed{)} \boxed{=}$$

The answer in the display should be 92.

Practice evaluating variable expressions by trying some of the problems in your text. Remember that a term with more than one factor must have parentheses around it.

Chapter Summary

Key Words   A *variable* is a letter that is used to stand for a quantity that is unknown.

A *variable expression* is an expression that contains one or more variables.

The *terms* of a variable expression are the addends of the expression.

A *variable term* is composed of a numerical coefficient and a variable part.

Like terms of a variable expression are the terms with the same variable part.

The *additive inverse* of a number is the opposite of the number.

The *multiplicative inverse* of a number is the reciprocal of the number.

Essential Rules

The Associative Property of Addition	If a, b, and c are real numbers, then $(a + b) + c = a + (b + c)$.
The Commutative Property of Addition	If a and b are real numbers, then $a + b = b + a$.
The Addition Property of Zero	If a is a real number, then $a + 0 = 0 + a = a$.
The Additive Inverse Property	If a is a real number, then $a + (-a) = (-a) + a = 0$.
The Associative Property of Multiplication	If a, b, and c are real numbers, then $(ab)c = a(bc)$.
The Commutative Property of Multiplication	If a and b are real numbers, then $ab = ba$.
The Multiplication Property of One	If a is a real number, then $1 \cdot a = a \cdot 1 = a$.
The Inverse Property of Multiplication	If a is a non-zero real number, then $a\left(\dfrac{1}{a}\right) = \left(\dfrac{1}{a}\right)a = 1$.
The Distributive Property	If a, b, and c are real numbers, then $a(b + c) = ab + ac$.

Chapter Review

SECTION 2.1

1. Evaluate $a^2 - b^2$ when $a = 3$ and $b = 4$.
-7

2. Evaluate $(b - a)^2 + c$ when $a = -2$, $b = 3$, and $c = 4$.
29

3. Evaluate $2bc \div (a + 7)$ when $a = 3$, $b = -5$, and $c = 4$.
-4

4. Evaluate $(5c - 4a)^2 - b$ when $a = -1$, $b = 2$, and $c = 1$.
79

SECTION 2.2

5. Simplify $7x + 4x$.
$11x$

6. Simplify $12y - 17y$.
$-5y$

7. Simplify $6a - 4b + 2a$.
$8a - 4b$

8. Simplify $4x - 3x^2 + 2x - x^2$.
$-4x^2 + 6x$

9. Simplify $5c + (-2d) - 3d - (-4c)$.
$9c - 5d$

10. Simplify $-9r + 2s - 6s + 12s$.
$-9r + 8s$

11. Simplify $5(4x)$.
$20x$

12. Simplify $-3(-12y)$.
$36y$

13. Simplify $\frac{1}{4}(-24a)$.
$-6a$

14. Simplify $-6(7x^2)$.
$-42x^2$

15. Simplify $(-50n)\left(\frac{1}{10}\right)$.
$-5n$

16. Simplify $5(2x - 7)$.
$10x - 35$

17. Simplify $2(6y^2 + 4y - 5)$.
$12y^2 + 8y - 10$

18. Simplify $-9(7 + 4x)$.
$-63 - 36x$

19. Simplify $3(x^2 - 8x - 7)$.
$3x^2 - 24x - 21$

20. Simplify $(7a^2 - 2a + 3)4$.
$28a^2 - 8a + 12$

21. Simplify $18 - (4x - 2)$.
$-4x + 20$

22. Simplify $-4(2x - 9) + 5(3x + 2)$.
$7x + 46$

23. Simplify $6(8y - 3) - 8(3y - 6)$.
$24y + 30$

24. Simplify $5[2 - 3(6x - 1)]$.
$-90x + 25$

SECTION 2.3

25. Translate "the product of 4 and x" into a variable expression.
$4x$

26. Translate "6 less than x" into a variable expression.
$x - 6$

27. Translate "two thirds of the total of x and 10" into a variable expression.
$\frac{2}{3}(x + 10)$

28. Translate "a number plus twice the number" into a variable expression. Then simplify.
$x + 2x; 3x$

29. Translate "three times a number plus the product of five and one less than the number" into a variable expression. Then simplify.
$3x + 5(x - 1); 8x - 5$

30. Translate "the difference between twice a number and one half of the number" into a variable expression. Then simplify.
$2x - \frac{1}{2}x; \frac{3}{2}x$

31. A candy bar contains eight more calories than twice the number of calories in an apple. Express the number of calories in a candy bar in terms of the number of calories in an apple.
number of calories in an apple: a
number of calories in a candy bar: $2a + 8$

32. A club treasurer has some five-dollar bills and some ten-dollar bills. The treasurer has a total of 35 bills. Express the number of five-dollar bills in terms of the number of ten-dollar bills.
number of ten-dollar bills: T
number of five-dollar bills: $35 - T$

33. The diameter of #8 copper wire is two mils less than 13 times the diameter of #30 copper wire. Express the diameter of #8 copper wire in terms of the diameter of #30 copper wire.
diameter of #30 copper wire: d
diameter of #8 copper wire: $13d - 2$

34. A baseball card collection contains five times as many National League players' cards as American League players' cards. Express the number of National League players' cards in terms of the number of American League players' cards.
number of American League cards: A
number of National League cards: $5A$

Chapter Test

1. Evaluate $b^2 - 3ab$ when $a = 3$ and $b = -2$.
   22   [2.1A]

2. Evaluate $\frac{-2ab}{2b-a}$ when $a = -4$ and $b = 6$.
   3   [2.1A]

3. Simplify $3x - 5x + 7x$.
   $5x$ [2.2A]

4. Simplify $-7y^2 + 6y^2 - (-2y^2)$.
   y^2 [2.2A]

5. Simplify $3x - 7y - 12x$.
   $-9x - 7y$ [2.2A]

6. Simplify $3x + (-12y) - 5x - (-7y)$.
   $-2x - 5y$ [2.2A]

7. Simplify $\frac{1}{5}(10x)$.
   $2x$ [2.2B]

8. Simplify $(12x)\left(\frac{1}{4}\right)$.
   $3x$ [2.2B]

9. Simplify $(-3)(-12y)$.
   $36y$ [2.2B]

10. Simplify $\frac{2}{3}(-15a)$.
    $-10a$ [2.2B]

11. Simplify $5(3 - 7b)$.
    $15 - 35b$ [2.2C]

12. Simplify $-2(2x - 4)$.
    $-4x + 8$ [2.2C]

13. Simplify $-3(2x^2 - 7y^2)$.
    $-6x^2 + 21y^2$ [2.2C]

14. Simplify $-5(2x^2 - 3x + 6)$.
    $-10x^2 + 15x - 30$ [2.2C]

15. Simplify $2x - 3(x - 2)$.
    $-x + 6$ [2.2D]

16. Simplify $5(2x + 4) - 3(x - 6)$.
    $7x + 38$ [2.2D]

17. Simplify $2x + 3[4 - (3x - 7)]$.
    $-7x + 33$ [2.2D]

18. Simplify $-2[x - 2(x - y)] + 5y$.
    $2x + y$ [2.2D]

19. Translate "the difference of the squares of *a* and *b*" into a variable expression.
$a^2 - b^2$ [2.3A]

20. Translate "the sum of a number and twice the square of the number" into a variable expression.
$x + 2x^2$ [2.3A]

21. Translate "three less than the quotient of six and a number" into a variable expression.
$\frac{6}{x} - 3$ [2.3A]

22. Translate "*b* decreased by the product of *b* and 7" into a variable expression. Then simplify.
$b - 7b$; $-6b$ [2.3B]

23. Translate "10 times the difference between *x* and 3" into a variable expression. Then simplify.
$10(x - 3)$; $10x - 30$ [2.3B]

24. The speed of a pitcher's fastball is twice the speed of the catcher's return throw. Express the speed of the fastball in terms of the speed of the return throw.
speed of return: *s*; speed of fastball: $2s$ [2.3C]

25. A wire is cut into two lengths. The length of the longer piece is three inches less than four times the length of the shorter piece. Express the length of the longer piece in terms of the length of the shorter piece.
shorter piece: *x*; longer piece: $4x - 3$ [2.3C]

Cumulative Review

1. Add: $-4 + 7 + (-10)$
   -7 [1.2A]

2. Subtract: $-16 - (-25) - 4$
   5 [1.2B]

3. Multiply: $(-2)(3)(-4)$
   24 [1.3A]

4. Divide: $(-60) \div 12$
   -5 [1.3B]

5. Write $1\frac{1}{4}$ as a decimal.
   1.25 [1.4A]

6. Simplify $\frac{7}{12} - \frac{11}{16} - \left(-\frac{1}{3}\right)$.
   $\frac{11}{48}$ [1.4B]

7. Simplify $\frac{5}{12} \div \left(2\frac{1}{2}\right)$.
   $\frac{1}{6}$ [1.4C]

8. Simplify $\left(-\frac{9}{16}\right) \cdot \left(\frac{8}{27}\right) \cdot \left(-\frac{3}{2}\right)$.
   $\frac{1}{4}$ [1.4C]

9. Simplify $-3^2 \cdot \left(-\frac{2}{3}\right)^3$.
   $\frac{8}{3}$ [1.5A]

10. Simplify $-2^5 \div (3-5)^2 - (-3)$.
    -5 [1.5B]

11. Simplify $\left(-\frac{3}{4}\right)^2 - \left(\frac{3}{8} - \frac{11}{12}\right)$.
    $\frac{53}{48}$ [1.5B]

12. Evaluate $a^2 - 3b$ when $a = 2$ and $b = -4$.
    16 [2.1A]

13. Simplify $-2x^2 - (-3x^2) + 4x^2$.
    $5x^2$ [2.2A]

14. Simplify $5a - 10b - 12a$.
    $-7a - 10b$ [2.2A]

15. Simplify $\frac{1}{2}(12a)$.
    $6a$ [2.2B]

16. Simplify $\left(-\frac{5}{6}\right)(-36b)$.
    $30b$ [2.2B]

17. Simplify $3(8 - 2x)$.
    $24 - 6x$ [2.2C]

18. Simplify $-2(-3y + 9)$.
    $6y - 18$ [2.2C]

19. Simplify $-4(2x^2 - 3y^2)$.
$-8x^2 + 12y^2$ [2.2C]

20. Simplify $-3(3y^2 - 3y - 7)$.
$-9y^2 + 9y + 21$ [2.2C]

21. Simplify $-3x - 2(2x - 7)$.
$-7x + 14$ [2.2D]

22. Simplify $4(3x - 2) - 7(x + 5)$.
$5x - 43$ [2.2D]

23. Simplify
$2x + 3[x - 2(4 - 2x)]$.
$17x - 24$ [2.2D]

24. Simplify
$3[2x - 3(x - 2y)] + 3y$.
$-3x + 21y$ [2.2D]

25. Translate "the sum of one-half of b and b" into a variable expression.
$\frac{1}{2}b + b$ [2.3A]

26. Translate "ten divided by the difference between y and two" into a variable expression.
$\frac{10}{y - 2}$ [2.3A]

27. Translate "the difference between eight and the quotient of a number and twelve" into a variable expression.
$8 - \frac{x}{12}$ [2.3B]

28. Translate "the sum of a number and two more than the number" into a variable expression. Then simplify.
$x + (x + 2)$; $2x + 2$ [2.3B]

29. Translate and simplify "the product of four and the sum of two consecutive integers."
$4[x + (x + 1)]$; $8x + 4$ [2.3B]

30. Translate and simplify "twelve more than the product of three plus a number and five."
$(3 + x)5 + 12$; $27 + 5x$ [2.3B]

3

Solving Equations

OBJECTIVES

▶ To determine whether a given number is a solution of an equation
▶ To solve an equation of the form $x + a = b$
▶ To solve an equation of the form $ax = b$
▶ To solve an equation of the form $ax + b = c$
▶ To solve application problems
▶ To solve an equation of the form $ax + b = cx + d$
▶ To solve an equation containing parentheses
▶ To solve application problems
▶ To translate a sentence into an equation and solve
▶ To solve application problems

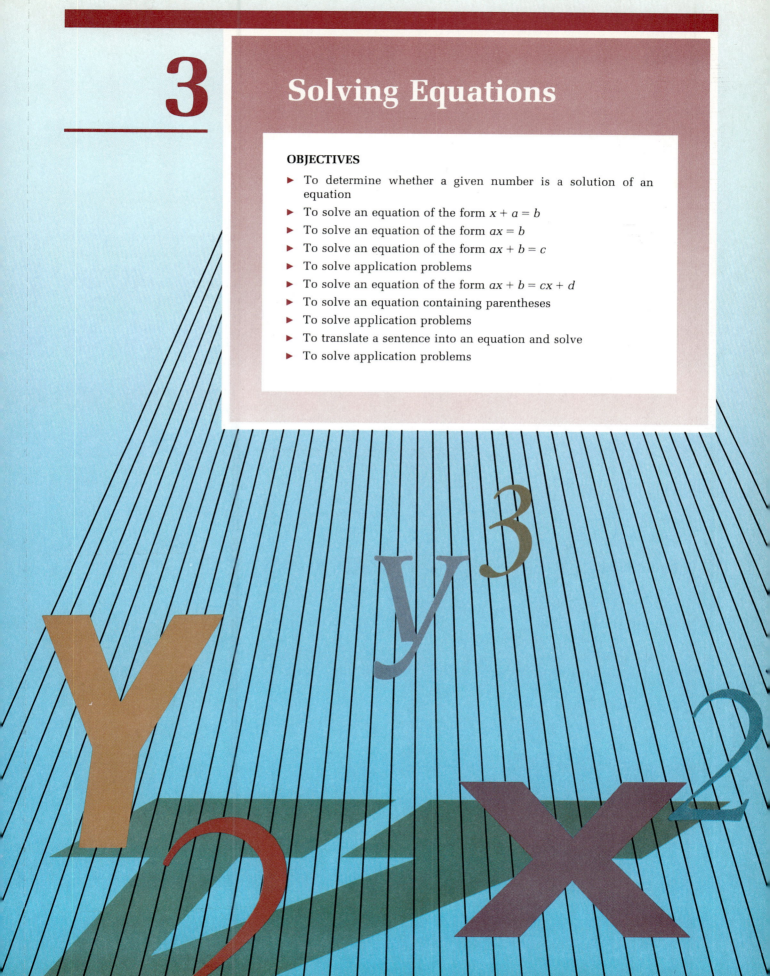

Mersenne Primes

A prime number that can be written in the form $2^n - 1$, where n is also prime, is called a Mersenne prime. The table below shows some Mersenne primes.

$$3 = 2^2 - 1$$
$$7 = 2^3 - 1$$
$$31 = 2^5 - 1$$
$$127 = 2^7 - 1$$

You might notice that not every prime number is a Mersenne prime. For example, 5 is a prime number but not a Mersenne prime. Also, not all numbers in the form $2^n - 1$, where n is prime, yield a prime number. For example, $2^{11} - 1 = 2047$ is not a prime number.

The search for Mersenne primes has been quite extensive, especially since the advent of the computer. One reason for the extensive research into large prime numbers (not only Mersenne primes) has to do with cryptology.

Cryptology is the study of making or breaking secret codes. One method for making a code that is difficult to break is called public key cryptology. For this method to work, it is necessary to use very large prime numbers. To keep anyone from breaking the code, each prime should have at least 200 digits.

Today the largest known Mersenne prime is $2^{216091} - 1$. This number has 65,050 digits in its representation.

Another Mersenne prime got special recognition in a postage-meter stamp. It is the number $2^{11213} - 1$. This number has 3276 digits in its representation.

SECTION 3.1   Introduction to Equations

Objective A   **To determine whether a given number is a solution of an equation**

An **equation** expresses the equality of two mathematical expressions. The expressions can be either numerical or variable expressions.

$$\left.\begin{array}{l} 9 + 3 = 12 \\ 3x - 2 = 10 \\ y^2 + 4 = 2y - 1 \\ z = 2 \end{array}\right\} \text{Equations}$$

The equation at the right is true if the variable is replaced by 5.

$x + 8 = 13$
$5 + 8 = 13$ A true equation

The equation is false if the variable is replaced by 7.

$7 + 8 = 13$ A false equation

A **solution** of an equation is a number that, when substituted for the variable, results in a true equation. 5 is a solution of the equation $x + 8 = 13$. 7 is not a solution of the equation $x + 8 = 13$.

Is -3 a solution of the equation $4x + 18 = x^2 - 3$?

Replace the variable by the given number, -3.

Evaluate the numerical expressions, using the Order of Operations Agreement.

$$\begin{array}{c|c} \multicolumn{2}{c}{4x + 18 = x^2 - 3} \\ \hline 4(-3) + 18 & (-3)^2 - 3 \\ -12 + 18 & 9 - 3 \\ 6 = & 6 \end{array}$$

Compare the results. If the results are equal, the given number is a solution. If the results are not equal, the given number is not a solution.

Yes, -3 is a solution of the equation $4x + 18 = x^2 - 3$.

Example 1   Is $\frac{2}{3}$ a solution of
$12x - 2 = 6x + 2$?

Solution
$$\begin{array}{c|c} \multicolumn{2}{c}{12x - 2 = 6x + 2} \\ \hline 12\left(\frac{2}{3}\right) - 2 & 6\left(\frac{2}{3}\right) + 2 \\ 8 - 2 & 4 + 2 \\ 6 = & 6 \end{array}$$

Yes, $\frac{2}{3}$ is a solution.

Example 2   Is $\frac{1}{4}$ a solution of
$5 - 4x = 8x + 2$?

Your solution   yes

Example 3   Is -4 a solution of
$4 + 5x = x^2 - 2x$?

Solution
$$\begin{array}{c|c} \multicolumn{2}{c}{4 + 5x = x^2 - 2x} \\ \hline 4 + 5(-4) & (-4)^2 - 2(-4) \\ 4 + (-20) & 16 - (-8) \\ -16 \neq & 24 \end{array}$$
($\neq$ means "is not equal to")

No, -4 is not a solution.

Example 4   Is 5 a solution of
$10x - x^2 = 3x - 10$?

Your solution   no

Solutions on p. A9

| Objective B | **To solve an equation of the form $x + a = b$** |

To **solve** an equation means to find a solution of the equation. The simplest equation to solve is an equation of the form **variable = constant,** because the constant is the solution.

If $x = 5$, then 5 is the solution of the equation, because $5 = 5$ is a true equation.

| The solution of the equation shown at the right is 7. | $x + 2 = 9$ | $7 + 2 = 9$ |

| Note that if 4 is added to each side of the equation, the solution is still 7. | $x + 2 + 4 = 9 + 4$
 $x + 6 = 13$ | $7 + 6 = 13$ |

| If -5 is added to each side of the equation, the solution is still 7. | $x + 2 + (-5) = 9 + (-5)$
 $x - 3 = 4$ | $7 - 3 = 4$ |

This illustrates the Addition Property of Equations.

Addition Property of Equations

> If a, b, and c are algebraic expressions, then the equation $a = b$ has the same solution as the equation $a + c = b + c$.

This property states that the same quantity can be added to each side of an equation without changing the solution of the equation.

In solving an equation, the goal is to rewrite the given equation in the form **variable = constant.** The Addition Property of Equations is used to remove a *term* from one side of the equation by adding the opposite of that term to each side of the equation.

Solve: $x - 2 = 9$

Add the opposite of the constant term -2 to each side of the equation. After simplifying, the equation will be in the form variable = constant.

$$x - 2 = 9$$
$$x - 2 + 2 = 9 + 2$$
$$x + 0 = 11$$
$$x = 11$$

$$\boxed{\text{variable}} = \boxed{\text{constant}}$$

Check: $x - 2 = 9$
$$\frac{11 - 2 \mid 9}{9 = 9} \text{ A true equation}$$

The solution is 11.

Because subtraction is defined in terms of addition, the Addition Property of Equations allows the same number to be subtracted from each side of an equation.

Solve: $y + \frac{1}{2} = \frac{5}{4}$

The goal is to rewrite the equation in the form *variable = constant.*

Add the opposite of the constant term $\frac{1}{2}$ to each side of the equation. This is equivalent to subtracting $\frac{1}{2}$ from each side of the equation.

$$y + \frac{1}{2} = \frac{5}{4}$$

$$y + \frac{1}{2} - \frac{1}{2} = \frac{5}{4} - \frac{1}{2}$$

$$y + 0 = \frac{5}{4} - \frac{2}{4}$$

$$y = \frac{3}{4}$$

You should check this solution.   The solution is $\frac{3}{4}$.

Example 5   Solve: $x + \frac{1}{3} = -\frac{3}{4}$

Solution $x + \frac{1}{3} = -\frac{3}{4}$

$$x + \frac{1}{3} - \frac{1}{3} = -\frac{3}{4} - \frac{1}{3}$$

$$x = -\frac{13}{12}$$

The solution is $-\frac{13}{12}$.

Example 6   Solve: $\frac{1}{2} = x - \frac{2}{3}$

Your solution $\frac{7}{6}$

Solution on p. A9

Objective C   **To solve an equation of the form $ax = b$**

The solution of the equation shown at the right is 3. Note that if each side of the equation is multiplied by 5, the solution is still 3.

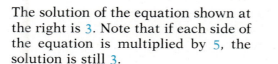

$$2x = 6 \qquad 2 \cdot 3 = 6$$
$$5 \cdot 2x = 5 \cdot 6$$
$$10x = 30 \qquad 10 \cdot 3 = 30$$

If each side of the equation is multiplied by -4, the solution is still 3

$$(-4) \cdot 2x = (-4) \cdot 6$$
$$-8x = -24 \qquad -8 \cdot 3 = -24$$

This illustrates the Multiplication Property of Equations.

Multiplication Property of Equations

If a, b, and c are algebraic expressions and $c \neq 0$, then the equation $a = b$ has the same solutions as the equation $ac = bc$.

This property states that each side of an equation can be multiplied by the same nonzero number without changing the solutions of the equation.

Recall that the goal of solving an equation is to rewrite the equation in the form *variable = constant.* The Multiplication Property of Equations is used to remove a coefficient by multiplying each side of the equation by the reciprocal of the coefficient.

Solve: $\frac{2}{3}x = 6$

Multiply each side of the equation by $\frac{3}{2}$, the reciprocal of $\frac{2}{3}$. After simplifying, the equation will be in the form *variable = constant*.

$$\frac{2}{3}x = 6$$

$$\frac{3}{2} \cdot \frac{2}{3}x = \frac{3}{2} \cdot 6$$

$$1x = 9$$

$$x = 9$$

Check: $\quad \frac{2}{3}x = 6$

$$\frac{2}{3}(9) \;\Big|\; 6$$

$$6 = 6 \quad \text{A true equation}$$

The solution is 9.

Because division is defined in terms of multiplication, the Multiplication Property of Equations allows each side of an equation to be divided by the same nonzero quantity.

Solve: $4x = 6$

The goal is to rewrite the equation in the form *variable = constant*.

Multiply each side of the equation by the reciprocal of 4. This is equivalent to dividing each side of the equation by 4.

$$4x = 6$$

$$\frac{4x}{4} = \frac{6}{4}$$

$$1x = \frac{3}{2}$$

$$x = \frac{3}{2}$$

The solution is $\frac{3}{2}$.

You should check this solution.

When using the Multiplication Property of Equations, multiply each side of the equation by the reciprocal of the coefficient when the coefficient is a fraction. Divide each side of the equation by the coefficient when the coefficient is an integer or decimal.

Example 7   Solve: $\frac{3x}{4} = -9$

Solution

$$\frac{3x}{4} = -9$$

$$\frac{4}{3} \cdot \frac{3}{4}x = \frac{4}{3}(-9) \qquad \left[\frac{3x}{4} = \frac{3}{4}x\right]$$

$$x = -12$$

The solution is -12.

Example 8   Solve: $-\frac{2x}{5} = 6$

Your solution -15

Example 9   Solve: $5x - 9x = 12$

Solution

$$5x - 9x = 12 \quad \text{Combine like}$$
$$-4x = 12 \quad \text{terms.}$$
$$\frac{-4x}{-4} = \frac{12}{-4}$$
$$x = -3$$

The solution is -3.

Example 10   Solve: $4x - 8x = 16$

Your solution -4

Solutions on p. A9

3.1   EXERCISES

▶ Objective A

1. Is 4 a solution of
$2x = 8$?
yes

2. Is 3 a solution of
$y + 4 = 7$?
yes

3. Is -1 a solution of
$2b - 1 = 3$?
no

4. Is -2 a solution of
$3a - 4 = 10$?
no

5. Is 1 a solution of
$4 - 2m = 3$?
no

6. Is 2 a solution of
$7 - 3n = 2$?
no

7. Is 5 a solution of
$2x + 5 = 3x$?
yes

8. Is 4 a solution of
$3y - 4 = 2y$?
yes

9. Is 0 a solution of
$4a + 5 = 3a + 5$?
yes

10. Is 0 a solution of
$4 - 3b = 4 - 5b$?
yes

11. Is -2 a solution of
$4 - 2n = n + 10$?
yes

12. Is -3 a solution of
$5 - m = 2 - 2m$?
yes

13. Is 3 a solution of
$z^2 + 1 = 4 + 3z$?
no

14. Is 2 a solution of
$2x^2 - 1 = 4x - 1$?
yes

15. Is -1 a solution of
$y^2 - 1 = 4y + 3$?
no

16. Is -2 a solution of
$m^2 - 4 = m + 3$?
no

17. Is 5 a solution of
$x^2 + 2x + 1 = (x + 1)^2$?
yes

18. Is -6 a solution of
$(n - 2)^2 = n^2 - 4n + 4$?
yes

19. Is 4 a solution of
$x(x + 1) = x^2 + 5$?
no

20. Is 3 a solution of
$2a(a - 1) = 3a + 3$?
yes

21. Is $-\frac{1}{4}$ a solution of
$8t + 1 = -1$?
yes

22. Is $\frac{1}{2}$ a solution of
$4y + 1 = 3$?
yes

23. Is $\frac{2}{5}$ a solution of
$5m + 1 = 10m - 3$?
no

24. Is $\frac{3}{4}$ a solution of
$8x - 1 = 12x + 3$?
no

25. Is $\frac{1}{3}$ a solution of
$2n + 2 = 5n - 1$?
no

26. Is $\frac{1}{4}$ a solution of
$5x + 4 = x + 5$?
yes

27. Is $-\frac{1}{3}$ a solution of
$3b(b - 1) = 2b + 2$?
yes

28. Is 2.1 a solution of
$x^2 - 4x = x + 1.89$?
no

29. Is 1.56 a solution of
$c^2 - 3c = 4c - 8.4864$?
yes

30. Is -1.8 a solution of
$a(a + 1) = 2.6 - 2a$?
no

▶ **Objective B**

Solve and check.

31. $x + 5 = 7$
2

32. $y + 3 = 9$
6

33. $b - 4 = 11$
15

34. $z - 6 = 10$
16

35. $2 + a = 8$
6

36. $5 + x = 12$
7

37. $m + 9 = 3$
-6

38. $t + 12 = 10$
-2

39. $n - 5 = -2$
3

40. $x - 6 = -5$
1

41. $b + 7 = 7$
0

42. $y - 5 = -5$
0

43. $a - 3 = -5$
-2

44. $x - 6 = -3$
3

45. $z + 9 = 2$
-7

46. $n + 11 = 1$
-10

47. $10 + m = 3$
-7

48. $8 + x = 5$
-3

49. $9 + x = -3$
-12

50. $10 + y = -4$
-14

51. $b - 5 = -3$
2

52. $t - 6 = -4$
2

53. $4 + x = 10$
6

54. $9 + a = 20$
11

55. $2 = x + 7$
-5

56. $-8 = n + 1$
-9

57. $4 = m - 11$
15

58. $-6 = y - 5$
-1

59. $12 = 3 + w$
9

60. $-9 = 5 + x$
-14

61. $4 = -10 + b$
14

62. $-7 = -2 + x$
-5

63. $13 = -6 + a$
19

64. $m + \dfrac{2}{3} = -\dfrac{1}{3}$
-1

65. $c + \dfrac{3}{4} = -\dfrac{1}{4}$
-1

66. $x - \dfrac{1}{2} = \dfrac{1}{2}$
1

Solve and check.

67. $x - \dfrac{2}{5} = \dfrac{3}{5}$
1

68. $\dfrac{5}{8} + y = \dfrac{1}{8}$
$-\dfrac{1}{2}$

69. $\dfrac{4}{9} + a = -\dfrac{2}{9}$
$-\dfrac{2}{3}$

70. $m + \dfrac{1}{2} = -\dfrac{1}{4}$
$-\dfrac{3}{4}$

71. $b + \dfrac{1}{6} = -\dfrac{1}{3}$
$-\dfrac{1}{2}$

72. $x + \dfrac{2}{3} = \dfrac{3}{4}$
$\dfrac{1}{12}$

73. $n + \dfrac{2}{5} = \dfrac{2}{3}$
$\dfrac{4}{15}$

74. $-\dfrac{5}{6} = x - \dfrac{1}{4}$
$-\dfrac{7}{12}$

75. $-\dfrac{1}{4} = c - \dfrac{2}{3}$
$\dfrac{5}{12}$

76. $d + 1.3619 = 2.0148$
0.6529

77. $w + 2.932 = 4.801$
1.869

78. $-0.813 + x = -1.096$
-0.283

79. $-1.926 + t = -1.042$
0.884

80. $6.149 = -3.108 + z$
9.257

81. $5.237 = -2.014 + x$
7.251

▶ **Objective C**

Solve and check.

82. $5x = 15$
3

83. $4y = 28$
7

84. $3b = -12$
-4

85. $2a = -14$
-7

86. $-3x = 6$
-2

87. $-5m = 20$
-4

88. $-3x = -27$
9

89. $-6n = -30$
5

90. $20 = 4c$
5

91. $18 = 2t$
9

92. $-32 = 8w$
-4

93. $-56 = 7x$
-8

94. $8d = 0$
0

95. $-5x = 0$
0

96. $36 = 9z$
4

Solve and check.

97. $35 = -5x$
 −7

98. $-64 = 8a$
 −8

99. $-32 = -4y$
 8

100. $-54 = 6c$
 −9

101. $49 = -7t$
 −7

102. $\dfrac{x}{3} = 2$
 6

103. $\dfrac{x}{4} = 3$
 12

104. $-\dfrac{y}{2} = 5$
 −10

105. $-\dfrac{b}{3} = 6$
 −18

106. $\dfrac{3}{4}y = 9$
 12

107. $\dfrac{2}{5}x = 6$
 15

108. $-\dfrac{2}{3}d = 8$
 −12

109. $-\dfrac{3}{5}m = 12$
 −20

110. $\dfrac{2n}{3} = 2$
 3

111. $\dfrac{5x}{6} = -10$
 −12

112. $\dfrac{-3z}{8} = 9$
 −24

113. $\dfrac{-4x}{5} = -12$
 15

114. $-6 = -\dfrac{2}{3}y$
 9

115. $-15 = -\dfrac{3}{5}x$
 25

116. $\dfrac{2}{5}a = 3$
 $\dfrac{15}{2}$

117. $\dfrac{3x}{4} = 2$
 $\dfrac{8}{3}$

118. $\dfrac{3}{4}c = \dfrac{3}{5}$
 $\dfrac{4}{5}$

119. $\dfrac{2}{9} = \dfrac{2}{3}y$
 $\dfrac{1}{3}$

120. $-\dfrac{6}{7} = -\dfrac{3}{4}b$
 $\dfrac{8}{7}$

121. $-\dfrac{2}{5}m = -\dfrac{6}{7}$
 $\dfrac{15}{7}$

122. $5x + 2x = 14$
 2

123. $3n + 2n = 20$
 4

124. $7d - 4d = 9$
 3

125. $10y - 3y = 21$
 3

126. $2x - 5x = 9$
 −3

127. $\dfrac{x}{1.46} = 3.25$
 4.745

128. $\dfrac{z}{2.95} = -7.88$
 −23.246

129. $3.47a = 7.1482$
 2.06

130. $2.31m = 2.4255$
 1.05

131. $-3.7x = 7.881$
 −2.13

132. $\dfrac{n}{2.65} = 9.08$
 24.062

SECTION 3.2   General Equations—Part I

Objective A   **To solve an equation of the form $ax + b = c$**

In solving an equation of the form $ax + b = c$, the goal is to rewrite the equation in the form *variable* = *constant*. This requires the application of both the Addition and the Multiplication Properties of Equations.

Solve: $\frac{2}{5}x - 3 = -7$

The goal is to rewrite the equation in the form *variable* = *constant*.

Add the opposite of the constant term -3 to each side of the equation. Then simplify.

$$\frac{2}{5}x - 3 = -7$$

$$\frac{2}{5}x - 3 + 3 = -7 + 3$$

$$\frac{2}{5}x = -4$$

Multiply each side of the equation by the reciprocal of the coefficient $\frac{2}{5}$. Simplify. The equation is in the form *variable* = *constant*.

$$\frac{5}{2} \cdot \frac{2}{5}x = \frac{5}{2}(-4)$$

$$x = -10$$

-10 checks as the solution.

The solution is -10.

Example 1   Solve: $3x - 7 = -5$

Solution
$$3x - 7 = -5$$
$$3x - 7 + 7 = -5 + 7$$
$$3x = 2$$
$$\frac{3x}{3} = \frac{2}{3}$$
$$x = \frac{2}{3}$$

The solution is $\frac{2}{3}$.

Example 2   Solve: $5x + 7 = 10$

Your solution $\frac{3}{5}$

Example 3   Solve: $5 = 9 - 2x$

Solution
$$5 = 9 - 2x$$
$$5 - 9 = 9 - 9 - 2x$$
$$-4 = -2x$$
$$\frac{-4}{-2} = \frac{-2x}{-2}$$
$$2 = x$$

The solution is 2.

Example 4   Solve: $2 = 11 + 3x$

Your solution -3

Solutions on p. A10

Example 5

Solve: $2x + 4 - 5x = 10$

Solution

$$\begin{aligned}
2x + 4 - 5x &= 10 \quad \text{Combine} \\
-3x + 4 &= 10 \quad \text{like terms.} \\
-3x + 4 - 4 &= 10 - 4 \\
-3x &= 6 \\
\frac{-3x}{-3} &= \frac{6}{-3} \\
x &= -2
\end{aligned}$$

The solution is -2.

Example 6

Solve: $x - 5 + 4x = 25$

Your solution

6

Solution on p. A10

Objective B   To solve application problems

Example 7

To determine the total cost of production, an economist uses the equation $T = U \cdot N + F$, where T is the total cost, U is the unit cost, N is the number of units made, and F is the fixed cost. Use this equation to find the number of units made during a month when the total cost was $9000, the unit cost was $25, and the fixed costs were $3000.

Strategy

To find the number of units made, replace each of the variables by their given value and solve for N.

Solution

$$\begin{aligned}
T &= U \cdot N + F \\
9000 &= 25N + 3000 \\
9000 - 3000 &= 25N + 3000 - 3000 \\
6000 &= 25N \\
\frac{6000}{25} &= \frac{25N}{25} \\
240 &= N
\end{aligned}$$

There were 240 units made.

Example 8

The pressure at a certain depth in the ocean can be approximated by the equation $P = 15 + \frac{1}{2}D$, where P is the pressure in pounds per square inch and D is the depth in feet. Use this equation to find the depth when the pressure is 45 pounds per square inch.

Your strategy

Your solution

60 ft

Solution on p. A10

3.2   **EXERCISES**

▶ **Objective A**

Solve and check.

1. $3x + 1 = 10$
3

2. $4y + 3 = 11$
2

3. $2a - 5 = 7$
6

4. $5m - 6 = 9$
3

5. $5 = 4x + 9$
−1

6. $2 = 5b + 12$
−2

7. $2x - 5 = -11$
−3

8. $3n - 7 = -19$
−4

9. $10 = 4 + 3d$
2

10. $13 = 9 + 4z$
1

11. $7 - c = 9$
−2

12. $2 - x = 11$
−9

13. $4 - 3w = -2$
2

14. $5 - 6x = -13$
3

15. $8 - 3t = 2$
2

16. $12 - 5x = 7$
1

17. $4a - 20 = 0$
5

18. $3y - 9 = 0$
3

19. $6 + 2b = 0$
−3

20. $10 + 5m = 0$
−2

21. $-2x + 5 = -7$
6

22. $-5d + 3 = -12$
3

23. $-12x + 30 = -6$
3

24. $-13 = -11y + 9$
2

25. $2 = 7 - 5a$
1

26. $3 = 11 - 4n$
2

27. $-35 = -6b + 1$
6

28. $-8x + 3 = -29$
4

29. $-3m - 21 = 0$
−7

30. $-5x - 30 = 0$
−6

31. $-4y + 15 = 15$
0

32. $-3x + 19 = 19$
0

33. $0 = 14 - 7c$
2

34. $0 = 24 - 6z$
4

35. $7x - 3 = 3$
$\frac{6}{7}$

36. $8y + 3 = 7$
$\frac{1}{2}$

37. $6a + 5 = 9$
$\frac{2}{3}$

38. $4 = 5b + 6$
$-\frac{2}{5}$

39. $11 = 15 + 4n$
−1

40. $4 = 2 - 3c$
$-\frac{2}{3}$

Solve and check.

41. $9 - 4x = 6$
$\dfrac{3}{4}$

42. $3t - 2 = 0$
$\dfrac{2}{3}$

43. $9x - 4 = 0$
$\dfrac{4}{9}$

44. $7 - 8z = 0$
$\dfrac{7}{8}$

45. $1 - 3x = 0$
$\dfrac{1}{3}$

46. $9d + 10 = 7$
$-\dfrac{1}{3}$

47. $12w + 11 = 5$
$-\dfrac{1}{2}$

48. $6y - 5 = -7$
$-\dfrac{1}{3}$

49. $8b - 3 = -9$
$-\dfrac{3}{4}$

50. $5 - 6m = 2$
$\dfrac{1}{2}$

51. $7 - 9a = 4$
$\dfrac{1}{3}$

52. $9 = -12c + 5$
$-\dfrac{1}{3}$

53. $10 = -18x + 7$
$-\dfrac{1}{6}$

54. $2y + \dfrac{1}{3} = \dfrac{7}{3}$
1

55. $4a + \dfrac{3}{4} = \dfrac{19}{4}$
1

56. $2n - \dfrac{3}{4} = \dfrac{13}{4}$
2

57. $3x - \dfrac{5}{6} = \dfrac{13}{6}$
1

58. $5y + \dfrac{3}{7} = \dfrac{3}{7}$
0

59. $9x + \dfrac{4}{5} = \dfrac{4}{5}$
0

60. $8 = 7d - 1$
$\dfrac{9}{7}$

61. $8 = 10x - 5$
$\dfrac{13}{10}$

62. $4 = 7 - 2w$
$\dfrac{3}{2}$

63. $7 = 9 - 5a$
$\dfrac{2}{5}$

64. $8t + 13 = 3$
$-\dfrac{5}{4}$

65. $12x + 19 = 3$
$-\dfrac{4}{3}$

66. $-6y + 5 = 13$
$-\dfrac{4}{3}$

67. $-4x + 3 = 9$
$-\dfrac{3}{2}$

68. $\dfrac{1}{2}a - 3 = 1$
8

69. $\dfrac{1}{3}m - 1 = 5$
18

70. $\dfrac{2}{5}y + 4 = 6$
5

71. $\dfrac{3}{4}n + 7 = 13$
8

72. $-\dfrac{2}{3}x + 1 = 7$
-9

73. $-\dfrac{3}{8}b + 4 = 10$
-16

74. $\dfrac{x}{4} - 6 = 1$
28

75. $\dfrac{y}{5} - 2 = 3$
25

76. $\dfrac{2x}{3} - 1 = 5$
9

77. $\dfrac{3c}{7} - 1 = 8$
21

78. $4 - \dfrac{3}{4}z = -2$
8

79. $3 - \dfrac{4}{5}w = -9$
15

80. $5 + \dfrac{2}{3}y = 3$
-3

81. $17 + \dfrac{5}{8}x = 7$
-16

82. $17 = 7 - \dfrac{5}{6}t$
-12

83. $9 = 3 - \dfrac{2x}{7}$
-21

84. $3 = \dfrac{3a}{4} + 1$
$\dfrac{8}{3}$

Solve and check.

85. $7 = \dfrac{2x}{5} + 4$

$\dfrac{15}{2}$

86. $5 - \dfrac{4c}{7} = 8$

$-\dfrac{21}{4}$

87. $7 - \dfrac{5}{9}y = 9$

$-\dfrac{18}{5}$

88. $6a + 3 + 2a = 11$

1

89. $5y + 9 + 2y = 23$

2

90. $7x - 4 - 2x = 6$

2

91. $11z - 3 - 7z = 9$

3

92. $2x - 6x + 1 = 9$

-2

93. $b - 8b + 1 = -6$

1

94. $3 = 7x + 9 - 4x$

-2

95. $-1 = 5m + 7 - m$

-2

96. $8 = 4n - 6 + 3n$

2

97. $0.135y + 0.0257 = -0.0742$

-0.74

98. $1.42x - 3.449 = 1.308$

3.35

99. $3.58 = 3.5686 + 0.076x$

0.15

100. $-6.54 = 1.48y - 3.062$

-2.35

101. If $2x - 3 = 7$, evaluate $3x + 4$.

19

102. If $3x + 5 = -4$, evaluate $2x - 5$.

-11

103. If $4 - 5x = -1$, evaluate $x^2 - 3x + 1$.

-1

104. If $2 - 3x = 11$, evaluate $x^2 + 2x - 3$.

0

105. If $5x + 3 - 2x = 12$, evaluate $4 - 5x$.

-11

106. If $2x - 4 - 7x = 16$, evaluate $x^2 + 1$.

17

▶ **Objective B**   *Application Problems*

The distance s that an object will fall in t seconds is given by $s = 16t^2 + vt$, where v is the initial velocity of the object.

107. Find the initial velocity of an object that falls 80 ft in 2 s.

8 ft/s

108. Find the initial velocity of an object that falls 144 ft in 3 s.

0 ft/s

A company uses the equation $V = C - 6000t$ to determine the depreciated value V, after t years, of a milling machine that originally cost C dollars. Equations like this are used in accounting for *straight-line depreciation*.

109. A milling machine originally cost $50,000. In how many years will the depreciated value be $38,000?
2 years

110. A milling machine originally cost $78,000. In how many years will the depreciated value be $48,000?
5 years

Anthropologists can approximate the height of a primate by the size of its *humerus* (the bone from the elbow to the shoulder) by using the equation $H = 1.2L + 27.8$, where L is the length of the humerus and H is the height of the primate.

111. An anthropologist estimates the height of a primate to be 66 in. What is the approximate length of the humerus of this primate to the nearest tenth of an inch?
31.8 in.

112. An anthropologist finds the humerus of a primate to be 28.4 in. long. Approximate the height of the primate to the nearest tenth of an inch.
61.9 in.

A telephone company estimates that the number N of phone calls per day between two cities of population P_1 and P_2 that are d miles apart is given by the equation $N = \frac{2.51P_1P_2}{d^2}$.

113. Estimate the population P_2, to the nearest thousand, given that the population of one city (P_1) is 48,000, the number of phone calls is 1,100,000, and the distance between the cities is 75 mi.
51,000 people

114. Estimate the population P_2, to the nearest thousand, given that the population of one city (P_1) is 125,000, the number of phone calls is 2,500,000, and the distance between the cities is 50 mi.
20,000 people

The equation $t = 17.08 - 0.0067y$ can be used to approximate the world record time for a 1-mile race, where t is the time and y is the year of the race.

115. Use this equation to approximate the year in which the first "4-minute mile" would be run. (The actual year was 1954.)
1952

116. In 1985, the world record for a 1-mile race was 3.77 minutes. For what year does the equation predict this record time?
1987

SECTION 3.3 General Equations—Part II

Objective A   **To solve an equation of the form $ax + b = cx + d$**

In solving an equation of the form $ax + b = cx + d$, the goal is to rewrite the equation in the form *variable = constant*. Begin by rewriting the equation so that there is only one variable term in the equation. Then rewrite the equation so that there is only one constant term.

Solve: $4x - 5 = 6x + 11$

The goal is to rewrite the equation in the form *variable = constant*.

Subtract $6x$ from each side of the equation. Simplify. Now there is only one variable term in the equation.

$$4x - 5 = 6x + 11$$
$$4x - 6x - 5 = 6x - 6x + 11$$
$$-2x - 5 = 11$$

Add 5 to each side of the equation. Simplify. Now there is only one constant term in the equation.

$$-2x - 5 + 5 = 11 + 5$$
$$-2x = 16$$

Divide each side of the equation by -2. Simplify. Now the equation is in the form *variable = constant*.

$$\frac{-2x}{-2} = \frac{16}{-2}$$
$$x = -8$$

-8 checks as the solution.

The solution is -8.

Example 1

Solve: $4x - 5 = 8x - 7$

Solution

$$4x - 5 = 8x - 7$$
$$4x - 8x - 5 = 8x - 8x - 7$$
$$-4x - 5 = -7$$
$$-4x - 5 + 5 = -7 + 5$$
$$-4x = -2$$
$$\frac{-4x}{-4} = \frac{-2}{-4}$$
$$x = \frac{1}{2}$$

The solution is $\frac{1}{2}$.

Example 2

Solve: $5x + 4 = 6 + 10x$

Your solution

$-\dfrac{2}{5}$

Solution on p. A10

Example 3

Solve: $3x + 4 - 5x = 2 - 4x$

Solution

$$
\begin{aligned}
3x + 4 - 5x &= 2 - 4x \\
-2x + 4 &= 2 - 4x \\
-2x + 4x + 4 &= 2 - 4x + 4x \\
2x + 4 &= 2 \\
2x + 4 - 4 &= 2 - 4 \\
2x &= -2 \\
\frac{2x}{2} &= \frac{-2}{2} \\
x &= -1
\end{aligned}
$$

The solution is -1.

Example 4

Solve: $5x - 10 - 3x = 6 - 4x$

Your solution

$\dfrac{8}{3}$

Solution on p. A10

Objective B   **To solve an equation containing parentheses**

When an equation contains parentheses, one of the steps in solving the equation requires the use of the Distributive Property. The Distributive Property is used to remove parentheses from a variable expression.

Solve: $4 + 5(2x - 3) = 3(4x - 1)$

The goal is to rewrite the equation in the form *variable* = *constant*.

Use the Distributive Property to remove parentheses. Simplify.

$$
\begin{aligned}
4 + 5(2x - 3) &= 3(4x - 1) \\
4 + 10x - 15 &= 12x - 3 \\
10x - 11 &= 12x - 3
\end{aligned}
$$

Subtract $12x$ from each side of the equation. Simplify. Now there is only one variable term in the equation.

$$
\begin{aligned}
10x - 12x - 11 &= 12x - 12x - 3 \\
-2x - 11 &= -3
\end{aligned}
$$

Add 11 to each side of the equation. Simplify. Now there is only one constant term in the equation.

$$
\begin{aligned}
-2x - 11 + 11 &= -3 + 11 \\
-2x &= 8
\end{aligned}
$$

Divide each side of the equation by -2. Simplify. Now the equation is in the form *variable* = *constant*.

$$
\begin{aligned}
\frac{-2x}{-2} &= \frac{8}{-2} \\
x &= -4
\end{aligned}
$$

-4 checks as the solution.

The solution is -4.

Content and Format © 1991 HMCo.

Example 5

Solve:
$3x - 4(2 - x) = 3(x - 2) - 4$

Solution

$$3x - 4(2 - x) = 3(x - 2) - 4$$
$$3x - 8 + 4x = 3x - 6 - 4$$
$$7x - 8 = 3x - 10$$
$$7x - 3x - 8 = 3x - 3x - 10$$
$$4x - 8 = -10$$
$$4x - 8 + 8 = -10 + 8$$
$$4x = -2$$
$$\frac{4x}{4} = \frac{-2}{4}$$
$$x = -\frac{1}{2}$$

The solution is $-\frac{1}{2}$.

Example 6

Solve:
$5x - 4(3 - 2x) = 2(3x - 2) + 6$

Your solution

2

Example 7

Solve:
$3[2 - 4(2x - 1)] = 4x - 10$

Solution

$$3[2 - 4(2x - 1)] = 4x - 10$$
$$3[2 - 8x + 4] = 4x - 10$$
$$3[6 - 8x] = 4x - 10$$
$$18 - 24x = 4x - 10$$
$$18 - 24x - 4x = 4x - 4x - 10$$
$$18 - 28x = -10$$
$$18 - 18 - 28x = -10 - 18$$
$$-28x = -28$$
$$\frac{-28x}{-28} = \frac{-28}{-28}$$
$$x = 1$$

The solution is 1.

Example 8

Solve:
$-2[3x - 5(2x - 3)] = 3x - 8$

Your solution

2

Solutions on p. A11

| Objective C | **To solve application problems** |

A lever system is shown at the right. It constists of a lever, or bar; a fulcrum; and two forces, F_1 and F_2. The distance d represents the length of the lever, x represents the distance from F_1 to the fulcrum, and $d - x$ represents the distance from F_2 to the fulcrum.

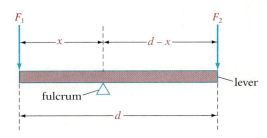

A principle of physics states that when the lever system balances, $F_1 \cdot x = F_2 \cdot (d - x)$.

Example 9

A lever is 15 ft long. A force of 50 lb is applied to one end of the lever, and a force of 100 lb is applied to the other end. Where is the fulcrum located when the system balances?

Example 10

A lever is 25 ft long. A force of 45 lb is applied to one end of the lever, and a force of 80 lb is applied to the other end. Where is the location of the fulcrum when the system balances?

Strategy

To find the location of the fulcrum when the system balances, replace the variables F_1, F_2, and d in the lever system equation by the given values, and solve for x.

Your strategy

Solution

$$F_1 \cdot x = F_2 \cdot (d - x)$$
$$50x = 100(15 - x)$$
$$50x = 1500 - 100x$$
$$50x + 100x = 1500 - 100x + 100x$$
$$150x = 1500$$
$$\frac{150x}{150} = \frac{1500}{150}$$
$$x = 10$$

The fulcrum is 10 ft from the 50-lb force.

Your solution

16 ft from the 45-lb force

Solution on p. A11

3.3 EXERCISES

▶ **Objective A**

Solve and check.

1. $8x + 5 = 4x + 13$
2

2. $6y + 2 = y + 17$
3

3. $7m + 4 = 6m + 7$
3

4. $11n + 3 = 10n + 11$
8

5. $5x - 4 = 2x + 5$
3

6. $9a - 10 = 3a + 2$
2

7. $12y - 4 = 9y - 7$
−1

8. $13b - 1 = 4b - 19$
−2

9. $15x - 2 = 4x - 13$
−1

10. $7a - 5 = 2a - 20$
−3

11. $3x + 1 = 11 - 2x$
2

12. $n - 2 = 6 - 3n$
2

13. $2x - 3 = -11 - 2x$
−2

14. $4y - 2 = -16 - 3y$
−2

15. $2b + 3 = 5b + 12$
−3

16. $m + 4 = 3m + 8$
−2

17. $4x - 7 = 5x + 1$
−8

18. $6d - 2 = 7d + 5$
−7

19. $4y - 8 = y - 8$
0

20. $5a + 7 = 2a + 7$
0

21. $6 - 5x = 8 - 3x$
−1

22. $10 - 4n = 16 - n$
−2

23. $5 + 7x = 11 + 9x$
−3

24. $3 - 2y = 15 + 4y$
−2

25. $2x - 4 = 6x$
−1

26. $2b - 10 = 7b$
−2

27. $8m = 3m + 20$
4

28. $9y = 5y + 16$
4

29. $-3x - 4 = 2x + 6$
−2

30. $-5a - 3 = 2a + 18$
−3

31. $8b + 5 = 5b + 7$
$\dfrac{2}{3}$

32. $6y - 1 = 2y + 2$
$\dfrac{3}{4}$

33. $7x - 8 = x - 3$
$\dfrac{5}{6}$

Solve and check.

34. $10x - 3 = 3x - 1$
$\dfrac{2}{7}$

35. $5n + 3 = 2n + 1$
$-\dfrac{2}{3}$

36. $8a - 2 = 4a - 5$
$-\dfrac{3}{4}$

37. $8.7y = 3.9y + 16.8$
3.5

38. $4.5x + 6.03 = 2.7x$
-3.35

39. $5.6x = 7.2x - 3.92$
2.45

40. If $5x = 3x - 8$, evaluate $4x + 2$.
-14

41. If $7x + 3 = 5x - 7$, evaluate $3x - 2$.
-17

42. If $2 - 6a = 5 - 3a$, evaluate $4a^2 - 2a + 1$.
7

43. If $1 - 5c = 4 - 4c$, evaluate $3c^2 - 4c + 2$.
41

44. If $2y + 3 = 5 - 4y$, evaluate $6y - 7$.
-5

45. If $3z + 1 = 1 - 5z$, evaluate $3z^2 - 7z + 8$.
8

▶ **Objective B**

Solve and check.

46. $5x + 2(x + 1) = 23$
3

47. $6y + 2(2y + 3) = 16$
1

48. $9n - 3(2n - 1) = 15$
4

49. $12x - 2(4x - 6) = 28$
4

50. $7a - (3a - 4) = 12$
2

51. $9m - 4(2m - 3) = 11$
-1

52. $2(3x + 1) - 4 = 16$
3

53. $4(2b + 3) - 9 = 11$
1

54. $5(3 - 2y) + 4y = 3$
2

55. $4(1 - 3x) + 7x = 9$
-1

56. $10x + 1 = 2(3x + 5) - 1$
2

57. $5y - 3 = 7 + 4(y - 2)$
2

58. $5b + 2(3b + 1) = 3b + 5$
$\dfrac{3}{8}$

59. $x + 3(4x - 2) = 7x - 1$
$\dfrac{5}{6}$

60. $3y - 7 = 5(2y - 3) + 4$
$\dfrac{4}{7}$

Solve and check.

61. $2a - 5 = 4(3a + 1) - 2$
$-\dfrac{7}{10}$

62. $5 - (9 - 6x) = 2x - 2$
$\dfrac{1}{2}$

63. $7 - (5 - 8x) = 4x + 3$
$\dfrac{1}{4}$

64. $3[2 - 4(y - 1)] = 3(2y + 8)$
$-\dfrac{1}{3}$

65. $5[2 - (2x - 4)] = 2(5 - 3x)$
5

66. $3a + 2[2 + 3(a - 1)] = 2(3a + 4)$
$\dfrac{10}{3}$

67. $5 + 3[1 + 2(2x - 3)] = 6(x + 5)$
$\dfrac{20}{3}$

68. $-2[4 - (3b + 2)] = 5 - 2(3b + 6)$
$-\dfrac{1}{4}$

69. $-4[x - 2(2x - 3)] + 1 = 2x - 3$
2

70. $0.36x - 2(1.63x) - 8.03 = 3(1.96x - 4.14)$
0.5

71. $0.593 - 0.14(2.1y - 3.15) = 0.26(2.9y - 6.1)$
2.5

72. If $4 - 3a = 7 - 2(2a + 5)$, evaluate $a^2 + 7a$.
0

73. If $9 - 5x = 12 - (6x + 7)$, evaluate $x^2 - 3x - 2$.
26

74. If $2z - 5 = 3(4z + 5)$, evaluate $\dfrac{z^2}{(z - 2)}$.
-1

75. If $3n - 7 = 5(2n + 7)$, evaluate $\dfrac{n^2}{(2n - 6)}$.
-2

▶ **Objective C** *Application Problems*

Use the lever system equation $F_1 x = F_2(d - x)$.

76. A lever 10 ft long is used to move a 100-lb rock. The fulcrum is placed 2 ft from the rock. What force must be applied to the other end of the lever to move the rock?
25 lb

77. A screwdriver 9 in. long is used as a lever to open a can of paint. The tip of the screwdriver is placed under the lip of the can with the fulcrum 0.15 in. from the lip. A force of 30 lb is applied to the other end of the screwdriver. Find the force on the lip of the can.
1770 lb

Use the lever system equation $F_1x = F_2(d - x)$.

78. In preparation for a stunt, two acrobats are standing on a plank 18 ft long. One acrobat weighs 128 lb and the second acrobat weighs 160 lb. How far from the 128-lb acrobat must the fulcrum be placed so that the acrobats are balanced on the plank?
10 ft

79. An adult and a child are on a see-saw 14 ft long. The adult weighs 175 lb and the child weighs 70 lb. How many feet from the child must the fulcrum be placed so that the see-saw balances?
10 ft

80. An 80-lb weight is applied to the left end of a see-saw 10 ft long with its fulcrum 5 ft from the 80-lb weight. A weight of 30 lb is applied to the right end of the see-saw. Is the 30-lb weight adequate to balance the see-saw?
No

81. Two children are sitting 8 ft apart on a see-saw. One child weighs 60 lb and the second child weighs 50 lb. The fulcrum is 3.5 ft from the 60-lb child. Is the see-saw balanced?
No

Economists use a *break-even point* to determine the number of units that must be sold so that no profit or loss occurs. The equation used is $Px = Cx + F$, where P is the selling price per unit, x is the number of units sold, C is the cost to make each unit, and F is the fixed cost.

82. An economist has determined the selling price for a television to be $250. The cost to make one television is $165 and the fixed costs are $29,750. Find the number of televisions that must be sold to break even.
350 televisions

83. A business analyst has determined the selling price for a cellular telephone to be $675. The cost to make one cellular telephone is $485 and the fixed costs are $36,100. Find the number of cellular telephones that must be sold to break even.
190 telephones

84. A manufacturing engineer determines the cost to make one compact disc to be $3.35 and the fixed costs to be $6180. The selling price for each compact disc is $8.50. Find the number of compact discs that must be sold to break even.
1200 compact discs

85. To manufacture a softball bat requires two steps. The first step is to cut a rough shape. The second step is to sand the bat to its final form. The cost to rough-shape a bat is $.45; and the cost to sand a bat to final form is $1.05. The total fixed costs for the two steps are $16,500. How many softball bats must be sold at a price of $7.00 to break even?
3000 softball bats

SECTION 3.4   Translating Sentences into Equations

Objective A   **To translate a sentence into an equation and solve**

An equation states that two mathematical expressions are equal. Therefore, to translate a sentence into an equation requires recognition of the words or phrases that mean "equals." Some of these phrases are listed below.

equals
is
is equal to   } translate to =
amounts to
represents

Once the sentence is translated into an equation, the equation can be solved by rewriting the equation in the form *variable = constant*.

Translate "five less than a number is thirteen" into an equation and solve.

Assign a variable to the unknown quantity.

The unknown number: n

Find two verbal expressions for the same value.

| Five less than a number | is | thirteen |

Write a mathematical expression for each verbal expression. Write the equals sign.

$$n - 5 = 13$$

Solve the equation.

$$n - 5 + 5 = 13 + 5$$
$$n = 18$$

The number is 18.

Example 1

Translate "three more than twice a number is the number plus six" into an equation and solve.

Solution

The unknown number: n

| Three more than twice a number | is | the number plus six |

$$2n + 3 = n + 6$$
$$2n - n + 3 = n - n + 6$$
$$n + 3 = 6$$
$$n + 3 - 3 = 6 - 3$$
$$n = 3$$

The number is 3.

Example 2

Translate "four less than one-third of a number equals five minus two-thirds of the number" into an equation and solve.

Your solution

$\frac{1}{3}x - 4 = 5 - \frac{2}{3}x;$

$x = 9$

Solution on p. A11

Example 3

Translate "four times the sum of a number and three is six less than twice the number" into an equation and solve.

Solution

The unknown number: n

Four times the sum of a number and three	is	six less than twice the number

$$4(n + 3) = 2n - 6$$
$$4n + 12 = 2n - 6$$
$$4n - 2n + 12 = 2n - 2n - 6$$
$$2n + 12 = -6$$
$$2n + 12 - 12 = -6 - 12$$
$$2n = -18$$
$$\frac{2n}{2} = \frac{-18}{2}$$
$$n = -9$$

The number is -9.

Example 4

Translate "two times the difference between a number and eight is equal to the sum of six times the number and eight" into an equation and solve.

Your solution

$2(n - 8) = 6n + 8$;
$n = -6$

Example 5

The sum of two numbers is sixteen. The difference between four times the smaller number and two is two more than twice the larger number. Find the two numbers.

Solution

The smaller number: n
The larger number: $16 - n$

The difference between four times the smaller and two	is	two more than twice the larger

$$4n - 2 = 2(16 - n) + 2$$
$$4n - 2 = 32 - 2n + 2$$
$$4n - 2 = 34 - 2n$$
$$4n + 2n - 2 = 34 - 2n + 2n$$
$$6n - 2 = 34$$
$$6n - 2 + 2 = 34 + 2$$
$$6n = 36$$
$$\frac{6n}{6} = \frac{36}{6}$$
$$n = 6 \quad 16 - n = 10$$

The smaller number is 6.
The larger number is 10.

Example 6

The sum of two numbers is twelve. The total of three times the smaller number and six amounts to seven less than the product of four and the larger number. Find the two numbers.

Your solution

5 and 7

Solutions on pp. A11–A12

Objective B   To solve application problems

Example 7

The temperature of the sun on the Kelvin scale is 6500° K (K is an abbreviation for kelvins, the units in which temperature is measured on the Kelvin scale). This is 454° more than the temperature on the Fahrenheit scale. Find the Fahrenheit temperature.

Strategy

To find the Fahrenheit temperature, write and solve an equation using F to represent the Fahrenheit temperature.

Solution

| 6500 | is | 454° more than Fahrenheit temperature |

$$6500 = F + 454$$
$$6500 - 454 = F + 454 - 454$$
$$6046 = F$$

The Fahrenheit temperature of the sun is 6046°.

Example 8

A molecule of octane gas has 8 carbon atoms. This represents twice the number of carbon atoms in a butane gas molecule. Find the number of carbon atoms in a butane molecule.

Your strategy

Your solution

4 carbon atoms

Example 9

The Fahrenheit temperature is 68°. This is 32° more than $\frac{9}{5}$ the Celsius temperature. Find the Celsius temperature.

Strategy

To find the Celsius temperature, write and solve an equation using C to represent the Celsius temperature.

Solution

| 68 | is | 32 degrees more than $\frac{9}{5}$ the Celsius temperature |

$$68 = \frac{9}{5}C + 32$$
$$68 - 32 = \frac{9}{5}C + 32 - 32$$
$$36 = \frac{9}{5}C$$
$$\frac{5}{9} \cdot 36 = \frac{5}{9} \cdot \frac{9}{5}C$$
$$20 = C$$

The Celsius temperature is 20°.

Example 10

The Celsius temperature is 20°. This is equal to $\frac{5}{9}$ of the difference between the Fahrenheit temperature and 32. Find the Fahrenheit temperature.

Your strategy

Your solution

68°F

Solutions on p. A12

Example 11

Primary and secondary waves are two types of waves that occur after an earthquake. The speed of a secondary wave is 8 mi/s. This is 12 mi/s less than twice the speed of a primary wave. Find the speed of a primary wave.

Strategy

To find the speed, write and solve an equation using s to represent the unknown speed.

Solution

| 8 | is | 12 mi/s less than twice the speed of a primary wave |

$$8 = 2s - 12$$
$$8 + 12 = 2s - 12 + 12$$
$$20 = 2s$$
$$\frac{20}{2} = \frac{2s}{2}$$
$$10 = s$$

The speed is 10 mi/s.

Example 13

A board 20 ft long is cut into two pieces. Five times the length of the smaller piece is 2 ft more than twice the length of the longer piece. Find the length of each piece.

Strategy

To find the length of each piece, write and solve an equation using x to represent the length of the shorter piece and $20 - x$ to represent the length of the longer piece.

Solution

| Five times the smaller piece | is | two feet more than twice the longer |

$$5x = 2(20 - x) + 2$$
$$5x = 40 - 2x + 2$$
$$5x = 42 - 2x$$
$$5x + 2x = 42 - 2x + 2x$$
$$7x = 42$$
$$\frac{7x}{7} = \frac{42}{7}$$
$$x = 6$$
$$20 - x = 14$$

The shorter piece is 6 ft.
The longer piece is 14 ft.

Example 12

The length of a wire that produces a C note on a piano is 10 in. This represents 6 in. less than half the length of the wire that produces an A note. Find the length of the wire that produces an A note.

Your strategy

Your solution

32 in.

Example 14

A company makes 140 televisions per day. Three times the number of black and white TV's made equals 20 less than the number of color TV's made. Find the number of color TV's made each day.

Your strategy

Your solution

110 color TV's

Solutions on pp. A12–A13

Content and Format © 1991 HMCo.

3.4   EXERCISES

▶ **Objective A**

Translate into an equation and solve.

1. The difference between a number and fifteen is seven. Find the number.
    $x - 15 = 7$; $x = 22$

2. The sum of five and a number is three. Find the number.
    $5 + x = 3$; $x = -2$

3. The product of seven and a number is negative twenty-one. Find the number.
    $7x = -21$; $x = -3$

4. The quotient of a number and four is two. Find the number.
    $\frac{x}{4} = 2$; $x = 8$

5. Four less than three times a number is five. Find the number.
    $3x - 4 = 5$; $x = 3$

6. The difference between five and twice a number is one. Find the number.
    $5 - 2x = 1$; $x = 2$

7. Four times the sum of twice a number and three is twelve. Find the number.
    $4(2x + 3) = 12$; $x = 0$

8. Seventeen is three times the difference between four times a number and 5. Find the number.
    $17 = 3(4x - 5)$; $x = \frac{8}{3}$

9. Twelve is six times the difference between a number and three. Find the number.
    $12 = 6(x - 3)$; $x = 5$

10. The difference between six times a number and four times the number is negative 14. Find the number.
    $6x - 4x = -14$; $x = -7$

11. The product of a number and four added to seven is equal to three. Find the number.
    $4x + 7 = 3$; $x = -1$

12. The total of six times a number and four times a number is one. Find the number.
    $6x + 4x = 1$; $x = \frac{1}{10}$

13. Twenty-two is two less than six times a number. Find the number.
    $22 = 6x - 2$; $x = 4$

14. Negative fifteen is three more than twice a number. Find the number.
    $-15 = 2x + 3$; $x = -9$

15. Three times a number is equal to the product of the number and two added to four. Find the number.
    $3x = 2x + 4$; $x = 4$

16. A number is equal to four more than five times the number. Find the number.
    $x = 5x + 4$; $x = -1$

Translate into an equation and solve.

17. Seven more than four times a number is three more than two times the number. Find the number.
$4x + 7 = 2x + 3; x = -2$

18. The difference between three times a number and four is five times the number. Find the number.
$3x - 4 = 5x; x = -2$

19. Eight less than five times a number is four more than eight times the number. Find the number.
$5x - 8 = 8x + 4; x = -4$

20. The sum of a number and six is four less than six times the number. Find the number.
$x + 6 = 6x - 4; x = 2$

21. Twice the difference between a number and twenty-five is three times the number. Find the number.
$2(x - 25) = 3x; x = -50$

22. Four times a number is three times the difference between thirty-five and the number. Find the number.
$4x = 3(35 - x); x = 15$

23. The sum of two numbers is twenty. Three times the smaller is equal to two times the larger. Find the two numbers.
$3x = 2(20 - x);$ 8 and 12

24. The sum of two numbers is fifteen. One less than three times the smaller is equal to the larger. Find the two numbers.
$3x - 1 = 15 - x;$ 4 and 11

25. The sum of two numbers is twenty-four. Four less than three times the smaller is twelve less than twice the larger. Find the two numbers.
$3x - 4 = 2(24 - x) - 12;$ 8 and 16

26. The sum of two numbers is fourteen. The difference between two times the smaller and the larger is one. Find the two numbers.
$2x - (14 - x) = 1;$ 5 and 9

27. The sum of two numbers is eighteen. The total of three times the smaller and twice the larger is forty-four. Find the two numbers.
$3x + 2(18 - x) = 44;$ 8 and 10

28. The sum of two numbers is 2. The difference between eight and twice the smaller number is two less than four times the larger. Find the two numbers.
$8 - 2x = 4(2 - x) - 2;$ -1 and 3.

▶ **Objective B**   *Application Problems*

Write an equation and solve.

29. The manufacturer's suggested retail price for a compact disc player is $238. This price is $96 more than the cost of the compact disc player. Find the cost of the compact disc player.
$142

30. The width of a football field is 20 yd more than $\frac{1}{3}$ the length. The length is 100 yd. Find the width of the field.
$53\frac{1}{3}$ yd

Write an equation and solve.

31. The operating speed of a personal computer is 8 megahertz. This is one-fourth the speed of a newer model personal computer. Find the speed of the newer personal computer.
32 megahertz

32. One measure of computer speed is *mips*, *m*illions of *i*nstructions *per s*econd. One computer has a rating of 10 mips. This is two-thirds the speed of a second computer. Find the mips rating of the second computer.
15 mips

33. The score for one team of a football game was 26. The team scored twice as many field goals (3 points each) as it did touchdowns with extra points (7 points each). Find the number of field goals that were scored.
4 field goals

34. An office furniture store maintains an inventory of twice as many chairs as desks. There are currently 36 desks and chairs in stock. How many chairs are in stock?
24 chairs

35. A university employs a total of 600 teaching assistants and research assistants. There are three times as many teaching assistants as research assistants. Find the number of research assistants employed by the university.
150 research assistants

36. A basketball team scored 105 points. There were as many two-point baskets as free throws (one point each). The number of three-point baskets was five less than the number of free throws. Find the number of three-point baskets.
15 three-point baskets

37. A 24-lb soil supplement contains iron, potassium, and a mulch. There is five times as much mulch as iron and twice as much potassium as iron. Find the amount of iron, potassium, and mulch in the soil supplement.
3 lb iron, 6 lb potassium, 15 lb mulch

38. A real estate agent sold two homes and received a total commission of $6000. The agent's commission on one home was one and one-half times the commission on the second home. Find the agent's commission for each home.
$3600 first home, $2400 second home

39. The purchase price of a new big-screen TV, including finance charges, was $3276. A down payment of $450 was made. The remainder was paid in 24 equal monthly installments. Find the monthly payment to the nearest cent.
$117.75

40. The purchase price of a new computer system, including finance charges, was $6150. A down payment of $250 was made. The remainder was paid in 36 equal monthly installments. Find the monthly payment to the nearest cent.
$163.89

41. The cost to replace a water pump in a sports car was $600. This included $375 for the water pump and $45 per hour for labor. Find the number of hours of labor.
5 h

Write an equation and solve.

42. The cost of electricity in a certain city is $.08 for each of the first 300 kWh (kilowatt-hours) and $.13 for each kWh over 300 kWh. Find the number of kilowatt hours used for a family with a $51.95 electric bill.
515 kWh

43. An investor deposits $5000 into two accounts. Two times the smaller deposit is $1000 more than the larger deposit. Find the amount deposited into each account.
$2000, $3000

44. A rectangular cement patio is constructed so that the length of the patio is 3 ft less than three times the width. The sum of the length and width of the patio is 21 ft. Find the length of the patio.
15 ft

45. Greek architects thought that a rectangle whose length was approximately 1.6 times the width yielded the most eye-pleasing proportions for the front of a building. The sum of the length and width of a certain rectangle constructed in this manner is 130 ft. Find the width and length of the rectangle.
width 50 ft, length 80 ft

46. The sum of the angles of a triangle, measured in degrees, is always 180°. One angle of a certain triangle is twice the smallest angle. The third angle is three times the smallest angle. Find each angle.
30°, 60°, 90°

47. A computer screen consists of tiny dots of light called pixels. In a certain graphics mode, there are 640 horizontal pixels. This is 40 more than three times the number of vertical pixels. Find the number of vertical pixels.
200 pixels

48. The velocity of a falling object equals the sum of its initial velocity and the product of 32 and the time it falls in seconds. How long has an object been falling whose velocity is 88 ft/s and whose initial velocity was 24 ft/s?
2 s

49. A wire 12 ft long is cut into two pieces. Each piece is bent into the shape of a square. The perimeter of the larger square is twice the perimeter of the smaller square. Find the perimeter of the larger square.
8 ft

50. Five thousand dollars is divided between two scholarships so that three times the smaller scholarship is twice the larger scholarship. Find the amount of the larger scholarship.
$3000

51. A carpenter is building a wood door frame whose height is 1 ft less than three times the width. What is the width of the largest door frame that can be constructed from a board 22 ft long?
3 ft

52. A board 20 ft long is cut into two pieces so that twice the length of the shorter piece is 4 ft more than the length of the longer piece. Find the length of the shorter piece.
8 ft

Calculators and Computers

Solving a First-Degree Equation

Chapter 3, Solving Equations, is a core chapter of this text. You will be using the equation-solving skills taught in this chapter throughout the remainder of your studies in algebra.

The order in which we apply the skills to solve equations is very important. The order in which these skills are used to solve an equation is as follows:

1. Remove parentheses.

2. Get all variable terms on one side of the equation and simplify.

3. Get all constant terms on the other side of the equation and simplify.

4. Multiply both sides of the equation by the reciprocal of the coefficient of the variable term.

Solving equations is a learned skill and requires practice. To provide you with additional practice, the program SOLVE A FIRST-DEGREE EQUATION on the Math ACE Disk will enable you to practice solving the following three types of equations:

1. $ax + b = c$

2. $ax + b = cx + d$

3. Equations with parentheses

1. After you select the type of equation you want to practice, a problem will be displayed on the screen.

2. Using paper and pencil, solve the problem.

3. When you are ready, press the RETURN key and the complete solution will be displayed.

4. Compare your solution with the displayed solution.

5. All answers are rounded to the nearest hundredth.

6. When you finish a problem, you may continue practicing the type of problem you have selected, return to the main menu and select a different type, or quit the program.

Chapter Summary

Key Words   An *equation* expresses the equality of two mathematical expressions.

A *solution* of an equation is a number that, when substituted for the variable, results in a true equation.

To *solve* an equation means to find a solution of the equation. The goal is to rewrite the equation in the form *variable = constant*.

To *translate* a sentence into an equation requires recognition of the words or phrases that mean "equals." Some of these phrases are *equals, is, is equal to, amounts to,* and *represents.*

Essential Rules   *Addition Property of Equations*

The same quantity can be added to each side of an equation without changing the solution of the equation.

If $a = b$, then
$a + c = b + c.$

Multiplication Property of Equations

Each side of an equation can be multiplied by the same nonzero number without changing the solution of the equation.

If $a = b$ and $c \neq 0$, then
$ac = bc.$

Chapter Review

SECTION 3.1

1. Is 3 a solution of $5x - 2 = 4x + 5$?
no

2. Is $-\dfrac{1}{6}$ a solution of $5 - 18x = 18x + 11$?
yes

3. Is 2 a solution of $x^2 + 4x + 1 = 3x + 7$?
yes

4. Is -4 a solution of $x^2 - 3x = (x + 2)^2$?
no

5. Solve $x + 3 = 24$.
21

6. Solve $a - \dfrac{1}{6} = \dfrac{2}{3}$.
$\dfrac{5}{6}$

7. Solve $4.6 = 2.1 + x$.
2.5

8. Solve $\dfrac{3}{5}a = 12$.
20

SECTION 3.2

9. Solve $5x - 6 = 29$.
7

10. Solve $-4x - 2 = 10$.
-3

11. Solve $32 = 9x - 4 - 3x$.
6

12. Solve $14x + 7x + 8 = -10$.
$-\dfrac{6}{7}$

13. Use the equation $I = Prt$, where I is the simple interest, P is the principal, r is the annual simple interest rate, and t is the time in years. If \$1500 is invested at the annual simple interest rate of 0.08, in how many years will the interest be \$1800?
15 years

14. Use the equation $T = c + rc$, where T is the total amount of a purchase including tax, r is the sales tax rate, and c is the cost before tax is added. Find the cost of a cookbook before tax is added if the sales tax rate is 0.07 and the total price of the book, including tax, is \$29.96.
\$28.00

15. Use the equation $F = \dfrac{9}{5}C + 32$, where F is the temperature in degrees Fahrenheit and C is the temperature in degrees Celsius. Find the temperature in degrees Celsius of a person whose temperature is 98.6 degrees Fahrenheit.
37° Celsius

16. Use the equation $S = R - D$, where S is the sale price, R is the regular price, and D is the discount. Find the discount on a mattress set if the sale price is \$220.75 and the regular price is \$300.00.
\$79.25

SECTION 3.3

17. Solve $5x + 3 = 10x - 17$.
4

18. Solve $12y - 1 = 3y + 2$.
$\frac{1}{3}$

19. Solve $-6x + 16 = -2x$.
4

20. Solve $-7x - 5 = 4x + 50$.
-5

21. Solve $6x + 3(2x - 1) = -27$.
-2

22. Solve $5 + 2(x + 1) = 13$.
3

23. Solve $x + 5(3x - 20) = 10(x - 4)$.
10

24. Solve $7 - [4 + 2(x - 3)] = 11(x + 2)$.
-1

25. A lever is 12 ft long. At a distance of 2 ft from the fulcrum a force of 120 lb is applied. How large a force must be applied to the other end of the lever so that the system will balance? Use the lever system equation $F_1 \cdot x = F_2 \cdot (d - x)$.
24 lb

26. A lever is 8 ft long. A force of 25 lb is applied to one end of the lever and a force of 15 lb is applied to the other end. Find the location of the fulcrum when the system balances. Use the lever system equation $F_1 \cdot x = F_2 \cdot (d - x)$.
3 ft from the 25 lb force

SECTION 3.4

27. Translate "four less than the product of five and a number is sixteen" into an equation and solve.
$5n - 4 = 16$; $n = 4$

28. Translate "the product of six and three more than a number is ten less than twice the number" into an equation and solve.
$6(n + 3) = 2n - 10$; $n = -7$

29. The sum of two numbers is twenty-one. Three times the smaller number is two less than twice the larger number. Translate into an equation and then find the two numbers.
$3x = 2(21 - x) - 2$; 8 and 13

30. A piano wire is 35 in. long. A fifth chord can be produced by dividing this wire into two parts so that three times the length of the shorter piece is twice the length of the longer piece. Find the length of each piece.
14 in. and 21 in.

31. An optical engineer's consulting fee was $600. This included $80 for supplies and $65 for each hour of consulting. Find the number of hours of consulting.
8 h

32. The Empire State Building is 1472 ft tall. This is 514 ft less than twice the height of the Eiffel Tower. Find the height of the Eiffel Tower.
993 ft

Chapter Test

1. Is -2 a solution of $x^2 - 3x = 2x - 6$?
   no   [3.1A]

2. Is $\frac{2}{3}$ a solution of $6x - 7 = 3 - 9x$?
   yes   [3.1A]

3. Solve: $x - 3 = -8$
   -5 [3.1B]

4. Solve: $\frac{3}{4}x = -9$
   -12 [3.1C]

5. Solve: $3x - 5 = -14$
   -3 [3.2A]

6. Solve: $7 - 4x = -13$
   5   [3.2A]

7. A financial manager has determined that the cost per unit for a calculator is $15 and that the fixed costs per month are $2000. Find the number of calculators produced during a month in which the total cost was $5000. Use the equation $T = U \cdot N + F$, where T is the total cost, U is the cost per unit, N is the number of units produced, and F is the fixed costs.
   200 calcluators   [3.2B]

8. Solve: $3x - 2 = 5x + 8$
   -5 [3.3A]

9. Solve: $6 - 5x = 5x + 11$
   $-\dfrac{1}{2}$ [3.3A]

10. Solve: $5x + 3 - 7x = 2x - 5$
    2   [3.3A]

11. Solve: $4 - 2(3 - 2x) = 2(5 - x)$
    2   [3.3B]

12. Solve: $5x - 2(4x - 3) = 6x + 9$
    $-\dfrac{1}{3}$ [3.3B]

13. Solve: $9 - 3(2x - 5) = 12 + 5x$
    $\dfrac{12}{11}$ [3.3B]

14. A chemist mixes 100 g of water at 80°C with 50 g of water at 20°C. Use the equation $m_1 \cdot (T_1 - T) = m_2 \cdot (T - T_2)$ to find the final temperature of the water after mixing. m_1 is the quantity of water at the hotter temperature, T_1 is the temperature of the hotter water, m_2 is the quantity of water at the cooler temperature, T_2 is the temperature of the cooler water, and T is the final temperature of the water after mixing.
60°C   [3.3C]

15. Translate "the difference between three times a number and fifteen is twenty-seven" into an equation and solve.
$3x - 15 = 27$; $x = 14$ [3.4A]

16. Translate "the sum of five times a number and six equals the product of the number plus twelve and three" into an equation and solve.
$5x + 6 = 3(x + 12)$; $x = 15$ [3.4A]

17. The sum of two numbers is 18. The difference between four times the smaller number and seven is equal to the sum of two times the larger number and five. Find the two numbers.
8 and 10   [3.4A]

18. A train travels between two cities in 26 h. This is 5 h more than the product of three and the time required for a plane to fly between the two cities. Find the number of hours required for the plane to fly between the two cities.
7 h   [3.4B]

19. A board 18 ft long is cut into two pieces. Two feet less than the product of five and the length of the smaller piece is equal to the difference between three times the length of the longer piece and eight. Find the length of each piece.
6 ft and 12 ft   [3.4B]

20. The time it takes a hard disk controller in a computer to access data is 28 milliseconds (ms). This is 2 ms less than twice the time a new hard disk controller can access the data. How many milliseconds are required for the new disk controller to access the data?
15 ms   [3.4B]

Cumulative Review

1. Subtract: $-6 - (-20) - 8$
   6   [1.2B]

2. Multiply: $(-2)(-6)(-4)$
   -48 [1.3A]

3. Simplify $-\dfrac{5}{6} - \left(-\dfrac{7}{16}\right)$.
   $-\dfrac{19}{48}$ [1.4B]

4. Simplify $-2\dfrac{1}{3} \div 1\dfrac{1}{6}$.
   -2 [1.4C]

5. Simplify $-4^2 \cdot \left(-\dfrac{3}{2}\right)^3$.
   54   [1.5A]

6. Simplify $25 - 3\dfrac{(5-2)^2}{2^3+1} - (-2)$.
   24   [1.5B]

7. Evaluate $3(a - c) - 2ab$ when $a = 2$, $b = 3$, and $c = -4$.
   6   [2.1A]

8. Simplify $3x - 8x + (-12x)$.
   $-17x$ [2.2A]

9. Simplify $2a - (-3b) - 7a - 5b$.
   $-5a - 2b$ [2.2A]

10. Simplify $(16x)\left(\dfrac{1}{8}\right)$.
    $2x$ [2.2B]

11. Simplify $-4(-9y)$.
    $36y$ [2.2B]

12. Simplify $-2(-x^2 - 3x + 2)$.
    $2x^2 + 6x - 4$ [2.2C]

13. Simplify $-2(x - 3) + 2(4 - x)$.
    $-4x + 14$ [2.2D]

14. Simplify $-3[2x - 4(x - 3)] + 2$.
    $6x - 34$ [2.2D]

15. Is -3 a solution of $x^2 + 6x + 9 = x + 3$?
    yes   [3.1A]

16. Is $\dfrac{1}{2}$ a solution of $3 - 8x = 12x - 2$?
    no   [3.1A]

17. Solve: $x - 4 = -9$
    -5 [3.1B]

18. Solve: $\dfrac{3}{5}x = -15$
    -25 [3.1C]

19. Solve: $7x - 8 = -29$
    -3 [3.2A]

20. Solve: $13 - 9x = -14$
    3   [3.2A]

21. Solve: $8x - 3(4x - 5) = -2x - 11$
13   [3.3B]

22. Solve: $6 - 2(5x - 8) = 3x - 4$
2   [3.3B]

23. Solve: $5x - 8 = 12x + 13$
−3   [3.3A]

24. Solve: $11 - 4x = 2x + 8$
$\frac{1}{2}$ [3.3A]

25. A business manager has determined that the cost per unit for a camera is $70 and that the fixed costs per month are $3500. Find the number of cameras that are produced during a month in which the total cost was $21,000. Use the equation $T = U \cdot N + F$, where T is the total cost, U is the cost per unit, N is the number of units produced, and F is the fixed cost.
250 cameras   [3.2B]

26. A chemist mixes 300 g of water at 75°C with 100 g of water at 15°C. Use the equation $m_1 \cdot (T_1 - T) = m_2 \cdot (T - T_2)$ to find the final temperature of the water. m_1 is the quantity of water at the hotter temperature, T_1 is the temperature of the hotter water, m_2 is the quantity of water at the cooler temperature, T_2 is the temperature of the cooler water, and T is the final temperature of the water after mixing.
60°C   [3.3C]

27. Translate "the difference between twelve and the product of five and a number is negative eighteen" into an equation and solve.
$12 - 5x = -18$; $x = 6$ [3.4A]

28. Translate "the sum of six times a number and thirteen is five less than the product of three and the number" into an equation and solve.
$6x + 13 = 3x - 5$; $x = -6$ [3.4A]

29. The area of the cement foundation of a house is 2000 ft². This is 200 ft² more than three times the area of the garage. Find the area of the garage.
600 ft²   [3.4B]

30. A board 16 ft long is cut into two pieces. Four feet more than the product of three and the length of the shorter piece is equal to three feet less than twice the length of the longer piece. Find the length of each piece.
5 ft, 11 ft   [3.4B]

4

Solving Equations: Applications

OBJECTIVES

▶ To write a percent as a fraction or a decimal
▶ To write a fraction or a decimal as a percent
▶ To solve the basic percent equation
▶ To solve application problems
▶ To solve markup problems
▶ To solve discount problems
▶ To solve investment problems
▶ To solve value mixture problems
▶ To solve percent mixture problems
▶ To solve uniform motion problems
▶ To solve perimeter problems
▶ To solve problems involving the angles of a triangle
▶ To solve consecutive integer problems
▶ To solve coin and stamp problems
▶ To solve age problems

Word Problems

Word problems have been challenging students of mathematics for a long time. Here are two types of problems you may have seen before:

A number added to $\frac{1}{7}$ of the number is 19. What is the number?

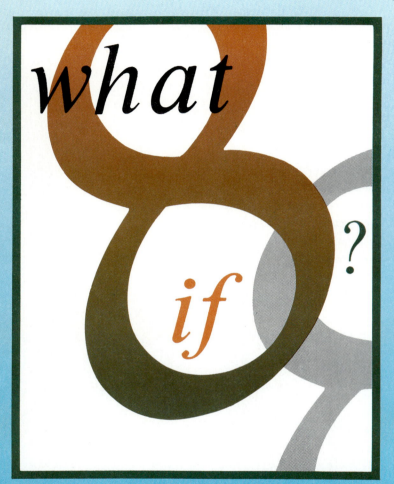

A dog is chasing a rabbit that has a head start of 150 feet.
The dog jumps 9 feet every time the rabbit jumps 7 feet.
In how many jumps will the dog catch up with the rabbit?

What is unusual about these problems is that the first one is about 4000 years old and occurred as Problem 1 in the Rhind Papyrus. The second problem is about 1500 years old and comes from a Latin algebra book written in A.D. 450.

These examples illustrate that word problems have been around for a long time. The long history of word problems also reflects the importance that each generation has placed on solving these problems. It is through word problems that the initial steps of applying mathematics are taken.

The answer to the first problem is $16\frac{5}{8}$.

The answer to the second problem is 75 jumps.

Content and Format © 1991 HMCo.

SECTION 4.1   Introduction to Percent

Objective A   **To write a percent as a fraction or a decimal**

"A population growth rate of 3%," "a manufacturer's discount of 25%," and "an 8% increase in pay" are typical examples of the many ways in which percent is used in applied problems. **Percent** means "parts of 100." Thus, 27% means 27 parts of 100.

In applied problems involving a percent, it is usually necessary either to rewrite the percent as a fraction or a decimal, or to rewrite a fraction or a decimal as a percent.

To write a percent as a fraction, remove the percent sign and multiply by $\frac{1}{100}$.

Write 27% as a fraction.

Remove the percent sign and multiply by $\frac{1}{100}$.

$$27\% = 27\left(\frac{1}{100}\right) = \frac{27}{100}$$

To write a percent as a decimal, remove the percent sign and multiply by 0.01.

Write 33% as a decimal.

Remove the percent sign. Then multiply by 0.01.

$$33\% = 33(0.01) = 0.33$$

Move the decimal point two places to the left. Then remove the percent sign.

Example 1
Write 130% as a fraction and as a decimal.

Solution
$$130\% = 130\left(\frac{1}{100}\right) = \frac{130}{100} = 1\frac{3}{10}$$

$$130\% = 130(0.01) = 1.30$$

Example 2
Write 125% as a fraction and as a decimal.

Your solution
$1\frac{1}{4}$,   1.25

Example 3
Write $33\frac{1}{3}\%$ as a fraction.

Solution
$$33\frac{1}{3}\% = 33\frac{1}{3}\left(\frac{1}{100}\right) = \frac{100}{3}\left(\frac{1}{100}\right) = \frac{1}{3}$$

Example 4
Write $16\frac{2}{3}\%$ as a fraction.

Your solution
$\frac{1}{6}$

Example 5
Write 0.25% as a decimal.

Solution
$$0.25\% = 0.25(0.01) = 0.0025$$

Example 6
Write 0.5% as a decimal.

Your solution
0.005

Solutions on p. A13

| **Objective B** | **To write a fraction or a decimal as a percent** |

A fraction or decimal can be written as a percent by multiplying by 100%.

Write $\frac{5}{8}$ as a percent.

Multiply by 100%.

$$\frac{5}{8} = \frac{5}{8}(100\%) = \frac{500}{8}\% = 62.5\% \text{ or } 62\frac{1}{2}\%$$

Write 0.82 as a percent.

Multiply by 100%.

$$0.82 = 0.82(100\%) = 82\%$$

Move the decimal point two places to the right. Then write the percent sign.

Example 7

Write 0.027 as a percent.

Solution

$0.027 = 0.027(100\%) = 2.7\%$

Example 8

Write 0.043 as a percent.

Your solution

4.3%

Example 9

Write 1.34 as a percent.

Solution

$1.34 = 1.34(100\%) = 134\%$

Example 10

Write 2.57 as a percent.

Your solution

257%

Example 11

Write $\frac{5}{6}$ as a percent. Round to the nearest tenth of a percent.

Solution

$\frac{5}{6} = \frac{5}{6}(100\%) = \frac{500}{6}\% \approx 83.3\%$

Example 12

Write $\frac{5}{9}$ as a percent. Round to the nearest tenth of a percent.

Your solution

55.6%

Example 13

Write $\frac{7}{16}$ as a percent. Write the remainder in fractional form.

Solution

$\frac{7}{16} = \frac{7}{16}(100\%) = \frac{700}{16}\% = 43\frac{3}{4}\%$

Example 14

Write $\frac{9}{16}$ as a percent. Write the remainder in fractional form.

Your solution

$56\frac{1}{4}\%$

Solutions on p. A13

| 4.1 | **EXERCISES** |

▶ **Objective A**

Write as a fraction and a decimal.

1. 75%
$\frac{3}{4}$; 0.75

2. 40%
$\frac{2}{5}$; 0.40

3. 50%
$\frac{1}{2}$; 0.50

4. 10%
$\frac{1}{10}$; 0.10

5. 64%
$\frac{16}{25}$; 0.64

6. 88%
$\frac{22}{25}$; 0.88

7. 125%
$1\frac{1}{4}$; 1.25

8. 160%
$1\frac{3}{5}$; 1.60

9. 19%
$\frac{19}{100}$; 0.19

10. 87%
$\frac{87}{100}$; 0.87

11. 5%
$\frac{1}{20}$; 0.05

12. 2%
$\frac{1}{50}$; 0.02

13. 450%
$4\frac{1}{2}$; 4.50

14. 380%
$3\frac{4}{5}$; 3.80

15. 8%
$\frac{2}{25}$; 0.08

16. 4%
$\frac{1}{25}$; 0.04

Write as a fraction.

17. $11\frac{1}{9}\%$
$\frac{1}{9}$

18. $4\frac{2}{7}\%$
$\frac{3}{70}$

19. $12\frac{1}{2}\%$
$\frac{1}{8}$

20. $37\frac{1}{2}\%$
$\frac{3}{8}$

21. $31\frac{1}{4}\%$
$\frac{5}{16}$

22. $66\frac{2}{3}\%$
$\frac{2}{3}$

23. $\frac{1}{4}\%$
$\frac{1}{400}$

24. $\frac{1}{2}\%$
$\frac{1}{200}$

25. $5\frac{3}{4}\%$
$\frac{23}{400}$

26. $68\frac{3}{4}\%$
$\frac{11}{16}$

27. $6\frac{1}{4}\%$
$\frac{1}{16}$

28. $83\frac{1}{3}\%$
$\frac{5}{6}$

Write as a decimal.

29. 7.3%
0.073

30. 9.1%
0.091

31. 15.8%
0.158

32. 16.7%
0.167

33. 0.3%
0.003

34. 0.9%
0.009

35. 9.15%
0.0915

36. 121.2%
1.212

37. 18.23%
0.1823

38. 62.14%
0.6214

39. 0.15%
0.0015

40. 0.27%
0.0027

▶ **Objective B**

Write as a percent.

41. 0.15
15%

42. 0.37
37%

43. 0.05
5%

44. 0.02
2%

45. 0.175
17.5%

46. 0.125
12.5%

47. 1.15
115%

48. 1.36
136%

49. 0.62
62%

50. 0.96
96%

51. 3.165
316.5%

52. 2.142
214.2%

53. 0.008
0.8%

54. 0.004
0.4%

55. 0.065
6.5%

56. 0.083
8.3%

Write as a percent. Round to the nearest tenth of a percent.

57. $\frac{27}{50}$

54%

58. $\frac{83}{100}$

83%

59. $\frac{1}{3}$

33.3%

60. $\frac{3}{8}$

37.5%

61. $\frac{5}{11}$

45.5%

62. $\frac{4}{9}$

44.4%

63. $\frac{7}{8}$

87.5%

64. $\frac{9}{20}$

45%

65. $1\frac{2}{3}$

166.7%

66. $2\frac{1}{2}$

250%

67. $1\frac{2}{7}$

128.6%

68. $1\frac{11}{12}$

191.7%

Write as a percent. Write the remainder in fractional form.

69. $\frac{17}{50}$

34%

70. $\frac{17}{25}$

68%

71. $\frac{3}{8}$

$37\frac{1}{2}\%$

72. $\frac{7}{16}$

$43\frac{3}{4}\%$

73. $\frac{5}{14}$

$35\frac{5}{7}\%$

74. $\frac{3}{19}$

$15\frac{15}{19}\%$

75. $\frac{3}{16}$

$18\frac{3}{4}\%$

76. $\frac{4}{7}$

$57\frac{1}{7}\%$

77. $1\frac{1}{4}$

125%

78. $2\frac{5}{8}$

$262\frac{1}{2}\%$

79. $1\frac{5}{9}$

$155\frac{5}{9}\%$

80. $1\frac{13}{16}$

$181\frac{1}{4}\%$

SECTION 4.2   ## The Percent Equation

| **Objective A** | **To solve the basic percent equation** |

4

Solving a problem that involves a percent requires solving the basic percent equation shown at the right.

Basic Percent Equation

Percent × Base = Amount
P × B = A

In any percent problem, two parts of the equation are given, and one is unknown.

To translate a problem involving a percent into an equation, remember that the word "of" translates to "multiply" and the word "is" translates to "=". The base usually follows the word "of."

20% of what number is 30?

Given: $P = 20\% = 0.20$
$\qquad A = 30$
Unknown: Base

$$P \times B = A$$
$$(0.20)B = 30$$
$$\frac{0.20B}{0.20} = \frac{30}{0.20}$$
$$B = 150$$

The number is 150.

Find 25% of 200.

Given: $P = 25\% = 0.25$
$\qquad B = 200$
Unknown: Amount

$$P \times B = A$$
$$0.25\,(200) = A$$
$$50 = A$$

25% of 200 is 50.

In most cases, the percent is written as a decimal before solving the basic percent equation. However, some percents are more easily written as a fraction. For example,

$$33\tfrac{1}{3}\% = \tfrac{1}{3} \qquad\qquad 66\tfrac{2}{3}\% = \tfrac{2}{3} \qquad\qquad 16\tfrac{2}{3}\% = \tfrac{1}{6} \qquad\qquad 83\tfrac{1}{3}\% = \tfrac{5}{6}$$

Example 1

12 is $33\tfrac{1}{3}\%$ of what number?

Solution

$$12 = \tfrac{1}{3}B \qquad \left(33\tfrac{1}{3}\% = \tfrac{1}{3}\right)$$
$$3 \cdot 12 = 3 \cdot \tfrac{1}{3}B$$
$$36 = B$$

The number is 36.

Example 2

27 is what percent of 60?

Your Solution

45%

Solution on p. A13

| Objective B | **To solve application problems** |

The key to solving a percent problem is identifying the percent, the base, and the amount. The base usually follows the word "of."

Example 3

A student answered 76 of the 80 questions on a test correctly. What percent of the questions were answered correctly?

Strategy

To find the percent of the questions answered correctly, solve the basic percent equation, using $B = 80$ and $A = 76$. The percent is unknown.

Solution

$$P \times B = A$$
$$P(80) = 76$$
$$\frac{P(80)}{80} = \frac{76}{80}$$
$$P = 0.95$$

95% of the questions were answered correctly.

Example 4

A quality control inspector found that 6 out of 200 wheel bearings inspected were defective. What percent of the wheel bearings were defective?

Your strategy

Your solution

3%

Example 5

A new labor contract increased an employee's hourly wage by 5%. What is the amount of increase for an employee who was making $9.60 an hour?

Strategy

To find the amount of increase, solve the basic percent equation, using $B = 9.60$ and $P = 5\% = 0.05$. The amount is unknown.

Solution

$$P \times B = A$$
$$(0.05)(9.60) = A$$
$$0.48 = A$$

The amount of increase is $.48.

Example 6

A company was producing 2500 gal of paint each week. Because of a decrease in demand for the paint, the company reduced its weekly production by 500 gal. What percent decrease does this represent?

Your strategy

Your solution

20%

Solutions on p. A14

4.2 EXERCISES

▶ Objective A

Solve.

1. 12 is what percent of 50?
24%

2. What percent of 125 is 50?
40%

3. Find 18% of 40.
7.2

4. What is 25% of 60?
15

5. 12% of what is 48?
400

6. 45% of what is 9?
20

7. What is $33\frac{1}{3}$% of 27?
9

8. Find $16\frac{2}{3}$% of 30.
5

9. What percent of 12 is 3?
25%

10. 10 is what percent of 15?
$66\frac{2}{3}$%

11. 60% of what is 3?
5

12. 75% of what is 6?
8

13. 12 is what percent of 6?
200%

14. 20 is what percent of 16?
125%

15. $5\frac{1}{4}$% of what is 21?
400

16. $37\frac{1}{2}$% of what is 15?

40

17. Find 15.4% of 50.
7.7

18. What is 18.5% of 46?
8.51

19. 1 is 0.5% of what?
200

20. 3 is 1.5% of what?
200

21. $\frac{3}{4}$% of what is 3?

400

22. $\frac{1}{2}$% of what is 3?

600

23. Find 125% of 16.
20

24. What is 250% of 12?
30

25. 16.43 is what percent of 20.45? Round to the nearest hundredth of a percent.
80.34%

26. Find 18.37% of 625.43. Round to the nearest hundredth.
114.89

▶ **Objective B**   *Application Problems*

Solve.

27. A university consists of three colleges: business, engineering, and fine arts. There are 2900 students in the business college, 1500 students in the engineering college, and 1000 students in the fine arts college. What percent of the total number of students in the university, to the nearest percent, are in the fine arts college?
19%

28. Approximately 21% of air is oxygen. Using this estimate, determine how many liters of oxygen there are in a room containing 21,600 liters of air.
4536 L

29. The baseball playoff series were increased from 5 games to 7 games. What percent increase does this represent?
40%

30. The cost of a 60-second television commercial during the 1967 Super Bowl was $80,000. In 1983, a 60-second commercial for the Super Bowl cost $800,000. What percent of the 1967 commercial cost is the 1983 commercial cost?
1000%

31. A ski vacation package regularly costs $850 for one week, including lift tickets. By making an early reservation, a person receives a 15% discount. How much is the early reservation discount?
$127.50

32. A football stadium increased its 60,000 seat capacity by 15%. How many seats were added to the stadium?
9000 seats

33. To override a presidential veto, at least $66\frac{2}{3}$% of the Senate must vote to override the veto. There are 100 senators in the Senate. What is the minimum number of votes needed to override a veto?
67 votes

34. To receive a B− grade in a history class, a student must give 75 correct responses on a test of 90 questions. What percent of the total number of questions must a student answer correctly to receive a B− grade?
$83\frac{1}{3}$%

35. An airline knowingly overbooks certain flights by selling 18% more tickets than there are available seats. How many tickets would this airline sell for an airplane that has 150 seats?
177 tickets

36. In a survey of 250 people to determine their soft drink preferences, 100 people preferred cola flavor, 60 people preferred lemon/lime flavor, 50 people preferred orange flavor, and 40 people preferred cherry flavor. What percent of the people surveyed preferred a non-cola flavored drink?
60%

SECTION **4.3**   **Markup and Discount**

| Objective A | **To solve markup problems** |

Cost is the price that a business pays for a product. **Selling price** is the price for which a business sells a product to a customer. The difference between selling price and cost is called **markup.** Markup is added to a retailer's cost to cover the expenses of operating a business. Markup is usually expressed as a percent of the retailer's cost. This percent is called the **markup rate.**

The basic markup equations used by a business are:

Selling price = cost + markup   Markup = markup rate × cost
$$S \quad = C \; + \; M \qquad\qquad M \; = \quad r \quad \times C$$

Substituting $r \times C$ for M in the first equation, selling price can also be written as
$$S = C + (r \times C) = C + rC$$

The manager of a clothing store buys a suit for \$80 and sells the suit for \$116. Find the markup rate.

Given: $C = \$80$
         $S = \$116$
Unknown: r
Use the equation $S = C + rC$.

$$S = C + rC$$
$$116 = 80 + 80r$$

$$\boxed{116 - 80 = 80 - 80 + 80r}$$ Do this step mentally.

$$36 = 80r$$

$$\boxed{\frac{36}{80} = \frac{80r}{80}}$$ Do this step mentally.

$$\frac{36}{80} = r$$
$$0.45 = r$$

The markup rate is 45%.

Example 1

A hardware store employee uses a markup rate of 40% on all items. The selling price of a lawn mower is \$105. Find the cost.

Strategy

Given: $r = 40\% = 0.40$
         $S = \$105$
Unknown: C
Use the equation $S = C + rC$.

Solution

$$S = C + rC$$
$$105 = C + 0.40C$$
$$105 = 1.40C$$
$$75 = C$$

The cost is \$75.

Example 2

The cost to the manager of a sporting goods store for a tennis racket is \$40. The selling price of the racket is \$60. Find the markup rate.

Your strategy

Your Solution

50%

Solution on p. A14

| **Objective B** | **To solve discount problems** |

Discount is the amount by which a retailer reduces the regular price of a product for a promotional sale. Discount is usually expressed as a percent of the regular price. This percent is called the **discount rate.**

The basic discount equations used by a business are

$$\begin{matrix} \text{Sale} \\ \text{price} \end{matrix} = \begin{matrix} \text{regular} \\ \text{price} \end{matrix} - \text{discount} \qquad \text{Discount} = \begin{matrix} \text{discount} \\ \text{rate} \end{matrix} \times \begin{matrix} \text{regular} \\ \text{price} \end{matrix}$$
$$\ \ S \ \ = \ \ R \ \ - \ \ D \qquad\qquad\qquad D \ \ = \ \ r \ \ \times \ \ R$$

Substituting $r \times R$ for D in the first equation, sale price can also be written as
$S = R - (r \times R) = R - rR$

In a garden supply store, the regular price of a 100-foot garden hose is \$32. During an "after-summer sale," the hose is being sold for \$24. Find the discount rate.

Given: $R = \$32$
$\qquad\quad S = \$24$
Unknown: r
Use the equation $S = R - rR$.

$$S = R - rR$$
$$24 = 32 - 32r$$

$\boxed{24 - 32 = 32 - 32 - 32r}$ Do this step mentally.

$$-8 = -32r$$

$\boxed{\dfrac{-8}{-32} = \dfrac{-32r}{-32}}$ Do this step mentally.

$$\frac{1}{4} = r$$
$$0.25 = r$$

The discount rate is 25%.

Example 3

The sale price for a chemical sprayer is \$23.40. This price is 35% off the regular price. Find the regular price.

Strategy

Given: $S = \$23.40$
$\qquad\quad r = 35\% = 0.35$
Unknown: R
Use the equation $S = R - rR$.

Solution

$$S = R - rR$$
$$23.40 = R - 0.35R$$
$$23.40 = 0.65R$$
$$36 = R$$

The regular price is \$36.

Example 4

A case of motor oil that regularly sells for \$27.60 is on sale for \$20.70. What is the discount rate?

Your strategy

Your Solution

25%

Solution on p. A14

4.3   EXERCISES

▶ **Objective A**   *Application Problems*

Solve.

1. A computer software retailer uses a markup rate of 40%. Find the selling price of a computer game that cost the retailer $25.
    $35

2. A car dealer advertises a 5% markup over cost. Find the selling price of a car that cost the dealer $12,000.
    $12,600

3. The pro in a golf shop purchases a one iron for $40 and fixes a selling price on the club of $75. Find the markup rate.
    87.5%

4. A jeweler purchases a diamond ring for $350 and fixes a selling price of $700 on the ring. Find the markup rate.
    100%

5. A leather jacket costs a clothing store manager $140. Find the selling price of the leather jacket when the markup rate is 40%.
    $196

6. The cost to a landscape architect for a 25-gallon tree is $65. Find the selling price of the tree when the markup rate used by the architect is 30%.
    $84.50

7. A digitally recorded compact disc costs the manager of a music store $8.50. The selling price of the disc is $11.90. Find the markup rate.
    40%

8. A grocer purchases a can of fruit juice for $.68. The selling price for the fruit juice is $.85. Find the markup rate.
    25%

9. A tire dealer uses a markup rate of 55% on steel-belted tires. Find the selling price of a steel-belted tire that costs the dealer $45.
    $69.75

10. A cobbler uses a markup rate of 52% on rubber heels for shoes. Find the selling price of a rubber heel that costs the cobbler $3.50.
    $5.32

11. An electronics store manager adds $50 to the cost of every 17-inch television, regardless of the cost of the television. Find the markup rate, to the nearest tenth of a percent, on a television that costs the manager $215.
    23.3%

12. A department store uses a markup rate of 40% on items that cost over $100 and a markup rate of 50% on items that cost less than $100. Find the selling price of a ceramic bowl that costs the department store $86.
    $129.00

▶ **Objective B** *Application Problems*

Solve.

13. A tennis racket that regularly sells for $55 is on sale for 25% off the regular price. Find the sale price.
$41.25

14. A fax machine that regularly sells for $975 is on sale for $33\frac{1}{3}$% off the regular price. Find the sale price.
$650

15. A car stereo system that regularly sells for $425 is on sale for $318.75. Find the discount rate.
25%

16. A car dealer is having a year-end clearance sale that offers $2500 off the regular price of a car. Find the discount rate for a car that regularly sells for $12,500.
20%

17. A college bookstore sells a used book at a discount of 30% off the regular price of a new book. Find the price of a used book that originally cost $35.
$24.50

18. An airline is offering a 35% discount on round-trip air fares. Find the sale price of a round-trip ticket that normally costs $385.
$250.25

19. A gold bracelet that regularly sells for $1250 is on sale for $750. Find the discount rate.
40%

20. A pair of skis that regularly sell for $325 are on sale for $250. Find the discount rate to the nearest percent.
23%

21. A clothing wholesaler offers a discount of 10% per shirt when 10 to 20 shirts are purchased and a discount of 15% per shirt when 21 to 50 shirts are purchased. For a shirt that regularly sells for $17, find the sale price per shirt when 35 shirts are purchased.
$14.45

22. A supplier of electrical equipment offers a 10% discount for a cash purchase or a 5% discount for a purchase that is paid for within 30 days. Find the discount price of a transformer that regularly sells for $230 and is paid for 10 days after the purchase.
$218.50

23. A service station offers a discount of $10 per tire when 2 tires are purchased and a discount of $25 per tire when 4 tires are purchased. Find the discount rate, to the nearest percent, when a customer purchases 4 tires that regularly sell for $95 each.
26%

24. A department store offers a discount of $3 per dinner plate when 5 or fewer plates are purchased and a discount of $5 per plate when more than 5 plates are purchased. Find the discount rate, to the nearest percent, when a customer purchases 3 dinner plates that regularly sell for $18 each.
17%

SECTION 4.4 Investment Problems

Objective A **To solve investment problems**

The annual simple interest that an investment earns is given by the equation $I = Pr$, where I is the simple interest, P is the principal, or the amount invested, and r is the simple interest rate.

The annual interest rate on a $2500 investment is 14%. Find the annual simple interest earned on the investment.

Given: $P = \$2500$
$\qquad r = 14\% = 0.14$
Unknown: I

$I = Pr$
$I = 2500(0.14)$
$I = 350$

The annual simple interest is $350.

An investor has a total of $10,000 to deposit into two simple interest accounts. On one account, the annual simple interest rate is 7%. On the second account, the annual simple interest rate is 11%. How much should be invested in each account so that the total annual interest earned is $1000?

Strategy for Solving a Problem Involving Money Deposited in Two Simple Interest Accounts

For each amount invested, write a numerical or variable expression for the principal, the interest rate, and the interest earned. The results can be recorded in a table.

The sum of the amounts at each interest rate is $10,000.

Amount invested at 7%: x
Amount invested 11%: $\$10,000 - x$

	Principal, P	$\cdot$	Interest rate, r	$=$	Interest earned, I
Amount at 7%	x	$\cdot$	0.07	$=$	$0.07x$
Amount at 11%	$10,000 - x$	$\cdot$	0.11	$=$	$0.11(10,000 - x)$

Determine how the amounts of interest earned on each amount are related. For example, the total interest earned by both accounts may be known, or it may be known that the interest earned on one account is equal to the interest earned on the other account.

The total annual interest earned is $1000.

$$0.07x + 0.11(10,000 - x) = 1000$$
$$0.07x + 1100 - 0.11x = 1000$$
$$-0.04x + 1100 = 1000$$
$$-0.04x = -100$$
$$x = 2500$$
$$10,000 - x = 10,000 - 2500 = 7500$$

The amount invested at 7% is $2500.
The amount invested at 11% is $7500.

Example 1

An investment counselor invested 75% of a client's money in a 13% annual simple interest money market fund. The remainder was invested in 9% annual simple interest government securities. Find the amount invested in each if the total annual interest earned is $5400.

Strategy

■ Amount invested: x
Amount invested at 9%: $0.25x$
Amount invested at 13%: $0.75x$

	Principal	Rate	Interest
Amount at 9%	$0.25x$	0.09	$0.0225x$
Amount at 13%	$0.75x$	0.13	$0.0975x$

■ The sum of the interest earned by the two investments equals the total annual interest earned ($5400).

Solution

$$0.0225x + 0.0975x = 5400$$
$$0.12x = 5400$$
$$x = 45,000$$

$$0.25x = 0.25(45,000) = 11,250$$

$$0.75x = 0.75(45,000) = 33,750$$

The amount invested at 9% is $11,250.
The amount invested at 13% is $33,750.

Example 2

An investment of $5000 is made at an annual simple interest rate of 8%. How much additional money must be invested at 14% so that the total interest earned will be 11% of the total investment?

Your strategy

Your solution

$5000

Solution on p. A14

4.4 EXERCISES

▶ **Objective A**   *Application Problems*

Solve.

1. An investment of $2500 is made at an annual simple interest rate of 7%. How much additional money must be invested at an annual simple interest rate of 11% so that the total interest earned is 9% of the total investment?
$2500

2. A total of $6000 is invested into two simple interest accounts. The annual simple interest rate on one account is 9%; on the second account, the annual simple interest rate is 6%. How much should be invested in each account so that both accounts earn the same amount of interest.
$3600 at 6%; $2400 at 9%.

3. An engineer invested a portion of $15,000 in a 7% annual simple interest account and the remainder in a 6.5% annual simple interest government bond. The amount of interest earned for one year was $1020. How much was invested in each account?
$9000 at 7%; $6000 at 6.5%

4. An investment club invested part of $20,000 in preferred stock that pays 8% annual simple interest and the remainder in a municipal bond that pays 7% annual simple interest. The amount of interest earned each year is $1520. How much was invested in each account?
$12,000 at 8%; $8000 at 7%

5. A grocery checker deposited an amount of money into a high-yield mutual fund that returns a 13% annual simple interest rate. A second deposit, $2500 more than the first, was placed in a certificate of deposit that returns a 7% annual simple interest rate. The total interest earned on both investments for one year was $475. How much money was deposited in the mutual fund?
$1500

6. A deposit was made into a 7% annual simple interest account. Another deposit, $1500 less than the first deposit, was placed in a 9% annual simple interest certificate of deposit. The total interest earned on both accounts for one year was $505. How much money was deposited in the certificate of deposit?
$2500

7. A corporation gave a university $300,000 to support product safety research. The university deposited some of the money in a 10% simple interest account and the remainder in an 8.5% simple interest account. How much should be deposited in each account so that the interest earned is $28,500?
$200,000 at 10%; $100,000 at 8.5%

8. A financial consultant advises a client to invest part of $30,000 in municipal bonds that earn 6.5% annual simple interest and the remainder of the money in 8.5% corporate bonds. How much should be invested in each account so that the total interest earned each year is $2190?
$18,000 at 6.5%; $12,000 at 8.5%

Solve.

9. To provide for retirement income, an auto mechanic purchases a $5000 bond that earns 7.5% annual simple interest. How much money must be invested in additional bonds that have an interest rate of 8% so that the total interest earned from the two investments is $615?
$3000

10. The portfolio manager for an investment group invested $40,000 in a certificate of deposit that earns 7.25% annual simple interest. How much money must be invested in additional certificates that have an interest rate of 8.5% so that the total interest earned from the two investments is $5025?
$25,000

11. A charity deposited a total of $54,000 into two simple interest accounts. The annual simple interest rate on one account is 8%; on the second account, the annual simple interest rate is 12%. How much should be invested in each account so that the total interest earned is 9% of the total investment?
$40,500 at 8%; $13,500 at 12%

12. A college sports foundation deposited a total of $24,000 into two simple interest accounts. The annual simple interest rate on one account is 7%; on the second account, the annual simple interest rate is 11%. How much should be invested in each account so that the total interest earned is 10% of the total investment?
$6000 at 7%; $18,000 at 11%

13. An investment banker decided to invest 55% of the bank's available cash in an account that earns 8.25% annual simple interest. The remainder of the cash was placed in an account that earns 10% annual simple interest. The interest earned in one year was $58,743.75. What was the total amount invested?
$650,000

14. A financial planner recommended that 40% of a client's cash account be placed in preferred stock that earns 9% annual simple interest. The remainder of the client's cash was placed in treasury bonds that earn 7% annual interest. The total annual interest earned from the two investments was $2496. What was the total amount invested?
$32,000

15. The manager of a mutual fund placed 30% of the fund's available cash in a 6% simple interest account, 25% in 8% corporate bonds, and the remainder in a money market fund that earns 7.5% annual simple interest. The total annual interest from the investments was $35,875. What was the total amount invested?
$500,000

16. The manager of a trust decided to invest 30% of a client's cash in government bonds that earn 6.5% annual simple interest. Another 30% was placed in utility stocks that earn 7% annual simple interest. The remainder of the cash was placed in an account earning 8% annual simple interest. The total interest earned from the investments was $5437.50. What was the total amount invested?
$75,000

SECTION 4.5   Mixture Problems

Objective A   To solve value mixture problems

A value mixture problem involves combining two ingredients that have different prices into a single blend. For example, a coffee merchant may blend two types of coffee into a single blend, or a candy manufacturer may combine two types of candy to sell as a "variety pack."

The solution of a value mixture problem is based on the equation $V = AC$, where V is the value of an ingredient, A is the amount of the ingredient, and C is the cost per unit of the ingredient.

A coffee merchant wants to make 6 lb of a blend of coffee to sell for $5 per pound. The blend is made using a $6 per pound grade and a $3 per pound grade of coffee. How many pounds of each of these grades should be used?

Strategy for Solving a Value Mixture Problem

For each ingredient in the mixture, write a numerical or variable expression for the amount of the ingredient used, the unit cost of the ingredient, and the value of the amount used. For the blend, write a numerical or variable expression for the amount, the unit cost of the blend, and the value of the amount. The results can be recorded in a table.

The sum of the amounts is 6 lb.

Amount of $6 coffee: x
Amount of $3 coffee: $6 - x$

	Amount, A	·	Unit cost, C	=	Value, V
$6 grade	x	·	$6	=	$6x$
$3 grade	$6 - x$	·	$3	=	$3(6 - x)$
$5 blend	6	·	$5	=	$5(6)$

Determine how the values of each ingredient are related. Use the fact that the sum of the values of each ingredient is equal to the value of the blend.

The sum of the values of the $6 grade and the $3 grade is equal to the value of the $5 blend.

$$6x + 3(6 - x) = 5(6)$$
$$6x + 18 - 3x = 30$$
$$3x + 18 = 30$$
$$3x = 12$$
$$x = 4$$

$$6 - x = 6 - 4 = 2$$

The merchant must use 4 lb of the $6 coffee and 2 lb of the $3 coffee.

Example 1

How many ounces of a silver alloy that costs $4 an ounce must be mixed with 10 oz of an alloy that costs $6 an ounce to make a mixture that costs $4.32 an ounce?

Example 2

A gardener has 20 lb of a lawn fertilizer that costs $.80 per pound. How many pounds of a fertilizer that costs $.55 per pound should be mixed with this 20 lb of lawn fertilizer to produce a mixture that costs $.75 per pound?

Strategy

- Ounces of $4 alloy: x

	Amount	Cost	Value
$4 alloy	x	$4	$4x$
$6 alloy	10	$6	6(10)
$4.32 mixture	$10 + x$	$4.32	4.32(10 + x)

- The sum of the values before mixing equals the value after mixing.

Your strategy

Solution

$$4x + 6(10) = 4.32(10 + x)$$
$$4x + 60 = 43.2 + 4.32x$$
$$-0.32x + 60 = 43.2$$
$$-0.32x = -16.8$$
$$x = 52.5$$

52.5 oz of the $4 silver alloy must be used.

Your solution

5 lb

Solution on p. A15

Objective B	**To solve percent mixture problems**

The amount of a substance in a solution can be given as a percent of the total solution. For example, a 5% salt water solution means that 5% of the total solution is salt. The remaining 95% is water.

The solution of a percent mixture problem is based on the equation $Q = Ar$, where Q is the quantity of a substance in the solution, r is the percent of concentration, and A is the amount of solution.

A 500-milliliter bottle contains a 4% solution of hydrogen peroxide. Find the amount of hydrogen peroxide in the solution.

Given: $A = 500$ $Q = Ar$
            $r = 4\% = 0.04$ $Q = 500(0.04)$
Unknown: Q $Q = 20$

The bottle contains 20 ml of hydrogen peroxide.

How many gallons of a 20% salt solution must be mixed with 6 gal of a 30% salt solution to make a 22% salt solution?

Strategy for Solving a Percent Mixture Problem

> For each solution, write a numerical or variable expression for the amount of solution, the percent of concentration, and the quantity of the substance in the solution. The results can be recorded in a table.

The unknown quantity of 20% solution: x

	Amount of solution, A	$\cdot$	Percent of concentration, r	$=$	Quantity of substance, Q
20% solution	x	$\cdot$	0.20	$=$	$0.20x$
30% solution	6	$\cdot$	0.30	$=$	$0.30(6)$
22% solution	$x + 6$	$\cdot$	0.22	$=$	$0.22(x + 6)$

> Determine how the quantities of the substances in each solution are related. Use the fact that the sum of the quantities of the substances being mixed is equal to the quantity of the substance after mixing.

The sum of the quantities of the substances in the 20% solution and the 30% solution is equal to the quantity of the substance in the 22% solution.

$$0.20x + 0.30(6) = 0.22(x + 6)$$
$$0.20x + 1.80 = 0.22x + 1.32$$
$$-0.02x + 1.80 = 1.32$$
$$-0.02x = -0.48$$
$$x = 24$$

24 gal of the 20% solution are required.

Example 3

A chemist wishes to make 2 L of an 8% acid solution by mixing a 10% acid solution and a 5% acid solution. How many liters of each solution should the chemist use?

Strategy

■ Liters of 10% solution: x
Liters of 5% solution: $2 - x$

	Amount	Percent	Quantity
10%	x	0.10	$0.10x$
5%	$2 - x$	0.05	$0.05(2 - x)$
8%	2	0.08	$0.08(2)$

■ The sum of the quantities before mixing is equal to the quantity after mixing.

Solution

$$0.10x + 0.05(2 - x) = 0.08(2)$$
$$0.10x + 0.10 - 0.05x = 0.16$$
$$0.05x + 0.10 = 0.16$$
$$0.05x = 0.06$$
$$x = 1.2$$

$$2 - x = 2 - 1.2 = 0.8$$

The chemist needs 1.2 L of the 10% solution and 0.8 L of the 5% solution.

Example 4

A pharmacist dilutes 5 L of a 12% solution by adding water. How many liters of water are added to make an 8% solution?

Your strategy

Your solution

2.5 L

Solution on p. A15

4.5   EXERCISES

▶ **Objective A**   *Application Problems*

Solve.

1. A high-protein diet supplement that costs $6.75 per pound is mixed with a vitamin supplement that costs $3.25 per pound. How many pounds of each should be used to make 5 lb of a mixture that sells for $4.65 per pound?
    2 lb of diet supplement; 3 lb of vitamin supplement

2. A 20-oz alloy of platinum that costs $220 per ounce is mixed with an alloy that costs $400 per ounce. How many ounces of the $400 alloy should be mixed with the 20-oz alloy to make an alloy that costs $300 per ounce?
    16 oz

3. Find the selling price per pound of a coffee mixture made from 8 lb of coffee that costs $9.20 per pound and 12 lb of coffee that costs $5.50 per pound.
    $6.98 per pound

4. How many pounds of tea that costs $4.20 per pound must be mixed with 12 lb of tea that costs $2.25 per pound to make a mixture that costs $3.40 per pound?
    17.25 lb

5. A goldsmith combined an alloy that costs $4.30 per ounce with an alloy that costs $1.80 per ounce. How many ounces of each were used to make a mixture of 200 oz to sell for $2.50 per ounce?
    56 oz at $4.30; 144 oz at $1.80

6. How many liters of a solvent that costs $80 per liter must be mixed with 6 L of a solvent that costs $25 per liter to make a solvent that sells for $36 per liter?
    1.5 L

7. Find the selling price per pound of a trail mix made from 40 pounds of raisins that cost $4.40 per pound and 100 pounds of granola that costs $2.30 per pound.
    $2.90 per lb

8. Find the selling price per ounce of a mixture of 200 oz of cologne that costs $5.50 per ounce and 500 oz of a cologne that costs $2.00 per ounce.
    $3.00 per ounce

9. How many kilograms of hard candy that costs $7.50 per kilogram must be mixed with 24 kg of jelly beans that cost $3.25 per kilogram to make a mixture that sells for $4.50 per kilogram?
    10 kg

10. A grocery store offers a cheese and fruit sampler that combines cheddar cheese that costs $8 per kilogram with kiwis that cost $3 per kilogram. How many kilograms of each were used to make a 5-kg mixture that costs $4.50 per kilogram?
    1.5 kg of cheese; 3.5 kg of kiwis

Solve.

11. A ground meat mixture is formed by combining meat that costs $2.20 per pound with meat that costs $4.20 per pound. How many pounds of each were used to make a 50-lb mixture that costs $3.00 per pound?
   30 lb at $2.20; 20 lb at $4.20

12. A lumber company combined oak wood chips that cost $3.10 per pound with pine wood chips that cost $2.50 per pound. How many pounds of each were used to make an 80-lb mixture that costs $2.65 per pound?
   60 lb pine; 20 lb oak

13. How many kilograms of a soil supplement that costs $7.00 per kilogram must be mixed with 20 kg of aluminum nitrate that costs $3.50 per kilogram to make a fertilizer that costs $4.50 per kilogram?
   8 kg

14. A caterer made an ice cream punch by combining fruit juice that costs $2.25 per gallon with ice cream that costs $3.25 per gallon. How many gallons of each should be used to make 100 gal of punch to sell for $2.50 per gallon?
   25 gal ice cream; 75 gal fruit juice

15. The manager of a specialty food store combined almonds that cost $4.50 per pound with walnuts that cost $2.50 per pound. How many pounds of each were used to make a 100-lb mixture that costs $3.24 per pound?
   63 lb walnuts; 37 lb almonds

16. Find the cost per gallon of a carbonated fruit drink made from 12 gal of fruit juice that costs $4.00 per gallon and 30 gal of carbonated water that costs $2.25 per gallon.
   $2.75 per gallon

17. Find the cost per pound of a sugar-coated breakfast cereal made from 40 pounds of sugar that costs $1.00 per pound and 120 pounds of corn flakes that cost $.60 per pound.
   $.70 per pound

18. Find the cost per ounce of a gold alloy made from 25 oz of pure gold that costs $482 per ounce and 40 oz of an alloy that costs $300 per ounce.
   $370 per ounce

19. How many pounds of lima beans that cost $.90 per pound must be mixed with 16 lb of corn that costs $.50 per pound to make a mixture of vegetables that costs $.65 per pound?
   9.6 lb

20. How many liters of a blue dye that costs $1.60 per liter must be mixed with 18 L of anil that costs $2.50 per liter to make a mixture that costs $1.90 per liter?
   36 L

▶ **Objective B**   *Applications*

Solve.

21. A chemist wants to make 50 ml of a 16% acid solution. How many milliliters each of a 13% acid solution and an 18% acid solution should be mixed to produce the desired solution?
     20 ml of 13%; 30 ml of 18%

22. A blend of coffee was made by combining some coffee that was 40% java beans with 80 lb of coffee that was 30% java beans to make a mixture that is 32% java. How many pounds of the 40% java coffee were used?
     20 lb

23. Thirty ounces of pure silver are added to 50 oz of a silver alloy that is 20% silver. What is the percent concentration of the silver in the resulting mixture?
     50%

24. A 200-liter punch that contains 35% fruit juice is mixed with 300 L of another punch. The resulting fruit punch is 20% fruit juice. Find the percent of fruit juice in the 300-liter punch.
     10%

25. The manager of a garden shop mixes grass seed that is 60% rye grass with 70 lb of grass seed that is 80% rye grass to make a mixture that is 74% rye grass. How much of the 60% mixture is used?
     30 lb

26. Ten grams of sugar are added to a 40-g serving of a breakfast cereal that is 30% sugar. What is the percent concentration of sugar in the resulting mixture?
     44%

27. A dermatologist mixes 50 g of a cream that is 0.5% hydrocortizone with 150 g of another hydrocortizone cream. The resulting mixture is 0.68% hydrocortizone. Find the percent of hydrocortizone in the 150-g cream.
     0.74%

28. A carpet manufacturer blends two fibers, one 20% wool and the second 50% wool. How many pounds of each fiber should be woven together to produce 500 lb of a fabric that is 35% wool?
     250 lb of 20%; 250 lb of 50%

29. A hair dye is made by blending some 7% hydrogen peroxide solution with some 4% hydrogen peroxide solution. How many milliliters of each should be mixed to make a 300-ml solution that is 5% hydrogen peroxide?
     100 ml of 7%; 200 ml of 4%

30. How many grams of pure salt must be added to 40 g of a 20% salt solution to make a solution that is 36% salt?
     10 g

Solve.

31. How many ounces of pure water must be added to 50 oz of a 15% saline solution to make a saline solution that is 10% salt?
25 oz

32. A paint is blended by using a paint that contains 21% green dye and a paint that contains 15% green dye. How many gallons of each must be mixed to produce 60 gal of paint that is 19% green dye?
40 gal of 21%; 20 gal of 15%

33. A goldsmith mixes 8 oz of a 30% gold alloy with 12 oz of a 25% gold alloy. What is the percent concentration of the resulting alloy?
27%

34. A physicist mixes 40 L of oxygen with 50 L of air that contains 64% oxygen. What is the percent concentration of the resulting air?
80% oxygen

35. A 50-oz box of cereal is 40% bran flakes. How many ounces of pure bran flakes must be added to this box to produce a mixture that is 50% bran flakes?
10 oz

36. A pastry chef has 150 ml of a chocolate topping that is 50% chocolate. How many milliliters of pure chocolate must be added to this topping to make a topping that is 75% chocolate?
150 ml

37. A cafe has some tea that is 20% jasmine and some tea that is 15% jasmine. How many pounds of each must be mixed to make 5 lb of tea that is 18% jasmine?
3 lb of 20%; 2 lb of 15%

38. A clothing manufacturer has some pure silk thread and some thread that is 85% silk. How many kilograms of each must be woven together to make 75 kg of a blend that is 96% silk?
20 kg of 85%; 55 kg of pure silk

39. How many ounces of dried apricots must be added to 18 oz of a snack mix that contains 20% dried apricots to make a mixture that is 25% dried apricots?
1.2 oz

40. A recipe for a rice dish calls for 12 oz of a rice mixture that is 20% wild rice and 8 oz of pure wild rice. What is the percent concentration of wild rice in the 20-oz mixture?
52%

SECTION 4.6

Uniform Motion Problems

Objective A	**To solve uniform motion problems**

A train that travels constantly in a straight line at 50 mph is in *uniform motion*. **Uniform motion** means that the speed or direction of an object does not change.

The solution of a uniform motion problem is based on the equation $d = rt$, where d is the distance traveled, r is the rate of travel, and t is the time traveled.

A car leaves a town traveling at 40 mph. Two hours later, a second car leaves the same town, on the same road, traveling at 60 mph. In how many hours will the second car pass the first car?

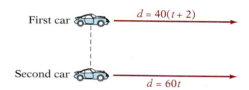

First car $\quad d = 40(t + 2)$

Second car $\quad d = 60t$

Strategy for Solving a Uniform Motion Problem

> For each object, write a numerical or variable expression for the distance, rate, and time. The results can be recorded in a table.

The first car traveled 2 h longer than the second car.

Unknown time for the second car: t
Time for the first car: $t + 2$

	Rate, r	$\cdot$	Time, t	$=$	Distance, d
First car	40	$\cdot$	$t + 2$	$=$	$40(t + 2)$
Second car	60	$\cdot$	t	$=$	$60t$

> Determine how the distances traveled by each object are related. For example, the total distance traveled by both objects may be known, or it may be known that the two objects traveled the same distance.

The two cars travel the same distance.

$$40(t + 2) = 60t$$
$$40t + 80 = 60t$$
$$80 = 20t$$
$$4 = t$$

The second car will pass the first car in 4 h.

Example 1

Two cars, one traveling 10 mph faster than the other car, start at the same time from the same point and travel in opposite directions. In 3 h they are 300 mi apart. Find the rate of each car.

Strategy

- Rate of 1st car: r
  Rate of 2nd car: $r + 10$

	Rate	Time	Distance
1st car	r	3	$3r$
2nd car	$r + 10$	3	$3(r + 10)$

- The total distance traveled by the two cars is 300 mi.

Solution

$$3r + 3(r + 10) = 300$$
$$3r + 3r + 30 = 300$$
$$6r + 30 = 300$$
$$6r = 270$$
$$r = 45$$

$r + 10 = 45 + 10 = 55$

The first car is traveling 45 mph.
The second car is traveling 55 mph.

Example 2

Two trains, one traveling at twice the speed of the other, start at the same time on parallel tracks from stations which are 288 mi apart and travel toward each other. In 3 h, the trains pass each other. Find the rate of each train.

Your strategy

Your solution

32 mph; 64 mph

Example 3

How far can a bicycling club ride out into the country at a speed of 12 mph and return over the same road at 8 mph if they travel a total of 10 h?

Strategy

- Time spend riding out: t
  Time spent riding back: $10 - t$

	Rate	Time	Distance
Out	12	t	$12t$
Back	8	$10 - t$	$8(10 - t)$

- The distance out equals the distance back.

Solution

$$12t = 8(10 - t)$$
$$12t = 80 - 8t$$
$$20t = 80$$
$$t = 4 \quad \text{(The time is 4 h.)}$$

The distance out $= 12t = 12(4) = 48$ mi.
The club can ride 48 mi into the country.

Example 4

A pilot flew out to a parcel of land and back in 5 h. The rate out was 150 mph and the rate returning was 100 mph. How far away was the parcel of land?

Your strategy

Your solution

300 mi

Content and Format © 1991 HMCo.

4.6   **EXERCISES**

▶ **Objective A**   *Application Problems*

Solve.

1. A 555-mile, 5-hour plane trip was flown at two speeds. For the first part of the trip, the average speed was 105 mph. For the remainder of the trip, the average speed was 115 mph. For how long did the plane fly at each speed?
   2 h at 105 mph; 3 h at 115 mph

2. An executive drove from home at an average speed of 30 mph to an airport where a helicopter was waiting. The executive boarded the helicopter and flew to the corporate offices at an average speed of 60 mph. The entire distance was 150 mi. The entire trip took 3 h. Find the distance from the airport to the corporate offices.
   120 mi

3. After a sailboat had been on the water for 3 h, a change in wind direction reduced the average speed of the boat by 5 mph. The entire distance sailed was 57 mi. The total time spent sailing was 6 h. How far did the sailboat travel in the first 3 h?
   36 mi

4. A car and a bus set out at 2 P.M. from the same point headed in the same direction. The average speed of the car is 30 mph slower than twice the speed of the bus. In 2 h the car is 20 mi ahead of the bus. Find the rate of the car.
   50 mph

5. A passenger train leaves a train depot 2 h after a freight train leaves the same depot. The freight train is traveling 20 mph slower than the passenger train. Find the rate of each train if the passenger train overtakes the freight train in 3 h.
   passenger train, 50 mph; freight train, 30 mph

6. Two cyclists start at the same time from opposite ends of a course that is 45 mi long. One cyclist is riding at 14 mph and the second cyclist is riding at 16 mph. How long after they begin will they meet?
   1.5 h

7. A cyclist and a jogger set out at 11 A.M. from the same point headed in the same direction. The average speed of the cyclist is twice the average speed of the jogger. In 1 h the cyclist is 8 mi ahead of the jogger. Find the rate of the cyclist.
   16 mph

8. Two cyclists start from the same point and ride in opposite directions. One cyclist rides twice as fast as the other. In 3 h they are 72 mi apart. Find the rate of each cyclist.
   8 mph; 16 mph

9. Two small planes start from the same point and fly in opposite directions. The first plane is flying 25 mph slower than the second plane. In 2 h the planes are 430 mi apart. Find the rate of each plane.
   95 mph; 120 mph

Solve.

10. A motorboat leaves a harbor and travels at an average speed of 8 mph toward a small island. Two hours later a cabin cruiser leaves the same harbor and travels at an average speed of 16 mph toward the same island. In how many hours after the cabin cruiser leaves will it be alongside the motorboat?
2 h

11. Two joggers start at the same time from opposite ends of a 10-mile course. One jogger is running at 4 mph and the other is running at 6 mph. How long after they begin will they meet?
1 h

12. On a 195-mile trip, a car traveled at an average speed of 45 mph and then reduced its speed to an average of 30 mph for the remainder of the trip. The trip took a total of 5 h. How long did the car travel at each speed?
3 h at 45 mph; 2 h at 30 mph

13. A long-distance runner started on a course running at an average speed of 6 mph. One hour later, a second runner began the same course at an average speed of 8 mph. How long after the second runner started will the second runner overtake the first runner?
3 h

14. A family drove to a resort at an average speed of 30 mph and later returned over the same road at an average speed of 50 mph. Find the distance to the resort if the total driving time was 8 h.
150 mi

15. Three campers left their campsite by canoe and paddled downstream at an average rate of 8 mph. They then turned around and paddled back upstream at an average rate of 4 mph to return to their campsite. How long did it take the campers to canoe downstream if the total trip took 1 h?
20 min

16. A car traveling at 48 mph overtakes a cyclist who, riding at 12 mph, had a 3-h head start. How far from the starting point does the car overtake the cyclist?
48 mi

17. As part of flight training, a student pilot was required to fly to an airport and then return. The average speed to the airport was 90 mph, and the average speed returning was 120 mph. Find the distance between the two airports if the total flying time was 7 h.
360 mi

18. Running at an average rate of 8 m/s, a sprinter ran to the end of a track and then jogged back to the starting point at an average rate of 3 m/s. The sprinter took 55 s to run to the end of the track and jog back. Find the length of the track.
120 m

19. A jet plane traveling at 600 mph overtakes a propeller-driven plane that had a 2-h head start. The propeller-driven plane is traveling at 200 mph. How far from the starting point does the jet overtake the propeller-driven plane?
600 mi

20. A bus traveled on a level road for 2 h at an average speed of 20 mph faster than it traveled on a winding road. The time spent on the winding road was 3 h. Find the average speed on the winding road if the total trip was 200 mi.
32 mph

SECTION 4.7

Geometry Problems

Objective A

To solve perimeter problems

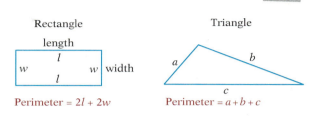

The **perimeter** of a geometric figure is a measure of the distance around the figure. The equations for the perimeters of a rectangle and a triangle are shown at the right.

Rectangle

length

w l w width

l

Perimeter $= 2l + 2w$

Triangle

a b

c

Perimeter $= a + b + c$

The perimeter of a rectangle is 26 ft. The length of the rectangle is 1 ft more than twice the width. Find the width of the rectangle.

Strategy for Solving a Perimeter Problem

Let a variable represent the measure of one of the unknown sides of the figure. Express the measures of the remaining sides in terms of that variable.

Width: w
Length: $2w + 1$

Determine which perimeter equation to use. Use the equation for the perimeter of a rectangle.

$$2l + 2w = P$$
$$2(2w + 1) + 2w = 26$$
$$4w + 2 + 2w = 26$$
$$6w + 2 = 26$$
$$6w = 24$$
$$w = 4$$

The width is 4 ft.

Example 1

The perimeter of a triangle is 25 ft. Two sides of the triangle are equal. The third side is 2 ft less than the length of one of the equal sides. Find the measure of one of the equal sides.

Strategy

- Each equal side: x
  The third side: $x - 2$
- Use the equation for the perimeter of a triangle.

Solution $a + b + c = P$
$$x + x + (x - 2) = 25$$
$$3x - 2 = 25$$
$$3x = 27$$
$$x = 9$$

Each of the equal sides measures 9 ft.

Example 2

The perimeter of a rectangle is 34 m. The width of the rectangle is 3 m less than the length. Find the measure of the width.

Your strategy

Your solution   7 m

Solution on p. A16

| Objective B | **To solve problems involving the angles of a triangle** |

In a triangle, the sum of the measures of all the angles is 180°.

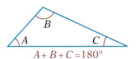

$$A + B + C = 180°$$

Two special types of triangles are shown at the right. A **right triangle** has one right angle (90°). An **isosceles triangle** has two equal angles.

Right triangle

Equal angles

Isosceles triangle

In a right triangle, the measure of one angle is twice the measure of the smallest angle. Find the measure of the smallest angle.

Strategy for Solving a Problem Involving the Angles of a Triangle

Let a variable represent one of the unknown angles. Express the other angles in terms of that variable.

Measure of smallest angle: x
Measure of second angle: $2x$
Measure of right angle: $90°$

Use the equation $A + B + C = 180°$

$$x + 2x + 90 = 180$$
$$3x + 90 = 180$$
$$3x = 90$$
$$x = 30$$

The measure of the smallest angle is 30°.

Example 3

In an isosceles triangle, the measure of one angle is 20° more than twice the measure of one of the equal angles. Find the measure of one of the equal angles.

Strategy

- Measure of one of the equal angles: x
  Measure of the second equal angle: x
  Measure of the third angle: $2x + 20$
- Use the equation $A + B + C = 180°$

Solution $x + x + (2x + 20) = 180$
$$4x + 20 = 180$$
$$4x = 160$$
$$x = 40$$

The measure of one of the equal angles is 40°.

Example 4

In a triangle, the measure of one angle is twice the measure of the second angle. The measure of the third angle is 4° less than the measure of the second angle. Find the measure of each angle.

Your strategy

Your solution   92°, 46°, 42°

Solution on p. A16

4.7	**EXERCISES**

▶ **Objective A**   *Application Problems*

Solve.

1. In an isosceles triangle, two sides are equal. The third side is 50% of the length of one of the equal sides. Find the length of each side when the perimeter is 125 ft.
50 ft; 50 ft; 25 ft

2. The width of a rectangle is 25% of the length. The perimeter is 250 cm. Find the length and width of the rectangle.
length 100 cm; width 25 cm

3. In an isosceles triangle, two sides are equal. The length of one of the equal sides is 3 times the length of the third side. The perimeter is 21 m. Find the length of each side.
9 m; 9 m; 3 m

4. The perimeter of a rectangle is 42 m. The length of the rectangle is 3 m less than twice the width. Find the length and width of the rectangle.
length 13 m; width 8 m

5. The perimeter of a rectangle is 120 ft. The length of the rectangle is twice the width. Find the length and width of the rectangle.
length 40 ft; width 20 ft

6. The perimeter of a triangle is 110 cm. One side is twice the second side. The third side is 30 cm more than the second side. Find the measure of each side.
40 cm; 20 cm; 50 cm

7. The perimeter of a triangle is 33 ft. One side of the triangle is 1 ft longer than the second side. The third side is 2 ft longer than the second side. Find the measure of each side.
11 ft; 10 ft; 12 ft

8. The perimeter of a rectangle is 50 m. The width of the rectangle is 5 m less than the length. Find the length and width of the rectangle.
length 15 m; width 10 m

9. The width of a rectangle is 30% of the length. The perimeter of the rectangle is 338 ft. Find the length and width of the rectangle.
length 130 ft; width 39 ft

10. The perimeter of a rectangle is 48 m. The width of the rectangle is 8 m less than the length. Find the length and width of the rectangle.
length 16 m; width 8 m

11. The perimeter of a triangle is 18.4 m. The first side is 0.1 m less than twice the second side. The third side is 1 m more than twice the second side. Find the measure of each side.
3.5 m; 6.9 m; 8 m

12. In an isosceles triangle, two sides are equal. The third side is 2.38 m less than one of the equal sides. The perimeter is 11.9 m. Find the length of each side.
2.38 m; 4.76 m; 4.76 m

▶ **Objective B**   *Application Problems*

Solve.

13. In an isosceles triangle, one angle is 5° less than three times the measure of one of the equal angles. Find the measure of each angle.
37°; 37°; 106°

14. The first angle of a triangle is twice the measure of the second angle. The third angle is 10° less than the measure of the first angle. Find the measure of each angle.
76°; 38°; 66°

15. One angle of a triangle is twice the measure of the second angle. The third angle is three times the measure of the first angle. Find the measure of each angle.
40°; 20°; 120°

16. In a triangle, one angle is twice the measure of the second angle. The third angle is three times the measure of the second angle. Find the measure of each angle.
60°; 30°; 90°

17. The first angle of a triangle is three times the measure of the second angle. The third angle is 33° more than the measure of the first angle. Find the measure of each angle.
63°; 21°; 96°

18. In an isosceles right triangle, two angles are equal and the third angle is 90°. Find the measures of the equal angles.
45°

19. In an isosceles triangle, one angle is three times the measure of one of the equal angles. Find the measure of each angle.
36°; 36°; 108°

20. In an isosceles triangle, one angle is 12° more than twice the measure of one of the equal angles. Find the measure of each angle.
42°; 42°; 96°

21. One angle of a right triangle is 3° less than twice the measure of the smallest angle. Find the measure of each angle.
31°; 59°; 90°

22. In an equiangular triangle, all three angles are equal. Find the measures of the equal angles.
60°

23. In a triangle, one angle is 5° more than the measure of the second angle. The third angle is 10° more than the measure of the second angle. Find the measure of each angle.
60°; 55°; 65°

24. One angle of a triangle is three times the measure of the third angle. The second angle is 5° less than the measure of the third angle. Find the measure of each angle.
111°; 32°; 37°

<table>
<tr><td>**SECTION 4.8**</td></tr>
</table>

Puzzle Problems

Objective A

To solve consecutive integer problems

Recall that the integers are the numbers . . . $-4, -3, -2, -1, 0, 1, 2, 3, 4, \ldots$. An **even integer** is an integer that is divisible by 2. Examples of even integers are -8, 0, and 22. An **odd integer** is an integer that is not divisible by 2. Examples of odd integers are -17, 1, and 39.

Consecutive integers are integers that follow one another in order. Examples of consecutive integers are shown at the right. (Assume that the variable n represents an integer.)	11, 12, 13 $-8, -7, -6$ $n, n + 1, n + 2$
Examples of **consecutive even integers** are shown at the right. (Assume that the variable n represents an even integer.)	24, 26, 28 $-10, -8, -6$ $n, n + 2, n + 4$
Examples of **consecutive odd integers** are shown at the right. (Assume that the variable n represents an odd integer.)	19, 21, 23 $-1, 1, 3$ $n, n + 2, n + 4$

The sum of three consecutive odd integers is 45. Find the integers.

Strategy for Solving a Consecutive Integer Problem

Let a variable represent one of the integers. Express each of the other integers in terms of that variable. Remember that consecutive integers differ by 1. Consecutive even or consecutive odd integers differ by 2.

Represent three consecutive odd integers.

First odd integer: n
Second odd integer: $n + 2$
Third odd integer: $n + 4$

Determine the relationship among the integers.

The sum of the three odd integers is 45.

$$n + (n + 2) + (n + 4) = 45$$
$$3n + 6 = 45$$
$$3n = 39$$
$$n = 13$$

$$n + 2 = 13 + 2 = 15$$

$$n + 4 = 13 + 4 = 17$$

The three consecutive odd integers are 13, 15, and 17.

Example 1

Find three consecutive even integers such that three times the second is four more than the sum of the first and third.

Strategy

- First even integer: n
  Second even integer: $n + 2$
  Third even integer: $n + 4$
- Three times the second equals four more than the sum of the first and third.

Solution

$$3(n + 2) = n + (n + 4) + 4$$
$$3n + 6 = 2n + 8$$
$$n + 6 = 8$$
$$n = 2$$

$$n + 2 = 2 + 2 = 4$$
$$n + 4 = 2 + 4 = 6$$

The three even integers are 2, 4, and 6.

Example 2

Find three consecutive integers whose sum is −6.

Your strategy

Your solution

−3; −2; −1

Solution on p. A16

Objective B

To solve coin and stamp problems

In solving problems dealing with coins or stamps of different values, it is necessary to represent the value of the coins or stamps in the same unit of money. The unit of money is frequently cents. For example:

The value of 3 quarters in cents is $3 \cdot 25$, or 75 cents.
The value of 4 nickels in cents is $4 \cdot 5$ or 20 cents.
The value of d dimes in cents is $d \cdot 10$, or $10d$ cents.

A coin bank contains $1.35 in dimes and quarters. In all, there are nine coins in the bank. Find the number of dimes and the number of quarters in the bank.

Strategy for Solving a Coin Problem

For each denomination of coin, write a numerical or variable expression for the number of coins, the value of the coin in cents, and the total value of the coins in cents. The results can be recorded in a table.

The total number of coins is 9.

Number of quarters: x
Number of dimes: $9 - x$

Coin	Number of coins	·	Value of coin in cents	=	Total value in cents
Quarter	x	·	25	=	$25x$
Dime	$9 - x$	·	10	=	$10(9 - x)$

Content and Format © 1991 HMCo.

> Determine the relationship between the total values of the different coins. Use the fact that the sum of the total values of each denomination of coin is equal to the total value of all the coins.

The sum of the total values of the different denominations of coins is equal to the total value of all the coins (135 cents).

$$25x + 10(9 - x) = 135$$
$$25x + 90 - 10x = 135$$
$$15x + 90 = 135$$
$$15x = 45$$
$$x = 3$$

$$9 - x = 9 - 3 = 6$$

There are 3 quarters and 6 dimes in the bank.

Example 3

A collection of stamps consists of 3¢ stamps and 8¢ stamps. The number of 8¢ stamps is five more than three times the number of 3¢ stamps. The total value of all the stamps is $1.48. Find the number of 3¢ stamps.

Strategy

- Number of 3¢ stamps: x
  Number of 8¢ stamps: $3x + 5$

Stamp	Number	Value	Total value
3¢	x	3	$3x$
8¢	$3x + 5$	8	$8(3x + 5)$

- The sum of the total values of the different types of stamps equals the total value of all the stamps (148 cents).

Solution

$$3x + 8(3x + 5) = 148$$
$$3x + 24x + 40 = 148$$
$$27x + 40 = 148$$
$$27x = 108$$
$$x = 4$$

There are four 3¢ stamps in the collection.

Example 4

A coin bank contains nickels, dimes, and quarters. There are four times as many nickels as dimes, and five more quarters than dimes. The total value of all the coins is $6.75. Find the number of each kind of coin in the bank.

Your strategy

Your solution

40 nickels; 10 dimes; 15 quarters

Solution on p. A16

Objective C   **To solve age problems**   4

The goal of an age problem is to determine the age of a person or an object.

A painting is 20 years old and a sculpture is 10 years old. How many years ago was the painting three times as old as the sculpture was then?

Strategy for Solving an Age Problem

Represent the ages in terms of numerical or variable expressions. To represent a past age, subtract from the present age. To represent a future age, add to the present age. The results can be recorded in a table.

The number of years ago: x

	Present age	*Past age*
Painting	20	$20 - x$
Sculpture	10	$10 - x$

Determine the relationship among the ages.

At a past age, the painting was three times as old as the sculpture was then.

$$20 - x = 3(10 - x)$$
$$20 - x = 30 - 3x$$
$$20 + 2x = 30$$
$$2x = 10$$
$$x = 5$$

Five years ago the painting was three times as old as the sculpture.

Example 5

A stamp collector has a 3¢ stamp that is 25 years older than a 5¢ stamp. In 18 years, the 3¢ stamp will be twice as old as the 5¢ stamp will be then. How old are the stamps now?

Strategy

■ Present age of 5¢ stamp: x

	Present	*Future*
3¢ stamp	$x + 25$	$x + 43$
5¢ stamp	x	$x + 18$

■ At a future age, the 3¢ stamp will be twice as old as the 5¢ stamp.

Solution

$$2(x + 18) = x + 43$$
$$2x + 36 = x + 43$$
$$x + 36 = 43$$
$$x = 7$$
$$x + 25 = 7 + 25 = 32$$

The 3¢ stamp is 32 years old and the 5¢ stamp is 7 years old.

Example 6

A half-dollar is now 25 years old. A dime is 15 years old. How many years ago was the half-dollar twice as old as the dime?

Your strategy

Your solution

5 years

Solution on p. A17

4.8   **EXERCISES**

▶ **Objective A**   *Application Problems*

Solve.

1. Find two consecutive even integers such that four times the first is three times the second.
    6, 8

2. The sum of three consecutive odd integers is 51. Find the integers.
    15, 17, 19

3. Find three consecutive odd integers such that three times the middle integer is one more than the sum of the first and third.
    −1, 1, 3

4. Twice the smallest of three consecutive odd integers is seven more than the largest. Find the integers.
    11, 13, 15

5. The sum of three consecutive integers is 60. Find the integers.
    19, 20, 21

6. Find two consecutive even integers such that three times the first equals twice the second.
    4, 6

7. Seven times the first of two consecutive odd integers is five times the second. Find the integers.
    5, 7

8. Find three consecutive even integers such that three times the middle integer is four more than the sum of the first and third.
    2, 4, 6

9. The sum of three consecutive even integers is 42. Find the integers.
    12, 14, 16

10. Five times the first of two consecutive odd integers equals three times the second. Find the integers.
    3, 5

11. Find three consecutive even integers whose sum is negative eighteen.
    −8, −6, −4

12. The sum of three consecutive even integers is 66. Find the integers.
    20, 22, 24

13. The sum of three consecutive odd integers is 75. Find the integers.
    23, 25, 27

14. Find three consecutive integers whose sum is negative twenty-one.
    −8, −7, −6

Solve.

15. Three times the smallest of three consecutive even integers is four more than twice the largest. Find the integers.
12, 14, 16

16. The sum of three consecutive integers is 48. Find the integers.
15, 16, 17

▶ **Objective B**   *Application Problems*

Solve.

17. A drawer contains 15¢ stamps and 18¢ stamps. The number of 15¢ stamps is two less than three times the number of 18¢ stamps. The total value of all the stamps is $.96. How many 15¢ stamps are in the drawer?
4 stamps

18. A coin bank contains pennies, nickels, and quarters. There are six times as many nickels as pennies and four times as many quarters as pennies. The total amount of money in the bank is $6.55. Find the number of pennies in the bank.
5 pennies

19. A bank contains 30 coins in dimes and quarters. The coins have a total value of $5.40. Find the number of dimes and quarters in the bank.
14 dimes; 16 quarters

20. The total value of the dimes and quarters in a bank is $6.50. There are five more quarters than dimes. Find the number of each type of coin in the bank.
15 dimes; 20 quarters

21. A collection of stamps consists of 3¢ stamps, 7¢ stamps, and 12¢ stamps. The number of 3¢ stamps is six less than the number of 7¢ stamps. The number of 12¢ stamps is half the number of 7¢ stamps. The total value of all the stamps is $3.02. Find the number of each type of stamp in the collection.
3¢ stamps, 14; 7¢ stamps, 20; 12¢ stamps, 10

22. A coin purse contains 16 coins in nickels and dimes. The coins have a total value of $1. Find the number of nickels and dimes in the coin purse.
12 nickels; 4 dimes

23. A business executive purchased 40 stamps for $7.70. The purchase included 15¢ stamps and 20¢ stamps. How many of each type of stamp were purchased?
15¢ stamps, 6; 20¢ stamps, 34

24. A collection of stamps consists of 2¢ stamps, 5¢ stamps, and 7¢ stamps. There are eight more 2¢ stamps than 5¢ stamps and twice as many 7¢ stamps as 5¢ stamps. The total value of the stamps is $1.63. Find the number of each type of stamp in the collection.
2¢ stamps, 15; 5¢ stamps, 7; 7¢ stamps, 14

Solve.

25. A collection of stamps consists of 6¢ stamps, 8¢ stamps, and 15¢ stamps. The number of 6¢ stamps is three times the number of 8¢ stamps. There are seven more 15¢ stamps than there are 6¢ stamps. The total value of all the stamps is $4.60. Find the number of each type of stamp.
6¢ stamps, 15; 8¢ stamps, 5; 15¢ stamps, 22

26. A postal clerk sold some 15¢ stamps and some 25¢ stamps. Altogether 10 stamps were sold for a total cost of $1.70. How many of each type of stamp were sold?
15¢ stamps, 8; 25¢ stamps, 2

27. A total of 30 bills are in a cash box. Some of the bills are one-dollar bills and the rest are five-dollar bills. The total amount of cash in the box is $50. Find the number of each type of bill in the cash box.
25 one-dollar bills; 5 five-dollar bills

28. A child's piggy bank contains nickels, dimes, and quarters. There are twice as many nickels as dimes and three more quarters than nickels. The total value of all the coins is $11.25. Find the number of each type of coin.
30 nickels; 15 dimes; 33 quarters

29. A coin bank contains pennies, nickels, and dimes. There are five times as many nickels as pennies and three times as many dimes as pennies. The total amount of money in the bank is $6.72. Find the number of pennies in the bank.
12 pennies

30. A bank teller cashed a check for $150 using twenty-dollar bills and ten-dollar bills. In all, nine bills were handed to the customer. Find the number of twenty-dollar bills and ten-dollar bills.
6 twenty-dollar bills; 3 ten-dollar bills

▶ **Objective C** *Application Problems*

Solve.

31. A gold coin is 84 years older than a silver coin. Twenty years ago, the gold coin was three times as old as the silver coin was then. Find the present ages of the two coins.
silver coin, 62 years; gold coin, 146 years

32. The sum of the ages of a 5¢ coin and a 10¢ coin is 12. Two years from now the age of the 5¢ coin will equal the age of the 10¢ coin two years ago. Find the present age of each coin.
5¢ coin, 4 years; 10¢ coin, 8 years

33. A collector of hand-woven rugs has an oval rug that is 42 years old and a circular rug that is 14 years old. How many years ago was the oval rug five times as old as the circular rug was then?
7 years

Solve.

34. The sum of the ages of an oil painting and a watercolor is 20 years. One year from now the oil painting will be nine times the age the watercolor was one year ago. Find the present age of each painting.
watercolor, 3 years; oil painting, 17 years

35. The sum of the ages of two cars is 8 years. Two years ago the age of the older car was three times the age the younger car was then. Find the present age of each car.
younger car, 3 years; older car, 5 years

36. An antique car is 45 years older than a replica of the car. In 13 years, the antique car will be four times as old as the replica will be then. Find the present ages of the two cars.
replica, 2 years; antique, 47 years

37. A stamp collector has a 2¢ stamp that is 20 years old and a 3¢ stamp that is 16 years old. How many years ago was the 2¢ stamp twice as old as the 3¢ stamp was then?
12 years

38. A book dealer has an autographed, first-edition book that is 35 years old and a reprint of the book that is 7 years old. In how many years will the autographed first edition be three times as old as the reprint will be then?
7 years

39. The sum of the ages of two children is 18. Six years from now the age of the older child will be twice the age of the younger child. Find the present ages of the two children.
younger child, 4 years; older child, 14 years

40. An antique butterchurn is 85 years old and an antique ice box is 75 years old. How many years ago was the butterchurn twice the age the ice box was then?
65 years

41. An oil painting is 10 years older than a lithograph. Five years ago the painting was twice as old as the lithograph was then. Find the present age of each.
lithograph, 15 years; painting, 25 years

42. An art collector has a porcelain vase that is 15 years old and a crystal vase that is 95 years old. In how many years will the crystal vase be three times as old as the porcelain vase will be then?
25 years

43. A diamond ring is 2 years old and a ruby ring is 22 years old. In how many years will the ruby ring be twice the age the diamond ring will be then?
18 years

44. A coin collector has a dime that is 24 years older than a nickel. In 8 years the dime will be twice as old as the nickel will be then. Find the present age of the dime and nickel.
nickel, 16 years; dime, 40 years

Calculators and Computers

The EEX Key on a Calculator

Many application problems require the use of very large or very small numbers. The EEX key, or on some calculators the EXP key, is used for entering these numbers. For example, the speed of light is roughly 29,800,000,000 cm/sec. This number cannot be directly entered on a calculator. The EEX key (which means exponent) is used.

To understand this key, consider the following table:

$$5489 = 548.9 \times 10^1$$
$$5489 = 54.89 \times 10^2$$
$$5489 = 5.489 \times 10^3$$

The number 5489 can be represented in various forms. Note that each time the decimal point is moved to the *left*, it is necessary to *multiply* by a power of 10.

Now study the following table:

$$0.004638 = 0.04638 \div 10^1$$
$$0.004638 = 0.4638 \div 10^2$$
$$0.004638 = 4.638 \div 10^3$$

The number 0.004638 can be represented in various forms. Note that each time the decimal point is moved to the *right*, it is necessary to *divide* by a power of 10. It is customary to express dividing by a power of 10 by using multiplication and a negative exponent. Thus the above table could be written as follows:

$$0.004638 = 0.04638 \times 10^{-1}$$
$$0.004638 = 0.4638 \times 10^{-2}$$
$$0.004638 = 4.638 \times 10^{-3}$$

To enter the speed of light on a calculator, rewrite the number by moving the decimal point 10 places to the left and then multiplying by 10^{10}.

Now enter the number.   2.98 EEX 10

The EEX key is used to enter the exponent on 10.

The wave length of an x-ray is approximately 0.00000000537 cm. To enter this number on your calculator, rewrite the number by moving the decimal point 9 places to the right and then dividing by 10^9 (or multiplying by 10^{-9}).

$$0.00000000537 = 5.37 \div 10^9 = 5.37 \times 10^{-9}$$

Now enter the number.   5.37 EEX 9 +/−

The +/− key is used to change the sign of the exponent and thus make it negative.

Chapter Summary

Key Words   *Percent* means "parts of 100."

Cost is the price that a business pays for a product.

Selling price is the price for which a business sells a product to a customer.

Markup is the difference between selling price and cost.

Discount is the amount by which a retailer reduces the regular price of a product.

Uniform motion means that an object at a constant speed moves in a straight line.

The *perimeter* of a geometric figure is a measure of the distance around the figure.

A *right angle* is an angle whose measure is 90 degrees.

A *right triangle* has one right angle.

An *isosceles triangle* has two equal angles and two equal sides.

Consecutive integers are integers that follow one another in order.

Essential Rules

Basic Percent Equation

Percent × base = amount
$$\mathbf{P} \times \mathbf{B} = \mathbf{A}$$

Basic Markup Equation

Selling price = cost + markup
$$\mathbf{S} = \mathbf{C} + \mathbf{M}$$

Markup = markup rate × cost
$$\mathbf{M} = \mathbf{r} \times \mathbf{C}$$

Basic Discount Equations

Sale price = regular price − discount
$$\mathbf{S} = \mathbf{R} - \mathbf{D}$$

Discount = discount rate × regular price
$$\mathbf{D} = \mathbf{r} \times \mathbf{R}$$

Annual Simple Interest Equation

Simple interest = principal × simple interest rate
$$\mathbf{I} = \mathbf{P} \times \mathbf{r}$$

Value Mixture Equation

Value = amount × unit cost
$$\mathbf{V} = \mathbf{A} \times \mathbf{C}$$

Percent Mixture Equation

Quantity = amount × percent concentration
$$\mathbf{Q} = \mathbf{A} \times \mathbf{r}$$

Uniform Motion Equation

Distance = rate × time
$$\mathbf{d} = \mathbf{r} \times \mathbf{t}$$

Triangle Equation

$$\mathbf{A + B + C = 180°}$$

Chapter Review

SECTION 4.1

1. Write $79\frac{1}{2}\%$ as a fraction.

   $\frac{159}{200}$

2. Write 6.2% as a decimal.

   0.062

3. Write $\frac{5}{8}$ as a percent. Round to the nearest tenth of a percent.
   62.5%

4. Write $\frac{19}{35}$ as a percent. Write the remainder in fractional form.

   $54\frac{2}{7}\%$

SECTION 4.2

5. What is $\frac{1}{2}\%$ of 3000?

   15

6. 27 is what percent of 40?

   67.5%

7. In 1987, the value of a Yogi Berra rookie card was $75. By 1990 the value of that card had increased by $300. What percent increase does this represent?
   400%

8. A person's weight on the moon is $16\frac{2}{3}\%$ of the person's weight on the earth. If an astronaut weighed 180 lb on the earth, how much would the astronaut weigh on the moon?

   30 lb

SECTION 4.3

9. A furniture store uses a markup rate of 60%. The store sells a solid oak curio cabinet for $1074. Find the cost of the curio cabinet.
   $671.25

10. A pair of athletic shoes, which regularly sell for $55, are on sale for 25% off the regular price. Find the sale price.
    $41.25

11. The sale price for a carpet sweeper is $26.56, which is 17% off the regular price. Find the regular price.
    $32

12. A ceiling fan, which regularly sells for $60, is on sale for $40. Find the discount rate.

    $33\frac{1}{3}\%$

SECTION 4.4

13. A club treasurer has $2400 to be deposited into two simple interest accounts. On one account the annual simple interest rate is 6.75%. The annual simple interest on the other account is 9.45%. How much should the treasurer deposit in each account so that the interest earned in each account is the same?
    $1400 at 6.75% $1000 at 9.45%

14. An engineering consultant invested $14,000 in an 8.15% annual simple interest Individual Retirement Account. How much additional money must be deposited into an account which pays 12% annual simple interest so that the total interest earned on both accounts is 9.25% of the total investment?
    $5600

SECTION 4.5

15. A health food store combined cranberry juice which cost $1.79 per quart with apple juice which cost $1.19 per quart. How many quarts of each were used to make 10 qt of a cranapple juice mixture to sell for $1.61 per quart?
7 qt cranberry;   3 qt apple

17. A dairy mixed five gallons of cream which is 30% butterfat with eight gallons of milk which is 4% butterfat. What is the percent of butterfat in the resulting mixture?
14%

16. Find the selling price per pound of a meatloaf mixture made from three pounds of ground beef which cost $1.99 per pound and one pound of ground turkey which cost $1.39 per pound.
$1.84 per pound

18. A pharmacist has 15 liters of an 80% alcohol solution. How many liters of pure water should be added to the alcohol solution to make a diluted alcohol solution which is 75% alcohol?
1 liter

SECTION 4.6

19. A motorcyclist and a bicyclist set out at 8:00 A.M. from the same point headed in the same direction. The average speed of the motorcyclist is three times the speed of the bicyclist. In two hours the motorcyclist is 60 miles ahead of the bicyclist. Find the rate of the motorcyclist.
45 mph

20. The largest swimming pool in the world, located in Casablanca, Morocco, is 480 meters long. Two swimmers start at the same time from opposite ends of the pool and start swimming toward each other. One swimmer's rate is 65 meters per minute and the other swimmer's rate is 55 meters per minute. How many minutes after they begin will they meet?
4 min

SECTION 4.7

21. The perimeter of a triangle is 35 in. The second side is 4 in. longer than the first side. The third side is 1 in. shorter than twice the first side. Find the measure of each side.
8 in., 12 in., 15 in.

23. In an isosceles triangle one angle is 25° less than half the measure of one of the equal angles. Find the measure of each angle.
16°, 82°, 82°

22. The length of a rectangle is four times the width of the rectangle. The perimeter is 200 ft. Find the length and the width of the rectangle.
length: 80 ft;   width: 20 ft

24. One angle of a triangle is 15° more than the measure of the second angle. The third angle is 15° less than the measure of the second angle. Find the measure of each angle.
75°, 60°, 45°

SECTION 4.8

25. Find two consecutive integers such that five times the first integer is 15 more than three times the second integer.
9, 10

27. An antique telephone is 85 years old and an antique typewriter is 160 years old. How many years ago was the typewriter twice the age the telephone was then?
10 years

26. A ticket seller at a baseball card show had $145 in one-dollar bills and five-dollar bills. In all, there were 53 bills. Find the number of one-dollar bills and the number of five-dollar bills.
30 one-dollar bills, 23 five-dollar bills

28. A grandparent is eight times the age of a child. One year from now the grandparent will be seven times the age the child will be then. Find the present ages of the grandparent and the child.
grandparent: 48 years;   child: 6 years

Chapter Test

1. Write 45% as a fraction and as a decimal.

   $\frac{9}{20}$; 0.45   [4.1A]

2. Write $37\frac{1}{2}\%$ as a fraction.

   $\frac{3}{8}$ [4.1A]

3. Write 1.025 as a percent.

   102.5%   [4.1B]

4. Write $\frac{5}{6}$ as a percent. Write the remainder in fractional form.

   $83\frac{1}{3}\%$ [4.1B]

5. Find 28% of 90.
   25.2   [4.2A]

6. 30 is what percent of 20?
   150%   [4.2A]

7. The value of a motorhome today is $33,000. This is 75% of the motorhome's value last year. Find the value of the motorhome last year.
   $44,000   [4.2B]

8. A survey of students indicated that 120 out of 200 favored a grading system that used plus and minus along with letter grades. What percent of the students favored the plus/minus system?
   60%   [4.2B]

9. The manager of a jewelry store uses a markup rate of 75%. The selling price for a gold ring is $612.50. Find the cost of the gold ring.
   $350   [4.3A]

10. A television that regularly sells for $450 is on sale for $360. Find the discount rate.
    20%   [4.3B]

11. The cost to a college bookstore for a textbook is $27. The bookstore sells the textbook for $36. Find the markup rate.

    $33\frac{1}{3}\%$ [4.3A]

12. A total of $5000 is deposited into two simple interest accounts. The annual simple interest rate on one account is 6%; on the second account, the simple interest rate is 9%. How much should be invested in each account so that the total annual interest earned is $345?
    $3500 at 6%; $1500 at 9%   [4.4A]

13. A baker wants to make a 15-lb blend of flour that costs $.60 per pound. The blend is made using a rye flour that costs $.70 per pound and a wheat flour that costs $.40 per pound. How many pounds of each flour should be used?

10 lb at $.70; 5 lb at $.40   [4.5A]

14. How many gallons of water must be mixed with 5 gal of a 20% salt solution to make a 16% salt solution?

1.25 gal   [4.5B]

15. A cross-country skier leaves a camp to explore a wilderness area. Two hours later a friend leaves the camp in a snowmobile, traveling 4 mph faster than the skier, and overtakes the skier 1 h later. Find the rate of the snowmobile.

6 mph   [4.6A]

16. The perimeter of a rectangular sandbox is 20 ft. The length of the sandbox is 2 ft less than twice the width. Find the length and width of the sandbox.

length 6 ft; width 4 ft   [4.7A]

17. In an isosceles triangle, two angles are equal. The third angle of the triangle is 30° less than one of the equal angles. Find the measure of one of the equal angles.

70°   [4.7B]

18. Find three consecutive even integers whose sum is 36.

10, 12, 14   [4.8A]

19. A coin bank contains 30 dimes and quarters. The total amount of money in the bank is $6.00. Find the number of quarters in the bank.

20 quarters   [4.8B]

20. A coin is seven times the age of a stamp. In six years, the coin will be three times the age the stamp will be then. Find the age of each now.

stamp, 3 years old; coin, 21 years old   [4.8C]

Cumulative Review

1. Simplify $8 \cdot (-6) - 4(-3)$.
-36 [1.5B]

2. Simplify $\left(-\frac{3}{8}\right)^2 \cdot \left(-\frac{4}{9}\right)$.
$-\frac{1}{16}$ [1.5A]

3. Simplify $\frac{5}{8} - \left(\frac{1}{2}\right)^2 \div \left(\frac{1}{3} - \frac{3}{4}\right)$
$\frac{49}{40}$ [1.5B]

4. Evaluate $a^2 - (2a - b^2)$ when $a = 2$ and $b = -3$.
9 [2.1A]

5. Simplify $6a - 4b - (-2a) - 5b$.
$8a - 9b$ [2.2A]

6. Simplify $-3(3x^2 + 4x - 7)$.
$-9x^2 - 12x + 21$ [2.2C]

7. Simplify $-2[4 - 2(2x + 1) - x]$.
$10x - 4$ [2.2D]

8. Is -1 a solution of $3 + 4x = x^2 - 2$?
Yes [3.1A]

9. Solve: $7 - x = 9$
-2 [3.1B]

10. Solve: $\frac{2}{3}x = -8$
-12 [3.1C]

11. Solve: $-2 = 6 - 4x$
2 [3.2A]

12. Solve: $3x - 4 = 4 - 2(x - 1)$
2 [3.3B]

13. Write 55% as a fraction.
$\frac{11}{20}$ [4.1A]

14. Write $66\frac{2}{3}\%$ as a fraction.
$\frac{2}{3}$ [4.1A]

15. The sum of two numbers is ten. The difference of four times the smaller number and eight equals ten less than the product of three and the larger number. Find the two numbers.
4 and 6 [3.4A]

16. The repair bill for a washing machine was $73. This includes $28 for parts and $30 for each hour of labor. Find the number of hours of labor.
1.5 h [3.4B]

17. Write 1.03 as a percent.

103%   [4.1B]

18. Write $\frac{9}{20}$ as a percent.

45%   [4.1B]

19. Find $16\frac{2}{3}\%$ of 90.

15   [4.2A]

20. 25% of what number is 30?

120   [4.2A]

21. The value of an investment today is $4400. This is a 10% increase over the value of the investment 1 year ago. Find the value of the investment last year.

$4000   [4.2B]

22. A deposit of $2400 is made into an account that earns 9% simple interest. How much additional money must be deposited into an account that pays 12% simple interest so that the total interest earned is 10% of the total investment?

$1200   [4.4A]

23. A tree farm buys a 5-gal fruit tree for $3.30 and sells it for $4.62. Find the markup rate.

40%   [4.3A]

24. The toner cartridge for a personal copier, which normally sells for $62.00, is on sale for 40% off the regular price. Find the sale price.

$37.20   [4.3B]

25. How many pounds of an oat flour that costs $.80 per pound must be mixed with 40 pounds of a wheat flour that costs $.50 per pound to make a blend that costs $.60 per pound?

20 lb   [4.5A]

26. How many grams of pure gold must be added to 100 g of a 20% gold alloy to make an alloy that is 36% gold?

25 g   [4.5B]

27. The perimeter of a rectangular office is 44 ft. The length of the office is 2 ft more than the width. Find the dimensions of the office.

length 12 ft; width 10 feet   [4.7A]

28. In an equilateral triangle, all angles are equal. Find the measure of one of the angles of an equilateral triangle.

60°   [4.7B]

29. Four times the second of three consecutive integers equals the sum of the first and third integers. Find the integers.

−1, 0, 1   [4.8A]

30. A coin bank contains nickels and dimes. The number of dimes is two more than twice the number of nickels. The total amount in the bank is $5.45. Find the number of dimes in the bank.

44 dimes   [4.8B]

5

Polynomials

OBJECTIVES

- ▶ To add polynomials
- ▶ To subtract polynomials
- ▶ To multiply monomials
- ▶ To simplify powers of monomials
- ▶ To multiply a polynomial by a monomial
- ▶ To multiply two polynomials
- ▶ To multiply two binomials
- ▶ To multiply binomials that have special products
- ▶ To solve application problems
- ▶ To divide monomials and simplify expressions containing negative exponents
- ▶ To divide a polynomial by a monomial
- ▶ To divide polynomials

Early Egyptian Arithmetic Operations

The early Egyptian arithmetic processes are recorded on the early Rhind Papyrus without showing the underlying principles. Scholars of today can only guess how these early developments were discovered.

Egyptian hieroglyphics used a base-ten system of numbers in which a vertical line represented 1, a heel bone, ∩, represented 10, and a scroll, 𝟗, represented 100.

The symbols at the right represent the number 237.
There are 7 vertical lines, 3 heel bones, and 2 scrolls.
Thus the symbols at the right represent 7 + 30 + 200, or 237.

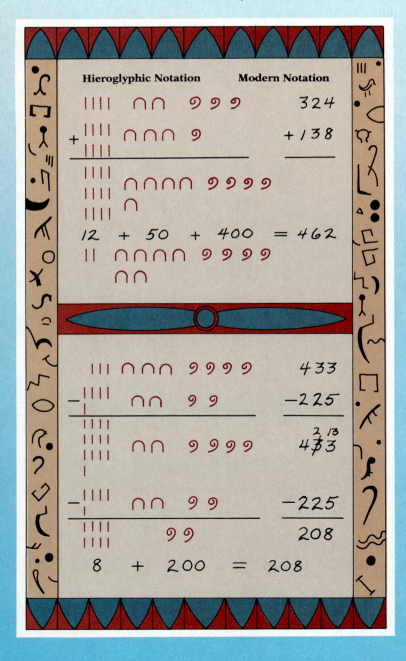

Addition in hieroglyphic notation does not require memorization of addition facts. Addition is done just by counting symbols.

Addition is a simple grouping operation.

Write down the total of each kind of symbol.

Group the 10 straight lines into one heel bone.

Subtraction in the hieroglyphic system is similar to making change. For example, what change do you get from a $1.00 bill when buying a $.55 item?

5 cannot be subtracted from 3, so a 10 is "borrowed" and 10 ones are added.

Note that no zero is provided in this number system. That place value symbol is just not used. As shown at the right, the heel bone is not used because there are no 10's necessary in 208.

Content and Format © 1991 HMCo.

SECTION 5.1   Addition and Subtraction of Polynomials

Objective A   To add polynomials

A **monomial** is a number, a variable, or a product of numbers and variables. A **polynomial** is a variable expression in which the terms are monomials.

A **monomial** is a polynomial of *one* term. $5x^3$ is a monomial.
A **binomial** is a polynomial of *two* terms. $5y^2 - 3x$ is a binomial.
A **trinomial** is a polynomial of *three* terms. $6xy - 2r^2s + 4r$ is a trinomial.

The terms of a polynomial in one variable are usually arranged so that the exponents of the variable decrease from left to right. This is called **descending order.**

$4x^3 - 3x^2 + 6x - 1$
$5y^4 - 2y^3 + y^2 - 7y + 8$

The **degree of a polynomial** in one variable is its largest exponent. The degree of $4x^3 - 3x^2 + 6x - 1$ is 3. The degree of $5y^4 - 2y^3 + y^2 - 7y + 8$ is 4.

Polynomials can be added, using either a vertical or a horizontal format, by combining like terms.

Simplify $(2x^2 + x - 1) + (3x^3 + 4x^2 - 5)$. Use a vertical format.

Arrange the terms of each polynomial in descending order with like terms in the same column. Combine the terms in each column.

$$\begin{array}{r} 2x^2 + x - 1 \\ +3x^3 + 4x^2 \quad\;\; - 5 \\ \hline 3x^3 + 6x^2 + x - 6 \end{array}$$

Simplify $(3x^3 - 7x + 2) + (7x^2 + 2x - 7)$. Use a horizontal format.

Use the Commutative and Associative Properties of Addition to rearrange and group like terms.
Combine like terms.
Write the polynomial in descending order.

$(3x^3 - 7x + 2) + (7x^2 + 2x - 7)$

$3x^3 + 7x^2 + (-7x + 2x) + (2 - 7)$ Do this step mentally.

$3x^3 + 7x^2 - 5x - 5$

Example 1
Simplify $(7y^2 - 6y + 9) + (-8y^2 - 2)$.
Use a vertical format.

Solution
$$\begin{array}{r} 7y^2 - 6y + 9 \\ -8y^2 \quad\;\; - 2 \\ \hline -y^2 - 6y + 7 \end{array}$$

Example 2
Simplify $(2x^2 + 4x - 3) + (5x^2 - 6x)$.
Use a vertical format.

Your solution
$7x^2 - 2x - 3$

Example 3
Simplify $(-4x^2 - 3xy + 2y^2) + (3x^2 - 4y^2)$.
Use a horizontal format.

Solution
$(-4x^2 - 3xy + 2y^2) + (3x^2 - 4y^2)$
$-x^2 - 3xy - 2y^2$

Example 4
Simplify $(-3x^2 + 2y^2) + (-8x^2 + 9xy)$.
Use a horizontal format.

Your solution
$-11x^2 + 9xy + 2y^2$

Solutions on p. A17

Objective B	**To subtract polynomials**

The **opposite** of the polynomial $x^2 - 2x + 3$ is $-(x^2 - 2x + 3)$.

To simplify the opposite of a polynomial, change the sign of every term inside the parentheses.

$$-(x^2 - 2x + 3) = -x^2 + 2x - 3$$

Polynomials can be subtracted using either a vertical or horizontal format. To subtract, add the opposite of the second polynomial to the first.

Simplify $(-3x^2 - 7) - (-8x^2 + 3x - 4)$. Use a vertical format.

Arrange the terms of each polynomial in descending order with like terms in the same column.
Rewrite subtraction as addition of the opposite.
Combine like terms in each column.

$$\begin{array}{r} -3x^2 \qquad -7 \\ -(-8x^2 + 3x - 4) \end{array} = \begin{array}{r} -3x^2 \qquad -7 \\ 8x^2 - 3x + 4 \\ \hline 5x^2 - 3x - 3 \end{array}$$

Simplify $(5x^2 - 3x + 4) - (-3x^3 - 2x + 8)$. Use a horizontal format.

Rewrite subtraction as addition of the opposite. Combine like terms.
Write the polynomial in descending order.

$$(5x^2 - 3x + 4) - (-3x^3 - 2x + 8)$$
$$(5x^2 - 3x + 4) + (3x^3 + 2x - 8)$$
$$3x^3 + 5x^2 - x - 4$$

Example 5

Simplify $(6y^2 - 3y - 1) - (7y^2 - y)$.
Use a vertical format.

Solution

$$\begin{array}{r} 6y^2 - 3y - 1 \\ -(7y^2 - y) \end{array} = \begin{array}{r} 6y^2 - 3y - 1 \\ -7y^2 + y \\ \hline -y^2 - 2y - 1 \end{array}$$

Example 6

Simplify
$(8y^2 - 4xy + x^2) - (2y^2 - xy + 5x^2)$.
Use a vertical format.

Your solution

$6y^2 - 3xy - 4x^2$

Example 7

Simplify $(4x^3 - 3x - 7) - (7x^2 - 4x - 2)$.
Use a horizontal format.

Solution

$(4x^3 - 3x - 7) - (7x^2 - 4x - 2)$
$(4x^3 - 3x - 7) + (-7x^2 + 4x + 2)$
$4x^3 - 7x^2 + x - 5$

Example 8

Simplify $(-3a^2 - 4a + 2) - (5a^3 + 2a - 6)$.
Use a horizontal format.

Your solution

$-5a^3 - 3a^2 - 6a + 8$

Solutions on p. A17

| **5.1** | **EXERCISES** |

▶ Objective A

Simplify. Use a vertical format.

1. $(x^2 + 7x) + (-3x^2 - 4x)$
$-2x^2 + 3x$

2. $(3y^2 - 2y) + (5y^2 + 6y)$
$8y^2 + 4y$

3. $(y^2 + 4y) + (-4y - 8)$
$y^2 - 8$

4. $(3x^2 + 9x) + (6x - 24)$
$3x^2 + 15x - 24$

5. $(2x^2 + 6x + 12) + (3x^2 + x + 8)$
$5x^2 + 7x + 20$

6. $(x^2 + x + 5) + (3x^2 - 10x + 4)$
$4x^2 - 9x + 9$

7. $(x^3 - 7x + 4) + (2x^2 + x - 10)$
$x^3 + 2x^2 - 6x - 6$

8. $(3y^3 + y^2 + 1) + (-4y^3 - 6y - 3)$
$-y^3 + y^2 - 6y - 2$

9. $(2a^3 - 7a + 1) + (-3a^2 - 4a + 1)$
$2a^3 - 3a^2 - 11a + 2$

10. $(5r^3 - 6r^2 + 3r) + (r^2 - 2r - 3)$
$5r^3 - 5r^2 + r - 3$

Simplify. Use a horizontal format.

11. $(4x^2 + 2x) + (x^2 + 6x)$
$5x^2 + 8x$

12. $(-3y^2 + y) + (4y^2 + 6y)$
$y^2 + 7y$

13. $(4x^2 - 5xy) + (3x^2 + 6xy - 4y^2)$
$7x^2 + xy - 4y^2$

14. $(2x^2 - 4y^2) + (6x^2 - 2xy + 4y^2)$
$8x^2 - 2xy$

15. $(2a^2 - 7a + 10) + (a^2 + 4a + 7)$
$3a^2 - 3a + 17$

16. $(-6x^2 + 7x + 3) + (3x^2 + x + 3)$
$-3x^2 + 8x + 6$

17. $(5x^3 + 7x - 7) + (10x^2 - 8x + 3)$
$5x^3 + 10x^2 - x - 4$

18. $(3y^3 + 4y + 9) + (2y^2 + 4y - 21)$
$3y^3 + 2y^2 + 8y - 12$

19. $(2r^2 - 5r + 7) + (3r^3 - 6r)$
$3r^3 + 2r^2 - 11r + 7$

20. $(3y^3 + 4y + 14) + (-4y^2 + 21)$
$3y^3 - 4y^2 + 4y + 35$

21. $(3x^2 + 7x + 10) + (-2x^3 + 3x + 1)$
$-2x^3 + 3x^2 + 10x + 11$

22. $(7x^3 + 4x - 1) + (2x^2 - 6x + 2)$
$7x^3 + 2x^2 - 2x + 1$

▶ **Objective B**

Simplify. Use a vertical format.

23. $(x^2 - 6x) - (x^2 - 10x)$
$4x$

24. $(y^2 + 4y) - (y^2 + 10y)$
$-6y$

25. $(2y^2 - 4y) - (-y^2 + 2)$
$3y^2 - 4y - 2$

26. $(-3a^2 - 2a) - (4a^2 - 4)$
$-7a^2 - 2a + 4$

27. $(x^2 - 2x + 1) - (x^2 + 5x + 8)$
$-7x - 7$

28. $(3x^2 + 2x - 2) - (5x^2 - 5x + 6)$
$-2x^2 + 7x - 8$

29. $(4x^3 + 5x + 2) - (-3x^2 + 2x + 1)$
$4x^3 + 3x^2 + 3x + 1$

30. $(5y^2 - y + 2) - (-2y^3 + 3y - 3)$
$2y^3 + 5y^2 - 4y + 5$

31. $(2y^3 + 6y - 2) - (y^3 + y^2 + 4)$
$y^3 - y^2 + 6y - 6$

32. $(-2x^2 - x + 4) - (-x^3 + 3x - 2)$
$x^3 - 2x^2 - 4x + 6$

Simplify. Use a horizontal format.

33. $(y^2 - 10xy) - (2y^2 + 3xy)$
$-y^2 - 13xy$

34. $(x^2 - 3xy) - (-2x^2 + xy)$
$3x^2 - 4xy$

35. $(3x^2 + x - 3) - (x^2 + 4x - 2)$
$2x^2 - 3x - 1$

36. $(5y^2 - 2y + 1) - (-3y^2 - y - 2)$
$8y^2 - y + 3$

37. $(-2x^3 + x - 1) - (-x^2 + x - 3)$
$-2x^3 + x^2 + 2$

38. $(2x^2 + 5x - 3) - (3x^3 + 2x - 5)$
$-3x^3 + 2x^2 + 3x + 2$

39. $(4a^3 - 2a + 1) - (a^3 - 2a + 3)$
$3a^3 - 2$

40. $(b^2 - 8b + 7) - (4b^3 - 7b - 8)$
$-4b^3 + b^2 - b + 15$

41. $(4y^3 - y - 1) - (2y^2 - 3y + 3)$
$4y^3 - 2y^2 + 2y - 4$

42. $(3x^2 - 2x - 3) - (2x^3 - 2x^2 + 4)$
$-2x^3 + 5x^2 - 2x - 7$

SECTION 5.2   Multiplication of Monomials

Objective A   To multiply monomials

Recall that in the exponential expression x^5, x is the base and 5 is the exponent. The exponent indicates the number of times the base occurs as a factor.

The product of exponential expressions with the *same* base can be simplified by writing each expression in factored form and writing the result with an exponent.

$$\overbrace{}^{\text{3 factors}}\ \overbrace{}^{\text{2 factors}}$$
$$x^3 \cdot x^2 = \underbrace{(x \cdot x \cdot x) \cdot (x \cdot x)}_{\text{5 factors}}$$

$$= x \cdot x \cdot x \cdot x \cdot x$$
$$= x^5$$

Note that adding the exponents results in the same product.

$$x^3 \cdot x^2 = x^{3+2} = x^5$$

Rule for Multiplying Exponential Expressions

If m and n are positive integers, then $x^m \cdot x^n = x^{m+n}$.

Simplify $a^2 \cdot a^6 \cdot a$.

The bases are the same.
Add the exponents.

$$a^2 \cdot a^6 \cdot a \boxed{= a^{2+6+1}}$$ Do this step mentally.

$$= a^9$$

Simplify $(2xy)(3x^2y)$.

Use the Commutative and Associative Properties of Multiplication to rearrange and group factors. Multiply variables with the same base by adding the exponents.

$$(2xy)(3x^2y) = (2 \cdot 3)(x \cdot x^2)(y \cdot y)$$

$$\boxed{= 6x^{1+2}y^{1+1}}$$ Do this step mentally.

$$= 6x^3y^2$$

Example 1
Simplify $(-4y)(5y^3)$.

Solution

$(-4y)(5y^3) = (-4 \cdot 5)(y \cdot y^3) = -20y^4$

Example 2
Simplify $(3x^2)(6x^3)$.

Your solution

$18x^5$

Example 3
Simplify $(2x^2y)(-5xy^4)$.

Solution

$(2x^2y)(-5xy^4) = [2(-5)](x^2 \cdot x)(y \cdot y^4)$
$$= -10x^3y^5$$

Example 4
Simplify $(-3xy^2)(-4x^2y^3)$.

Your solution

$12x^3y^5$

Solutions on p. A17

| Objective B | **To simplify powers of monomials** |

A power of a monomial can be simplified by rewriting the expression in factored form and then using the Rule for Multiplying Exponential Expressions.

$$(x^2)^3 = x^2 \cdot x^2 \cdot x^2$$
$$= x^{2+2+2}$$
$$= x^6$$

$$(x^4y^3)^2 = (x^4y^3)(x^4y^3)$$
$$= x^4 \cdot y^3 \cdot x^4 \cdot y^3$$
$$= (x^4 \cdot x^4)(y^3 \cdot y^3)$$
$$= x^{4+4}y^{3+3}$$
$$= x^8y^6$$

Note that multiplying each exponent inside the parentheses by the exponent outside the parentheses gives the same result.

$$(x^2)^3 = x^{2 \cdot 3} = x^6$$

$$(x^4y^3)^2 = x^{4 \cdot 2}y^{3 \cdot 2} = x^8y^6$$

Rule for Simplifying Powers of Exponential Expressions

If m and n are positive integers, then $(x^m)^n = x^{m \cdot n}$.

Rule for Simplifying Powers of Products

If m, n, and p are positive integers, then $(x^m \cdot y^n)^p = x^{m \cdot p}y^{n \cdot p}$.

Simplify $(x^5)^2$.

Multiply the exponents.

$(x^5)^2 \boxed{= x^{5 \cdot 2}}$ Do this step mentally.

$$= x^{10}$$

Simplify $(3a^2b)^3$.

Multiply each exponent inside the parentheses by the exponent outside the parentheses.

$(3a^2b)^3 \boxed{= 3^{1 \cdot 3}a^{2 \cdot 3}b^{1 \cdot 3}}$ Do this step mentally.

$$= 3^3a^6b^3$$
$$= 27a^6b^3$$

Example 5

Simplify $(2xy^3)^4$.

Solution

$(2xy^3)^4 = 2^4x^4y^{12} = 16x^4y^{12}$

Example 6

Simplify $(3x)(2x^2y)^3$.

Your solution

$24x^7y^3$

Example 7

Simplify $(-2x)(-3xy^2)^3$.

Solution

$$(-2x)(-3xy^2)^3 = (-2x)(-3)^3x^3y^6$$
$$= (-2x)(-27)x^3y^6$$
$$= [-2(-27)](x \cdot x^3)y^6 = 54x^4y^6$$

Example 8

Simplify $(3x^2)^2(-2xy^2)^3$.

Your solution

$-72x^7y^6$

Solutions on p. A17

5.2   **EXERCISES**

▶ **Objective A**

Simplify.

1. $(x)(2x)$
$2x^2$

2. $(-3y)(y)$
$-3y^2$

3. $(3x)(4x)$
$12x^2$

4. $(7y^3)(7y^2)$
$49y^5$

5. $(-2a^3)(-3a^4)$
$6a^7$

6. $(5a^6)(-2a^5)$
$-10a^{11}$

7. $(x^2y)(xy^4)$
x^3y^5

8. $(x^2y^4)(xy^7)$
x^3y^{11}

9. $(-2x^4)(5x^5y)$
$-10x^9y$

10. $(-3a^3)(2a^2b^4)$
$-6a^5b^4$

11. $(x^2y^4)(x^5y^4)$
x^7y^8

12. $(a^2b^4)(ab^3)$
a^3b^7

13. $(2xy)(-3x^2y^4)$
$-6x^3y^5$

14. $(-3a^2b)(-2ab^3)$
$6a^3b^4$

15. $(x^2yz)(x^2y^4)$
x^4y^5z

16. $(-ab^2c)(a^2b^5)$
$-a^3b^7c$

17. $(a^2b^3)(ab^2c^4)$
$a^3b^5c^4$

18. $(x^2y^3z)(x^3y^4)$
x^5y^7z

19. $(-a^2b^2)(a^3b^6)$
$-a^5b^8$

20. $(xy^4)(-xy^3)$
$-x^2y^7$

21. $(-6a^3)(a^2b)$
$-6a^5b$

22. $(2a^2b^3)(-4ab^2)$
$-8a^3b^5$

23. $(-5y^4z)(-8y^6z^5)$
$40y^{10}z^6$

24. $(3x^2y)(-4xy^2)$
$-12x^3y^3$

25. $(10ab^2)(-2ab)$
$-20a^2b^3$

26. $(x^2y)(yz)(xyz)$
$x^3y^3z^2$

27. $(xy^2z)(x^2y)(z^2y^2)$
$x^3y^5z^3$

28. $(-2x^2y^3)(3xy)(-5x^3y^4)$
$30x^6y^8$

29. $(4a^2b)(-3a^3b^4)(a^5b^2)$
$-12a^{10}b^7$

30. $(3ab^2)(-2abc)(4ac^2)$
$-24a^3b^3c^3$

31. $(3a^2b)(-6bc)(2ac^2)$
$-36a^3b^2c^3$

▶ **Objective B**

Simplify.

32. $(2^2)^3$
    64

33. $(3^2)^2$
    81

34. $(-2)^2$
    4

35. $(-3)^3$
    -27

36. $(-2^2)^3$
    -64

37. $(-2^3)^3$
    -512

38. $(x^3)^3$
    x^9

39. $(y^4)^2$
    y^8

40. $(x^7)^2$
    x^{14}

41. $(y^5)^3$
    y^{15}

42. $(-x^2)^2$
    x^4

43. $(-x^2)^3$
    $-x^6$

44. $(2x)^2$
    $4x^2$

45. $(3y)^3$
    $27y^3$

46. $(-2x^2)^3$
    $-8x^6$

47. $(-3y^3)^2$
    $9y^6$

48. $(x^2y^3)^2$
    x^4y^6

49. $(x^3y^4)^5$
    $x^{15}y^{20}$

50. $(3x^2y)^2$
    $9x^4y^2$

51. $(-2ab^3)^4$
    $16a^4b^{12}$

52. $(a^2)(3a^2)^3$
    $27a^8$

53. $(b^2)(2a^3)^4$
    $16a^{12}b^2$

54. $(-2x)(2x^3)^2$
    $-8x^7$

55. $(2y)(-3y^4)^3$
    $-54y^{13}$

56. $(x^2y)(x^2y)^3$
    x^8y^4

57. $(a^3b)(ab)^3$
    a^6b^4

58. $(ab^2)^2(ab)^2$
    a^4b^6

59. $(x^2y)^2(x^3y)^3$
    $x^{13}y^5$

60. $(-2x)(-2x^3y)^3$
    $16x^{10}y^3$

61. $(-3y)(-4x^2y^3)^3$
    $192x^6y^{10}$

62. $(-2x)(-3xy^2)^2$
    $-18x^3y^4$

63. $(-3y)(-2x^2y)^3$
    $24x^6y^4$

64. $(ab^2)(-2a^2b)^3$
    $-8a^7b^5$

65. $(a^2b^2)(-3ab^4)^2$
    $9a^4b^{10}$

66. $(-2a^3)(3a^2b)^3$
    $-54a^9b^3$

67. $(-3b^2)(2ab^2)^3$
    $-24a^3b^8$

68. $(-3ab)^2(-2ab)^3$
    $-72a^5b^5$

69. $(-3a^2b)^3(-3ab)^3$
    $729a^9b^6$

SECTION 5.3   Multiplication of Polynomials

Objective A   To multiply a polynomial by a monomial

To multiply a polynomial by a monomial, use the Distributive Property and the Rule for Multiplying Exponential Expressions.

Simplify $-2x(x^2 - 4x - 3)$.

$$-2x(x^2 - 4x - 3)$$

Use the Distributive Property. $\boxed{-2x(x^2) - (-2x)(4x) - (-2x)(3)}$ Do this step mentally.

Use the Rule for Multiplying Exponential Expressions. $-2x^3 + 8x^2 + 6x$

Example 1

Simplify $(5x + 4)(-2x)$.

Solution

$(5x + 4)(-2x) = -10x^2 - 8x$

Example 3

Simplify $x^3(2x^2 - 3x + 2)$.

Solution

$x^3(2x^2 - 3x + 2) = 2x^5 - 3x^4 + 2x^3$

Example 2

Simplify $(-2y + 3)(-4y)$.

Your solution

$8y^2 - 12y$

Example 4

Simplify $-a^2(3a^2 + 2a - 7)$.

Your solution

$-3a^4 - 2a^3 + 7a^2$

Solutions on p. A17

Objective B   To multiply two polynomials

Multiplication of two polynomials requires the repeated application of the Distributive Property.

$$(y - 2)(y^2 + 3y + 1) = (y - 2)(y^2) + (y - 2)(3y) + (y - 2)(1)$$
$$= y^3 - 2y^2 + 3y^2 - 6y + y - 2$$
$$= y^3 + y^2 - 5y - 2$$

A more convenient method of multiplying two polynomials is to use a vertical format similar to that used for multiplication of whole numbers.

Multiply each term in the trinomial by -2.
Multiply each term in the trinomial by y.
Like terms must be in the same column.
Add the terms in each column.

$$
\begin{array}{r}
y^2 + 3y + 1 \\
y - 2 \\
\hline
-2y^2 - 6y - 2 \\
y^3 + 3y^2 + y \\
\hline
y^3 + y^2 - 5y - 2
\end{array}
$$

Simplify $(a^2 - 3)(a + 5)$.

Multiply each term of $a^2 - 3$ by 5.
Multiply each term of $a^2 - 3$ by a.
Arrange the terms in descending order.
Add the terms in each column.

$$
\begin{array}{r}
a^2 - 3 \\
a + 5 \\
\hline
5a^2 \qquad - 15 \\
a^3 \qquad - 3a \qquad \\
\hline
a^3 + 5a^2 - 3a - 15
\end{array}
$$

Example 5

Simplify $(2b^3 - b + 1)(2b + 3)$.

Solution

$$
\begin{array}{r}
2b^3 - b + 1 \\
2b + 3 \\
\hline
6b^3 \qquad - 3b + 3 \\
4b^4 + \qquad - 2b^2 + 2b \qquad \\
\hline
4b^4 + 6b^3 - 2b^2 - b + 3
\end{array}
$$

Example 6

Simplify $(2y^3 + 2y^2 - 3)(3y - 1)$.

Your solution

$6y^4 + 4y^3 - 2y^2 - 9y + 3$

Solution on p. A18

Objective C

To multiply two binomials

It is frequently necessary to find the product of two binomials. The product can be found using a method called **FOIL,** which is based on the Distributive Property. The letters of FOIL stand for **F**irst, **O**uter, **I**nner, and **L**ast.

Simplify $(2x + 3)(x + 5)$.

Multiply the **F**irst terms.	$(2x + 3)(x + 5)$	$2x \cdot x = 2x^2$
Multiply the **O**uter terms.	$(2x + 3)(x + 5)$	$2x \cdot 5 = 10x$
Multiply the **I**nner terms.	$(2x + 3)(x + 5)$	$3 \cdot x = 3x$
Multiply the **L**ast terms.	$(2x + 3)(x + 5)$	$3 \cdot 5 = 15$

$$\qquad\qquad\qquad\qquad\qquad\quad \text{F}\qquad \text{O}\qquad \text{I}\quad \text{L}$$

Add the products. $\qquad (2x + 3)(x + 5)\qquad = 2x^2 + 10x + 3x + 15$

Combine like terms. $\qquad\qquad\qquad\qquad\qquad\; = 2x^2 + 13x + 15$

Simplify $(4x - 3)(3x - 2)$.

$(4x - 3)(3x - 2)\quad \boxed{= 4x(3x) + 4x(-2) + (-3)(3x) + (-3)(-2)}$ \quad Do this step mentally.

$$= 12x^2 - 8x - 9x + 6$$
$$= 12x^2 - 17x + 6$$

Simplify $(3x - 2y)(x + 4y)$.

$(3x - 2y)(x + 4y)\quad \boxed{= 3x(x) + 3x(4y) + (-2y)(x) + (-2y)(4y)}$ \quad Do this step mentally.

$$= 3x^2 + 12xy - 2xy - 8y^2$$
$$= 3x^2 + 10xy - 8y^2$$

Example 7

Simplify $(2a - 1)(3a - 2)$.

Solution

$(2a - 1)(3a - 2) = 6a^2 - 4a - 3a + 2$
$= 6a^2 - 7a + 2$

Example 8

Simplify $(4y - 5)(2y - 3)$.

Your solution

$8y^2 - 22y + 15$

Example 9

Simplify $(3x - 2)(4x + 3)$.

Solution

$(3x - 2)(4x + 3) = 12x^2 + 9x - 8x - 6$
$= 12x^2 + x - 6$

Example 10

Simplify $(3b + 2)(3b - 5)$.

Your solution

$9b^2 - 9b - 10$

Solutions on p. A18

Objective D   **To multiply binomials that have special products**

Using FOIL, a pattern for the product of the sum and difference of two terms and for the square of a binomial can be found.

The Sum and Difference of Two Terms

$$(a + b)(a - b) = a^2 - ab + ab - b^2$$
$$= a^2 - b^2$$

Square of first term
Square of second term

The Square of a Binomial

$$(a + b)^2 = (a + b)(a + b) = a^2 + ab + ab + b^2$$
$$= a^2 + 2ab + b^2$$

Square of first term
Twice the product of the two terms
Square of the last term

Simplify $(2x + 3)(2x - 3)$.

$(2x + 3)(2x - 3)$ is the sum and difference of two terms.

$(2x + 3)(2x - 3)$ $\boxed{= (2x)^2 - 3^2}$

$= 4x^2 - 9$

Do this step mentally.

Simplify $(3x - 2)^2$.

$(3x - 2)^2$ is the square of a binomial.

$(3x - 2)^2$ $\boxed{= (3x)^2 + 2(3x)(-2) + (-2)^2}$

$= 9x^2 - 12x + 4$

Do this step mentally.

Example 11

Simplify $(4z - 2w)(4z + 2w)$.

Solution

$(4z - 2w)(4z + 2w) = 16z^2 - 4w^2$

Example 12

Simplify $(2a + 5c)(2a - 5c)$.

Your solution

$4a^2 - 25c^2$

Example 13

Simplify $(2r - 3s)^2$.

Solution

$(2r - 3s)^2 = 4r^2 - 12rs + 9s^2$

Example 14

Simplify $(3x + 2y)^2$.

Your solution

$9x^2 + 12xy + 4y^2$

Solutions on p. A18

Objective E   To solve application problems

Example 15

The length of a rectangle is $x + 7$. The width is $x - 4$. Find the area of the rectangle in terms of the variable x.

$x - 4$ [ ]
$x + 7$

Strategy

To find the area, replace the variables l and w in the equation $A = l \cdot w$ by the given values and solve for A.

Solution

$A = l \cdot w$
$A = (x + 7)(x - 4)$
$A = x^2 - 4x + 7x - 28$
$A = x^2 + 3x - 28$

The area is $x^2 + 3x - 28$.

Example 16

The radius of a circle is $x - 4$. Use the equation $A = \pi r^2$, where r is the radius, to find the area of the circle in terms of x. Use $\pi \approx 3.14$.

Your strategy

Your solution

$3.14x^2 - 25.12x + 50.24$

Solution on p. A18

5.3 EXERCISES

▶ Objective A

Simplify.

1. $x(x - 2)$
$x^2 - 2x$

2. $y(3 - y)$
$-y^2 + 3y$

3. $-x(x + 7)$
$-x^2 - 7x$

4. $-y(7 - y)$
$y^2 - 7y$

5. $3a^2(a - 2)$
$3a^3 - 6a^2$

6. $4b^2(b + 8)$
$4b^3 + 32b^2$

7. $-5x^2(x^2 - x)$
$-5x^4 + 5x^3$

8. $-6y^2(y + 2y^2)$
$-12y^4 - 6y^3$

9. $-x^3(3x^2 - 7)$
$-3x^5 + 7x^3$

10. $-y^4(2y^2 - y^6)$
$y^{10} - 2y^6$

11. $2x(6x^2 - 3x)$
$12x^3 - 6x^2$

12. $3y(4y - y^2)$
$-3y^3 + 12y^2$

13. $(2x - 4)3x$
$6x^2 - 12x$

14. $(3y - 2)y$
$3y^2 - 2y$

15. $(3x + 4)x$
$3x^2 + 4x$

16. $(2x + 1)2x$
$4x^2 + 2x$

17. $-xy(x^2 - y^2)$
$-x^3y + xy^3$

18. $-x^2y(2xy - y^2)$
$-2x^3y^2 + x^2y^3$

19. $x(2x^3 - 3x + 2)$
$2x^4 - 3x^2 + 2x$

20. $y(-3y^2 - 2y + 6)$
$-3y^3 - 2y^2 + 6y$

21. $-a(-2a^2 - 3a - 2)$
$2a^3 + 3a^2 + 2a$

22. $-b(5b^2 + 7b - 35)$
$-5b^3 - 7b^2 + 35b$

23. $x^2(3x^4 - 3x^2 - 2)$
$3x^6 - 3x^4 - 2x^2$

24. $y^3(-4y^3 - 6y + 7)$
$-4y^6 - 6y^4 + 7y^3$

25. $2y^2(-3y^2 - 6y + 7)$
$-6y^4 - 12y^3 + 14y^2$

26. $4x^2(3x^2 - 2x + 6)$
$12x^4 - 8x^3 + 24x^2$

27. $(a^2 + 3a - 4)(-2a)$
$-2a^3 - 6a^2 + 8a$

28. $(b^3 - 2b + 2)(-5b)$
$-5b^4 + 10b^2 - 10b$

29. $-3y^2(-2y^2 + y - 2)$
$6y^4 - 3y^3 + 6y^2$

30. $-5x^2(3x^2 - 3x - 7)$
$-15x^4 + 15x^3 + 35x^2$

31. $xy(x^2 - 3xy + y^2)$
$x^3y - 3x^2y^2 + xy^3$

32. $ab(2a^2 - 4ab - 6b^2)$
$2a^3b - 4a^2b^2 - 6ab^3$

▶ Objective B

Simplify.

33. $(x^2 + 3x + 2)(x + 1)$
$x^3 + 4x^2 + 5x + 2$

34. $(x^2 - 2x + 7)(x - 2)$
$x^3 - 4x^2 + 11x - 14$

35. $(a^2 - 3a + 4)(a - 3)$
$a^3 - 6a^2 + 13a - 12$

Simplify.

36. $(x^2 - 3x + 5)(2x - 3)$
$2x^3 - 9x^2 + 19x - 15$

37. $(-2b^2 - 3b + 4)(b - 5)$
$-2b^3 + 7b^2 + 19b - 20$

38. $(-a^2 + 3a - 2)(2a - 1)$
$-2a^3 + 7a^2 - 7a + 2$

39. $(-2x^2 + 7x - 2)(3x - 5)$
$-6x^3 + 31x^2 - 41x + 10$

40. $(-a^2 - 2a + 3)(2a - 1)$
$-2a^3 - 3a^2 + 8a - 3$

41. $(x^2 + 5)(x - 3)$
$x^3 - 3x^2 + 5x - 15$

42. $(y^2 - 2y)(2y + 5)$
$2y^3 + y^2 - 10y$

43. $(x^3 - 3x + 2)(x - 4)$
$x^4 - 4x^3 - 3x^2 + 14x - 8$

44. $(y^3 + 4y^2 - 8)(2y - 1)$
$2y^4 + 7y^3 - 4y^2 - 16y + 8$

45. $(5y^2 + 8y - 2)(3y - 8)$
$15y^3 - 16y^2 - 70y + 16$

46. $(3y^2 + 3y - 5)(4y - 3)$
$12y^3 + 3y^2 - 29y + 15$

47. $(5a^3 - 5a + 2)(a - 4)$
$5a^4 - 20a^3 - 5a^2 + 22a - 8$

48. $(3b^3 - 5b^2 + 7)(6b - 1)$
$18b^4 - 33b^3 + 5b^2 + 42b - 7$

49. $(y^3 + 2y^2 - 3y + 1)(y + 2)$
$y^4 + 4y^3 + y^2 - 5y + 2$

50. $(2a^3 - 3a^2 + 2a - 1)(2a - 3)$
$4a^4 - 12a^3 + 13a^2 - 8a + 3$

▶ **Objective C**

Simplify.

51. $(x + 1)(x + 3)$
$x^2 + 4x + 3$

52. $(y + 2)(y + 5)$
$y^2 + 7y + 10$

53. $(a - 3)(a + 4)$
$a^2 + a - 12$

54. $(b - 6)(b + 3)$
$b^2 - 3b - 18$

55. $(y + 3)(y - 8)$
$y^2 - 5y - 24$

56. $(x + 10)(x - 5)$
$x^2 + 5x - 50$

57. $(y - 7)(y - 3)$
$y^2 - 10y + 21$

58. $(a - 8)(a - 9)$
$a^2 - 17a + 72$

59. $(2x + 1)(x + 7)$
$2x^2 + 15x + 7$

60. $(y + 2)(5y + 1)$
$5y^2 + 11y + 2$

61. $(3x - 1)(x + 4)$
$3x^2 + 11x - 4$

62. $(7x - 2)(x + 4)$
$7x^2 + 26x - 8$

63. $(4x - 3)(x - 7)$
$4x^2 - 31x + 21$

64. $(2x - 3)(4x - 7)$
$8x^2 - 26x + 21$

65. $(3y - 8)(y + 2)$
$3y^2 - 2y - 16$

66. $(5y - 9)(y + 5)$
$5y^2 + 16y - 45$

67. $(3x + 7)(3x + 11)$
$9x^2 + 54x + 77$

68. $(5a + 6)(6a + 5)$
$30a^2 + 61a + 30$

69. $(7a - 16)(3a - 5)$
$21a^2 - 83a + 80$

70. $(5a - 12)(3a - 7)$
$15a^2 - 71a + 84$

Simplify.

71. $(3b + 13)(5b - 6)$
$15b^2 + 47b - 78$

72. $(x + y)(2x + y)$
$2x^2 + 3xy + y^2$

73. $(2a + b)(a + 3b)$
$2a^2 + 7ab + 3b^2$

74. $(3x - 4y)(x - 2y)$
$3x^2 - 10xy + 8y^2$

75. $(2a - b)(3a + 2b)$
$6a^2 + ab - 2b^2$

76. $(5a - 3b)(2a + 4b)$
$10a^2 + 14ab - 12b^2$

77. $(2x + y)(x - 2y)$
$2x^2 - 3xy - 2y^2$

78. $(3x - 7y)(3x + 5y)$
$9x^2 - 6xy - 35y^2$

79. $(2x + 3y)(5x + 7y)$
$10x^2 + 29xy + 21y^2$

80. $(5x + 3y)(7x + 2y)$
$35x^2 + 31xy + 6y^2$

81. $(3a - 2b)(2a - 7b)$
$6a^2 - 25ab + 14b^2$

82. $(5a - b)(7a - b)$
$35a^2 - 12ab + b^2$

83. $(a - 9b)(2a + 7b)$
$2a^2 - 11ab - 63b^2$

84. $(2a + 5b)(7a - 2b)$
$14a^2 + 31ab - 10b^2$

85. $(10a - 3b)(10a - 7b)$
$100a^2 - 100ab + 21b^2$

86. $(12a - 5b)(3a - 4b)$
$36a^2 - 63ab + 20b^2$

87. $(5x + 12y)(3x + 4y)$
$15x^2 + 56xy + 48y^2$

88. $(11x + 2y)(3x + 7y)$
$33x^2 + 83xy + 14y^2$

89. $(2x - 15y)(7x + 4y)$
$14x^2 - 97xy - 60y^2$

90. $(5x + 2y)(2x - 5y)$
$10x^2 - 21xy - 10y^2$

91. $(8x - 3y)(7x - 5y)$
$56x^2 - 61xy + 15y^2$

92. $(2x - 9y)(8x - 3y)$
$16x^2 - 78xy + 27y^2$

▶ **Objective D**

Simplify.

93. $(y - 5)(y + 5)$
$y^2 - 25$

94. $(y + 6)(y - 6)$
$y^2 - 36$

95. $(2x + 3)(2x - 3)$
$4x^2 - 9$

96. $(4x - 7)(4x + 7)$
$16x^2 - 49$

97. $(x + 1)^2$
$x^2 + 2x + 1$

98. $(y - 3)^2$
$y^2 - 6y + 9$

99. $(3a - 5)^2$
$9a^2 - 30a + 25$

100. $(6x - 5)^2$
$36x^2 - 60x + 25$

101. $(3x - 7)(3x + 7)$
$9x^2 - 49$

102. $(9x - 2)(9x + 2)$
$81x^2 - 4$

103. $(2a + b)^2$
$4a^2 + 4ab + b^2$

104. $(x + 3y)^2$
$x^2 + 6xy + 9y^2$

Simplify.

105. $(x - 2y)^2$
$x^2 - 4xy + 4y^2$

106. $(2x - 3y)^2$
$4x^2 - 12xy + 9y^2$

107. $(4 - 3y)(4 + 3y)$
$16 - 9y^2$

108. $(4x - 9y)(4x + 9y)$
$16x^2 - 81y^2$

109. $(5x + 2y)^2$
$25x^2 + 20xy + 4y^2$

110. $(2a - 9b)^2$
$4a^2 - 36ab + 81b^2$

▶ **Objective E** *Application Problems*

Solve.

111. Let L represent the length of a rectangle. The width of the rectangle is 2 ft less than twice the length. Express the area of the rectangle in terms of the variable L.
$(2L^2 - 2L)$ ft^2

112. The side of a square is $(2x + 1)$ ft. Express the area of the square in terms of the variable x.
$(4x^2 + 4x + 1)$ ft^2

113. A softball diamond has dimensions 45 ft by 45 ft. A base path border x ft wide lies on both the first-base side and the third-base side of the diamond. Express the total area of the softball diamond and the base path in terms of the variable x.
$(90x + 2025)$ ft^2

114. A field has dimensions 30 yd by 100 yd. An endzone that is w yd wide borders each end of the field. Express the total area of the field and the endzone in terms of the variable w.
$(60w + 3000)$ yd^2

115. The radius of a circle is $x + 3$. Use the equation $A = \pi r^2$, where r is the radius, to find the area of the circle in terms of the variable x. Use $\pi \approx 3.14$.
$(3.14x^2 + 18.84x + 28.26)$

116. The radius of a circle is $x - 2$. Use the equation $A = \pi r^2$, where r is the radius, to find the area of the circle in terms of the variable x. Use $\pi \approx 3.14$.
$(3.14x^2 - 12.56x + 12.56)$

| SECTION 5.4 | **Integer Exponents** |

Objective A

To divide monomials and simplify expressions containing negative exponents

The quotient of two exponential expressions with the same base can be simplified by writing each expression in factored form, dividing by the common factors, and then writing the result with an exponent.

$$\frac{x^5}{x^2} = \frac{\overset{1}{\cancel{x}} \cdot \overset{1}{\cancel{x}} \cdot x \cdot x \cdot x}{\underset{1}{\cancel{x}} \cdot \underset{1}{\cancel{x}}} = x^3$$

Note that subtracting the exponents gives the same result.

$$\frac{x^5}{x^2} = x^{5-2} = x^3$$

To divide two monomials with the same base, subtract the exponents of the like bases.

Simplify $\frac{a^7}{a^3}$.

The bases are the same. Subtract the exponents.

$$\frac{a^7}{a^3} \boxed{= a^{7-3}}$$ Do this step mentally.

$$= a^4$$

Simplify $\frac{r^8 s^6}{r^7 s}$.

Subtract the exponents of the like bases.

$$\frac{r^8 s^6}{r^7 s} \boxed{= r^{8-7} s^{6-1}}$$ Do this step mentally.

$$= r s^5$$

Recall that for any number $a \neq 0$, $\frac{a}{a} = 1$. This property is true for exponential expressions as well.

For example, for $x \neq 0$, $\frac{x^4}{x^4} = 1$. This expression also can be simplified using the rules of dividing exponential expressions with the same base.

$$\frac{x^4}{x^4} = x^{4-4} = x^0$$

Because $\frac{x^4}{x^4} = 1$ and $\frac{x^4}{x^4} = x^0$, the following definition of zero as an exponent is used.

Zero as an Exponent

If $x \neq 0$, then $x^0 = 1$.

Note in this definition that $x \neq 0$. The expression 0^0 is not defined.

Simplify $(12a^3)^0$, $a \neq 0$.

Any nonzero expression to the zero power is 1.

$$(12a^3)^0 = 1$$

Simplify $-(xy^4)^0$, $x \neq 0$, $y \neq 0$

Any nonzero expression to the zero power is 1. Because the negative sign is outside the parenthesis, the answer is -1.

$$-(xy^4)^0 = -(1) = -1$$

Now consider the meaning of a negative exponent. Examine the quotient $\frac{x^4}{x^6}$.

Write the numerator and denominator in factored form. Divide by the common factors.

$$\frac{x^4}{x^6} = \frac{\overset{1}{\cancel{x}} \cdot \overset{1}{\cancel{x}} \cdot \overset{1}{\cancel{x}} \cdot \overset{1}{\cancel{x}}}{\underset{1}{\cancel{x}} \cdot \underset{1}{\cancel{x}} \cdot \underset{1}{\cancel{x}} \cdot \underset{1}{\cancel{x}} \cdot x \cdot x} = \frac{1}{x^2}$$

Now simplify the same expression by subtracting the exponents on the like bases.

$$\frac{x^4}{x^6} = x^{4-6} = x^{-2}$$

Because $\frac{x^4}{x^6} = \frac{1}{x^2}$ and $\frac{x^4}{x^6} = x^{-2}$, the following definition of a negative exponent is used.

Rule for Negative Exponents

If n is a positive integer and $x \neq 0$, then $x^{-n} = \frac{1}{x^n}$ and $x^n = \frac{1}{x^{-n}}$.

Write 2^{-4} with a positive exponent and then evaluate.

Write the expression with a positive exponent.

$$2^{-4} = \frac{1}{2^4}$$

Evaluate.

$$= \frac{1}{16}$$

Now that negative exponents have been defined, the rule for dividing exponential expressions can be stated.

Rule for Dividing Exponential Expressions

If m and n are integers and $x \neq 0$, then $\frac{x^m}{x^n} = x^{m-n}$.

Write $\frac{4^{-2}}{4}$ with a positive exponent and then evaluate.

Write the expression with a positive exponent.

$$\frac{4^{-2}}{4} \quad \boxed{= 4^{-2-1}} \qquad \text{Do this step mentally.}$$

$$= 4^{-3} = \frac{1}{4^3}$$

Evaluate.

$$= \frac{1}{64}$$

An exponential expression is in simplest form when it is written with only positive exponents.

Simplify $\frac{x^{-4}y^6}{xy^2}$.

Divide variables with like bases by subtracting the exponents.

$$\frac{x^{-4}y^6}{xy^2} \quad \boxed{= x^{-4-1}y^{6-2}} \qquad \text{Do this step mentally.}$$

$$= x^{-5}y^4 = \frac{y^4}{x^5}$$

Simplify $\frac{b^8}{a^{-5}b^6}$.

Divide variables with like bases by subtracting the exponents.

$$\frac{b^8}{a^{-5}b^6} \quad \boxed{= a^5 b^{8-6}} \quad \text{Do this step mentally.}$$

$$= a^5 b^2$$

The rules for multiplying exponential expressions and powers of exponential expressions are true for all integers. These rules are restated here.

Rules of Exponents

If m, n, and p are integers, then

$$x^m \cdot x^n = x^{m+n} \qquad (x^m)^n = x^{mn} \qquad (x^m y^n)^p = x^{mp} y^{np}$$

$$\frac{x^m}{x^n} = x^{m-n},\ x \neq 0 \qquad x^0 = 1,\ x \neq 0 \qquad \left(\frac{x^m}{y^n}\right)^p = \frac{x^{mp}}{y^{np}},\ y \neq 0$$

Simplify $(-3ab)(2a^3 b^{-2})^{-3}$.

Use the Rules of Exponents.

$$(-3ab)(2a^3 b^{-2})^{-3} = (-3ab)(2^{-3} a^{-9} b^6)$$

$$\boxed{= -3 \cdot 2^{-3} a^{1-9} b^{1+6}} \quad \text{Do this step mentally.}$$

$$= -3 \cdot 2^{-3} a^{-8} b^7$$

$$= -\frac{3b^7}{2^3 a^8} = -\frac{3b^7}{8a^8}$$

Simplify $\frac{(2a^{-2}b^3)^{-3}}{(a^2 b^{-2})^{-4}}$.

Use the Rules of Exponents.

$$\frac{(2a^{-2}b^3)^{-3}}{(a^2 b^{-2})^{-4}} = \frac{2^{-3} a^6 b^{-9}}{a^{-8} b^8}$$

$$= 2^{-3} a^{14} b^{-17}$$

$$= \frac{a^{14}}{2^3 b^{17}} = \frac{a^{14}}{8 b^{17}}$$

Example 1

Write $\frac{3^{-3}}{3^2}$ with a positive exponent. Then evaluate.

Solution

$$\frac{3^{-3}}{3^2} = 3^{-5} = \frac{1}{3^5} = \frac{1}{243}$$

Example 2

Write $\frac{2^2}{2^{-3}}$ with a positive exponent. Then evaluate.

Your solution

32

Solution on p. A18

Example 3

Simplify $(-2x)(3x^{-2})^{-3}$.

Solution

$$(-2x)(3x^{-2})^{-3} = (-2x)(3^{-3}x^6)$$
$$= \frac{-2x \cdot x^6}{3^3} = -\frac{2x^7}{27}$$

Example 4

Simplify $(-2x^2)(x^{-3}y^{-4})^{-2}$.

Your solution

$-2x^8y^8$

Example 5

Simplify $\frac{-35y^5}{25y^9}$.

Solution

$$\frac{-35y^5}{25y^9} = \frac{-5 \cdot 7y^{5-9}}{5 \cdot 5} = -\frac{7}{5y^4}$$

Example 6

Simplify $\frac{18y^3}{-27y^7}$.

Your solution

$-\dfrac{2}{3y^4}$

Example 7

Simplify $\frac{-42a^6b^{-2}}{28a^{-2}b^5}$.

Solution

$$\frac{-42a^6b^{-2}}{28a^{-2}b^5} = -\frac{14 \cdot 3a^{6-(-2)}b^{-2-5}}{14 \cdot 2}$$
$$= -\frac{3a^8b^{-7}}{2} = -\frac{3a^8}{2b^7}$$

Example 8

Simplify $\frac{12x^{-8}y^4}{-16xz^{-2}}$.

Your solution

$-\dfrac{3y^4z^2}{4x^9}$

Example 9

Simplify $\frac{(2r^2t^{-1})^{-3}}{(r^{-3}t^4)^2}$.

Solution

$$\frac{(2r^2t^{-1})^{-3}}{(r^{-3}t^4)^2} = \frac{2^{-3}r^{-6}t^3}{r^{-6}t^8} = 2^{-3}r^0t^{-5}$$
$$= \frac{1}{8t^5}$$

Example 10

Simplify $\frac{(6a^{-2}b^3)^{-1}}{(4a^3b^{-2})^{-2}}$.

Your solution

$\dfrac{8a^8}{3b^7}$

Example 11

Simplify $\left[\frac{4a^{-2}b^3}{6a^4b^{-2}}\right]^{-3}$.

Solution

$$\left[\frac{4a^{-2}b^3}{6a^4b^{-2}}\right]^{-3} = \left[\frac{2a^{-6}b^5}{3}\right]^{-3} = \frac{2^{-3}a^{18}b^{-15}}{3^{-3}}$$
$$= \frac{27a^{18}}{8b^{15}}$$

Example 12

Simplify $\left[\frac{6r^3s^{-3}}{9r^3s^{-1}}\right]^{-2}$.

Your solution

$\dfrac{9s^4}{4}$

Solutions on p. A18

Content and Format © 1991 HMCo.

| Objective B | **To divide a polynomial by a monomial** |

Note that $\frac{8+4}{2}$ can be simplified by first adding the terms in the numerator and then dividing the result. It can also be simplified by first dividing each term in the numerator by the denominator and then adding the result.

$$\frac{8+4}{2} = \frac{12}{2} = 6$$

$$\frac{8+4}{2} = \frac{8}{2} + \frac{4}{2} = 4 + 2 = 6$$

To divide a polynomial by a monomial, divide each term in the numerator by the denominator and write the sum of the quotients.

Simplify $\frac{6x^2 + 4x}{2x}$.

Divide each term of the polynomial by the monomial.
Simplify each expression.

$$\frac{6x^2 + 4x}{2x} = \frac{6x^2}{2x} + \frac{4x}{2x}$$
$$= 3x + 2$$

Check: $2x(3x + 2) = 6x^2 + 4x$

Example 13

Simplify $\frac{6x^3 - 3x^2 + 9x}{3x}$.

Solution

$$\frac{6x^3 - 3x^2 + 9x}{3x} = \frac{6x^3}{3x} - \frac{3x^2}{3x} + \frac{9x}{3x}$$
$$= 2x^2 - x + 3$$

Example 14

Simplify $\frac{4x^3y + 8x^2y^2 - 4xy^3}{2xy}$.

Your solution

$2x^2 + 4xy - 2y^2$

Example 15

Simplify $\frac{12x^2y - 6xy + 4x^2}{2xy}$.

Solution

$$\frac{12x^2y - 6xy + 4x^2}{2xy} = \frac{12x^2y}{2xy} - \frac{6xy}{2xy} + \frac{4x^2}{2xy}$$
$$= 6x - 3 + \frac{2x}{y}$$

Example 16

Simplify $\frac{24x^2y^2 - 18xy + 6y}{6xy}$.

Your solution

$4xy - 3 + \frac{1}{x}$

Solutions on p. A18

| Objective C | **To divide polynomials** |

To divide polynomials, use a method similar to that used for division of whole numbers. The same equation used to check division of whole numbers is used to check division of polynomials.

Dividend = (quotient × divisor) + remainder

Simplify $(x^2 - 5x + 8) \div (x - 3)$.

Step 1

$$\begin{array}{r} x \\ x - 3 \overline{\smash{)} x^2 - 5x + 8} \\ \underline{x^2 - 3x} \quad \downarrow \\ -2x + 8 \end{array}$$

Think: $x\overline{\smash{)}x^2} = \dfrac{x^2}{x} = x$

Multiply: $x(x - 3) = x^2 - 3x$

Subtract: $(x^2 - 5x) - (x^2 - 3x) = -2x$

Step 2

$$\begin{array}{r} x \;\; - 2 \\ x - 3 \overline{\smash{)} x^2 - 5x + 8} \\ \underline{x^2 - 3x} \\ -2x + 8 \\ \underline{-2x + 6} \\ 2 \end{array}$$

Think: $x\overline{\smash{)}-2x} = \dfrac{-2x}{x} = -2$

Multiply: $-2(x - 3) = -2x + 6$

Subtract: $(-2x + 8) - (-2x + 6) = 2$

The remainder is 2.

Check: $(x - 2)(x - 3) + 2 = x^2 - 3x - 2x + 6 + 2 = x^2 - 5x + 8$

$(x^2 - 5x + 8) \div (x - 3) = x - 2 + \dfrac{2}{x - 3}$

Simplify $(6x + 2x^3 + 26) \div (x + 2)$.

Arrange the terms in descending order.
There is no term of x^2 in $2x^3 + 6x + 26$.
Insert a zero for the missing term so
that like terms will be in columns.

$$\begin{array}{r} 2x^2 - 4x + 14 \\ x + 2 \overline{\smash{)} 2x^3 + 0 \quad + \;\; 6x + 26} \\ \underline{2x^3 + 4x^2} \\ -4x^2 + \;\; 6x \\ \underline{-4x^2 - \;\; 8x} \\ 14x + 26 \\ \underline{14x + 28} \\ -2 \end{array}$$

$(2x^3 + 6x + 26) \div (x + 2) = 2x^2 - 4x + 14 - \dfrac{2}{x + 2}$

Example 17

Simplify $(x^2 - 1) \div (x + 1)$.

Solution

Insert a zero for
the missing term.

$$\begin{array}{r} x \;\; - 1 \\ x + 1 \overline{\smash{)} x^2 + 0 - 1} \\ \underline{x^2 + x} \\ -x - 1 \\ \underline{-x - 1} \\ 0 \end{array}$$

$(x^2 - 1) \div (x + 1) = x - 1$

Example 18

Simplify $(2x^3 + x^2 - 8x - 3) \div (2x - 3)$.

Your solution

$x^2 + 2x - 1 - \dfrac{6}{2x - 3}$

Solution on p. A18

5.4 EXERCISES

▶ **Objective A**

Write with a positive or zero exponent. Then evaluate.

1. 5^{-2}

$\dfrac{1}{5^2} = \dfrac{1}{25}$

2. 3^{-3}

$\dfrac{1}{3^3} = \dfrac{1}{27}$

3. $\dfrac{1}{8^{-2}}$

$8^2 = 64$

4. $\dfrac{1}{12^{-1}}$

$12^1 = 12$

5. $\dfrac{3^{-2}}{3}$

$\dfrac{1}{3^3} = \dfrac{1}{27}$

6. $\dfrac{5^{-3}}{5}$

$\dfrac{1}{5^4} = \dfrac{1}{625}$

7. $\dfrac{2^{-3}}{2^{-3}}$

$2^0 = 1$

8. $\dfrac{3^2}{3^2}$

$3^0 = 1$

Simplify.

9. x^{-2}

$\dfrac{1}{x^2}$

10. y^{-10}

$\dfrac{1}{y^{10}}$

11. $\dfrac{1}{a^{-6}}$

a^6

12. $\dfrac{1}{b^{-4}}$

b^4

13. $\dfrac{x^{-2}y}{x}$

$\dfrac{y}{x^3}$

14. $\dfrac{x^4 y^{-3}}{x^2}$

$\dfrac{x^2}{y^3}$

15. $\dfrac{a^{-2}b}{b^0}$

$\dfrac{b}{a^2}$

16. $\dfrac{a^2 b^{-4}}{b^0}$

$\dfrac{a^2}{b^4}$

17. $(ab^{-1})^0$

1

18. $(a^2 b)^0$

1

19. $(x^{-2}y^2)^2$

$\dfrac{y^4}{x^4}$

20. $(x^{-3}y^{-1})^2$

$\dfrac{1}{x^6 y^2}$

21. $(-2xy^{-2})^3$

$-\dfrac{8x^3}{y^6}$

22. $(-3x^{-1}y^2)^2$

$\dfrac{9y^4}{x^2}$

23. $(3x^{-1}y^{-2})^2$

$\dfrac{9}{x^2 y^4}$

24. $(5xy^{-3})^{-2}$

$\dfrac{y^6}{25x^2}$

25. $(2x^{-1})(x^{-3})$

$\dfrac{2}{x^4}$

26. $(-2x^{-5})x^7$

$-2x^2$

27. $(-5a^2)(a^{-5})^2$

$-\dfrac{5}{a^8}$

28. $(2a^{-3})(a^7 b^{-1})^3$

$\dfrac{2a^{18}}{b^3}$

29. $(-2ab^{-2})(4a^{-2}b)^{-2}$

$-\dfrac{a^5}{8b^4}$

30. $(3ab^{-2})(2a^{-1}b)^{-3}$

$\dfrac{3a^4}{8b^5}$

31. $(-5x^{-2}y)(-2x^{-2}y^2)$

$\dfrac{10y^3}{x^4}$

32. $\dfrac{a^{-3}b^{-4}}{a^2 b^2}$

$\dfrac{1}{a^5 b^6}$

Simplify.

33. $\dfrac{3x^{-2}y^2}{6xy^2}$

$\dfrac{1}{2x^3}$

34. $\dfrac{2x^{-2}y}{8xy}$

$\dfrac{1}{4x^3}$

35. $\dfrac{3x^{-2}y}{xy}$

$\dfrac{3}{x^3}$

36. $\dfrac{2x^{-1}y^4}{x^2y^3}$

$\dfrac{2y}{x^3}$

37. $\dfrac{2x^{-1}y^{-4}}{4xy^2}$

$\dfrac{1}{2x^2y^6}$

38. $\dfrac{(x^{-1}y)^2}{xy^2}$

$\dfrac{1}{x^3}$

39. $\dfrac{(x^{-2}y)^2}{x^2y^3}$

$\dfrac{1}{x^6y}$

40. $\dfrac{(x^{-3}y^{-2})^2}{x^6y^8}$

$\dfrac{1}{x^{12}y^{12}}$

41. $\dfrac{(a^{-2}y^3)^{-3}}{a^2y}$

$\dfrac{a^4}{y^{10}}$

42. $\dfrac{12a^2b^3}{-27a^2b^2}$

$-\dfrac{4b}{9}$

43. $\dfrac{-16xy^4}{96x^4y^4}$

$-\dfrac{1}{6x^3}$

44. $\dfrac{-8x^2y^4}{44y^2z^5}$

$-\dfrac{2x^2y^2}{11z^5}$

45. $\dfrac{22a^2b^4}{-132b^3c^2}$

$-\dfrac{a^2b}{6c^2}$

46. $\dfrac{-(8a^2b^4)^3}{64a^3b^8}$

$-8a^3b^4$

47. $\dfrac{-(14ab^4)^2}{28a^4b^2}$

$-\dfrac{7b^6}{a^2}$

48. $\dfrac{(2a^{-2}b^3)^{-2}}{(4a^2b^{-4})^{-1}}$

$\dfrac{a^6}{b^{10}}$

49. $\dfrac{(3^{-1}r^4s^{-3})^{-2}}{(6r^2t^{-2}s^{-1})^2}$

$\dfrac{t^4s^8}{4r^{12}}$

50. $\left(\dfrac{6x^{-4}yz^{-1}}{14xy^{-4}z^2}\right)^{-3}$

$\dfrac{343x^{15}z^9}{27y^{15}}$

51. $\left(\dfrac{15m^3n^{-2}p^{-1}}{25m^{-2}n^{-4}}\right)^{-3}$

$\dfrac{125p^3}{27m^{15}n^6}$

52. $\left(\dfrac{18a^4b^{-2}c^4}{12ab^{-3}d^2}\right)^{-2}$

$\dfrac{4d^4}{9a^6b^2c^8}$

▶ **Objective B**

Simplify.

53. $\dfrac{2x+2}{2}$

$x+1$

54. $\dfrac{5y+5}{5}$

$y+1$

55. $\dfrac{10a-25}{5}$

$2a-5$

56. $\dfrac{16b-40}{8}$

$2b-5$

57. $\dfrac{3a^2+2a}{a}$

$3a+2$

58. $\dfrac{6y^2+4y}{y}$

$6y+4$

59. $\dfrac{4b^3-3b}{b}$

$4b^2-3$

60. $\dfrac{12x^2-7x}{x}$

$12x-7$

61. $\dfrac{3x^2-6x}{3x}$

$x-2$

62. $\dfrac{10y^2-6y}{2y}$

$5y-3$

63. $\dfrac{5x^2-10x}{-5x}$

$-x+2$

64. $\dfrac{3y^2-27y}{-3y}$

$-y+9$

65. $\dfrac{x^3+3x^2-5x}{x}$

x^2+3x-5

66. $\dfrac{a^3-5a^2+7a}{a}$

a^2-5a+7

67. $\dfrac{x^6-3x^4-x^2}{x^2}$

x^4-3x^2-1

68. $\dfrac{a^8-5a^5-3a^3}{a^2}$

a^6-5a^3-3a

Simplify.

69. $\dfrac{5x^2y^2 + 10xy}{5xy}$

$xy + 2$

70. $\dfrac{8x^2y^2 - 24xy}{8xy}$

$xy - 3$

71. $\dfrac{9y^6 - 15y^3}{-3y^3}$

$-3y^3 + 5$

72. $\dfrac{4x^4 - 6x^2}{-2x^2}$

$-2x^2 + 3$

73. $\dfrac{3x^2 - 2x + 1}{x}$

$3x - 2 + \dfrac{1}{x}$

74. $\dfrac{8y^2 + 2y - 3}{y}$

$8y + 2 - \dfrac{3}{y}$

75. $\dfrac{-3x^2 + 7x - 6}{x}$

$-3x + 7 - \dfrac{6}{x}$

76. $\dfrac{2y^2 - 6y + 9}{y}$

$2y - 6 + \dfrac{9}{y}$

77. $\dfrac{16a^2b - 20ab + 24ab^2}{4ab}$

$4a - 5 + 6b$

78. $\dfrac{22a^2b + 11ab - 33ab^2}{11ab}$

$2a + 1 - 3b$

79. $\dfrac{9x^2y + 6xy - 3xy^2}{xy}$

$9x + 6 - 3y$

80. $\dfrac{5a^2b - 15ab + 30ab^2}{5ab}$

$a - 3 + 6b$

▶ **Objective C**

Simplify.

81. $(x^2 + 2x + 1) \div (x + 1)$
$x + 1$

82. $(x^2 + 10x + 25) \div (x + 5)$
$x + 5$

83. $(a^2 - 6a + 9) \div (a - 3)$
$a - 3$

84. $(b^2 - 14b + 49) \div (b - 7)$
$b - 7$

85. $(x^2 - x - 6) \div (x - 3)$
$x + 2$

86. $(y^2 + 2y - 35) \div (y + 7)$
$y - 5$

87. $(2x^2 + 5x + 2) \div (x + 2)$
$2x + 1$

88. $(2y^2 - 13y + 21) \div (y - 3)$
$2y - 7$

89. $(4x^2 - 16) \div (2x + 4)$
$2x - 4$

90. $(2y^2 + 7) \div (y - 3)$

$2y + 6 + \dfrac{25}{y - 3}$

91. $(x^2 + 1) \div (x - 1)$

$x + 1 + \dfrac{2}{x - 1}$

92. $(x^2 + 4) \div (x + 2)$

$x - 2 + \dfrac{8}{x + 2}$

93. $(6x^2 - 7x) \div (3x - 2)$

$2x - 1 - \dfrac{2}{3x - 2}$

94. $(6y^2 + 2y) \div (2y + 4)$

$3y - 5 + \dfrac{20}{2y + 4}$

95. $(5x^2 + 7x) \div (x - 1)$

$5x + 12 + \dfrac{12}{x - 1}$

Simplify.

96. $(6x^2 - 5) \div (x + 2)$

$6x - 12 + \dfrac{19}{x + 2}$

97. $(a^2 + 5a + 10) \div (a + 2)$

$a + 3 + \dfrac{4}{a + 2}$

98. $(b^2 - 8b - 9) \div (b - 3)$

$b - 5 - \dfrac{24}{b - 3}$

99. $(2y^2 - 9y + 8) \div (2y + 3)$

$y - 6 + \dfrac{26}{2y + 3}$

100. $(3x^2 + 5x - 4) \div (x - 4)$

$3x + 17 + \dfrac{64}{x - 4}$

101. $(4x^2 + 8x + 3) \div (2x - 1)$

$2x + 5 + \dfrac{8}{2x - 1}$

102. $(10y^2 + 21y + 10) \div (2y + 3)$

$5y + 3 + \dfrac{1}{2y + 3}$

103. $(15a^2 - 8a - 8) \div (3a + 2)$

$5a - 6 + \dfrac{4}{3a + 2}$

104. $(12a^2 - 25a - 7) \div (3a - 7)$

$4a + 1$

105. $(12x^2 - 23x + 5) \div (4x - 1)$

$3x - 5$

106. $(6a^2 + 25a + 24) \div (3a - 1)$

$2a + 9 + \dfrac{33}{3a - 1}$

107. $(x^3 + 3x^2 + 5x + 3) \div (x + 1)$
$x^2 + 2x + 3$

108. $(x^3 - 6x^2 + 7x - 2) \div (x - 1)$
$x^2 - 5x + 2$

109. $(y^3 + 6y^2 + 4y - 5) \div (y + 3)$

$y^2 + 3y - 5 + \dfrac{10}{y + 3}$

110. $(4a^3 + 8a^2 + 5a + 9) \div (2a + 3)$

$2a^2 + a + 1 + \dfrac{6}{2a + 3}$

111. $(6a^3 + 5a^2 + 5) \div (3a - 2)$

$2a^2 + 3a + 2 + \dfrac{9}{3a - 2}$

112. $(4b^3 - b - 5) \div (2b + 3)$

$2b^2 - 3b + 4 - \dfrac{17}{2b + 3}$

113. $(2a^3 - 5a^2 + 2) \div (2a + 3)$

$a^2 - 4a + 6 - \dfrac{16}{2a + 3}$

114. $(5x^3 - 7x^2 - 3x) \div (x - 2)$

$5x^2 + 3x + 3 + \dfrac{6}{x - 2}$

115. $(x^4 - x^2 - 6) \div (x^2 + 2)$
$x^2 - 3$

116. $(x^4 + 3x^2 - 10) \div (x^2 - 2)$
$x^2 + 5$

Calculators and Computers

Evaluating Polynomials

One way to evaluate a polynomial is to first express the polynomial in a form that suggests a sequence of steps on the calculator. To illustrate this method, consider the polynomial $4x^2 - 5x + 2$. First the polynomial is rewritten as

$$4x^2 - 5x + 2 = (4x - 5)x + 2$$

To evaluate the polynomial, work through the rewritten expression from left to right, substituting the appropriate value for x.

Here are some examples:

Evaluate $5x^2 - 2x + 4$ when $x = 3$.

Rewrite the polynomial.

$$5x^2 - 2x + 4 = (5x - 2)x + 4$$

Replace x in the rewritten expression by the given value.

$$(5 \cdot 3 - 2) \cdot 3 + 4$$

Work through the expression from left to right.

$\boxed{(}\ 5\ \boxed{\times}\ 3\ \boxed{-}\ 2\ \boxed{)}\ \boxed{\times}\ 3\ \boxed{+}\ 4\ \boxed{=}$

The result in the display should be 43.

Evaluate $2x^3 - 4x^2 + 7x - 12$ when $x = 4$.

Rewrite the polynomial.

$$2x^3 - 4x^2 + 7x - 12 = [(2x - 4)x + 7]x - 12$$

Replace x in the rewritten expression by the given value.

$$[(2 \cdot 4 - 4) \cdot 4 + 7] \cdot 4 - 12$$

Work through the expression from left to right.

$\boxed{(}\ \boxed{(}\ 2\ \boxed{\times}\ 4\ \boxed{-}\ 4\ \boxed{)}\ \boxed{\times}\ 4\ \boxed{+}\ 7\ \boxed{)}\ \boxed{\times}\ 4\ \boxed{-}\ 12\ \boxed{=}$

The result in the display should be 80.

Evaluate $4x^2 - 3x + 5$ when $x = -2$.

Rewrite the polynomial.

$$4x^2 - 3x + 5 = (4x - 3)x + 5$$

Replace x in the given expression by the given value.

$$[4 \cdot (-2) - 3] \cdot (-2) + 5$$

Work through the expression from left to right.

$\boxed{(}\ 4\ \boxed{\times}\ 2\ \boxed{+/-}\ \boxed{-}\ 3\ \boxed{)}\ \boxed{\times}\ 2\ \boxed{+/-}\ \boxed{+}\ 5\ \boxed{=}$

The result in the display should be 27.

Here are some practice exercises. The answers are on page 194.
Evaluate for the given value.

1. $2x^2 - 3x + 7$; $x = 4$
2. $3x^2 + 7x - 12$; $x = -3$
3. $3x^3 - 2x^2 + 6x - 8$; $x = 3$
4. $2x^3 + 4x^2 - x - 2$; $x = -2$
5. $x^4 - 3x^3 + 6x^2 + 5x - 1$; $x = 2$
6. $2x^3 - 4x + 8$; $x = 2$
   Hint: $2x^3 - 4x + 8 = 2x^3 + 0x^2 - 4x + 8$

Chapter Summary

Key Words   A *monomial* is a number, a variable, or a product of numbers and variables.

A *polynomial* is a variable expression in which the terms are monomials.

A *monomial* is a polynomial of *one* term.

A *binomial* is a polynomial of *two* terms.

A *trinomial* is a polynomial of *three* terms.

The *degree of a polynomial* in one variable is the largest exponent on a variable.

Essential Rules

Rule for Multiplying Exponential Expressions   If m and n are integers, then $x^m \cdot x^n = x^{m+n}$.

Rule for Simplifying Powers of Exponential Expressions   If m and n are integers, then $(x^m)^n = x^{m \cdot n}$.

Rule for Simplifying Powers of Products   If m, n, and p are integers, then $(x^m \cdot y^n)^p = x^{m \cdot p} y^{n \cdot p}$.

Rule for Zero as an Exponent   If $x \neq 0$, then $x^0 = 1$.

Rule for Negative Exponents   If n is a positive integer and $x \neq 0$, then $x^{-n} = \dfrac{1}{x^n}$.

Rule for Simplifying Powers of Quotients   If m, n, and p are integers, then $\left(\dfrac{x^m}{y^n}\right)^p = \dfrac{x^{m \cdot p}}{y^{n \cdot p}}$, $y \neq 0$.

Rule for Dividing Exponential Expressions   If m and n are integers and $x \neq 0$, then $\dfrac{x^m}{x^n} = x^{m-n}$.

The Sum and Difference of Two Terms $(a + b)(a - b) = a^2 - b^2$

The Square of a Binomial $(a + b)^2 = a^2 + 2ab + b^2$
$(a - b)^2 = a^2 - 2ab + b^2$

Dividend = (quotient × divisor) + remainder

Answers to Practice Exercises on p. 193
1. 27   **2.** −6   **3.** 73   **4.** 0   **5.** 25   **6.** 16

Chapter Review

SECTION 5.1

1. Simplify $(12y^2 + 17y - 4) + (9y^2 - 13y + 3)$.
$21y^2 + 4y - 1$

2. Simplify $(2x^3 + 7x^2 + x) + (2x^2 - 4x - 12)$.
$2x^3 + 9x^2 - 3x - 12$

3. Simplify $(5a^2 + 6a - 11) + (5a^2 + 6a - 11)$.
$10a^2 + 12a - 22$

4. Simplify $(5x^2 - 2x - 1) - (3x^2 - 5x + 7)$.
$2x^2 + 3x - 8$

5. Simplify $(13y^3 - 7y - 2) - (12y^2 - 2y - 1)$.
$13y^3 - 12y^2 - 5y - 1$

6. Simplify $(6y^2 + 2y + 7) - (8y^2 + y + 12)$.
$-2y^2 + y - 5$

SECTION 5.2

7. Simplify $(5xy^2)(-4x^2y^3)$.
$-20x^3y^5$

8. Simplify $(xy^5z^3)(x^3y^3z)$.
$x^4y^8z^4$

9. Simplify $(2a^{12}b^3)(-9b^2c^6)(3ac)$.
$-54a^{13}b^5c^7$

10. Simplify $(2^3)^2$.
64

11. Simplify $(-3x^2y^3)^2$.
$9x^4y^6$

12. Simplify $(5a^7b^6)^2 (4ab)$.
$100a^{15}b^{13}$

SECTION 5.3

13. Simplify $-2x(4x^2 + 7x - 9)$.
$-8x^3 - 14x^2 + 18x$

14. Simplify $2ab^3(4a^2 - 2ab + 3b^2)$.
$8a^3b^3 - 4a^2b^4 + 6ab^5$

15. Simplify $(3y^2 + 4y - 7)(2y + 3)$.
$6y^3 + 17y^2 - 2y - 21$

16. Simplify $(6b^3 - 2b^2 - 5)(2b^2 - 1)$.
$12b^5 - 4b^4 - 6b^3 - 8b^2 + 5$

17. Simplify $(5a - 7)(2a + 9)$.
$10a^2 + 31a - 63$

18. Simplify $(2b - 3)(4b + 5)$.
$8b^2 - 2b - 15$

19. Simplify $(a + 7)(a - 7)$.
$a^2 - 49$

20. Simplify $(5y - 7)^2$.
$25y^2 - 70y + 49$

21. The length of a Ping-Pong table is 1 ft less than twice the width of the table. Let w represent the width of the Ping-Pong table. Express the area of the Ping-Pong table in terms of the variable w.
$(2w^2 - w)$ ft^2

22. The side of a square checkerboard is $(3x - 2)$ inches. Express the area of the checkerboard in terms of the variable x.
$(9x^2 - 12x + 4)$ in.^{2}

SECTION 5.4

23. Simplify $(-3x^{-2}y^{-3})^{-2}$.
$\frac{x^4y^6}{9}$

24. Simplify $\frac{(4a^{-2}b^{-3})^2}{(2a^{-1}b^{-2})^4}$.
b^2

25. Simplify $\frac{16y^2 - 32y}{-4y}$.
$-4y + 8$

26. Simplify $\frac{12b^7 + 36b^5 - 3b^3}{3b^3}$.
$4b^4 + 12b^2 - 1$

27. Simplify $(6y^2 - 35y + 36) \div (3y - 4)$.
$2y - 9$

28. Simplify $(b^3 - 2b^2 - 33b - 7) \div (b - 7)$.
$b^2 + 5b + 2 + \frac{7}{b - 7}$

Content and Format © 1991 HMCo.

Chapter Test

1. Simplify
   $(3x^3 - 2x^2 - 4) + (8x^2 - 8x + 7)$.
   $3x^3 + 6x^2 - 8x + 3$ [5.1A]

2. Simplify
   $(3a^2 - 2a - 7) - (5a^3 + 2a - 10)$.
   $-5a^3 + 3a^2 - 4a + 3$ [5.1B]

3. Simplify $(ab^2)(a^3b^5)$.
   a^4b^7 [5.2A]

4. Simplify $(-2xy^2)(3x^2y^4)$.
   $-6x^3y^6$ [5.2A]

5. Simplify $(x^2y^3)^4$.
   x^8y^{12} [5.2B]

6. Simplify $(-2a^2b)^3$.
   $-8a^6b^3$ [5.2B]

7. Simplify $2x(2x^2 - 3x)$.
   $4x^3 - 6x^2$ [5.3A]

8. Simplify $-3y^2(-2y^2 + 3y - 6)$.
   $6y^4 - 9y^3 + 18y^2$ [5.3A]

9. Simplify $(x - 3)(x^2 - 4x + 5)$.
   $x^3 - 7x^2 + 17x - 15$ [5.3B]

10. Simplify $(-2x^3 + x^2 - 7)(2x - 3)$.
    $-4x^4 + 8x^3 - 3x^2 - 14x + 21$ [5.3B]

11. Simplify $(a - 2b)(a + 5b)$.
    $a^2 + 3ab - 10b^2$ [5.3C]

12. Simplify $(2x - 7y)(5x - 4y)$.
    $10x^2 - 43xy + 28y^2$ [5.3C]

13. Simplify $(4y - 3)(4y + 3)$.
    $16y^2 - 9$ [5.3D]

14. Simplify $(2x - 5)^2$.
    $4x^2 - 20x + 25$ [5.3D]

15. The radius of a circle is $x - 5$. Use the equation $A = \pi r^2$, where r is the radius, to find the area of the circle in terms of the variable x. Use $\pi \approx 3.14$.
($3.14\,x^2 - 31.4x + 78.5$)   [5.3E]

16. Simplify $\frac{12x^2}{-3x^8}$.
$-\frac{4}{x^6}$ [5.4A]

17. Simplify $\frac{-(2x^2y)^3}{4x^3y^3}$.
$-2x^3$ [5.4A]

18. Simplify $\frac{2a^{-1}b}{2^{-2}a^{-2}b^{-3}}$.
$8ab^4$ [5.4A]

19. Simplify $\frac{(3x^{-2}y^3)^3}{3x^4y^{-1}}$.
$\frac{9y^{10}}{x^{10}}$ [5.4A]

20. Simplify $\frac{20a - 35}{5}$.
$4a - 7$ [5.4B]

21. Simplify $\frac{16x^5 - 8x^3 + 20x}{4x}$.
$4x^4 - 2x^2 + 5$ [5.4B]

22. Simplify $\frac{12x^3 - 3x^2 + 9}{3x^2}$.
$4x - 1 + \frac{3}{x^2}$ [5.4B]

23. Simplify $(x^2 + 6x - 7) \div (x - 1)$.
$x + 7$ [5.4C]

24. Simplify $(x^2 + 1) \div (x + 1)$.
$x - 1 + \frac{2}{x + 1}$ [5.4C]

25. Simplify $(4x^2 - 7) \div (2x - 3)$.
$2x + 3 + \frac{2}{2x - 3}$ [5.4C]

Cumulative Review

1. Simplify $\frac{3}{16} - \left(-\frac{5}{8}\right) - \frac{7}{9}$.
   $\frac{5}{144}$ [1.4B]

2. Evaluate $-3^2 \cdot \left(\frac{2}{3}\right)^3 \cdot \left(-\frac{5}{8}\right)$.
   $\frac{5}{3}$ [1.5A]

3. Simplify $\left(-\frac{1}{2}\right)^3 \div \left(\frac{3}{8} - \frac{5}{6}\right) + 2$.
   $\frac{25}{11}$ [1.5B]

4. Evaluate $\frac{b - (a - b)^2}{b^2}$ when $a = -2$ and $b = 3$.
   $-\frac{22}{9}$ [2.1A]

5. Simplify $-2x - (-xy) + 7x - 4xy$.
   $5x - 3xy$ [2.2A]

6. Simplify $(12x)\left(-\frac{3}{4}\right)$.
   $-9x$ [2.2B]

7. Simplify $-2[3x - 2(4 - 3x) + 2]$.
   $-18x + 12$ [2.2D]

8. Solve: $12 = -\frac{3}{4}x$
   -16 [3.1C]

9. Solve: $2x - 9 = 3x + 7$
   -16 [3.3A]

10. Solve: $2 - 3(4 - x) = 2x + 5$
    15 [3.3B]

11. 35.2 is what percent of 160?
    22% [4.2A]

12. Simplify $(4b^3 - 7b^2 - 7) + (3b^2 - 8b + 3)$.
    $4b^3 - 4b^2 - 8b - 4$ [5.1A]

13. Simplify $(3y^3 - 5y + 8) - (-2y^2 + 5y + 8)$.
    $3y^3 + 2y^2 - 10y$ [5.1B]

14. Simplify $(a^3b^5)^3$.
    a^9b^{15} [5.2B]

15. Simplify $(4xy^3)(-2x^2y^3)$.
    $-8x^3y^6$ [5.2A]

16. Simplify $-2y^2(-3y^2 - 4y + 8)$.
    $6y^4 + 8y^3 - 16y^2$ [5.3A]

17. Simplify $(2a - 7)(5a^2 - 2a + 3)$.
$10a^3 - 39a^2 + 20a - 21$ [5.3B]

18. Simplify $(3b - 2)(5b - 7)$.
$15b^2 - 31b + 14$ [5.3C]

19. Simplify $(3b + 2)^2$.
$9b^2 + 12b + 4$ [5.3D]

20. Simplify $\frac{(-2a^2b^3)^2}{8a^4b^8}$.
$\frac{1}{2b^2}$ [5.4A]

21. Simplify $\frac{-18a^3 + 12a^2 - 6}{-3a^2}$.
$6a - 4 + \frac{2}{a^2}$ [5.4B]

22. Simplify $(a^2 - 4a - 21) \div (a + 3)$.
$a - 7$ [5.4C]

23. Simplify $(-2x^{-2}y)(-3x^{-4}y)$.
$\frac{6y^2}{x^6}$ [5.4A]

24. Translate "the difference between eight times a number and twice a number is eighteen" into an equation and solve.
$8x - 2x = 18$; $x = 3$ [3.4A]

25. A calculator costs a retailer $24. Find the selling price when the markup rate is 80%.
$43.20   [4.3A]

26. Fifty ounces of pure orange juice are added to 200 oz of a fruit punch which is 10% orange juice. What is the percent concentration of orange juice in the resulting mixture?
28%   [4.5B]

27. A car traveling at 50 mph overtakes a cyclist who, riding at 10 mph, has had a 2-h head start. How far from the starting point does the car overtake the cyclist?
25 mi   [4.6A]

28. The width of a rectangle is 40% of the length. The perimeter of the rectangle is 42 m. Find the length and width of the rectangle.
length 15 m; width 6 m   [4.7A]

29. The age of a gold coin is 60 years, and the age of a silver coin is 40 years. How many years ago was the gold coin twice the age the silver coin was then?
20 years ago   [4.8C]

30. The length of a side of a square is $2x + 3$. Use the equation $A = s^2$, where s is the length of the side of a square, to find the area of the square in terms of the variable x.
$4x^2 + 12x + 9$ [5.3E]

6

Factoring

OBJECTIVES

▶ To factor a monomial from a polynomial
▶ To factor by grouping
▶ To factor a trinomial of the form $x^2 + bx + c$
▶ To factor completely
▶ To factor $ax^2 + bx + c$ by using trial factors
▶ To factor $ax^2 + bx + c$ by grouping
▶ To factor the difference of two squares and perfect square trinomials
▶ To factor completely
▶ To solve equations by factoring
▶ To solve application problems

Algebra from Geometry

The early Babylonians made substantial progress in both algebra and geometry. Often the progress they made in algebra was based on geometric concepts.

Here are some geometric proofs of algebraic identities.

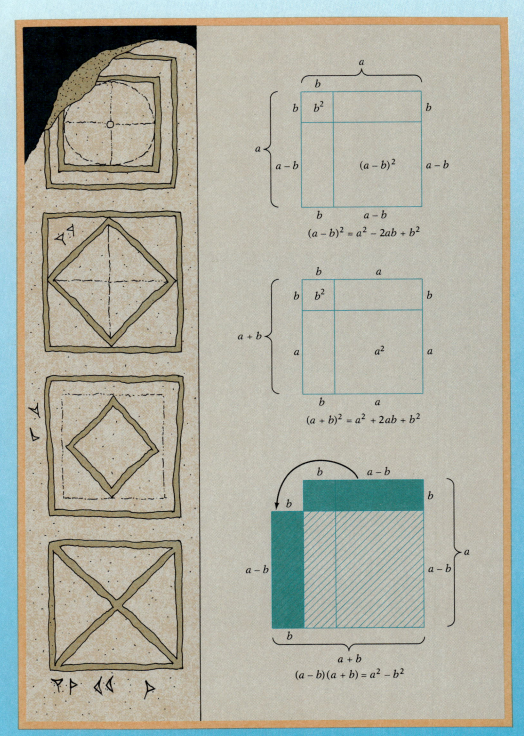

$$(a - b)^2 = a^2 - 2ab + b^2$$

$$(a + b)^2 = a^2 + 2ab + b^2$$

$$(a - b)(a + b) = a^2 - b^2$$

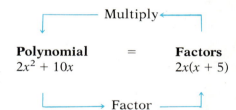

SECTION 6.1

Common Factors

Objective A

To factor a monomial from a polynomial

The **greatest common factor (GCF)** of two or more monomials is the product of the GCF of the coefficients and the common variable factors.

$$6x^3y = 2 \cdot 3 \cdot x \cdot x \cdot x \cdot y$$
$$8x^2y^2 = 2 \cdot 2 \cdot 2 \cdot x \cdot x \cdot y \cdot y$$
$$GCF = 2 \cdot x \cdot x \cdot y = 2x^2y$$

Note that the exponent of each variable in the GCF is the same as the *smallest* exponent of that variable in either of the monomials.

The GCF of $6x^3y$ and $8x^2y^2$ is $2x^2y$.

Find the GCF of $12a^4b$ and $18a^2b^2c$.

The common variable factors are a^2 and b. c is not a common variable factor.

$$12a^4b = 2 \cdot 2 \cdot 3 \cdot a^4 \cdot b$$
$$18a^2b^2c = 2 \cdot 3 \cdot 3 \cdot a^2 \cdot b^2 \cdot c$$
$$GCF = 2 \cdot 3 \cdot a^2 \cdot b = 6a^2b$$

To **factor** a polynomial means to write the polynomial as a product of other polynomials.

Multiply ←

Polynomial = **Factors**
$2x^2 + 10x$ $2x(x + 5)$

→ Factor

In the example above, $2x$ is the GCF of the terms $2x^2$ and $10x$. It is a **common monomial factor** of the terms. $x + 5$ is a **binomial factor** of $2x^2 + 10x$.

Factor $5x^3 - 35x^2 + 10x$.

Find the GCF of the terms of the polynomial.

$$5x^3 = 5 \cdot x^3$$
$$35x^2 = 5 \cdot 7 \cdot x^2$$
$$10x = 2 \cdot 5 \cdot x$$
The GCF is $5x$.

Divide each term of the polynomial by the GCF.

$$\frac{5x^3}{5x} = x^2 \quad \frac{-35x^2}{5x} = -7x \quad \frac{10x}{5x} = 2$$

Do this step mentally.

Use the quotients to rewrite the polynomial, expressing each term as a product with the GCF as one of the factors.

$$5x^3 - 35x^2 + 10x = 5x(x^2) + 5x(-7x) + 5x(2)$$

Use the Distributive Property to write the polynomial as a product of factors.

$$= 5x(x^2 - 7x + 2)$$

Example 1

Factor $8x^2 + 2xy$.

Solution

The GCF is $2x$.

$8x^2 + 2xy = 2x(4x) + 2x(y) = 2x(4x + y)$

Example 2

Factor $14a^2 - 21a^4b$.

Your solution

$7a^2(2 - 3a^2b)$

Example 3

Factor $n^3 - 5n^2 + 2n$.

Solution

The GCF is n.

$n^3 - 5n^2 + 2n =$
$n(n^2) + n(-5n) + n(2) =$
$n(n^2 - 5n + 2)$

Example 4

Factor $27b^2 + 18b + 9$.

Your solution

$9(3b^2 + 2b + 1)$

Example 5

Factor $16x^2y + 8x^4y^2 - 12x^4y^5$.

Solution

The GCF is $4x^2y$.

$16x^2y + 8x^4y^2 - 12x^4y^5 =$
$4x^2y(4) + 4x^2y(2x^2y) + 4x^2y(-3x^2y^4) =$
$4x^2y(4 + 2x^2y - 3x^2y^4)$

Example 6

Factor $6x^4y^2 - 9x^3y^2 + 12x^2y^4$.

Your solution

$3x^2y^2(2x^2 - 3x + 4y^2)$

Solutions on p. A19

Content and Format © 1991 HMCo.

| **Objective B** | **To factor by grouping** | |

In the examples at the right, the binomials in parentheses are called binomial factors.

$$2a(a + b)^2$$
$$3xy(x - y)$$

The Distributive Property is used to factor a common binomial factor from an expression.

Factor $6(x - 3) + y^2(x - 3)$.

The common binomial factor is $(x - 3)$. Use the Distributive Property to write the expression as a product of factors.

$$6(x - 3) + y^2(x - 3)$$
$$(x - 3)(6 + y^2)$$

Consider the binomial $b - a$.
Because $-(a - b) = -1(a - b) = -a + b = b - a$, the following equation is true.

$$b - a = -(a - b).$$

This equation is sometimes used to factor a common binomial from an expression.

Factor $2x(x - y) + 5(y - x)$.

Rewrite the expression as a difference of terms that have a common binomial factor.
Use $y - x = -(x - y)$.

$$2x(x - y) + 5(y - x)$$
$$2x(x - y) - 5(x - y)$$
$$(x - y)(2x - 5)$$

Some polynomials can be factored by grouping terms so that a common binomial factor is found.

Factor $ax + bx - ay - by$.

Group the first two terms and the last two terms.
Note that $-ay - by = -(ay + by)$.

$$ax + by - ay - by$$
$$(ax + bx) - (ay + by)$$

Factor the GCF from each group.

$$x(a + b) - y(a + b)$$

Write the expression as the product of factors.

$$(a + b)(x - y)$$

Factor $6x^2 - 9x - 4xy + 6y$.

Group the first two terms and the last two terms.
Note that $-4xy + 6y = -(4xy - 6y)$.

$$6x^2 - 9x - 4xy + 6y$$
$$(6x^2 - 9x) - (4xy - 6y)$$

Factor the GCF from each group.

$$3x(2x - 3) - 2y(2x - 3)$$

Write the expression as the product of factors.

$$(2x - 3)(3x - 2y)$$

Example 7

Factor $4x(3x - 2) - 7(3x - 2)$.

Solution

$4x(3x - 2) - 7(3x - 2) =$
$(3x - 2)(4x - 7)$

Example 8

Factor $2y(5x - 2) - 3(2 - 5x)$.

Your solution

$(5x - 2)(2y + 3)$

Example 9

Factor $9x^2 - 15x - 6xy + 10y$.

Solution

$9x^2 - 15x - 6xy + 10y =$
$(9x^2 - 15x) - (6xy - 10y) =$
$3x(3x - 5) - 2y(3x - 5) =$
$(3x - 5)(3x - 2y)$

Example 10

Factor $a^2 - 3a + 2ab - 6b$.

Your solution

$(a - 3)(a + 2b)$

Example 11

Factor $3x^2y - 4x - 15xy + 20$.

Solution

$3x^2y - 4x - 15xy + 20 =$
$(3x^2y - 4x) - (15xy - 20) =$
$x(3xy - 4) - 5(3xy - 4) =$
$(3xy - 4)(x - 5)$

Example 12

Factor $2mn^2 - n + 8mn - 4$.

Your solution

$(2mn - 1)(n + 4)$

Solutions on p. A19

6.1 EXERCISES

▶ **Objective A**

Factor.

1. $5a + 5$
$5(a + 1)$

2. $7b - 7$
$7(b - 1)$

3. $16 - 8a^2$
$8(2 - a^2)$

4. $12 + 12y^2$
$12(1 + y^2)$

5. $8x + 12$
$4(2x + 3)$

6. $16a - 24$
$8(2a - 3)$

7. $30a - 6$
$6(5a - 1)$

8. $20b + 5$
$5(4b + 1)$

9. $7x^2 - 3x$
$x(7x - 3)$

10. $12y^2 - 5y$
$y(12y - 5)$

11. $3a^2 + 5a^5$
$a^2(3 + 5a^3)$

12. $9x - 5x^2$
$x(9 - 5x)$

13. $14y^2 + 11y$
$y(14y + 11)$

14. $6b^3 - 5b^2$
$b^2(6b - 5)$

15. $2x^4 - 4x$
$2x(x^3 - 2)$

16. $3y^4 - 9y$
$3y(y^3 - 3)$

17. $10x^4 - 12x^2$
$2x^2(5x^2 - 6)$

18. $12a^5 - 32a^2$
$4a^2(3a^3 - 8)$

19. $8a^8 - 4a^5$
$4a^5(2a^3 - 1)$

20. $16y^4 - 8y^7$
$8y^4(2 - y^3)$

21. $x^2y^2 - xy$
$xy(xy - 1)$

22. $a^2b^2 + ab$
$ab(ab + 1)$

23. $3x^2y^4 - 6xy$
$3xy(xy^3 - 2)$

24. $12a^2b^5 - 9ab$
$3ab(4ab^4 - 3)$

25. $x^2y - xy^3$
$xy(x - y^2)$

26. $a^2b + a^4b^2$
$a^2b(1 + a^2b)$

27. $2a^5b + 3xy^3$
No common factor other than 1

28. $5x^2y - 7ab^3$
No common factor other than 1

29. $6a^2b^3 - 12b^2$
$6b^2(a^2b - 2)$

30. $8x^2y^3 - 4x^2$
$4x^2(2y^3 - 1)$

31. $a^3 - 3a^2 + 5a$
$a(a^2 - 3a + 5)$

32. $b^3 - 5b^2 - 7b$
$b(b^2 - 5b - 7)$

33. $5x^2 - 15x + 35$
$5(x^2 - 3x + 7)$

34. $8y^2 - 12y + 32$
$4(2y^2 - 3y + 8)$

35. $3x^3 + 6x^2 + 9x$
$3x(x^2 + 2x + 3)$

36. $5y^3 - 20y^2 + 10y$
$5y(y^2 - 4y + 2)$

37. $2x^4 - 4x^3 + 6x^2$
$2x^2(x^2 - 2x + 3)$

38. $3y^4 - 9y^3 - 6y^2$
$3y^2(y^2 - 3y - 2)$

39. $2x^3 + 6x^2 - 14x$
$2x(x^2 + 3x - 7)$

40. $3y^3 - 9y^2 + 24y$
$3y(y^2 - 3y + 8)$

41. $2y^5 - 3y^4 + 7y^3$
$y^3(2y^2 - 3y + 7)$

42. $6a^5 - 3a^3 - 2a^2$
$a^2(6a^3 - 3a - 2)$

43. $x^3y - 3x^2y^2 + 7xy^3$
$xy(x^2 - 3xy + 7y^2)$

44. $2a^2b - 5a^2b^2 + 7ab^2$
$ab(2a - 5ab + 7b)$

45. $5y^3 + 10y^2 - 25y$
$5y(y^2 + 2y - 5)$

46. $4b^5 + 6b^3 - 12b$
$2b(2b^4 + 3b^2 - 6)$

47. $3a^2b^2 - 9ab^2 + 15b^2$
$3b^2(a^2 - 3a + 5)$

48. $8x^2y^2 - 4x^2y + x^2$
$x^2(8y^2 - 4y + 1)$

▶ **Objective B**

Factor.

49. $x(b + 4) + 3(b + 4)$
$(b + 4)(x + 3)$

50. $y(a + z) + 7(a + z)$
$(a + z)(y + 7)$

51. $a(y - x) - b(y - x)$
$(y - x)(a - b)$

52. $3r(a - b) + s(a - b)$
$(a - b)(3r + s)$

53. $x(x - 2) + y(2 - x)$
$(x - 2)(x - y)$

54. $t(m - 7) + 7(7 - m)$
$(m - 7)(t - 7)$

55. $4x(2r - s) - 3y(s - 2r)$
$(2r - s)(4x + 3y)$

56. $5n(3p - 2) - m(2 - 3p)$
$(3p - 2)(5n + m)$

57. $7p(4a + 3b) - 2r(3b + 4a)$
$(4a + 3b)(7p - 2r)$

58. $2x(7 + b) - y(b + 7)$
$(b + 7)(2x - y)$

59. $2y(4a - b) - (b - 4a)$
$(4a - b)(2y + 1)$

60. $8c(2m - 3n) + (3n - 2m)$
$(2m - 3n)(8c - 1)$

61. $x^2 + 2x + 2xy + 4y$
$(x + 2)(x + 2y)$

62. $x^2 - 3x + 4ax - 12a$
$(x - 3)(x + 4a)$

63. $p^2 - 2p - 3rp + 6r$
$(p - 2)(p - 3r)$

64. $t^2 + 4t - st - 4s$
$(t + 4)(t - s)$

65. $ab + 6b - 4a - 24$
$(a + 6)(b - 4)$

66. $xy - 5y - 2x + 10$
$(x - 5)(y - 2)$

67. $2z^2 - z + 2yz - y$
$(2z - 1)(z + y)$

68. $2y^2 - 10y + 7xy - 35x$
$(y - 5)(2y + 7x)$

69. $8v^2 - 12vy + 14v - 21y$
$(2v - 3y)(4v + 7)$

70. $21x^2 + 6xy - 49x - 14y$
$(7x + 2y)(3x - 7)$

71. $2x^2 - 5x - 6xy + 15y$
$(2x - 5)(x - 3y)$

72. $4a^2 + 5ab - 10b - 8a$
$(4a + 5b)(a - 2)$

73. $3y^2 - 6y - ay + 2a$
$(y - 2)(3y - a)$

74. $2ra + a^2 - 2r - a$
$(2r + a)(a - 1)$

75. $3xy - y^2 - y + 3x$
$(3x - y)(y + 1)$

76. $2ab - 3b^2 - 3b + 2a$
$(2a - 3b)(b + 1)$

77. $3st + t^2 - 2t - 6s$
$(3s + t)(t - 2)$

78. $4x^2 + 3xy - 12y - 16x$
$(4x + 3y)(x - 4)$

79. $6x + 3a - 4ax - 2a^2$
$(2x + a)(3 - 2a)$

80. $8s + 12r - 6s^2 - 9rs$
$(2s + 3r)(4 - 3s)$

81. $6ab^2 - 5b + 12ab - 10$
$(6ab - 5)(b + 2)$

| **SECTION 6.2** | **Factoring Polynomials of the Form $x^2 + bx + c$** |

Objective A   To factor a trinomial of the form $x^2 + bx + c$

Trinomials of the form $x^2 + bx + c$, where b and c are integers, are shown at the right.

$x^2 + 8x + 12; \, b = 8, \quad c = 12$
$x^2 - 7x + 12; \, b = -7, \, c = 12$
$x^2 - 2x - 15; \, b = -2, \, c = -15$

To **factor** a trinomial of this form means to express the trinomial as the product of two binomials.

Trinomials expressed as the product of binomials are shown at the right.

$x^2 + 8x + 12 = (x + 6)(x + 2)$
$x^2 - 7x + 12 = (x - 3)(x - 4)$
$x^2 - 2x - 15 = (x + 3)(x - 5)$

The method by which factors of a trinomial are found is based on FOIL. Consider the following binomial products, noting the relationship between the constant terms of the binomials and the terms of the trinomials.

Signs in the binomials are the same

$(x + 6)(x + 2) = x^2 + 2x + 6x + (6)(2) \quad = x^2 + 8x + 12$
sum of 6 and 2
product of 6 and 2

$(x - 3)(x - 4) = x^2 - 4x - 3x + (-3)(-4) = x^2 - 7x + 12$
sum of -3 and -4
product of -3 and -4

Signs in the binomials are opposite

$(x + 3)(x - 5) = x^2 - 5x + 3x + (3)(-5) \quad = x^2 - 2x - 15$
sum of 3 and -5
product of 3 and -5

$(x - 4)(x + 6) = x^2 + 6x - 4x + (-4)(6) \quad = x^2 + 2x - 24$
sum of -4 and 6
product of -4 and 6

Important Relationships

1. When the constant term of the trinomial is positive, the constant terms of the binomials have the same sign. They are both positive when the coefficient of the x term in the trinomial is positive. They are both negative when the coefficient of the x term in the trinomial is negative.

2. When the constant term of the trinomial is negative, the constant terms of the binomials have opposite signs.

3. In the trinomial, the coefficient of x is the sum of the constant terms of the binomials.

4. In the trinomial, the constant term is the product of the constant terms of the binomials.

Factor $x^2 - 7x + 10$.

Because the constant term is positive and the coefficient of x is negative, the binomial constants will be negative.

Find two negative factors of 10 whose sum is -7.

Factors	Sum
$-1, -10$	-11
$-2, -5$	-7

Write the trinomial as a product of its factors. $(x - 2)(x - 5)$

You can always check that the proposed factorization is correct by multiplying the two binomials using FOIL.

$$\text{Check:} \quad (x - 2)(x - 5) = x^2 - 5x - 2x + 10$$
$$= x^2 - 7x + 10$$

Factor $x^2 - 9x - 36$.

Because the constant term is negative, the binomial constants will have opposite signs.

Find two factors of -36 whose sum is -9.

Factors	Sum
$+1, -36$	-35
$-1, +36$	35
$+2, -18$	-16
$-2, +18$	16
$+3, -12$	-9

Write the trinomial as a product of its factors. $(x + 3)(x - 12)$

In this example, once the correct factors are found, it is not necessary to try the remaining factors.

Factor $x^2 + 7x + 8$.

Because the constant term is positive and the coefficient of x is positive, the binomial constants will be positive.

Find two positive factors of 8 whose sum is 7.

Factors	Sum
1, 8	8
2, 4	6

There are no positive integer factors of 8 whose sum is 7. The trinomial $x^2 + 7x + 8$ is said to be **nonfactorable over the integers.** Just as 17 is a prime number, $x^2 + 7x + 8$ is a **prime polynomial.** Binomials of the form $x - a$ and $x + a$ are also prime polynomials.

Example 1

Factor $x^2 - 8x + 15$.

Solution

Find two negative factors of 15 whose sum is -8.

Factors	Sum
$-1, -15$	-16
$-3, -5$	-8

$x^2 - 8x + 15 = (x - 3)(x - 5)$

Example 2

Factor $x^2 + 9x + 20$.

Your solution

$(x + 4)(x + 5)$

Example 3

Factor $x^2 + 6x - 27$.

Solution

Find two factors of -27 whose sum is 6.

Factors	Sum
$+1, -27$	-26
$-1, +27$	26
$+3, -9$	-6
$-3, +9$	6

$x^2 + 6x - 27 = (x - 3)(x + 9)$

Example 4

Factor $x^2 + 7x - 18$.

Your solution

$(x + 9)(x - 2)$

Solutions on p. A19

Objective B   **To factor completely**

A polynomial is factored completely when it is written as a product of factors that are nonfactorable over the integers.

Factor $3x^3 + 15x^2 + 18x$.

Find the GCF of the terms of the polynomial.

The GCF is $3x$.

Factor out the GCF.

$3x^3 + 15x^2 + 18x =$

$3x(x^2) + 3x(5x) + 3x(6) =$ Do this step mentally.

$3x(x^2 + 5x + 6)$

Factor the trinomial. Find two positive factors of 6 whose sum is 5.

Factors	Sum
$+1, +6$	7
$+2, +3$	5

Write the product of the GCF and the binomial factors of the trinomial.

$3x(x + 2)(x + 3)$

Check: $3x(x + 2)(x + 3) = 3x(x^2 + 3x + 2x + 6)$
$= 3x(x^2 + 5x + 6)$
$= 3x^3 + 15x^2 + 18x$

Factor $x^2 + 9xy + 20y^2$.

The terms have no common factor.
There are two variables.
Find two positive factors of 20
whose sum is 9.

Factors	Sum
+1, +20	21
+2, +10	12
+4, +5	9

Write the factors of the trinomial. $x^2 + 9xy + 20y^2 = (x + 4y)(x + 5y)$

Check: $(x + 4y)(x + 5y) = x^2 + 5xy + 4xy + 20y^2$
$= x^2 + 9xy + 20y^2$

Example 5

Factor $2x^2y + 12xy - 14y$.

Solution

The GCF is $2y$.
$2x^2y + 12xy - 14y = 2y(x^2 + 6x - 7)$
Factor the trinomial.
Find two factors
of -7 whose sum is 6.

Factors	Sum
+1, −7	−6
−1, +7	6

$2x^2y + 12xy - 14y = 2y(x + 7)(x - 1)$

Example 6

Factor $3a^2b - 18ab - 81b$.

Your solution

$3b(a + 3)(a - 9)$

Example 7

Factor $4x^2 - 40xy + 84y^2$.

Solution

The GCF is 4.
$4x^2 - 40xy + 84y^2 = 4(x^2 - 10xy + 21y^2)$
Factor the trinomial.
Find two negative
factors of 21
whose sum is -10.

Factors	Sum
−1, −21	−22
−3, −7	−10

$4x^2 - 40xy + 84y^2 = 4(x - 3y)(x - 7y)$

Example 8

Factor $3x^2 - 9xy - 12y^2$.

Your solution

$3(x + y)(x - 4y)$

Solutions on p. A19

Content and Format © 1991 HMCo.

6.2   EXERCISES

▶ **Objective A**

Factor.

1. $x^2 + 3x + 2$
$(x + 1)(x + 2)$

2. $x^2 + 5x + 6$
$(x + 2)(x + 3)$

3. $x^2 - x - 2$
$(x + 1)(x - 2)$

4. $x^2 + x - 6$
$(x + 3)(x - 2)$

5. $a^2 + a - 12$
$(a + 4)(a - 3)$

6. $a^2 - 2a - 35$
$(a + 5)(a - 7)$

7. $a^2 - 3a + 2$
$(a - 1)(a - 2)$

8. $a^2 - 5a + 4$
$(a - 1)(a - 4)$

9. $a^2 + a - 2$
$(a + 2)(a - 1)$

10. $a^2 - 2a - 3$
$(a + 1)(a - 3)$

11. $b^2 - 6b + 9$
$(b - 3)(b - 3)$

12. $b^2 + 8b + 16$
$(b + 4)(b + 4)$

13. $b^2 + 7b - 8$
$(b + 8)(b - 1)$

14. $y^2 - y - 6$
$(y + 2)(y - 3)$

15. $y^2 + 6y - 55$
$(y + 11)(y - 5)$

16. $z^2 - 4z - 45$
$(z + 5)(z - 9)$

17. $y^2 - 5y + 6$
$(y - 2)(y - 3)$

18. $y^2 - 8y + 15$
$(y - 3)(y - 5)$

19. $z^2 - 14z + 45$
$(z - 5)(z - 9)$

20. $z^2 - 14z + 49$
$(z - 7)(z - 7)$

21. $z^2 - 12z - 160$
$(z + 8)(z - 20)$

22. $p^2 + 2p - 35$
$(p + 7)(p - 5)$

23. $p^2 + 12p + 27$
$(p + 3)(p + 9)$

24. $p^2 - 6p + 8$
$(p - 2)(p - 4)$

25. $x^2 + 20x + 100$
$(x + 10)(x + 10)$

26. $x^2 + 18x + 81$
$(x + 9)(x + 9)$

27. $b^2 + 9b + 20$
$(b + 4)(b + 5)$

28. $b^2 + 13b + 40$
$(b + 5)(b + 8)$

29. $x^2 - 11x - 42$
$(x + 3)(x - 14)$

30. $x^2 + 9x - 70$
$(x + 14)(x - 5)$

31. $b^2 - b - 20$
$(b + 4)(b - 5)$

32. $b^2 + 3b - 40$
$(b + 8)(b - 5)$

33. $y^2 - 14y - 51$
$(y + 3)(y - 17)$

34. $y^2 - y - 72$
$(y + 8)(y - 9)$

35. $p^2 - 4p - 21$
$(p + 3)(p - 7)$

36. $p^2 + 16p + 39$
$(p + 3)(p + 13)$

37. $y^2 - 8y + 32$
Nonfactorable over the
integers

38. $y^2 - 9y + 81$
Nonfactorable over the
integers

39. $x^2 - 20x + 75$
$(x - 5)(x - 15)$

Factor.

40. $p^2 + 24p + 63$
$(p + 3)(p + 21)$

41. $x^2 - 15x + 56$
$(x - 7)(x - 8)$

42. $x^2 + 21x + 38$
$(x + 2)(x + 19)$

43. $x^2 + x - 56$
$(x + 8)(x - 7)$

44. $x^2 + 5x - 36$
$(x + 9)(x - 4)$

45. $a^2 - 21a - 72$
$(a + 3)(a - 24)$

46. $a^2 - 7a - 44$
$(a + 4)(a - 11)$

47. $a^2 - 15a + 36$
$(a - 3)(a - 12)$

48. $a^2 - 21a + 54$
$(a - 3)(a - 18)$

49. $z^2 - 9z - 136$
$(z + 8)(z - 17)$

50. $z^2 + 14z - 147$
$(z + 21)(z - 7)$

51. $c^2 - c - 90$
$(c + 9)(c - 10)$

52. $c^2 - 3c - 180$
$(c + 12)(c - 15)$

53. $z^2 + 15z + 44$
$(z + 4)(z + 11)$

54. $p^2 + 24p + 135$
$(p + 9)(p + 15)$

55. $c^2 + 19c + 34$
$(c + 2)(c + 17)$

56. $c^2 + 11c + 18$
$(c + 2)(c + 9)$

57. $x^2 - 4x - 96$
$(x + 8)(x - 12)$

58. $x^2 + 10x - 75$
$(x + 15)(x - 5)$

59. $x^2 - 22x + 112$
$(x - 8)(x - 14)$

60. $x^2 + 21x - 100$
$(x + 25)(x - 4)$

61. $b^2 + 8b - 105$
$(b + 15)(b - 7)$

62. $b^2 - 22b + 72$
$(b - 4)(b - 18)$

63. $a^2 - 9a - 36$
$(a + 3)(a - 12)$

64. $a^2 + 42a - 135$
$(a + 45)(a - 3)$

65. $b^2 - 23b + 102$
$(b - 6)(b - 17)$

66. $b^2 - 25b + 126$
$(b - 7)(b - 18)$

67. $a^2 + 27a + 72$
$(a + 3)(a + 24)$

68. $z^2 + 24z + 144$
$(z + 12)(z + 12)$

69. $x^2 + 25x + 156$
$(x + 12)(x + 13)$

70. $x^2 - 29x + 100$
$(x - 4)(x - 25)$

71. $x^2 - 10x - 96$
$(x + 6)(x - 16)$

72. $x^2 + 9x - 112$
$(x + 16)(x - 7)$

▶ **Objective B**

Factor.

73. $2x^2 + 6x + 4$
$2(x + 1)(x + 2)$

74. $3x^2 + 15x + 18$
$3(x + 2)(x + 3)$

75. $3a^2 + 3a - 18$
$3(a + 3)(a - 2)$

76. $4x^2 - 4x - 8$
$4(x + 1)(x - 2)$

77. $ab^2 + 2ab - 15a$
$a(b + 5)(b - 3)$

78. $ab^2 + 7ab - 8a$
$a(b + 8)(b - 1)$

79. $xy^2 - 5xy + 6x$
$x(y - 2)(y - 3)$

80. $xy^2 + 8xy + 15x$
$x(y + 3)(y + 5)$

81. $z^3 - 7z^2 + 12z$
$z(z - 3)(z - 4)$

82. $2a^3 + 6a^2 + 4a$
$2a(a + 1)(a + 2)$

83. $3y^3 - 15y^2 + 18y$
$3y(y - 2)(y - 3)$

84. $4y^3 + 12y^2 - 72y$
$4y(y + 6)(y - 3)$

85. $3x^2 + 3x - 36$
$3(x + 4)(x - 3)$

86. $2x^3 - 2x^2 + 4x$
$2x(x + 1)(x - 2)$

87. $5z^2 - 15z - 140$
$5(z + 4)(z - 7)$

88. $6z^2 + 12z - 90$
$6(z + 5)(z - 3)$

89. $2a^3 + 8a^2 - 64a$
$2a(a + 8)(a - 4)$

90. $3a^3 - 9a^2 - 54a$
$3a(a + 3)(a - 6)$

91. $x^2 - 5xy + 6y^2$
$(x - 2y)(x - 3y)$

92. $x^2 + 4xy - 21y^2$
$(x + 7y)(x - 3y)$

93. $a^2 - 9ab + 20b^2$
$(a - 4b)(a - 5b)$

94. $a^2 - 15ab + 50b^2$
$(a - 5b)(a - 10b)$

95. $x^2 - 3xy - 28y^2$
$(x + 4y)(x - 7y)$

96. $s^2 + 2st - 48t^2$
$(s + 8t)(s - 6t)$

97. $y^2 - 15yz - 41z^2$
Nonfactorable over the
integers

98. $y^2 + 85yz + 36z^2$
Nonfactorable over the
integers

99. $z^4 - 12z^3 + 35z^2$
$z^2(z - 5)(z - 7)$

100. $z^4 + 2z^3 - 80z^2$
$z^2(z + 10)(z - 8)$

101. $b^4 - 22b^3 + 120b^2$
$b^2(b - 10)(b - 12)$

102. $b^4 - 3b^3 - 10b^2$
$b^2(b + 2)(b - 5)$

103. $2y^4 - 26y^3 - 96y^2$
$2y^2(y + 3)(y - 16)$

104. $3y^4 + 54y^3 + 135y^2$
$3y^2(y + 3)(y + 15)$

105. $x^4 + 7x^3 - 8x^2$
$x^2(x + 8)(x - 1)$

106. $x^4 - 11x^3 - 12x^2$
$x^2(x + 1)(x - 12)$

107. $4x^2y + 20xy - 56y$
$4y(x + 7)(x - 2)$

108. $3x^2y - 6xy - 45y$
$3y(x + 3)(x - 5)$

Factor.

109. $8y^2 - 32y + 24$
$8(y - 1)(y - 3)$

110. $10y^2 - 100y + 90$
$10(y - 1)(y - 9)$

111. $c^3 + 13c^2 + 30c$
$c(c + 3)(c + 10)$

112. $c^3 + 18c^2 - 40c$
$c(c + 20)(c - 2)$

113. $3x^3 - 36x^2 + 81x$
$3x(x - 3)(x - 9)$

114. $4x^3 + 4x^2 - 24x$
$4x(x + 3)(x - 2)$

115. $x^2 - 8xy + 15y^2$
$(x - 3y)(x - 5y)$

116. $y^2 - 7xy - 8x^2$
$(y + x)(y - 8x)$

117. $a^2 - 13ab + 42b^2$
$(a - 6b)(a - 7b)$

118. $y^2 + 4yz - 21z^2$
$(y + 7z)(y - 3z)$

119. $y^2 + 8yz + 7z^2$
$(y + z)(y + 7z)$

120. $y^2 - 16yz + 15z^2$
$(y - z)(y - 15z)$

121. $3x^2y + 60xy - 63y$
$3y(x + 21)(x - 1)$

122. $4x^2y - 68xy - 72y$
$4y(x + 1)(x - 18)$

123. $3x^3 + 3x^2 - 36x$
$3x(x + 4)(x - 3)$

124. $4x^3 + 12x^2 - 160x$
$4x(x + 8)(x - 5)$

125. $4z^3 + 32z^2 - 132z$
$4z(z + 11)(z - 3)$

126. $5z^3 - 50z^2 - 120z$
$5z(z + 2)(z - 12)$

127. $4x^3 + 8x^2 - 12x$
$4x(x + 3)(x - 1)$

128. $5x^3 + 30x^2 + 40x$
$5x(x + 2)(x + 4)$

129. $5p^2 + 25p - 420$
$5(p + 12)(p - 7)$

130. $4p^2 - 28p - 480$
$4(p + 8)(p - 15)$

131. $p^4 + 9p^3 - 36p^2$
$p^2(p + 12)(p - 3)$

132. $p^4 + p^3 - 56p^2$
$p^2(p + 8)(p - 7)$

133. $t^2 - 12ts + 35s^2$
$(t - 5s)(t - 7s)$

134. $a^2 - 10ab + 25b^2$
$(a - 5b)(a - 5b)$

135. $a^2 - 8ab - 33b^2$
$(a + 3b)(a - 11b)$

136. $x^2 + 4xy - 60y^2$
$(x + 10y)(x - 6y)$

137. $5x^4 - 30x^3 + 40x^2$
$5x^2(x - 2)(x - 4)$

138. $6x^3 - 6x^2 - 120x$
$6x(x + 4)(x - 5)$

139. $15ab^2 + 45ab - 60a$
$15a(b + 4)(b - 1)$

140. $20a^2b - 100ab + 120b$
$20b(a - 2)(a - 3)$

141. $3x^2y + 36xy - 135y$
$3y(x + 15)(x - 3)$

142. $4yz^2 - 52yz + 88y$
$4y(z - 2)(z - 11)$

SECTION 6.3   Factoring Polynomials of the Form $ax^2 + bx + c$

Objective A

To factor a trinomial of the form $ax^2 + bx + c$ by using trial factors

Trinomials of the form $ax^2 + bx + c$, where a, b, and c are integers, are shown at the right.

$3x^2 - x + 4$; $a = 3$, $b = -1$, $c = 4$
$6x^2 + 2x - 3$; $a = 6$, $b = 2$, $c = -3$

These trinomials differ from those in the previous section in that the coefficient of x^2 is not 1. There are various methods of factoring these trinomials. The method described in this objective is factoring polynomials using trial factors.

To reduce the number of trial factors that must be considered, remember the following:

1. Use the signs of the constant term and the coefficient of x in the trinomial to determine the signs of the binomial factors. If the constant term is positive, the signs of the binomial factors will be the same as the sign of the coefficient of x in the trinomial. If the sign of the constant term is negative, the constant terms in the binomials have opposite signs.

2. If the terms of the trinomial do not have a common factor, then the terms of neither of the binomial factors will have a common factor.

Factor $2x^2 - 7x + 3$.

The terms have no common factor. The constant term is positive. The coefficient of x is negative. The binomial constants will be negative.

Factors of 2 (coefficient of x^2)	Factors of 3 (constant term)
1, 2	-1, -3

Write trial factors. Use the **O**uter and **I**nner products of FOIL to determine the middle term, $-7x$, of the trinomial.

Trial Factors	Middle Term
$(x - 1)(2x - 3)$	$-3x - 2x = -5x$
$(x - 3)(2x - 1)$	$-x - 6x = -7x$

Write the factors of the trinomial.

$2x^2 - 7x + 3 = (x - 3)(2x - 1)$

Factor $15 - 2x - x^2$.

The terms have no common factor. The coefficient of x^2 is -1.

Factors of 15 (constant term)	Factors of -1 (coefficient of x^2)
1, 15	1, -1
3, 5	

Write trial factors. Use the **O**uter and **I**nner products of FOIL to determine the middle term, $-2x$, of the trinomial.

Trial Factors	Middle Term
$(1 + x)(15 - x)$	$-x + 15x = 14x$
$(1 - x)(15 + x)$	$x - 15x = -14x$
$(3 + x)(5 - x)$	$-3x + 5x = 2x$
$(3 - x)(5 + x)$	$3x - 5x = -2x$

Write the factors of the trinomial.

$15 - 2x - x^2 = (3 - x)(5 + x)$

Factor $6x^3 + 14x - 12x$

Factor the GCF, $2x$, from the terms.	$6x^3 + 14x^2 - 12x = 2x(3x^2 + 7x - 6)$

Factor the trinomial. The constant term is negative. The binomial constants will have opposite signs.

Factors of 3	Factors of -6
1, 3	$-1,\ \ 6$
	$1, -6$
	$-2,\ \ 3$
	$2, -3$

Write trial factors. Use the **O**uter and **I**nner products of FOIL to determine the middle term, $7x$, of the trinomial.

It is not necessary to test trial factors that have a common factor.

Trial Factors	Middle Term
$(x - 1)(3x + 6)$	Common factor
$(x + 6)(3x - 1)$	$-x + 18x = 17x$
$(x + 1)(3x - 6)$	Common factor
$(x - 6)(3x + 1)$	$x - 18x = -17x$
$(x - 2)(3x + 3)$	Common factor
$(x + 3)(3x - 2)$	$-2x + 9x = 7x$
$(x + 2)(3x - 3)$	Common factor
$(x - 3)(3x + 2)$	$2x - 9x = -7x$

Write the factors of the trinomial.

$6x^3 + 14x^2 - 12x = 2x(x + 3)(3x - 2)$

For this example, all the trial factors were listed. Once the correct factors have been found, the remaining trial factors can be omitted. For the examples and solutions in this text, all trial factors except those that have a common factor will be listed.

Example 1

Factor $3x^2 + x - 2$.

Solution

Factors of 3: 1, 3   Factors of -2:   1, -2
                                            $-1, 2$

Trial Factors	Middle Term
$(1x + 1)(3x - 2)$	$-2x + 3x = x$
$(1x - 2)(3x + 1)$	$x - 6x = -5x$
$(1x - 1)(3x + 2)$	$2x - 3x = -x$
$(1x + 2)(3x - 1)$	$-x + 6x = 5x$

$3x^2 + x - 2 = (x + 1)(3x - 2)$

Example 2

Factor $2x^2 - x - 3$.

Your solution

$(x + 1)(2x - 3)$

Example 3

Factor $12x - 32x^2 - 12x^3$.

Solution

The GCF is $4x$.

$12x - 32x^2 - 12x^3 = 4x(3 - 8x - 3x^2)$
Factor the trinomial.

Factors of 3: 1, 3   Factors of -3:   1, -3
                                            $-1, 3$

Trial Factors	Middle Term
$(1 + 3x)(3 - 1x)$	$-x + 9x = 8x$
$(1 - 3x)(3 + 1x)$	$x - 9x = -8x$

$12x - 32x^2 - 12x^3 = 4x(1 - 3x)(3 + x)$

Example 4

Factor $12y + 12y^2 - 45y^3$.

Your solution

$3y(2 - 3y)(2 + 5y)$

Solutions on p. A20

| **Objective B** | **To factor a trinomial of the form $ax^2 + bx + c$ by grouping** |

In the previous objective, trinomials of the form $ax^2 + bx + c$ were factored by using trial factors. In this objective, these trinomials will be factored by factoring by grouping.

Recall that to factor $x^2 - 6x + 8$, first find two negative factors of 8 whose sum is -6. The factors are -2 and -4. Thus,

$$x^2 - 6x + 8 = (x - 2)(x - 4)$$

To factor $ax^2 + bx + c$, first find two factors of $a \cdot c$ whose sum is b. Then use factoring by grouping to write the factorization of the trinomial.

Factor $2x^2 + 13x + 15$.

Find two positive factors of 30 $(2 \cdot 15)$ whose sum is 13.

When the required sum has been found, the remaining factors need not be checked.

Positive Factors of 30	Sum
1, 30	31
2, 15	17
3, 10	13
5, 6	11

Use the factors of 30 whose sum is 13 to write $13x$ as $3x + 10x$. Factor by grouping.

$$
\begin{aligned}
2x^2 + 13x + 15 &= 2x^2 + 3x + 10x + 15 \\
&= (2x^2 + 3x) + (10x + 15) \\
&= x(2x + 3) + 5(2x + 3) \\
&= (2x + 3)(x + 5)
\end{aligned}
$$

Check your answer.

$$
\begin{aligned}
(2x + 3)(x + 5) &= 2x^2 + 10x + 3x + 15 \\
&= 2x^2 + 13x + 15
\end{aligned}
$$

Factor $6x^2 - 11x - 10$.

Find two factors of -60 $[6 \cdot (-10)]$ whose sum is -11.

Factors of -60	Sum
1, -60	-59
-1, 60	59
2, -30	-28
-2, 30	28
3, -20	-17
-3, 20	17
4, -15	-11

Use the factors of -60 whose sum is -11 to write $-11x$ as $4x - 15x$. Factor by grouping. Recall that $-15x - 10 = -(15x + 10)$.

$$
\begin{aligned}
6x^2 - 11x - 10 &= 6x^2 + 4x - 15x - 10 \\
&= (6x^2 + 4x) - (15x + 10) \\
&= 2x(3x + 2) - 5(3x + 2) \\
&= (3x + 2)(2x - 5)
\end{aligned}
$$

Factor $3x^2 - 2x - 4$.

Find two factors of -12 $[3 \cdot (-4)]$ whose sum is -2.

Because no integer factors of -12 have a sum of -2, $3x^2 - 2x - 4$ is **nonfactorable over the integers**. $3x^2 - 2x - 4$ is a **prime polynomial**.

Factors of -12	Sum
1, -12	-11
-1, 12	11
2, -6	-4
-2, 6	4
3, -4	-1
-3, 4	1

Example 5

Factor $2x^2 + 19x - 10$.

Solution

Factors of -20 $[2(-10)]$	Sum
-1, 20	19

$$\begin{aligned} 2x^2 + 19x - 10 &= 2x^2 - x + 20x - 10 \\ &= (2x^2 - x) + (20x - 10) \\ &= x(2x - 1) + 10(2x - 1) \\ &= (2x - 1)(x + 10) \end{aligned}$$

Example 6

Factor $2a^2 + 13a - 7$.

Your solution

$(2a - 1)(a + 7)$

Example 7

Factor $24x^2y - 76xy + 40y$.

Solution

The GCF is $4y$.
$24x^2y - 76xy + 40y = 4y(6x^2 - 19x + 10)$

Negative factors of 60 $[6(10)]$	Sum
-1, -60	-61
-2, -30	-32
-3, -20	-23
-4, -15	-19

$$\begin{aligned} 6x^2 - 19x + 10 &= 6x^2 - 4x - 15x + 10 \\ &= (6x^2 - 4x) - (15x - 10) \\ &= 2x(3x - 2) - 5(3x - 2) \\ &= (3x - 2)(2x - 5) \end{aligned}$$

$$\begin{aligned} 24x^2y - 76xy + 40y &= 4y(6x^2 - 19x + 10) \\ &= 4y(3x - 2)(2x - 5) \end{aligned}$$

Example 8

Factor $15x^3 + 40x^2 - 80x$.

Your solution

$5x(3x - 4)(x + 4)$

Solutions on p. A20

Content and Format © 1991 HMCo.

6.3   EXERCISES

▶ **Objective A**

Factor by using trial factors.

1. $2x^2 + 3x + 1$
$(x + 1)(2x + 1)$

2. $5x^2 + 6x + 1$
$(x + 1)(5x + 1)$

3. $2y^2 + 7y + 3$
$(y + 3)(2y + 1)$

4. $3y^2 + 7y + 2$
$(y + 2)(3y + 1)$

5. $2a^2 - 3a + 1$
$(a - 1)(2a - 1)$

6. $3a^2 - 4a + 1$
$(a - 1)(3a - 1)$

7. $2b^2 - 11b + 5$
$(b - 5)(2b - 1)$

8. $3b^2 - 13b + 4$
$(b - 4)(3b - 1)$

9. $2x^2 + x - 1$
$(x + 1)(2x - 1)$

10. $4x^2 - 3x - 1$
$(x - 1)(4x + 1)$

11. $2x^2 - 5x - 3$
$(x - 3)(2x + 1)$

12. $3x^2 + 5x - 2$
$(x + 2)(3x - 1)$

13. $2t^2 - t - 10$
$(t + 2)(2t - 5)$

14. $2t^2 + 5t - 12$
$(t + 4)(2t - 3)$

15. $3p^2 - 16p + 5$
$(p - 5)(3p - 1)$

16. $6p^2 + 5p + 1$
$(2p + 1)(3p + 1)$

17. $12y^2 - 7y + 1$
$(3y - 1)(4y - 1)$

18. $6y^2 - 5y + 1$
$(2y - 1)(3y - 1)$

19. $6z^2 - 7z + 3$
Nonfactorable over the integers

20. $9z^2 + 3z + 2$
Nonfactorable over the integers

21. $6t^2 - 11t + 4$
$(2t - 1)(3t - 4)$

22. $10t^2 + 11t + 3$
$(2t + 1)(5t + 3)$

23. $8x^2 + 33x + 4$
$(x + 4)(8x + 1)$

24. $7x^2 + 50x + 7$
$(x + 7)(7x + 1)$

25. $5x^2 - 62x - 7$
Nonfactorable over the integers

26. $9x^2 - 13x - 4$
Nonfactorable over the integers

27. $12y^2 + 19y + 5$
$(3y + 1)(4y + 5)$

28. $5y^2 - 22y + 8$
$(y - 4)(5y - 2)$

29. $7a^2 + 47a - 14$
$(a + 7)(7a - 2)$

30. $11a^2 - 54a - 5$
$(a - 5)(11a + 1)$

31. $3b^2 - 16b + 16$
$(b - 4)(3b - 4)$

32. $6b^2 - 19b + 15$
$(2b - 3)(3b - 5)$

33. $2z^2 - 27z - 14$
$(z - 14)(2z + 1)$

34. $4z^2 + 5z - 6$
$(z + 2)(4z - 3)$

35. $3p^2 + 22p - 16$
$(p + 8)(3p - 2)$

36. $7p^2 + 19p + 10$
$(p + 2)(7p + 5)$

Factor by using trial factors.

37. $4x^2 + 6x + 2$
$2(x + 1)(2x + 1)$

38. $12x^2 + 33x - 9$
$3(x + 3)(4x - 1)$

39. $15y^2 - 50y + 35$
$5(y - 1)(3y - 7)$

40. $30y^2 + 10y - 20$
$10(y + 1)(3y - 2)$

41. $2x^3 - 11x^2 + 5x$
$x(x - 5)(2x - 1)$

42. $2x^3 - 3x^2 - 5x$
$x(x + 1)(2x - 5)$

43. $3a^2b - 16ab + 16b$
$b(a - 4)(3a - 4)$

44. $2a^2b - ab - 21b$
$b(a + 3)(2a - 7)$

45. $3z^2 + 95z + 10$
Nonfactorable over
the integers

46. $8z^2 - 36z + 1$
Nonfactorable over
the integers

47. $12 - x - x^2$
$(3 - x)(4 + x)$

48. $2 + x - x^2$
$(1 + x)(2 - x)$

49. $80y^2 - 36y + 4$
$4(4y - 1)(5y - 1)$

50. $24y^2 - 24y - 18$
$6(2y + 1)(2y - 3)$

51. $8z^3 + 14z^2 + 3z$
$z(2z + 3)(4z + 1)$

52. $6z^3 - 23z^2 + 20z$
$z(2z - 5)(3z - 4)$

53. $6x^2y - 11xy - 10y$
$y(2x - 5)(3x + 2)$

54. $8x^2y - 27xy + 9y$
$y(x - 3)(8x - 3)$

55. $10t^2 - 5t - 50$
$5(t + 2)(2t - 5)$

56. $16t^2 + 40t - 96$
$8(t + 4)(2t - 3)$

57. $3p^3 - 16p^2 + 5p$
$p(p - 5)(3p - 1)$

58. $6p^3 + 5p^2 + p$
$p(2p + 1)(3p + 1)$

59. $26z^2 + 98z - 24$
$2(z + 4)(13z - 3)$

60. $30z^2 - 87z + 30$
$3(2z - 5)(5z - 2)$

61. $10y^3 - 44y^2 + 16y$
$2y(y - 4)(5y - 2)$

62. $14y^3 + 94y^2 - 28y$
$2y(y + 7)(7y - 2)$

63. $4yz^3 + 5yz^2 - 6yz$
$yz(z + 2)(4z - 3)$

64. $12a^3 + 14a^2 - 48a$
$2a(2a - 3)(3a + 8)$

65. $42a^3 + 45a^2 - 27a$
$3a(2a + 3)(7a - 3)$

66. $36p^2 - 9p^3 - p^4$
$p^2(3 - p)(12 + p)$

67. $9x^2y - 30xy^2 + 25y^3$
$y(3x - 5y)(3x - 5y)$

68. $8x^2y - 38xy^2 + 35y^3$
$y(2x - 7y)(4x - 5y)$

69. $9x^3y - 24x^2y^2 + 16xy^3$
$xy(3x - 4y)(3x - 4y)$

70. $9x^3y + 12x^2y + 4xy$
$xy(3x + 2)(3x + 2)$

▶ **Objective B**

Factor by grouping.

71. $6x^2 - 17x + 12$
$(2x - 3)(3x - 4)$

72. $15x^2 - 19x + 6$
$(3x - 2)(5x - 3)$

73. $5b^2 + 33b - 14$
$(b + 7)(5b - 2)$

74. $8x^2 - 30x + 25$
$(2x - 5)(4x - 5)$

75. $6a^2 + 7a - 24$
$(2a - 3)(3a + 8)$

76. $14a^2 + 15a - 9$
$(2a + 3)(7a - 3)$

77. $4z^2 + 11z + 6$
$(z + 2)(4z + 3)$

78. $6z^2 - 25z + 14$
$(2z - 7)(3z - 2)$

79. $22p^2 + 51p - 10$
$(2p + 5)(11p - 2)$

80. $14p^2 - 41p + 15$
$(2p - 5)(7p - 3)$

81. $8y^2 + 17y + 9$
$(y + 1)(8y + 9)$

82. $12y^2 - 145y + 12$
$(y - 12)(12y - 1)$

83. $18t^2 - 9t - 5$
$(3t + 1)(6t - 5)$

84. $12t^2 + 28t - 5$
$(2t + 5)(6t - 1)$

85. $6b^2 + 71b - 12$
$(b + 12)(6b - 1)$

86. $8b^2 + 65b + 8$
$(b + 8)(8b + 1)$

87. $9x^2 + 12x + 4$
$(3x + 2)(3x + 2)$

88. $25x^2 - 30x + 9$
$(5x - 3)(5x - 3)$

89. $6b^2 - 13b + 6$
$(2b - 3)(3b - 2)$

90. $20b^2 + 37b + 15$
$(4b + 5)(5b + 3)$

91. $33b^2 + 34b - 35$
$(3b + 5)(11b - 7)$

92. $15b^2 - 43b + 22$
$(3b - 2)(5b - 11)$

93. $18y^2 - 39y + 20$
$(3y - 4)(6y - 5)$

94. $24y^2 + 41y + 12$
$(3y + 4)(8y + 3)$

95. $15a^2 + 26a - 21$
$(3a + 7)(5a - 3)$

96. $6a^2 + 23a + 21$
$(2a + 3)(3a + 7)$

97. $8y^2 - 26y + 15$
$(2y - 5)(4y - 3)$

98. $18y^2 - 27y + 4$
$(3y - 4)(6y - 1)$

99. $8z^2 + 2z - 15$
$(2z + 3)(4z - 5)$

100. $10z^2 + 3z - 4$
$(2z - 1)(5z + 4)$

101. $15x^2 - 82x + 24$
Nonfactorable over
the integers

102. $13z^2 + 49z - 8$
Nonfactorable over
the integers

103. $10z^2 - 29z + 10$
$(2z - 5)(5z - 2)$

104. $15z^2 - 44z + 32$
$(3z - 4)(5z - 8)$

105. $36z^2 + 72z + 35$
$(6z + 5)(6z + 7)$

106. $16z^2 + 8z - 35$
$(4z + 7)(4z - 5)$

107. $3x^2 + xy - 2y^2$
$(x + y)(3x - 2y)$

108. $6x^2 + 10xy + 4y^2$
$2(x + y)(3x + 2y)$

109. $3a^2 + 5ab - 2b^2$
$(a + 2b)(3a - b)$

110. $2a^2 - 9ab + 9b^2$
$(a - 3b)(2a - 3b)$

Factor by grouping.

111. $4y^2 - 11yz + 6z^2$
$(y - 2z)(4y - 3z)$

112. $2y^2 + 7yz + 5z^2$
$(y + z)(2y + 5z)$

113. $28 + 3z - z^2$
$(7 - z)(4 + z)$

114. $15 - 2z - z^2$
$(3 - z)(5 + z)$

115. $8 - 7x - x^2$
$(1 - x)(8 + x)$

116. $12 + 11x - x^2$
$(1 + x)(12 - x)$

117. $9x^2 + 33x - 60$
$3(x + 5)(3x - 4)$

118. $16x^2 - 16x - 12$
$4(2x + 1)(2x - 3)$

119. $24x^2 - 52x + 24$
$4(2x - 3)(3x - 2)$

120. $60x^2 + 95x + 20$
$5(3x + 4)(4x + 1)$

121. $35a^4 + 9a^3 - 2a^2$
$a^2(5a + 2)(7a - 1)$

122. $15a^4 + 26a^3 + 7a^2$
$a^2(3a + 1)(5a + 7)$

123. $15b^2 - 115b + 70$
$5(b - 7)(3b - 2)$

124. $25b^2 + 35b - 30$
$5(b + 2)(5b - 3)$

125. $3x^2 - 26xy + 35y^2$
$(x - 7y)(3x - 5y)$

126. $4x^2 + 16xy + 15y^2$
$(2x + 3y)(2x + 5y)$

127. $216y^2 - 3y - 3$
$3(8y - 1)(9y + 1)$

128. $360y^2 + 4y - 4$
$4(9y + 1)(10y - 1)$

129. $21 - 20x - x^2$
$(1 - x)(21 + x)$

130. $18 + 17x - x^2$
$(1 + x)(18 - x)$

131. $15a^2 + 11ab - 14b^2$
$(3a - 2b)(5a + 7b)$

132. $15a^2 - 31ab + 10b^2$
$(3a - 5b)(5a - 2b)$

133. $33z - 8z^2 - z^3$
$z(3 - z)(11 + z)$

134. $24z + 10z^2 - z^3$
$z(2 + z)(12 - z)$

135. $10x^3 + 12x^2 + 2x$
$2x(x + 1)(5x + 1)$

136. $9x^3 - 39x^2 + 12x$
$3x(x - 4)(3x - 1)$

137. $2yz^3 - 17yz^2 + 8yz$
$yz(z - 8)(2z - 1)$

138. $20b^4 + 41b^3 + 20b^2$
$b^2(4b + 5)(5b + 4)$

139. $6b^4 - 13b^3 + 6b^2$
$b^2(2b - 3)(3b - 2)$

140. $9a^3b - 9a^2b^2 - 10ab^3$
$ab(3a + 2b)(3a - 5b)$

141. $2a^3b - 11a^2b^2 + 5ab^3$
$ab(a - 5b)(2a - b)$

SECTION 6.4   Special Factoring

Objective A

To factor the difference of two squares and perfect square trinomials

Recall from Objective D in Section 5.3 that the product of the sum and difference of the same terms equals the square of the first term minus the square of the second term.

Sum and difference of two terms   Difference of two squares

$$(a + b)(a - b) \qquad = \qquad a^2 - b^2$$

This suggests that the difference of two squares can be factored as

$$a^2 - b^2 = (a + b)(a - b)$$

The polynomial $x^2 + y^2$ is the sum of two squares. The sum of two squares is nonfactorable over the integers.

Factor $x^2 - 16$.

Write $x^2 - 16$ as the difference
of two squares. Then factor.

$$\begin{aligned} x^2 - 16 &= (x)^2 - (4)^2 \\ &= (x - 4)(x + 4) \end{aligned}$$
$$\text{Check:} \quad (x - 4)(x + 4) = x^2 + 4x - 4x - 16 \\ = x^2 - 16$$

Factor $8x^3 - 18x$

The GCF is $2x$.
Factor the difference of two squares.
You should check the factorization.

$$\begin{aligned} 8x^3 - 18x &= 2x(4x^2 - 9) \\ &= 2x[(2x)^2 - 3^2] \\ &= 2x(2x - 3)(2x + 3) \end{aligned}$$

Factor $x^2 - 10$.

Because 10 cannot be written as the square of an integer, $x^2 - 10$ is nonfactorable over the integers.

Recall from Objective D in Section 5.3 the relationship between the terms of the trinomial and the terms of the binomial.

The Square of a Binomial

$$(a + b)^2 = (a + b)(a + b) = a^2 + ab + ab + b^2$$
$$= a^2 + 2ab + b^2$$

Square of first term
Twice the product of the two terms
Square of the last term

This relationship is used to factor a perfect square trinomial.

Factor $4x^2 - 20x + 25$.

Because the first and last terms are squares $[(2x)^2 = 4x^2;\ 5^2 = 25]$, try to factor as the square of a binomial. Check the factorization. The factorization is correct.

$4x^2 - 20x + 25 \overset{?}{=} (2x - 5)^2$
Check:
$(2x - 5)^2 = (2x)^2 + 2(2x)(-5) + 5^2$
$\qquad\qquad = 4x^2 - 20x + 25$
$4x^2 - 20x + 25 = (2x - 5)^2$

Factor $4x^2 + 37x + 9$.

Because the first and last terms are squares $[(2x)^2 = 4x^2;\ 3^2 = 9]$, try to factor as the square of a binomial. Check the proposed factorization.

$4x^2 + 37x + 9 \overset{?}{=} (2x + 3)^2$
Check:
$(2x + 3)^2 = (2x)^2 + 2(2x)(3) + 3^2$
$\qquad\qquad = 4x^2 + 12x + 9$

Because $4x^2 + 12x + 9 \neq 4x^2 + 37x + 9$, the proposed factorization is not correct. In this case, the polynomial is not a perfect square trinomial. It may however, still factor. In fact, $4x^2 + 37x + 9 = (4x + 1)(x + 9)$.

Example 1

Factor $16x^2 - y^2$.

Solution

$16x^2 - y^2 = (4x)^2 - y^2 = (4x + y)(4x - y)$

Example 2

Factor $25a^2 - b^2$.

Your solution

$(5a + b)(5a - b)$

Example 3

Factor $z^4 - 16$.

Solution

$z^4 - 16 = (z^2)^2 - 4^2 = (z^2 + 4)(z^2 - 4)$
$\qquad\quad = (z^2 + 4)(z - 2)(z + 2)$

Example 4

Factor $n^4 - 81$.

Your solution

$(n^2 + 9)(n - 3)(n + 3)$

Example 5

Factor $9x^2 - 30x + 25$.

Solution

Because $9x^2 = (3x)^2$, $25 = (-5)^2$, and $-30x = 2(3x)(-5)$, the trinomial is a perfect square trinomial.
$9x^2 - 30x + 25 = (3x - 5)^2$

Example 6

Factor $16y^2 + 8y + 1$.

Your solution

$(4y + 1)^2$

Solutions on p. A21

Example 7

Factor $9x^2 + 16x + 16$.

Solution

Because $9x^2 = (3x)^2$, $16 = 4^2$, and $16x \neq 2(3x)(4)$, the trinomial is not a perfect square trinomial.

Try to factor by another method.
$9x^2 + 16x + 16 = (3x + 4)(x + 4)$

Example 8

Factor $x^2 + 15x + 36$.

Your solution

$(x + 3)(x + 12)$

Example 9

Factor $(r + 2)^2 - 4$.

Solution

$$(r + 2)^2 - 4 = (r + 2)^2 - 2^2$$
$$= (r + 2 - 2)(r + 2 + 2)$$
$$= r(r + 4)$$

Example 10

Factor $(x^2 - 6x + 9) - y^2$.

Your solution

$(x - 3 - y)(x - 3 + y)$

Solutions on p. A21

Objective B   To factor completely

When factoring a polynomial completely, ask the following questions about the polynomial.

1. Is there a common factor? If so, factor out the common factor.
2. Is the polynomial the difference of two perfect squares? If so, factor.
3. Is the polynomial a perfect square trinomial? If so, factor.
4. Is the polynomial a trinomial that is the product of two binomials? If so, factor.
5. Does the polynomial contain four terms? If so, try factoring by grouping.
6. Is each binomial factor a prime polynomial over the integers? If not, factor.

Example 11

Factor $3x^2 - 48$.

Solution

The GCF is 3.

$$3x^2 - 48 = 3(x^2 - 16)$$
$$= 3(x + 4)(x - 4)$$

$$3x^2 - 48 = 3(x + 4)(x - 4)$$

Example 12

Factor $12x^3 - 75x$.

Your solution

$3x(2x + 5)(2x - 5)$

Example 13

Factor $x^3 - 3x^2 - 4x + 12$

Solution

Factor by grouping.

$$x^3 - 3x^2 - 4x + 12 = (x^3 - 3x^2) - (4x - 12)$$
$$= x^2(x - 3) - 4(x - 3)$$
$$= (x - 3)(x^2 - 4)$$
$$= (x - 3)(x + 2)(x - 2)$$

$$x^3 - 3x^2 - 4x + 12 = (x - 3)(x + 2)(x - 2)$$

Example 14

Factor $a^2b - 7a^2 - b + 7$

Your solution

$(b - 7)(a + 1)(a - 1)$

Example 15

Factor $4x^2y^2 + 12xy^2 + 9y^2$.

Solution

The GCF is y^2.

$$4x^2y^2 + 12xy^2 + 9y^2 = y^2(4x^2 + 12x + 9)$$
$$= y^2(2x + 3)^2$$

$$4x^2y^2 + 12xy^2 + 9y^2 = y^2(2x + 3)^2$$

Example 16

Factor $4x^3 + 28x^2 - 120x$.

Your solution

$4x(x + 10)(x - 3)$

Solutions on p. A21

6.4 EXERCISES

▶ **Objective A**

Factor.

1. $x^2 - 4$
$(x + 2)(x - 2)$

2. $x^2 - 9$
$(x + 3)(x - 3)$

3. $a^2 - 81$
$(a + 9)(a - 9)$

4. $a^2 - 49$
$(a + 7)(a - 7)$

5. $y^2 + 2y + 1$
$(y + 1)^2$

6. $y^2 + 14y + 49$
$(y + 7)^2$

7. $a^2 - 2a + 1$
$(a - 1)^2$

8. $x^2 - 12x + 36$
$(x - 6)^2$

9. $4x^2 - 1$
$(2x + 1)(2x - 1)$

10. $9x^2 - 16$
$(3x + 4)(3x - 4)$

11. $x^6 - 9$
$(x^3 + 3)(x^3 - 3)$

12. $y^{12} - 64$
$(y^6 + 8)(y^6 - 8)$

13. $x^2 + 8x - 16$
Nonfactorable
over the integers

14. $z^2 - 18z - 81$
Nonfactorable
over the integers

15. $x^2 + 2xy + y^2$
$(x + y)^2$

16. $x^2 + 6xy + 9y^2$
$(x + 3y)^2$

17. $4a^2 + 4a + 1$
$(2a + 1)^2$

18. $25x^2 + 10x + 1$
$(5x + 1)^2$

19. $9x^2 - 1$
$(3x + 1)(3x - 1)$

20. $1 - 49x^2$
$(1 + 7x)(1 - 7x)$

21. $1 - 64x^2$
$(1 + 8x)(1 - 8x)$

22. $t^2 + 36$
Nonfactorable
over the integers

23. $x^2 + 64$
Nonfactorable
over the integers

24. $64a^2 - 16a + 1$
$(8a - 1)^2$

25. $9a^2 + 6a + 1$
$(3a + 1)^2$

26. $x^4 - y^2$
$(x^2 + y)(x^2 - y)$

27. $b^4 - 16a^2$
$(b^2 + 4a)(b^2 - 4a)$

28. $16b^2 + 8b + 1$
$(4b + 1)^2$

29. $4a^2 - 20a + 25$
$(2a - 5)^2$

30. $4b^2 + 28b + 49$
$(2b + 7)^2$

31. $9a^2 - 42a + 49$
$(3a - 7)^2$

32. $9x^2 - 16y^2$
$(3x + 4y)(3x - 4y)$

33. $25z^2 - y^2$
$(5z + y)(5z - y)$

34. $x^2y^2 - 4$
$(xy + 2)(xy - 2)$

35. $a^2b^2 - 25$
$(ab + 5)(ab - 5)$

36. $16 - x^2y^2$
$(4 + xy)(4 - xy)$

Factor.

37. $25x^2 - 1$
$(5x + 1)(5x - 1)$

38. $25a^2 + 30ab + 9b^2$
$(5a + 3b)^2$

39. $4a^2 - 12ab + 9b^2$
$(2a - 3b)^2$

40. $49x^2 + 28xy + 4y^2$
$(7x + 2y)^2$

41. $4y^2 - 36yz + 81z^2$
$(2y - 9z)^2$

42. $64y^2 - 48yz + 9z^2$
$(8y - 3z)^2$

43. $\left[\dfrac{1}{x^2} - 4\right]$
$\left(\dfrac{1}{x} - 2\right)\left(\dfrac{1}{x} + 2\right)$

44. $\left[\dfrac{9}{a^2} - 16\right]$
$\left(\dfrac{3}{a} - 4\right)\left(\dfrac{3}{a} + 4\right)$

45. $9a^2b^2 - 6ab + 1$
$(3ab - 1)^2$

46. $16x^2y^2 - 24xy + 9$
$(4xy - 3)^2$

▶ **Objective B**

Factor.

47. $8y^2 - 2$
$2(2y - 1)(2y + 1)$

48. $12n^2 - 48$
$12(n - 2)(n + 2)$

49. $3a^3 + 6a^2 + 3a$
$3a(a + 1)^2$

50. $4rs^2 - 4rs + r$
$r(2s - 1)^2$

51. $m^4 - 256$
$(m^2 + 16)(m - 4)(m + 4)$

52. $81 - t^4$
$(9 + t^2)(3 - t)(3 + t)$

53. $9x^2 + 13x + 4$
$(9x + 4)(x + 1)$

54. $x^2 + 10x + 16$
$(x + 2)(x + 8)$

55. $16y^4 + 48y^3 + 36y^2$
$4y^2(2y + 3)^2$

56. $36c^4 - 48c^3 + 16c^2$
$4c^2(3c - 2)^2$

57. $y^8 - 81$
$(y^4 + 9)(y^2 - 3)(y^2 + 3)$

58. $32s^4 - 2$
$2(4s^2 + 1)(2s - 1)(2s + 1)$

59. $25 - 20p + 4p^2$
$(5 - 2p)^2$

60. $9 + 24a + 16a^2$
$(3 + 4a)^2$

61. $(4x - 3)^2 - y^2$
$(4x - 3 - y)(4x - 3 + y)$

62. $(2x + 5)^2 - 25$
$4x(x + 5)$

63. $(x^2 - 4x + 4) - y^2$
$(x - 2 - y)(x - 2 + y)$

64. $(4x^2 + 12x + 9) - 4y^2$
$(2x + 3 - 2y)(2x + 3 + 2y)$

Factor.

65. $5x^2 - 5$
$5(x + 1)(x - 1)$

66. $2x^2 - 18$
$2(x + 3)(x - 3)$

67. $x^3 + 4x^2 + 4x$
$x(x + 2)^2$

68. $y^3 - 10y^2 + 25y$
$y(y - 5)^2$

69. $x^4 + 2x^3 - 35x^2$
$x^2(x + 7)(x - 5)$

70. $a^4 - 11a^3 + 24a^2$
$a^2(a - 3)(a - 8)$

71. $5b^2 + 75b + 180$
$5(b + 3)(b + 12)$

72. $6y^2 - 48y + 72$
$6(y - 2)(y - 6)$

73. $3a^2 + 36a + 10$
Nonfactorable over the integers

74. $5a^2 - 30a + 4$
Nonfactorable over the integers

75. $2x^2y + 16xy - 66y$
$2y(x + 11)(x - 3)$

76. $3a^2b + 21ab - 54b$
$3b(a + 9)(a - 2)$

77. $x^3 - 6x^2 - 5x$
$x(x^2 - 6x - 5)$

78. $b^3 - 8b^2 - 7b$
$b(b^2 - 8b - 7)$

79. $3y^2 - 36$
$3(y^2 - 12)$

80. $3y^2 - 147$
$3(y + 7)(y - 7)$

81. $20a^2 + 12a + 1$
$(2a + 1)(10a + 1)$

82. $12a^2 - 36a + 27$
$3(2a - 3)^2$

83. $x^2y^2 - 7xy^2 - 8y^2$
$y^2(x + 1)(x - 8)$

84. $a^2b^2 + 3a^2b - 88a^2$
$a^2(b + 11)(b - 8)$

85. $10a^2 - 5ab - 15b^2$
$5(a + b)(2a - 3b)$

86. $16x^2 - 32xy + 12y^2$
$4(2x - y)(2x - 3y)$

87. $50 - 2x^2$
$2(5 + x)(5 - x)$

88. $72 - 2x^2$
$2(6 + x)(6 - x)$

89. $a^2b^2 - 10ab^2 + 25b^2$
$b^2(a - 5)^2$

90. $a^2b^2 + 6ab^2 + 9b^2$
$b^2(a + 3)^2$

91. $12a^3b - a^2b^2 - ab^3$
$ab(4a + b)(3a - b)$

92. $2x^3y - 7x^2y^2 + 6xy^3$
$xy(x - 2y)(2x - 3y)$

93. $12a^3 - 12a^2 + 3a$
$3a(2a - 1)^2$

94. $18a^3 + 24a^2 + 8a$
$2a(3a + 2)^2$

95. $243 + 3a^2$
$3(81 + a^2)$

96. $75 + 27y^2$
$3(25 + 9y^2)$

97. $12a^3 - 46a^2 + 40a$
$2a(2a - 5)(3a - 4)$

Factor.

98. $24x^3 - 66x^2 + 15x$
$3x(2x - 5)(4x - 1)$

99. $4a^3 + 20a^2 + 25a$
$a(2a + 5)^2$

100. $2a^3 - 8a^2b + 8ab^2$
$2a(a - 2b)^2$

101. $27a^2b - 18ab + 3b$
$3b(3a - 1)^2$

102. $a^2b^2 - 6ab^2 + 9b^2$
$b^2(a - 3)^2$

103. $48 - 12x - 6x^2$
$6(4 + x)(2 - x)$

104. $21x^2 - 11x^3 - 2x^4$
$x^2(7 + x)(3 - 2x)$

105. $x^4 - x^2y^2$
$x^2(x + y)(x - y)$

106. $b^4 - a^2b^2$
$b^2(b + a)(b - a)$

107. $18a^3 + 24a^2 + 8a$
$2a(3a + 2)^2$

108. $32xy^2 - 48xy + 18x$
$2x(4y - 3)^2$

109. $2b + ab - 6a^2b$
$b(2 - 3a)(1 + 2a)$

110. $20x - 11xy - 3xy^2$
$x(5 + y)(4 - 3y)$

111. $72xy^2 + 48xy + 8x$
$8x(3y + 1)^2$

112. $4x^2y + 8xy + 4y$
$4y(x + 1)^2$

113. $15y^2 - 2xy^2 - x^2y^2$
$y^2(5 + x)(3 - x)$

114. $4x^4 - 38x^3 + 48x^2$
$2x^2(x - 8)(2x - 3)$

115. $3x^2 - 27y^2$
$3(x + 3y)(x - 3y)$

116. $x^4 - 25x^2$
$x^2(x + 5)(x - 5)$

117. $y^3 - 9y$
$y(y + 3)(y - 3)$

118. $a^4 - 16$
$(a^2 + 4)(a + 2)(a - 2)$

119. $15x^4y^2 - 13x^3y^3 - 20x^2y^4$
$x^2y^2(5x + 4y)(3x - 5y)$

120. $45y^2 - 42y^3 - 24y^4$
$3y^2(5 + 2y)(3 - 4y)$

121. $a(2x - 2) + b(2x - 2)$
$2(x - 1)(a + b)$

122. $4a(x - 3) - 2b(x - 3)$
$2(x - 3)(2a - b)$

123. $x^2(x - 2) - (x - 2)$
$(x - 2)(x + 1)(x - 1)$

124. $y^2(a - b) - (a - b)$
$(a - b)(y + 1)(y - 1)$

125. $a(x^2 - 4) + b(x^2 - 4)$
$(x + 2)(x - 2)(a + b)$

126. $x(a^2 - b^2) - y(a^2 - b^2)$
$(a + b)(a - b)(x - y)$

127. $4(x - 5) - x^2(x - 5)$
$(x - 5)(2 + x)(2 - x)$

128. $y^2(a - b) - 9(a - b)$
$(a - b)(y + 3)(y - 3)$

129. $x^2(x - 2) + 4(2 - x)$
$(x + 2)(x - 2)^2$

130. $a(2y^2 - 4) - b(4 - 2y^2)$
$2(y^2 - 2)(a + b)$

SECTION 6.5   Solving Equations

Objective A	To solve equations by factoring	

Recall that the Multiplication Property of Zero states that the product of a number and zero is zero.

If a is a real number, then $a \cdot 0 = 0 \cdot a = 0$.

Consider $x \cdot y = 0$. If this is a true equation, then either $x = 0$ or $y = 0$.

Principle of Zero Products

If the product of two factors is zero, then at least one of the factors must be zero.

$$\text{If } a \cdot b = 0, \text{ then } a = 0 \text{ or } b = 0.$$

The Principle of Zero Products is used to solve an equation.

Solve: $(x - 2)(x - 3) = 0$

$$(x - 2)(x - 3) = 0$$

Let each factor equal zero (the Principle of Zero Products).

$$x - 2 = 0 \qquad x - 3 = 0$$

Rewrite each equation in the form *variable = constant*.

$$x = 2 \qquad x = 3$$

Write the solution.

The solutions are 2 and 3.

Check:

$$(x - 2)(x - 3) = 0 \quad (x - 2)(x - 3) = 0$$
$$(2 - 2)(2 - 3) = 0 \quad (3 - 2)(3 - 3) = 0$$
$$0(-1) = 0 \qquad\qquad (1)(0) = 0$$
$$0 = 0 \qquad\qquad\qquad 0 = 0$$

A true equation   A true equation

An equation of the form $ax^2 + bx + c = 0$, $a \neq 0$, is a **quadratic equation.** A quadratic equation is in **standard form** when the polynomial is in descending order and equal to zero.

$$3x^2 + 2x + 1 = 0$$

$$4x^2 - 3x + 2 = 0$$

Solve: $2x^2 + x = 6$

Write the equation in standard form.

$2x^2 + x = 6$
$2x^2 + x - 6 = 0$

Factor.

$(2x - 3)(x + 2) = 0$

Let each factor equal zero (the Principle of Zero Products).

$2x - 3 = 0 \qquad x + 2 = 0$

Rewrite each equation in the form *variable = constant*.

$2x = 3 \qquad\qquad x = -2$
$x = \dfrac{3}{2}$

$\dfrac{3}{2}$ and -2 check as solutions.

Write the solution.

The solutions are $\dfrac{3}{2}$ and -2.

Example 1

Solve: $x(x - 3) = 0$

Solution

$x(x - 3) = 0$
$x = 0 \qquad\qquad x - 3 = 0$
$\qquad\qquad\qquad\quad x = 3$

The solutions are 0 and 3.

Example 2

Solve: $2x(x + 7) = 0$

Your solution

0, −7

Example 3

Solve: $2x^2 - 50 = 0$

Solution

$2x^2 - 50 = 0$
$2(x^2 - 25) = 0$
$2(x + 5)(x - 5) = 0$
$x + 5 = 0 \qquad x - 5 = 0$
$\quad x = -5 \qquad\quad x = 5$

The solutions are -5 and 5.

Example 4

Solve: $4x^2 - 9 = 0$

Your solution

$\dfrac{3}{2}, -\dfrac{3}{2}$

Example 5

Solve: $(x - 3)(x - 10) = -10$

Solution

$(x - 3)(x - 10) = -10$

Write the equation in standard form.

$x^2 - 13x + 30 = -10$
$x^2 - 13x + 40 = 0$
$(x - 8)(x - 5) = 0$
$x - 8 = 0 \quad x - 5 = 0$
$\quad x = 8 \qquad\quad x = 5$

The solutions are 8 and 5.

Example 6

Solve: $(x + 2)(x - 7) = 52$

Your solution

−6, 11

Solutions on p. A21

Objective B   **To solve application problems**

6

Example 7

The sum of the squares of two consecutive positive even integers is equal to 100. Find the two integers.

Example 8

The sum of the squares of two consecutive positive integers is 61. Find the two integers.

Strategy

First positive even integer: n
Second positive even integer: $n + 2$

The sum of the square of the first positive even integer and the square of the second positive even integer is 100.

Your strategy

Solution

$n^2 + (n + 2)^2 = 100$
$n^2 + n^2 + 4n + 4 = 100$
$2n^2 + 4n + 4 = 100$
$2n^2 + 4n - 96 = 0$
$2(n^2 + 2n - 48) = 0$
$2(n - 6)(n + 8) = 0$

$n - 6 = 0 \quad\quad n + 8 = 0$
$\quad n = 6 \quad\quad\quad n = -8$

Because -8 is not a positive even integer, it is not a solution.

$n = 6$
$n + 2 = 6 + 2 = 8$

The two integers are 6 and 8.

Your solution

5 and 6

Solution on p. A22

Example 9

A stone is thrown into a well with an initial speed of 4 ft/s. The well is 420 ft deep. How many seconds later will the stone hit the bottom of the well? Use the equation $d = vt + 16t^2$, where d is the distance in feet, v is the initial speed, and t is the time in seconds.

Example 10

The length of a rectangle is 4 in. longer than twice the width. The area of the rectangle is 96 in.^{2}. Find the length and width of the rectangle.

Strategy

To find the time for the stone to drop to the bottom of the well, replace the variables d and v by their given values and solve for t.

Your strategy

Solution

$d = vt + 16t^2$

$420 = 4t + 16t^2$

$0 = -420 + 4t + 16t^2$

$16t^2 + 4t - 420 = 0$

$4(4t^2 + t - 105) = 0$

$4(4t + 21)(t - 5) = 0$

$4t + 21 = 0 \qquad t - 5 = 0$

$\quad 4t = -21 \qquad\quad t = 5$

$\quad\ t = -\dfrac{21}{4}$

Because the time cannot be a negative number, $-\frac{21}{4}$ is not a solution.

The time is 5 s.

Your solution

length 16 in.; width 6 in.

Solution on p. A22

6.5 EXERCISES

▶ **Objective A**

Solve.

1. $(y + 3)(y + 2) = 0$
$-3, -2$

2. $(y - 3)(y - 5) = 0$
$3, 5$

3. $(z - 7)(z - 3) = 0$
$7, 3$

4. $(z + 8)(z - 9) = 0$
$-8, 9$

5. $x(x - 5) = 0$
$0, 5$

6. $x(x + 2) = 0$
$0, -2$

7. $a(a - 9) = 0$
$0, 9$

8. $a(a + 12) = 0$
$0, -12$

9. $y(2y + 3) = 0$
$0, -\dfrac{3}{2}$

10. $t(4t - 7) = 0$
$0, \dfrac{7}{4}$

11. $2a(3a - 2) = 0$
$0, \dfrac{2}{3}$

12. $4b(2b + 5) = 0$
$0, -\dfrac{5}{2}$

13. $(b + 2)(b - 5) = 0$
$-2, 5$

14. $(b - 8)(b + 3) = 0$
$8, -3$

15. $x^2 - 81 = 0$
$9, -9$

16. $x^2 - 121 = 0$
$11, -11$

17. $4x^2 - 49 = 0$
$\dfrac{7}{2}, -\dfrac{7}{2}$

18. $16x^2 - 1 = 0$
$\dfrac{1}{4}, -\dfrac{1}{4}$

19. $9x^2 - 1 = 0$
$\dfrac{1}{3}, -\dfrac{1}{3}$

20. $16x^2 - 49 = 0$
$\dfrac{7}{4}, -\dfrac{7}{4}$

21. $x^2 + 6x + 8 = 0$
$-4, -2$

22. $x^2 - 8x + 15 = 0$
$3, 5$

23. $z^2 + 5z - 14 = 0$
$2, -7$

24. $z^2 + z - 72 = 0$
$8, -9$

25. $x^2 - 5x + 6 = 0$
$2, 3$

26. $x^2 - 3x - 10 = 0$
$-2, 5$

27. $y^2 + 4y - 21 = 0$
$3, -7$

28. $2y^2 - y - 1 = 0$
$-\dfrac{1}{2}, 1$

29. $2a^2 - 9a - 5 = 0$
$-\dfrac{1}{2}, 5$

30. $3a^2 + 14a + 8 = 0$
$-\dfrac{2}{3}, -4$

31. $6z^2 + 5z + 1 = 0$
$-\dfrac{1}{3}, -\dfrac{1}{2}$

32. $6y^2 - 19y + 15 = 0$
$\dfrac{5}{3}, \dfrac{3}{2}$

33. $x^2 - 3x = 0$
$0, 3$

34. $a^2 - 5a = 0$
$0, 5$

35. $x^2 - 7x = 0$
$0, 7$

36. $2a^2 - 8a = 0$
$0, 4$

Solve.

37. $a^2 + 5a = -4$
$-1, -4$

38. $a^2 - 5a = 24$
$-3, 8$

39. $y^2 - 5y = -6$
$2, 3$

40. $y^2 - 7y = 8$
$-1, 8$

41. $2t^2 + 7t = 4$
$\dfrac{1}{2}, -4$

42. $3t^2 + t = 10$
$\dfrac{5}{3}, -2$

43. $3t^2 - 13t = -4$
$\dfrac{1}{3}, 4$

44. $5t^2 - 16t = -12$
$\dfrac{6}{5}, 2$

45. $x(x - 12) = -27$
$3, 9$

46. $x(x - 11) = 12$
$12, -1$

47. $y(y - 7) = 18$
$9, -2$

48. $y(y + 8) = -15$
$-3, -5$

49. $p(p + 3) = -2$
$-1, -2$

50. $p(p - 1) = 20$
$5, -4$

51. $y(y + 4) = 45$
$5, -9$

52. $y(y - 8) = -15$
$3, 5$

53. $x(x + 3) = 28$
$4, -7$

54. $p(p - 14) = 15$
$15, -1$

55. $(x + 8)(x - 3) = -30$
$-2, -3$

56. $(x + 4)(x - 1) = 14$
$-6, 3$

57. $(y + 3)(y + 10) = -10$
$-8, -5$

58. $(z - 5)(z + 4) = 52$
$-8, 9$

59. $(z - 8)(z + 4) = -35$
$1, 3$

60. $(z - 6)(z + 1) = -10$
$1, 4$

61. $(a + 3)(a + 4) = 72$
$-12, 5$

62. $(a - 4)(a + 7) = -18$
$-5, 2$

63. $(2x + 5)(x + 1) = -1$
$-\dfrac{3}{2}, -2$

64. $(z + 3)(z - 10) = -42$
$3, 4$

65. $(y + 3)(2y + 3) = 5$
$-\dfrac{1}{2}, -4$

66. $(y + 5)(3y - 2) = -14$
$-\dfrac{1}{3}, -4$

▶ **Objective B**   *Application Problems*

Solve.

67. The square of a positive number is six more than five times the positive number. Find the number.
6

68. The square of a negative number is sixteen more than six times the negative number. Find the number.
-2

69. The sum of two numbers is six. The sum of the squares of the two numbers is twenty. Find the two numbers.
2, 4

Solve.

70. The sum of two numbers is eight. The sum of the squares of the two numbers is thirty-four. Find the two numbers.
3, 5

71. The sum of the squares of two consecutive positive integers is eighty-five. Find the two integers.
6, 7

72. The sum of the squares of two consecutive positive even integers is one hundred. Find the two integers.
6, 8

73. The sum of two numbers is ten. The product of the two numbers is twenty-one. Find the two numbers.
3, 7

74. The sum of two numbers is twenty-three. The product of the two numbers is one hundred twenty. Find the two numbers.
8, 15

The formula $S = \frac{n^2 + n}{2}$ gives the sum S of the first n natural numbers. Use this formula for Problems 75 and 76.

75. How many consecutive natural numbers beginning with 1 will give a sum of 78?
12

76. How many consecutive natural numbers beginning with 1 will give a sum of 120?
15

The formula $N = \frac{t^2 - t}{2}$ gives the number N of football games that must be scheduled in a league with t teams if each team is to play every other team once. Use this formula for Problems 77 and 78.

77. How many teams are in a league that schedules 28 games in such a way that each team plays every other team once?
8 teams

78. How many teams are in a league that schedules 45 games in such a way that each team plays every other team once.
10 teams

The distance s that an object will fall (neglecting air resistance) in t seconds is given by $s = vt + 16t^2$, where v is the initial velocity of the object. Use this formula for Problems 79 and 80.

79. An object is released from a plane at an altitude of 1600 ft. The initial velocity is 0 ft/s, and air resistance is neglected. How many seconds later will the object hit the ground?
10 s

80. An object is released from the top of a building 320 ft high. The initial velocity is 16 ft/s, and air resistance is neglected. How many seconds later will the object hit the ground?
4 s

Solve.

The height h an object will attain (neglecting air resistance) in t seconds is given by $h = vt - 16t^2$, where v is the initial velocity of the object. Use this formula for Problems 81 and 82.

81. A softball leaves a bat with an initial velocity of 64 ft/s. How many seconds later will the ball be 64 ft above the ground?
    2 s

82. A golf ball is thrown onto a cement surface and rebounds straight up. The initial velocity of the rebound is 96 ft/s. How many seconds later will the golf ball return to the ground?
    6 s

83. The length of a rectangle is 5 in. more than twice the width. The area is 75 in^2. Find the length and width of the rectangle.
    length 15 in.; width 5 in.

84. The width of a rectangle is 5 ft less than the length. The area of the rectangle is 176 ft^2. Find the length and width of the rectangle.
    length 16 ft; width 11 ft

85. The length of each side of a square is extended 2 in. The area of the resulting square is 144 in^2. Find the length of a side of the original square.
    10 in.

86. The length of each side of a square is extended 5 in. The area of the resulting square is 64 in^2. Find the length of a side of the original square.
    3 in.

87. The page of a book measures 6 in. by 9 in. A uniform border around the page leaves 28 in^2 for type. What are the dimensions of the type area?
    4 in. by 7 in.

88. A small garden measures 8 ft by 10 ft. A uniform border around the garden increases the total area to 168 ft^2. What is the width of the border?
    2 ft

89. The radius of a circle is increased by 3 in., increasing the area by 100 in^2. Find the radius of the original circle. Use $\pi \approx 3.14$.
    3.8078556 in.

90. A circle has a radius of 10 in. Find the increase in area when the radius is increased by 2 in. Use $\pi \approx 3.14$.
    138.16 in^2

Calculators and Computers

Factoring   Remember, when factoring a polynomial, ask the following questions about the polynomial.

1. Is there a common factor? If so, factor out the common factor.

2. Is the polynomial the difference of two perfect squares? If so, factor.

3. Is the polynomial a perfect square trinomial? If so, factor.

4. Is the polynomial a trinomial that is the product of two binomials? If so, factor.

5. Is each factor nonfactorable over the integers? If not, factor.

Factoring polynomials is an important part of the study of algebra and is a *learned* skill requiring practice. To provide you with additional practice, the program FACTORING on the Math ACE Disk will give you additional practice factoring a quadratic polynomial. You may choose to practice polynomials of the form

$$x^2 + bx + c \qquad \text{or}$$
$$ax^2 + bx + c$$

These choices, along with the option of quitting the program, are given on a menu screen. You may practice as long as you like with any type of problem. At the end of each problem, you may choose to return to the menu screen or to continue practicing.

The program will present you with a quadratic polynomial to factor. When you have tried to factor the polynomial using paper and pencil, press the RETURN key on the keyboard. The correct factorization will then be displayed.

Chapter Summary

Key Words   The *greatest common factor* (GCF) of two or more integers is the greatest integer that is a factor of all the integers.

To *factor* a polynomial means to write the polynomial as a product of other polynomials.

To *factor* a trinomial of the form $ax^2 + bx + c$ means to express the trinomial as the product of two binomials.

A polynomial that does not factor using only integers is *nonfactorable over the integers*.

A product of a term and itself is a *perfect square*.

An equation of the form $ax^2 + bx + c = 0$ is a *quadratic equation*.

A quadratic equation is in *standard form* when the polynomial is in descending order and equal to zero. $ax^2 + bx + c = 0$ is in standard form.

Essential Rules

Principle of Zero Products

If the product of two factors is zero, then at least one of the factors must be zero.

If $a \cdot b = 0$, then $a = 0$ or $b = 0$.

General Factoring Strategy

1. Is there a common factor? If so, factor out the common factor.

2. Is the polynomial the difference of two perfect squares? If so, factor.

3. Is the polynomial a perfect square trinomial? If so, factor.

4. Is the polynomial a trinomial that is the product of two binomials? If so, factor.

5. Does the polynomial contain four terms? If so, try factoring by grouping.

6. Is each binomial factor a prime polynomial over the integers? If not, factor.

Chapter Review

SECTION 6.1

1. Factor $5x^3 + 10x^2 + 35x$.
$5x(x^2 + 2x + 7)$

2. Factor $12a^2b + 3ab^2$.
$3ab(4a + b)$

3. Factor $14y^9 - 49y^6 + 7y^3$.
$7y^3(2y^6 - 7y^3 + 1)$

4. Factor $4x(x - 3) - 5(3 - x)$.
$(x - 3)(4x + 5)$

5. Factor $10x^2 + 25x + 4xy + 10y$.
$(2x + 5)(5x + 2y)$

6. Factor $21ax - 35bx - 10by + 6ay$.
$(3a - 5b)(7x + 2y)$

SECTION 6.2

7. Factor $b^2 - 13b + 30$.
$(b - 3)(b - 10)$

8. Factor $c^2 + 8c + 12$.
$(c + 6)(c + 2)$

9. Factor $y^2 + 5y - 36$.
$(y - 4)(y + 9)$

10. Factor completely $3a^2 - 15a - 42$.
$3(a + 2)(a - 7)$

11. Factor completely $4x^3 - 20x^2 - 24x$.
$4x(x - 6)(x + 1)$

12. Factor completely $n^4 - 2n^3 - 3n^2$.
$n^2(n + 1)(n - 3)$

SECTION 6.3

13. Factor $6x^2 - 29x + 28$ by using trial factors.
$(2x - 7)(3x - 4)$

14. Factor $12y^2 + 16y - 3$ by using trial factors.
$(6y - 1)(2y + 3)$

15. Factor $2x^2 - 5x + 6$ by using trial factors.
Nonfactorable over the integers

16. Factor $3x^2 - 17x + 10$ by grouping.
$(3x - 2)(x - 5)$

17. Factor $2a^2 - 19a - 60$ by grouping.
$(2a + 5)(a - 12)$

18. Factor $18a^2 - 3a - 10$ by grouping.
$(6a - 5)(3a + 2)$

SECTION 6.4

19. Factor $a^6 - 100$.
$(a^3 + 10)(a^3 - 10)$

20. Factor $9y^4 - 25z^2$.
$(3y^2 + 5z)(3y^2 - 5z)$

21. Factor $a^2b^2 - 1$.
$(ab + 1)(ab - 1)$

22. Factor $5x^2 - 5x - 30$.
$5(x + 2)(x - 3)$

23. Factor $3x^2 + 36x + 108$.
$3(x + 6)^2$

24. Factor $12b^3 - 58b^2 + 56b$.
$2b(3b - 4)(2b - 7)$

SECTION 6.5

25. Solve $4x^2 + 27x = 7$.
$\frac{1}{4}, -7$

26. Solve $(x + 1)(x - 5) = 16$.
$-3, 7$

27. The length of a hockey field is 20 yds less than twice the width of the hockey field. The area of the hockey field is 6000 yd². Find the length and width of the hockey field.
length: 100 yd; width: 60 yd

28. The size of a motion picture on the screen is given by $S = d^2$ where d is the distance between the projector and the screen. Find the distance between the projector and the screen when the size of the picture is 400 ft².
20 ft

29. A rectangular photograph has dimensions 15 in. by 12 in. A picture frame around the photograph increases the total area to 270 in.². What is the width of the frame?
$\frac{3}{2}$ in.

30. The length of each side of a square garden plot is extended 4 ft. The area of the resulting square is 576 ft². Find the length of a side of the original garden plot.
20 ft

Chapter Test

1. Factor $6x^3 - 8x^2 + 10x$.
$2x(3x^2 - 4x + 5)$ [6.1A]

2. Factor $ab + 6a - 3b - 18$.
$(b + 6)(a - 3)$ [6.1B]

3. Factor $p^2 + 5p + 6$.
$(p + 2)(p + 3)$ [6.2A]

4. Factor $a^2 - 19a + 48$.
$(a - 3)(a - 16)$ [6.2A]

5. Factor $x^2 + 2x - 15$.
$(x + 5)(x - 3)$ [6.2A]

6. Factor $x^2 - 9x - 36$.
$(x + 3)(x - 12)$ [6.2A]

7. Factor $5x^2 - 45x - 15$.
$5(x^2 - 9x - 3)$ [6.1A]

8. Factor $2y^4 - 14y^3 - 16y^2$.
$2y^2(y + 1)(y - 8)$ [6.2B]

9. Factor $2x^2 + 4x - 5$ by using trial factors.
Nonfactorable over the integers   [6.3A]

10. Factor $6x^2 + 19x + 8$ by using trial factors.
$(2x + 1)(3x + 8)$ [6.3A]

11. Factor $8x^2 + 20x - 48$ by grouping.
$4(x + 4)(2x - 3)$ [6.3B]

12. Factor $6x^2y^2 + 9xy^2 + 3y^2$ by grouping.
$3y^2(2x + 1)(x + 1)$ [6.3B]

13. Factor $a(x - 2) + b(x - 2)$.
$(x - 2)(a + b)$ [6.3B]

14. Factor $x(p + 1) - (p + 1)$.
$(p + 1)(x - 1)$ [6.3B]

15. Factor $b^2 - 16$.
$(b + 4)(b - 4)$ [6.4A]

16. Factor $4x^2 - 49y^2$.
$(2x + 7y)(2x - 7y)$ [6.4A]

17. Factor $p^2 + 12p + 36$.
$(p + 6)^2$ [6.4A]

18. Factor $4a^2 - 12ab + 9b^2$.
$(2a - 3b)^2$ [6.4A]

19. Factor $3a^2 - 75$.
$3(a + 5)(a - 5)$ [6.4B]

20. Factor $3x^2 + 12xy + 12y^2$.
$3(x + 2y)^2$ [6.4B]

21. Solve: $(2a - 3)(a + 7) = 0$
$\dfrac{3}{2}, -7$ [6.5A]

22. Solve: $4x^2 - 1 = 0$
$\dfrac{1}{2}, -\dfrac{1}{2}$ [6.5A]

23. Solve: $x(x - 8) = -15$
3, 5 [6.5A]

24. The sum of two numbers is 10. The sum of the squares of the two numbers is 58. Find the two numbers.
3, 7 [6.5B]

25. The length of a rectangle is 3 cm longer than twice the width. The area of the rectangle is 90 cm². Find the length and width of the rectangle.
length 15 cm; width 6 cm [6.5B]

Cumulative Review

1. Subtract: $-2 - (-3) - 5 - (-11)$
    7   [1.2B]

2. Simplify $(3 - 7)^2 \div (-2) - 3 \cdot (-4)$.
    4   [1.5B]

3. Evaluate $-2a^2 \div 2b - c$ when $a = -4$, $b = 2$, and $c = -1$.
    -31 [2.1A]

4. Simplify $-\frac{3}{4}(-20x^2)$.
    $15x^2$ [2.2B]

5. Simplify $-2[4x - 2(3 - 2x) - 8x]$.
    12   [2.2D]

6. Solve: $-\frac{5}{7}x = -\frac{10}{21}$
    $\frac{2}{3}$ [3.1C]

7. Solve: $3x - 2 = 12 - 5x$
    $\frac{7}{4}$ [3.3A]

8. Solve:
    $-2 + 4[3x - 2(4 - x) - 3] = 4x + 2$
    3   [3.3B]

9. 120% of what number is 54?
    45   [4.2A]

10. Simplify $(-3a^3b^2)^2$.
    $9a^6b^4$ [5.2B]

11. Simplify $(x + 2)(x^2 - 5x + 4)$.
    $x^3 - 3x^2 - 6x + 8$ [5.3B]

12. Simplify $(8x^2 + 4x - 3) \div (2x - 3)$.
    $4x + 8 + \dfrac{21}{2x - 3}$ [5.4C]

13. Simplify $(x^{-4}y^3)^2$.
    $\dfrac{y^6}{x^8}$ [5.4A]

14. Factor $3a - 3b - ax + bx$.
    $(a - b)(3 - x)$ [6.1B]

15. Factor $15xy^2 - 20xy^4$.
    $5xy^2(3 - 4y^2)$ [6.1A]

16. Factor $x^2 - 5xy - 14y^2$.
    $(x - 7y)(x + 2y)$ [6.2A]

17. Factor $p^2 - 9p - 10$.
    $(p - 10)(p + 1)$ [6.2A]

18. Factor $18a^3 + 57a^2 + 30a$.
    $3a(2a + 5)(3a + 2)$ [6.3B]

19. Factor $36a^2 - 49b^2$.
$(6a - 7b)(6a + 7b)$ [6.4A]

20. Factor $4x^2 + 28xy + 49y^2$.
$(2x + 7y)^2$ [6.4A]

21. Factor $9x^2 + 15x - 14$.
$(3x - 2)(3x + 7)$ [6.3A]

22. Factor $18x^2 - 48xy + 32y^2$.
$2(3x - 4y)^2$ [6.4B]

23. Factor $3y(x - 3) - 2(x - 3)$.
$(x - 3)(3y - 2)$ [6.1B]

24. Solve: $3x^2 + 19x - 14 = 0$
$\frac{2}{3}, -7$ [6.5A]

25. A board 10 ft long is cut into two pieces. Four times the length of the shorter piece is 2 ft less than three times the length of the longer piece. Find the length of each piece.
4 ft and 6 ft   [3.4B]

26. A stereo that regularly sells for $165 is on sale for $99. Find the discount rate.
40%   [4.3B]

27. An investment of $4000 is made at an annual simple interest rate of 8%. How much additional money must be invested at an annual simple interest rate of 11% so that the total interest earned is $1035?
$6500   [4.4A]

28. A family drove to a resort at an average speed of 42 mph and later returned over the same road at an average speed of 56 mph. Find the distance to the resort if the total driving time was 7 h.
168 mi   [4.6A]

29. Find three consecutive even integers such that five times the middle integer is twelve more than twice the sum of the first and third.
10, 12, and 14   [4.8A]

30. The length of the base of a triangle is three times the height. The area of the triangle is 24 in². Find the length of the base of the triangle.
12 in.   [6.5B]

7

Algebraic Fractions

OBJECTIVES

▶ To simplify an algebraic fraction
▶ To multiply algebraic fractions
▶ To divide algebraic fractions
▶ To find the least common multiple (LCM) of two or more polynomials
▶ To express two fractions in terms of the LCM of their denominators
▶ To add or subtract algebraic fractions with like denominators
▶ To add or subtract algebraic fractions with unlike denominators
▶ To simplify a complex fraction
▶ To solve an equation containing fractions
▶ To solve a proportion
▶ To solve application problems
▶ To solve a literal equation for one of the variables
▶ To solve work problems
▶ To solve uniform motion problems

Measurement of the Circumference of the Earth

Distances on the earth, the circumference of the earth, and the distance to the moon and stars are known to great precision. Eratosthenes, the fifth librarian of Alexandria (230 B.C.), laid the foundation of scientific geography with his determination of the circumference of the earth.

Eratosthenes was familiar with certain astronomical data that enabled him to calculate the circumference of the earth by using a proportion statement.

Eratosthenes knew that on a mid-summer day, the sun was directly overhead at Syrene, as shown in the diagram. At the same time, at Alexandria the sun was at a $7\frac{1}{2}°$ angle from the zenith. The distance from Syrene to Alexandria was 5000 stadia (about 520 mi).

Knowing that the ratio of the $7\frac{1}{2}°$ angle to one revolution (360°) is equal to the ratio of the arc length (520 mi) to the circumference, Eratosthenes was able to write and solve a proportion.

This result, calculated over 2000 years ago, is very close to the accepted value of 24,800 miles.

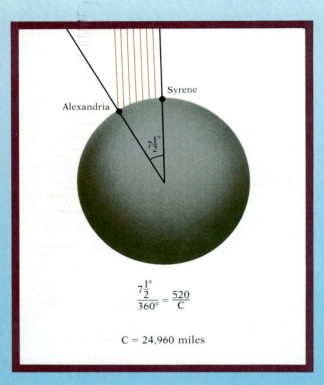

$$\frac{7\frac{1}{2}°}{360°} = \frac{520}{C}$$

$$C = 24{,}960 \text{ miles}$$

Content and Format © 1991 HMCo.

SECTION 7.1   Multiplication and Division of Algebraic Fractions

| **Objective A** | **To simplify an algebraic fraction** |

A fraction in which the numerator or denominator is a variable expression is called an **algebraic fraction.** Examples of algebraic fractions are shown at the right.

$$\frac{5}{z}, \quad \frac{x^2 + 1}{2x - 1}, \quad \frac{y^2 - 3}{3xy + 1}$$

Care must be exercised with algebraic fractions to ensure that when the variables are replaced with numbers, the resulting denominator is not zero.

Consider the algebraic fraction at the right. The value of x cannot be 2, because the denominator would then be zero.

$$\frac{3x^2 + 1}{x^2 - 4}$$

$$\frac{3 \cdot 2^2 + 1}{2^2 - 4} = \frac{13}{0} \quad \text{Not a real number}$$

An algebraic fraction is in simplest form when the numerator and denominator have no common factors.

Simplify: $\dfrac{x^2 - 4}{x^2 - 2x - 8}$

Factor the numerator and denominator.

$$\frac{x^2 - 4}{x^2 - 2x - 8} = \frac{(x - 2)(x + 2)}{(x - 4)(x + 2)}$$

$$= \frac{(x - 2)\overset{1}{(\cancel{x + 2})}}{(x - 4)\underset{1}{(\cancel{x + 2})}}$$

Write the answer in simplest form.

$$= \frac{x - 2}{x - 4}$$

Simplify: $\dfrac{10 + 3x - x^2}{x^2 - 4x - 5}$

Factor the numerator and denominator.

$$\frac{10 + 3x - x^2}{x^2 - 4x - 5} = \frac{(5 - x)(2 + x)}{(x - 5)(x + 1)}$$

Remember that $5 - x = -(x - 5)$.
Therefore, $\frac{5 - x}{x - 5} = \frac{-(x - 5)}{x - 5} = \frac{-1}{1} = -1$.

$$= \frac{\overset{-1}{(\cancel{5 - x})}(2 + x)}{\underset{1}{(\cancel{x - 5})}(x + 1)}$$

Write the answer in simplest form.

$$= \frac{x + 2}{x + 1}$$

Example 1

Simplify: $\frac{4x^3y^4}{6x^4y}$

Solution

$\frac{4x^3y^4}{6x^4y} = \frac{2y^3}{3x}$ — Use rules of exponents.

Example 2

Simplify: $\frac{6x^5y}{12x^2y^3}$

Your solution

$\frac{x^3}{2y^2}$

Example 3

Simplify: $\frac{9 - x^2}{x^2 + x - 12}$

Solution

$\frac{9 - x^2}{x^2 + x - 12} = \frac{\overset{-1}{\cancel{(3 - x)}}(3 + x)}{\underset{1}{\cancel{(x - 3)}}(x + 4)} = -\frac{x + 3}{x + 4}$

Example 4

Simplify: $\frac{x^2 + 2x - 24}{16 - x^2}$

Your solution

$-\frac{x + 6}{x + 4}$

Example 5

Simplify: $\frac{x^2 + 2x - 15}{x^2 - 7x + 12}$

Solution

$\frac{x^2 + 2x - 15}{x^2 - 7x + 12} = \frac{(x + 5)\overset{1}{\cancel{(x - 3)}}}{\underset{1}{\cancel{(x - 3)}}(x - 4)} = \frac{x + 5}{x - 4}$

Example 6

Simplify: $\frac{x^2 + 4x - 12}{x^2 - 3x + 2}$

Your solution

$\frac{x + 6}{x - 1}$

Solutions on p. A22

Objective B   **To multiply algebraic fractions**

The product of two fractions is a fraction whose numerator is the product of the numerators of the two fractions and whose denominator is the product of the denominators of the two fractions.

> If $\frac{a}{b}$ and $\frac{c}{d}$ are rational numbers, then $\frac{a}{b} \cdot \frac{c}{d} = \frac{ac}{bd}$

$$\frac{2}{3} \cdot \frac{4}{5} = \frac{8}{15} \qquad \frac{3x}{y} \cdot \frac{2}{z} = \frac{6x}{yz} \qquad \frac{x + 2}{x} \cdot \frac{3}{x - 2} = \frac{3x + 6}{x^2 - 2x}$$

Simplify: $\dfrac{x^2 + 3x}{x^2 - 3x - 4} \cdot \dfrac{x^2 - 5x + 4}{x^2 + 2x - 3}$

$$\dfrac{x^2 + 3x}{x^2 - 3x - 4} \cdot \dfrac{x^2 - 5x + 4}{x^2 + 2x - 3}$$

Factor the numerator and denominator of each fraction.

$$= \dfrac{x(x + 3)}{(x - 4)(x + 1)} \cdot \dfrac{(x - 4)(x - 1)}{(x + 3)(x - 1)}$$

Multiply.

$$= \dfrac{x(x + 3)(x - 4)(x - 1)}{(x - 4)(x + 1)(x + 3)(x - 1)}$$

Write the answer in simplest form.

$$= \dfrac{x}{x + 1}$$

Example 7

Simplify: $\dfrac{10x^2 - 15x}{12x - 8} \cdot \dfrac{3x - 2}{20x - 25}$

Solution

$$\dfrac{10x^2 - 15x}{12x - 8} \cdot \dfrac{3x - 2}{20x - 25}$$

$$= \dfrac{5x(2x - 3)}{4(3x - 2)} \cdot \dfrac{(3x - 2)}{5(4x - 5)}$$

$$= \dfrac{5x(2x - 3)(3x - 2)}{4(3x - 2)5(4x - 5)} = \dfrac{x(2x - 3)}{4(4x - 5)}$$

Example 8

Simplify: $\dfrac{12x^2 + 3x}{10x - 15} \cdot \dfrac{8x - 12}{9x + 18}$

Your solution

$$\dfrac{4x(4x + 1)}{15(x + 2)}$$

Example 9

Simplify: $\dfrac{x^2 + x - 6}{x^2 + 7x + 12} \cdot \dfrac{x^2 + 3x - 4}{4 - x^2}$

Solution

$$\dfrac{x^2 + x - 6}{x^2 + 7x + 12} \cdot \dfrac{x^2 + 3x - 4}{4 - x^2}$$

$$= \dfrac{(x + 3)(x - 2)}{(x + 3)(x + 4)} \cdot \dfrac{(x + 4)(x - 1)}{(2 - x)(2 + x)}$$

$$= \dfrac{(x + 3)(x - 2)(x + 4)(x - 1)}{(x + 3)(x + 4)(2 - x)(2 + x)} = -\dfrac{x - 1}{x + 2}$$

Example 10

Simplify: $\dfrac{x^2 + 2x - 15}{9 - x^2} \cdot \dfrac{x^2 - 3x - 18}{x^2 - 7x + 6}$

Your solution

$$-\dfrac{x + 5}{x - 1}$$

Solutions on p. A22

| **Objective C** | **To divide algebraic fractions** | | 7 |

The **reciprocal** of a fraction is a fraction with the numerator and denominator interchanged.

$$\text{Fraction}\left\{\begin{array}{cc} \dfrac{a}{b} & \dfrac{b}{a} \\[2mm] x^2 = \dfrac{x^2}{1} & \dfrac{1}{x^2} \\[2mm] \dfrac{x+2}{x} & \dfrac{x}{x+2} \end{array}\right\}\text{Reciprocal}$$

To divide two fractions, multiply by the reciprocal of the divisor.

$$\frac{a}{b} \div \frac{c}{d} = \frac{a}{b} \cdot \frac{d}{c} = \frac{ad}{bc}$$

$$\frac{4}{x} \div \frac{y}{5} = \frac{4}{x} \cdot \frac{5}{y} = \frac{20}{xy} \qquad \frac{x+4}{x} \div \frac{x-2}{4} = \frac{x+4}{x} \cdot \frac{4}{x-2} = \frac{4(x+4)}{x(x-2)}$$

The basis for the division rule is shown at the right.

$$\frac{a}{b} \div \frac{c}{d} = \frac{\dfrac{a}{b} \cdot \dfrac{d}{c}}{\dfrac{c}{d} \cdot \dfrac{d}{c}} = \frac{\dfrac{a}{b} \cdot \dfrac{d}{c}}{\dfrac{c}{d} \cdot \dfrac{d}{c}} = \frac{\dfrac{a}{b} \cdot \dfrac{d}{c}}{1} = \frac{a}{b} \cdot \frac{d}{c}$$

Example 11

Simplify: $\dfrac{xy^2 - 3x^2y}{z^2} \div \dfrac{6x^2 - 2xy}{z^3}$

Solution

$$\frac{xy^2 - 3x^2y}{z^2} \div \frac{6x^2 - 2xy}{z^3}$$

$$= \frac{xy^2 - 3x^2y}{z^2} \cdot \frac{z^3}{6x^2 - 2xy}$$

$$= \frac{xy(y - 3x) \cdot z^3}{z^2 \cdot 2x(3x - y)} = -\frac{yz}{2}$$

Example 12

Simplify: $\dfrac{a^2}{4bc^2 - 2b^2c} \div \dfrac{a}{6bc - 3b^2}$

Your solution

$\dfrac{3a}{2c}$

Example 13

Simplify: $\dfrac{2x^2 + 5x + 2}{2x^2 + 3x - 2} \div \dfrac{3x^2 + 13x + 4}{2x^2 + 7x - 4}$

Solution

$$\frac{2x^2 + 5x + 2}{2x^2 + 3x - 2} \div \frac{3x^2 + 13x + 4}{2x^2 + 7x - 4}$$

$$= \frac{2x^2 + 5x + 2}{2x^2 + 3x - 2} \cdot \frac{2x^2 + 7x - 4}{3x^2 + 13x + 4}$$

$$= \frac{(2x+1)(x+2) \cdot (2x-1)(x+4)}{(2x-1)(x+2) \cdot (3x+1)(x+4)} = \frac{2x+1}{3x+1}$$

Example 14

Simplify: $\dfrac{3x^2 + 26x + 16}{3x^2 - 7x - 6} \div \dfrac{2x^2 + 9x - 5}{x^2 + 2x - 15}$

Your solution

$\dfrac{x+8}{2x-1}$

Solutions on p. A22

7.1   EXERCISES

▶ Objective A

Simplify.

1. $\dfrac{9x^3}{12x^4}$

$\dfrac{3}{4x}$

2. $\dfrac{16x^2y}{24xy^3}$

$\dfrac{2x}{3y^2}$

3. $\dfrac{(x+3)^2}{(x+3)^3}$

$\dfrac{1}{x+3}$

4. $\dfrac{(2x-1)^5}{(2x-1)^4}$

$2x-1$

5. $\dfrac{3n-4}{4-3n}$

-1

6. $\dfrac{5-2x}{2x-5}$

-1

7. $\dfrac{6y(y+2)}{9y^2(y+2)}$

$\dfrac{2}{3y}$

8. $\dfrac{12x^2(3-x)}{18x(3-x)}$

$\dfrac{2x}{3}$

9. $\dfrac{6x(x-5)}{8x^2(5-x)}$

$-\dfrac{3}{4x}$

10. $\dfrac{14x^3(7-3x)}{21x(3x-7)}$

$-\dfrac{2x^2}{3}$

11. $\dfrac{a^2+4a}{ab+4b}$

$\dfrac{a}{b}$

12. $\dfrac{x^2-3x}{2x-6}$

$\dfrac{x}{2}$

13. $\dfrac{4-6x}{3x^2-2x}$

$-\dfrac{2}{x}$

14. $\dfrac{5xy-3y}{9-15x}$

$-\dfrac{y}{3}$

15. $\dfrac{y^2-3y+2}{y^2-4y+3}$

$\dfrac{y-2}{y-3}$

16. $\dfrac{x^2+5x+6}{x^2+8x+15}$

$\dfrac{x+2}{x+5}$

17. $\dfrac{x^2+3x-10}{x^2+2x-8}$

$\dfrac{x+5}{x+4}$

18. $\dfrac{a^2+7a-8}{a^2+6a-7}$

$\dfrac{a+8}{a+7}$

19. $\dfrac{x^2+x-12}{x^2-6x+9}$

$\dfrac{x+4}{x-3}$

20. $\dfrac{x^2+8x+16}{x^2-2x-24}$

$\dfrac{x+4}{x-6}$

21. $\dfrac{x^2-3x-10}{25-x^2}$

$-\dfrac{x+2}{x+5}$

22. $\dfrac{4-y^2}{y^2-3y-10}$

$\dfrac{2-y}{y-5}$

23. $\dfrac{2x^3+2x^2-4x}{x^3+2x^2-3x}$

$\dfrac{2(x+2)}{x+3}$

24. $\dfrac{3x^3-12x}{6x^3-24x^2+24x}$

$\dfrac{x+2}{2(x-2)}$

25. $\dfrac{6x^2-7x+2}{6x^2+5x-6}$

$\dfrac{2x-1}{2x+3}$

26. $\dfrac{2n^2-9n+4}{2n^2-5n-12}$

$\dfrac{2n-1}{2n+3}$

27. $\dfrac{x^2+3x-28}{24-2x-x^2}$

$\dfrac{x+7}{x+6}$

▶ **Objective B**

Simplify.

28. $\dfrac{8x^2}{9y^3} \cdot \dfrac{3y^2}{4x^3}$

$\dfrac{2}{3xy}$

29. $\dfrac{4a^2b^3}{15x^5y^2} \cdot \dfrac{25x^3y}{16ab}$

$\dfrac{5ab^2}{12x^2y}$

30. $\dfrac{12x^3y^4}{7a^2b^3} \cdot \dfrac{14a^3b^4}{9x^2y^2}$

$\dfrac{8xy^2ab}{3}$

31. $\dfrac{18a^4b^2}{25x^2y^3} \cdot \dfrac{50x^5y^6}{27a^6b^2}$

$\dfrac{4x^3y^3}{3a^2}$

32. $\dfrac{3x - 6}{5x - 20} \cdot \dfrac{10x - 40}{27x - 54}$

$\dfrac{2}{9}$

33. $\dfrac{8x - 12}{14x + 7} \cdot \dfrac{42x + 21}{32x - 48}$

$\dfrac{3}{4}$

34. $\dfrac{3x^2 + 2x}{2xy - 3y} \cdot \dfrac{2xy^3 - 3y^3}{3x^3 + 2x^2}$

$\dfrac{y^2}{x}$

35. $\dfrac{4a^2x - 3a^2}{2by + 5b} \cdot \dfrac{2b^3y + 5b^3}{4ax - 3a}$

ab^2

36. $\dfrac{x^2 + 5x + 4}{x^3y^2} \cdot \dfrac{x^2y^3}{x^2 + 2x + 1}$

$\dfrac{y(x + 4)}{x(x + 1)}$

37. $\dfrac{x^2 + x - 2}{xy^2} \cdot \dfrac{x^3y}{x^2 + 5x + 6}$

$\dfrac{x^2(x - 1)}{y(x + 3)}$

38. $\dfrac{x^4y^2}{x^2 + 3x - 28} \cdot \dfrac{x^2 - 49}{xy^4}$

$\dfrac{x^3(x - 7)}{y^2(x - 4)}$

39. $\dfrac{x^5y^3}{x^2 + 13x + 30} \cdot \dfrac{x^2 + 2x - 3}{x^7y^2}$

$\dfrac{y(x - 1)}{x^2(x + 10)}$

40. $\dfrac{2x^2 - 5x}{2xy + y} \cdot \dfrac{2xy^2 + y^2}{5x^2 - 2x^3}$

$-\dfrac{y}{x}$

41. $\dfrac{3a^3 + 4a^2}{5ab - 3b} \cdot \dfrac{3b^3 - 5ab^3}{3a^2 + 4a}$

$-ab^2$

42. $\dfrac{x^2 - 2x - 24}{x^2 - 5x - 6} \cdot \dfrac{x^2 + 5x + 6}{x^2 + 6x + 8}$

$\dfrac{x + 3}{x + 1}$

43. $\dfrac{x^2 - 8x + 7}{x^2 + 3x - 4} \cdot \dfrac{x^2 + 3x - 10}{x^2 - 9x + 14}$

$\dfrac{x + 5}{x + 4}$

44. $\dfrac{x^2 + 2x - 35}{x^2 + 4x - 21} \cdot \dfrac{x^2 + 3x - 18}{x^2 + 9x + 18}$

$\dfrac{x - 5}{x + 3}$

45. $\dfrac{y^2 + y - 20}{y^2 + 2y - 15} \cdot \dfrac{y^2 + 4y - 21}{y^2 + 3y - 28}$

1

Simplify.

46. $\dfrac{x^2 - 3x - 4}{x^2 + 6x + 5} \cdot \dfrac{x^2 + 5x + 6}{8 + 2x - x^2}$

$-\dfrac{x + 3}{x + 5}$

47. $\dfrac{25 - n^2}{n^2 - 2n - 35} \cdot \dfrac{n^2 - 8n - 20}{n^2 - 3n - 10}$

$-\dfrac{n - 10}{n - 7}$

48. $\dfrac{12x^2 - 6x}{x^2 + 6x + 5} \cdot \dfrac{2x^4 + 10x^3}{4x^2 - 1}$

$\dfrac{12x^4}{(x + 1)(2x + 1)}$

49. $\dfrac{8x^3 + 4x^2}{x^2 - 3x + 2} \cdot \dfrac{x^2 - 4}{16x^2 + 8x}$

$\dfrac{x(x + 2)}{2(x - 1)}$

50. $\dfrac{16 + 6x - x^2}{x^2 - 10x - 24} \cdot \dfrac{x^2 - 6x - 27}{x^2 - 17x + 72}$

$-\dfrac{x + 3}{x - 12}$

51. $\dfrac{x^2 - 11x + 28}{x^2 - 13x + 42} \cdot \dfrac{x^2 + 7x + 10}{20 - x - x^2}$

$-\dfrac{x + 2}{x - 6}$

52. $\dfrac{2x^2 + 5x + 2}{2x^2 + 7x + 3} \cdot \dfrac{x^2 - 7x - 30}{x^2 - 6x - 40}$

$\dfrac{x + 2}{x + 4}$

53. $\dfrac{x^2 - 4x - 32}{x^2 - 8x - 48} \cdot \dfrac{3x^2 + 17x + 10}{3x^2 - 22x - 16}$

$\dfrac{x + 5}{x - 12}$

54. $\dfrac{2x^2 + x - 3}{2x^2 - x - 6} \cdot \dfrac{2x^2 - 9x + 10}{2x^2 - 3x + 1}$

$\dfrac{2x - 5}{2x - 1}$

55. $\dfrac{3y^2 + 14y + 8}{2y^2 + 7y - 4} \cdot \dfrac{2y^2 + 9y - 5}{3y^2 + 16y + 5}$

$\dfrac{3y + 2}{3y + 1}$

56. $\dfrac{6x^2 - 11x + 4}{6x^2 + x - 2} \cdot \dfrac{12x^2 + 11x + 2}{8x^2 + 14x + 3}$

$\dfrac{3x - 4}{2x + 3}$

57. $\dfrac{6 - x - 2x^2}{4x^2 + 3x - 10} \cdot \dfrac{3x^2 + 7x - 20}{2x^2 + 5x - 12}$

$-\dfrac{3x - 5}{4x - 5}$

▶ **Objective C**

Simplify.

58. $\dfrac{4x^2y^3}{15a^2b^3} \div \dfrac{6xy}{5a^3b^5}$

$\dfrac{2xy^2ab^2}{9}$

59. $\dfrac{9x^3y^4}{16a^4b^2} \div \dfrac{45x^4y^2}{14a^7b}$

$\dfrac{7a^3y^2}{40bx}$

60. $\dfrac{6x - 12}{8x + 32} \div \dfrac{18x - 36}{10x + 40}$

$\dfrac{5}{12}$

61. $\dfrac{28x + 14}{45x - 30} \div \dfrac{14x + 7}{30x - 20}$

$\dfrac{4}{3}$

62. $\dfrac{6x^3 + 7x^2}{12x - 3} \div \dfrac{6x^2 + 7x}{36x - 9}$

$3x$

63. $\dfrac{5a^2y + 3a^2}{2x^3 + 5x^2} \div \dfrac{10ay + 6a}{6x^3 + 15x^2}$

$\dfrac{3a}{2}$

Simplify.

64. $\dfrac{x^2 + 4x + 3}{x^2y} \div \dfrac{x^2 + 2x + 1}{xy^2}$

$\dfrac{y(x + 3)}{x(x + 1)}$

65. $\dfrac{x^3y^2}{x^2 - 3x - 10} \div \dfrac{xy^4}{x^2 - x - 20}$

$\dfrac{x^2(x + 4)}{y^2(x + 2)}$

66. $\dfrac{x^2 - 49}{x^4y^3} \div \dfrac{x^2 - 14x + 49}{x^4y^3}$

$\dfrac{x + 7}{x - 7}$

67. $\dfrac{x^2y^5}{x^2 - 11x + 30} \div \dfrac{xy^6}{x^2 - 7x + 10}$

$\dfrac{x(x - 2)}{y(x - 6)}$

68. $\dfrac{4ax - 8a}{c^2} \div \dfrac{2y - xy}{c^3}$

$-\dfrac{4ac}{y}$

69. $\dfrac{3x^2y - 9xy}{a^2b} \div \dfrac{3x^2 - x^3}{ab^2}$

$-\dfrac{3by}{ax}$

70. $\dfrac{x^2 - 5x + 6}{x^2 - 9x + 18} \div \dfrac{x^2 - 6x + 8}{x^2 - 9x + 20}$

$\dfrac{x - 5}{x - 6}$

71. $\dfrac{x^2 + 3x - 40}{x^2 + 2x - 35} \div \dfrac{x^2 + 2x - 48}{x^2 + 3x - 18}$

$\dfrac{(x - 3)(x + 6)}{(x + 7)(x - 6)}$

72. $\dfrac{x^2 + 2x - 15}{x^2 - 4x - 45} \div \dfrac{x^2 + x - 12}{x^2 - 5x - 36}$

1

73. $\dfrac{y^2 - y - 56}{y^2 + 8y + 7} \div \dfrac{y^2 - 13y + 40}{y^2 - 4y - 5}$

1

74. $\dfrac{8 + 2x - x^2}{x^2 + 7x + 10} \div \dfrac{x^2 - 11x + 28}{x^2 - x - 42}$

$-\dfrac{x + 6}{x + 5}$

75. $\dfrac{x^2 - x - 2}{x^2 - 7x + 10} \div \dfrac{x^2 - 3x - 4}{40 - 3x - x^2}$

$-\dfrac{x + 8}{x - 4}$

76. $\dfrac{2x^2 - 3x - 20}{2x^2 - 7x - 30} \div \dfrac{2x^2 - 5x - 12}{4x^2 + 12x + 9}$

$\dfrac{2x + 3}{x - 6}$

77. $\dfrac{6n^2 + 13n + 6}{4n^2 - 9} \div \dfrac{6n^2 + n - 2}{4n^2 - 1}$

$\dfrac{2n + 1}{2n - 3}$

78. $\dfrac{9x^2 - 16}{6x^2 - 11x + 4} \div \dfrac{6x^2 + 11x + 4}{8x^2 + 10x + 3}$

$\dfrac{4x + 3}{2x - 1}$

79. $\dfrac{15 - 14x - 8x^2}{4x^2 + 4x - 15} \div \dfrac{4x^2 + 13x - 12}{3x^2 + 13x + 4}$

$\dfrac{3x + 1}{2x - 3}$

80. $\dfrac{8x^2 + 18x - 5}{10x^2 - 9x + 2} \div \dfrac{8x^2 + 22x + 15}{10x^2 + 11x - 6}$

$\dfrac{(2x + 5)(4x - 1)}{(2x - 1)(4x + 5)}$

81. $\dfrac{10 + 7x - 12x^2}{8x^2 - 2x - 15} \div \dfrac{6x^2 - 13x + 5}{10x^2 - 13x + 4}$

$\dfrac{(3x + 2)(5 - 4x)(5x - 4)}{(2x - 3)(4x + 5)(3x - 5)}$

SECTION 7.2 Expressing Fractions in Terms of the Least Common Multiple (LCM)

Objective A

To find the least common multiple (LCM) of two or more polynomials

The **least common multiple (LCM)** of two or more numbers is the smallest number that contains the prime factorization of each number.

The LCM of 12 and 18 is 36.
36 contains the prime factors of 12 and the prime factors of 18.

$12 = 2 \cdot 2 \cdot 3$
$18 = 2 \cdot 3 \cdot 3$

Factors of 12

$$LCM = 36 = \overbrace{2 \cdot 2 \cdot 3 \cdot 3}$$

Factors of 18

The least common multiple of two or more polynomials is the polynomial that contains the factors of each polynomial.

To find the LCM of two or more polynomials, first factor each polynomial completely. The LCM is the product of each factor the greatest number of times it occurs in any one factorization.

Find the LCM of $4x^2 + 4x$ and $x^2 + 2x + 1$.

The LCM of the polynomials is the product of the LCM of the numerical coefficients and each variable factor the greatest number of times it occurs in any one factorization.

$4x^2 + 4x = 4x(x + 1) = 2 \cdot 2 \cdot x(x + 1)$
$x^2 + 2x + 1 = (x + 1)(x + 1)$

Factors of $4x^2 + 4x$

$$LCM = \overbrace{2 \cdot 2 \cdot x(x + 1)(x + 1)}^{} = 4x(x + 1)(x + 1)$$

Factors of $x^2 + 2x + 1$

Example 1
Find the LCM of $4x^2y$ and $6xy^2$.

Solution
$4x^2y = 2 \cdot 2 \cdot x \cdot x \cdot y \qquad 6xy^2 = 2 \cdot 3 \cdot x \cdot y \cdot y$
$LCM = 2 \cdot 2 \cdot 3 \cdot x \cdot x \cdot y \cdot y = 12x^2y^2$

Example 2
Find the LCM of $8uv^2$ and $12uw$.

Your solution
$24uv^2w$

Example 3
Find the LCM of $x^2 - x - 6$ and $9 - x^2$.

Solution
$x^2 - x - 6 = (x - 3)(x + 2)$
$9 - x^2 = -(x^2 - 9) = -(x + 3)(x - 3)$
$LCM = (x - 3)(x + 2)(x + 3)$

Example 4
Find the LCM of $m^2 - 6m + 9$ and $m^2 - 2m - 3$.

Your solution
$(m - 3)(m - 3)(m + 1)$

Solutions on p. A23

| **Objective B** | **To express two fractions in terms of the LCM of their denominators** |

When adding and subtracting fractions, it is frequently necessary to express two or more fractions in terms of a common denominator. This common denominator is the LCM of the denominators of the fractions.

Write the fractions $\frac{x+1}{4x^2}$ and $\frac{x-3}{6x^2-12x}$ in terms of the LCM of the denominators.

Find the LCM of the denominators.

The LCM is $12x^2(x-2)$.

For each fraction, multiply the numerator and denominator by the factors whose product with the denominator is the LCM.

$$\frac{x+1}{4x^2} = \frac{x+1}{4x^2} \cdot \frac{3(x-2)}{3(x-2)} = \frac{3x^2-3x-6}{12x^2(x-2)}$$

$$\frac{x-3}{6x^2-12x} = \frac{x-3}{6x(x-2)} \cdot \frac{2x}{2x} = \frac{2x^2-6x}{12x^2(x-2)}$$

LCM

Example 5

Write the fractions $\frac{x+2}{3x^2}$ and $\frac{x-1}{8xy}$ in terms of the LCM of the denominators.

Solution

The LCM is $24x^2y$.

$$\frac{x+2}{3x^2} = \frac{x+2}{3x^2} \cdot \frac{8y}{8y} = \frac{8xy+16y}{24x^2y}$$

$$\frac{x-1}{8xy} = \frac{x-1}{8xy} \cdot \frac{3x}{3x} = \frac{3x^2-3x}{24x^2y}$$

Example 6

Write the fractions $\frac{x-3}{4xy^2}$ and $\frac{2x+1}{9y^2z}$ in terms of the LCM of the denominators.

Your solution

$$\frac{9xz-27z}{36xy^2z} \ , \ \frac{8x^2+4x}{36xy^2z}$$

Example 7

Write the fractions $\frac{2x-1}{2x-x^2}$ and $\frac{x}{x^2+x-6}$ in terms of the LCM of the denominators.

Solution

$$\frac{2x-1}{2x-x^2} = \frac{2x-1}{-(x^2-2x)} = -\frac{2x-1}{x^2-2x}$$

The LCM is $x(x-2)(x+3)$.

$$\frac{2x-1}{2x-x^2} = -\frac{2x-1}{x(x-2)} \cdot \frac{x+3}{x+3} = -\frac{2x^2+5x-3}{x(x-2)(x+3)}$$

$$\frac{x}{x^2+x-6} = \frac{x}{(x-2)(x+3)} \cdot \frac{x}{x} = \frac{x^2}{x(x-2)(x+3)}$$

Example 8

Write the fractions $\frac{x+4}{x^2-3x-10}$ and $\frac{2x}{25-x^2}$ in terms of the LCM of the denominators.

Your solution

$$\frac{x^2+9x+20}{(x+2)(x-5)(x+5)}, \ -\frac{2x^2+4x}{(x+2)(x-5)(x+5)}$$

Solutions on p. A23

7.2 EXERCISES

▶ **Objective A**

Find the LCM of the expressions.

1. $8x^3y$
$12xy^2$
$24x^3y^2$

2. $6ab^2$
$18ab^3$
$18ab^3$

3. $10x^4y^2$
$15x^3y$
$30x^4y^2$

4. $12a^2b$
$18ab^3$
$36a^2b^3$

5. $8x^2$
$4x^2 + 8x$
$8x^2(x + 2)$

6. $6y^2$
$4y + 12$
$12y^2(y + 3)$

7. $2x^2y$
$3x^2 + 12x$
$6x^2y(x + 4)$

8. $4xy^2$
$6xy^2 + 12y^2$
$12xy^2(x + 2)$

9. $9x(x + 2)$
$12(x + 2)^2$
$36x(x + 2)^2$

10. $8x^2(x - 1)^2$
$10x^3(x - 1)$
$40x^3(x - 1)^2$

11. $3x + 3$
$2x^2 + 4x + 2$
$6(x + 1)^2$

12. $4x - 12$
$2x^2 - 12x + 18$
$4(x - 3)^2$

13. $(x - 1)(x + 2)$
$(x - 1)(x + 3)$
$(x - 1)(x + 2)(x + 3)$

14. $(2x - 1)(x + 4)$
$(2x + 1)(x + 4)$
$(2x - 1)(x + 4)(2x + 1)$

15. $(2x + 3)^2$
$(2x + 3)(x - 5)$
$(2x + 3)^2(x - 5)$

16. $(x - 7)(x + 2)$
$(x - 7)^2$
$(x - 7)^2(x + 2)$

17. $(x - 1)$
$(x - 2)$
$(x - 1)(x - 2)$
$(x - 1)(x - 2)$

18. $(x + 4)(x - 3)$
$x + 4$
$x - 3$
$(x + 4)(x - 3)$

19. $x^2 - x - 6$
$x^2 + x - 12$
$(x - 3)(x + 2)(x + 4)$

20. $x^2 + 3x - 10$
$x^2 + 5x - 14$
$(x + 5)(x + 7)(x - 2)$

21. $x^2 + 5x + 4$
$x^2 - 3x - 28$
$(x + 4)(x + 1)(x - 7)$

22. $x^2 - 10x + 21$
$x^2 - 8x + 15$
$(x - 7)(x - 3)(x - 5)$

23. $x^2 - 2x - 24$
$x^2 - 36$
$(x - 6)(x + 6)(x + 4)$

24. $x^2 + 7x + 10$
$x^2 - 25$
$(x + 5)(x - 5)(x + 2)$

25. $x^2 - 7x - 30$
$x^2 - 5x - 24$
$(x - 10)(x - 8)(x + 3)$

26. $2x^2 - 7x + 3$
$2x^2 + x - 1$
$(2x - 1)(x - 3)(x + 1)$

27. $3x^2 - 11x + 6$
$3x^2 + 4x - 4$
$(3x - 2)(x - 3)(x + 2)$

28. $2x^2 - 9x + 10$
$2x^2 + x - 15$
$(2x - 5)(x - 2)(x + 3)$

29. $6 + x - x^2$
$x + 2$
$x - 3$
$(x + 2)(x - 3)$

30. $15 + 2x - x^2$
$x - 5$
$x + 3$
$(x + 3)(x - 5)$

31. $5 + 4x - x^2$
$x - 5$
$x + 1$
$(x - 5)(x + 1)$

32. $x^2 + 3x - 18$
$3 - x$
$x + 6$
$(x + 6)(x - 3)$

33. $x^2 - 5x + 6$
$1 - x$
$x - 6$
$(x - 3)(x - 2)(x - 1)(x - 6)$

▶ **Objective B**

Write each fraction in terms of the LCM of the denominators.

34. $\dfrac{4}{x}, \dfrac{3}{x^2}$

$\dfrac{4x}{x^2}, \dfrac{3}{x^2}$

35. $\dfrac{5}{ab^2}, \dfrac{6}{ab}$

$\dfrac{5}{ab^2}, \dfrac{6b}{ab^2}$

36. $\dfrac{x}{3y^2}, \dfrac{z}{4y}$

$\dfrac{4x}{12y^2}, \dfrac{3yz}{12y^2}$

37. $\dfrac{5y}{6x^2}, \dfrac{7}{9xy}$

$\dfrac{15y^2}{18x^2y}, \dfrac{14x}{18x^2y}$

38. $\dfrac{y}{x(x-3)}, \dfrac{6}{x^2}$

$\dfrac{xy}{x^2(x-3)}, \dfrac{6x-18}{x^2(x-3)}$

39. $\dfrac{a}{y^2}, \dfrac{6}{y(y+5)}$

$\dfrac{ay+5a}{y^2(y+5)}, \dfrac{6y}{y^2(y+5)}$

40. $\dfrac{9}{(x-1)^2}, \dfrac{6}{x(x-1)}$

$\dfrac{9x}{x(x-1)^2}, \dfrac{6x-6}{x(x-1)^2}$

41. $\dfrac{a^2}{y(y+7)}, \dfrac{a}{(y+7)^2}$

$\dfrac{a^2y+7a^2}{y(y+7)^2}, \dfrac{ay}{y(y+7)^2}$

42. $\dfrac{3}{x-3}, \dfrac{5}{x(3-x)}$

$\dfrac{3x}{x(x-3)}, -\dfrac{5}{x(x-3)}$

43. $\dfrac{b}{y(y-4)}, \dfrac{b^2}{4-y}$

$\dfrac{b}{y(y-4)}, -\dfrac{b^2y}{y(y-4)}$

44. $\dfrac{3}{(x-5)^2}, \dfrac{2}{5-x}$

$\dfrac{3}{(x-5)^2}, -\dfrac{2x-10}{(x-5)^2}$

45. $\dfrac{3}{7-y}, \dfrac{2}{(y-7)^2}$

$\dfrac{3y-21}{(y-7)^2}, \dfrac{2}{(y-7)^2}$

46. $\dfrac{3}{x^2+2x}, \dfrac{4}{x^2}$

$\dfrac{3x}{x^2(x+2)}, \dfrac{4x+8}{x^2(x+2)}$

47. $\dfrac{2}{y-3}, \dfrac{3}{y^3-3y^2}$

$\dfrac{2y^2}{y^2(y-3)}, \dfrac{3}{y^2(y-3)}$

48. $\dfrac{x-2}{x+3}, \dfrac{x}{x-4}$

$\dfrac{x^2-6x+8}{(x+3)(x-4)}, \dfrac{x^2+3x}{(x+3)(x-4)}$

49. $\dfrac{x^2}{2x-1}, \dfrac{x+1}{x+4}$

$\dfrac{x^3+4x^2}{(2x-1)(x+4)}, \dfrac{2x^2+x-1}{(2x-1)(x+4)}$

50. $\dfrac{3}{x^2+x-2}, \dfrac{x}{x+2}$

$\dfrac{3}{(x+2)(x-1)}, \dfrac{x^2-x}{(x+2)(x-1)}$

51. $\dfrac{3x}{x-5}, \dfrac{4}{x^2-25}$

$\dfrac{3x^2+15x}{(x-5)(x+5)}, \dfrac{4}{(x-5)(x+5)}$

52. $\dfrac{5}{2x^2-9x+10}, \dfrac{x-1}{2x-5}$

$\dfrac{5}{(2x-5)(x-2)}, \dfrac{x^2-3x+2}{(2x-5)(x-2)}$

53. $\dfrac{x-3}{3x^2+4x-4}, \dfrac{2}{x+2}$

$\dfrac{x-3}{(3x-2)(x+2)}, \dfrac{6x-4}{(3x-2)(x+2)}$

54. $\dfrac{x}{x^2+x-6}, \dfrac{2x}{x^2-9}$

$\dfrac{x^2-3x}{(x+3)(x-3)(x-2)}, \dfrac{2x^2-4x}{(x+3)(x-3)(x-2)}$

55. $\dfrac{x-1}{x^2+2x-15}, \dfrac{x}{x^2+6x+5}$

$\dfrac{x^2-1}{(x+5)(x-3)(x+1)}, \dfrac{x^2-3x}{(x+5)(x-3)(x+1)}$

56. $\dfrac{x}{9-x^2}, \dfrac{x-1}{x^2-6x+9}$

$-\dfrac{x^2-3x}{(x-3)^2(x+3)}, \dfrac{x^2+2x-3}{(x-3)^2(x+3)}$

57. $\dfrac{2x}{10+3x-x^2}, \dfrac{x+2}{x^2-8x+15}$

$-\dfrac{2x^2-6x}{(x-5)(x-3)(x+2)}, \dfrac{x^2+4x+4}{(x-5)(x-3)(x+2)}$

58. $\dfrac{3x}{x-5}, \dfrac{x}{x+4}, \dfrac{3}{20+x-x^2}$

$\dfrac{3x^2+12x}{(x-5)(x+4)}, \dfrac{x^2-5x}{(x-5)(x+4)}, -\dfrac{3}{(x-5)(x+4)}$

59. $\dfrac{x+1}{x+5}, \dfrac{x+2}{x-7}, \dfrac{3}{35+2x-x^2}$

$\dfrac{x^2-6x-7}{(x+5)(x-7)}, \dfrac{x^2+7x+10}{(x+5)(x-7)}, -\dfrac{3}{(x+5)(x-7)}$

SECTION 7.3   Addition and Subtraction of Algebraic Fractions

Objective A   **To add or subtract algebraic fractions with like denominators**

When adding algebraic fractions in which the denominators are the same, add the numerators. The denominator of the sum is the common denominator.

$$\frac{a}{b} + \frac{c}{b} = \frac{a+c}{b}$$

$$\frac{5x}{18} + \frac{7x}{18} = \frac{5x + 7x}{18} = \frac{12x}{18} = \frac{2x}{3}$$

$$\frac{x}{x^2 - 1} + \frac{1}{x^2 - 1} = \frac{x+1}{x^2 - 1} = \frac{\overset{1}{\cancel{(x+1)}}}{(x-1)\cancel{(x+1)}} = \frac{1}{x-1}$$

Note that the sum is written in simplest form.

When subtracting algebraic fractions with like denominators, subtract the numerators. The denominator of the difference is the common denominator. Write the answer in simplest form.

$$\frac{2x}{x-2} - \frac{4}{x-2} = \frac{2x - 4}{x-2} = \frac{2\overset{1}{\cancel{(x-2)}}}{\underset{1}{\cancel{x-2}}} = 2$$

$$\frac{3x - 1}{x^2 - 5x + 4} - \frac{2x + 3}{x^2 - 5x + 4} = \frac{(3x - 1) - (2x + 3)}{x^2 - 5x + 4}$$

$$= \frac{x - 4}{x^2 - 5x + 4} = \frac{\overset{1}{\cancel{(x-4)}}}{\cancel{(x-4)}(x - 1)} = \frac{1}{x - 1}$$

Example 1

Simplify: $\frac{7}{x^2} + \frac{9}{x^2}$

Solution

$$\frac{7}{x^2} + \frac{9}{x^2} = \frac{7 + 9}{x^2} = \frac{16}{x^2}$$

Example 2

Simplify: $\frac{3}{xy} + \frac{12}{xy}$

Your solution

$$\frac{15}{xy}$$

Example 3

Simplify: $\frac{3x^2}{x^2 - 1} - \frac{x + 4}{x^2 - 1}$

Solution

$$\frac{3x^2}{x^2 - 1} - \frac{x + 4}{x^2 - 1} = \frac{3x^2 - (x + 4)}{x^2 - 1} = \frac{3x^2 - x - 4}{x^2 - 1}$$

$$= \frac{(3x - 4)\overset{1}{\cancel{(x+1)}}}{(x - 1)\cancel{(x+1)}} = \frac{3x - 4}{x - 1}$$

Example 4

Simplify: $\frac{2x^2}{x^2 - x - 12} - \frac{7x + 4}{x^2 - x - 12}$

Your solution

$$\frac{2x + 1}{x + 3}$$

Solutions on p. A23

Example 5

Simplify:

$$\frac{2x^2 + 5}{x^2 + 2x - 3} - \frac{x^2 - 3x}{x^2 + 2x - 3} + \frac{x - 2}{x^2 + 2x - 3}$$

Solution

$$\frac{2x^2 + 5}{x^2 + 2x - 3} - \frac{x^2 - 3x}{x^2 + 2x - 3} + \frac{x - 2}{x^2 + 2x - 3}$$

$$= \frac{(2x^2 + 5) - (x^2 - 3x) + (x - 2)}{x^2 + 2x - 3}$$

$$= \frac{2x^2 + 5 - x^2 + 3x + x - 2}{x^2 + 2x - 3}$$

$$= \frac{x^2 + 4x + 3}{x^2 + 2x - 3} = \frac{\overset{1}{\cancel{(x + 3)}}(x + 1)}{\underset{1}{\cancel{(x + 3)}}(x - 1)} = \frac{x + 1}{x - 1}$$

Example 6

Simplify:

$$\frac{x^2 - 1}{x^2 - 8x + 12} - \frac{2x + 1}{x^2 - 8x + 12} + \frac{x}{x^2 - 8x + 12}$$

Your solution

$$\frac{x + 1}{x - 6}$$

Solution on p. A23

Objective B   **To add or subtract algebraic fractions with unlike denominators**

Before two fractions with unlike denominators can be added or subtracted, each fraction must be expressed in terms of a common denominator. This common denominator is the LCM of the denominators of the fractions.

Simplify: $\frac{x - 3}{x^2 - 2x} + \frac{6}{x^2 - 4}$

Find the LCM of the denominators.   The LCM is $x(x - 2)(x + 2)$

Write each fraction in terms of the LCM. Multiply the factors in the numerator and then add the fractions.

$$\frac{x - 3}{x^2 - 2x} + \frac{6}{x^2 - 4} = \frac{x - 3}{x(x - 2)} \cdot \frac{x + 2}{x + 2} + \frac{6}{(x - 2)(x + 2)} \cdot \frac{x}{x}$$

$$= \frac{x^2 - x - 6}{x(x - 2)(x + 2)} + \frac{6x}{x(x - 2)(x + 2)}$$

$$= \frac{(x^2 - x - 6) + 6x}{x(x - 2)(x + 2)}$$

$$= \frac{x^2 + 5x - 6}{x(x - 2)(x + 2)}$$

$$= \frac{(x + 6)(x - 1)}{x(x - 2)(x + 2)}$$

The last step is to factor the numerator to determine whether there are common factors in the numerator and denominator.

Example 7

Simplify: $\dfrac{y}{x} - \dfrac{4y}{3x} + \dfrac{3y}{4x}$

Solution

The LCM of the denominators is $12x$.

$$\dfrac{y}{x} - \dfrac{4y}{3x} + \dfrac{3y}{4x} = \dfrac{y}{x} \cdot \dfrac{12}{12} - \dfrac{4y}{3x} \cdot \dfrac{4}{4} + \dfrac{3y}{4x} \cdot \dfrac{3}{3}$$

$$= \dfrac{12y}{12x} - \dfrac{16y}{12x} + \dfrac{9y}{12x}$$

$$= \dfrac{12y - 16y + 9y}{12x} = \dfrac{5y}{12x}$$

Example 8

Simplify: $\dfrac{z}{8y} - \dfrac{4z}{3y} + \dfrac{5z}{4y}$

Your solution

$\dfrac{z}{24y}$

Example 9

Simplify: $\dfrac{2x}{x-3} - \dfrac{5}{3-x}$

Solution

Remember: $3 - x = -(x - 3)$.

Therefore, $\dfrac{5}{3-x} = \dfrac{-5}{x-3}$.

$$\dfrac{2x}{x-3} - \dfrac{5}{3-x} = \dfrac{2x}{x-3} - \dfrac{-5}{x-3}$$

$$= \dfrac{2x - (-5)}{x-3} = \dfrac{2x+5}{x-3}$$

Example 10

Simplify: $\dfrac{5x}{x-2} - \dfrac{3}{2-x}$

Your solution

$\dfrac{5x+3}{x-2}$

Example 11

Simplify: $\dfrac{2x}{2x-3} - \dfrac{1}{x+1}$

Solution

The LCM is $(2x - 3)(x + 1)$.

$$\dfrac{2x}{2x-3} - \dfrac{1}{x+1}$$

$$= \dfrac{2x}{2x-3} \cdot \dfrac{x+1}{x+1} - \dfrac{1}{x+1} \cdot \dfrac{2x-3}{2x-3}$$

$$= \dfrac{2x^2 + 2x}{(2x-3)(x+1)} - \dfrac{2x-3}{(2x-3)(x+1)}$$

$$= \dfrac{(2x^2 + 2x) - (2x - 3)}{(2x-3)(x+1)} = \dfrac{2x^2 + 3}{(2x-3)(x+1)}$$

Example 12

Simplify: $\dfrac{4x}{3x-1} - \dfrac{9}{x+4}$

Your solution

$\dfrac{4x^2 - 11x + 9}{(3x-1)(x+4)}$

Solutions on p. A23

Example 13

Simplify: $\dfrac{x+3}{x^2-2x-8} + \dfrac{3}{4-x}$

Solution

The LCM is $(x-4)(x+2)$.

Recall: $\dfrac{3}{4-x} = \dfrac{-3}{x-4}$

$\dfrac{x+3}{x^2-2x-8} + \dfrac{3}{4-x}$

$= \dfrac{x+3}{(x-4)(x+2)} + \dfrac{-3}{x-4}$

$= \dfrac{x+3}{(x-4)(x+2)} + \dfrac{-3}{x-4} \cdot \dfrac{x+2}{x+2}$

$= \dfrac{x+3}{(x-4)(x+2)} + \dfrac{-3(x+2)}{(x-4)(x+2)}$

$= \dfrac{(x+3)+(-3)(x+2)}{(x-4)(x+2)}$

$= \dfrac{x+3-3x-6}{(x-4)(x+2)} = \dfrac{-2x-3}{(x-4)(x+2)}$

Example 14

Simplify: $\dfrac{2x-1}{x^2-25} + \dfrac{2}{5-x}$

Your solution

$-\dfrac{11}{(x+5)(x-5)}$

Example 15

Simplify: $\dfrac{3x+2}{2x^2-x-1} - \dfrac{3}{2x+1} + \dfrac{4}{x-1}$

Solution

The LCM is $(2x+1)(x-1)$.

$\dfrac{3x+2}{2x^2-x-1} - \dfrac{3}{2x+1} + \dfrac{4}{x-1}$

$= \dfrac{3x+2}{(2x+1)(x-1)} - \dfrac{3}{2x+1} \cdot \dfrac{x-1}{x-1} + \dfrac{4}{x-1} \cdot \dfrac{2x+1}{2x+1}$

$= \dfrac{3x+2}{(2x+1)(x-1)} - \dfrac{3x-3}{(2x+1)(x-1)} + \dfrac{8x+4}{(2x+1)(x-1)}$

$= \dfrac{(3x+2)-(3x-3)+(8x+4)}{(2x+1)(x-1)}$

$= \dfrac{3x+2-3x+3+8x+4}{(2x+1)(x-1)} = \dfrac{8x+9}{(2x+1)(x-1)}$

Example 16

Simplify: $\dfrac{2x-3}{3x^2-x-2} + \dfrac{5}{3x+2} - \dfrac{1}{x-1}$

Your solution

$\dfrac{2(2x-5)}{(3x+2)(x-1)}$

Solutions on p. A24

7.3   **EXERCISES**

▶ **Objective A**

Simplify.

1. $\dfrac{3}{y^2} + \dfrac{8}{y^2}$

$\dfrac{11}{y^2}$

2. $\dfrac{6}{ab} - \dfrac{2}{ab}$

$\dfrac{4}{ab}$

3. $\dfrac{3}{x+4} - \dfrac{10}{x+4}$

$-\dfrac{7}{x+4}$

4. $\dfrac{x}{x+6} - \dfrac{2}{x+6}$

$\dfrac{x-2}{x+6}$

5. $\dfrac{3x}{2x+3} + \dfrac{5x}{2x+3}$

$\dfrac{8x}{2x+3}$

6. $\dfrac{6y}{4y+1} - \dfrac{11y}{4y+1}$

$-\dfrac{5y}{4y+1}$

7. $\dfrac{2x+1}{x-3} + \dfrac{3x+6}{x-3}$

$\dfrac{5x+7}{x-3}$

8. $\dfrac{4x+3}{2x-7} + \dfrac{3x-8}{2x-7}$

$\dfrac{7x-5}{2x-7}$

9. $\dfrac{5x-1}{x+9} - \dfrac{3x+4}{x+9}$

$\dfrac{2x-5}{x+9}$

10. $\dfrac{6x-5}{x-10} - \dfrac{3x-4}{x-10}$

$\dfrac{3x-1}{x-10}$

11. $\dfrac{x-7}{2x+7} - \dfrac{4x-3}{2x+7}$

$\dfrac{-3x-4}{2x+7}$

12. $\dfrac{2n}{3n+4} - \dfrac{5n-3}{3n+4}$

$-\dfrac{3(n-1)}{3n+4}$

13. $\dfrac{x}{x^2+2x-15} - \dfrac{3}{x^2+2x-15}$

$\dfrac{1}{x+5}$

14. $\dfrac{3x}{x^2+3x-10} - \dfrac{6}{x^2+3x-10}$

$\dfrac{3}{x+5}$

15. $\dfrac{2x+3}{x^2-x-30} - \dfrac{x-2}{x^2-x-30}$

$\dfrac{1}{x-6}$

16. $\dfrac{3x-1}{x^2+5x-6} - \dfrac{2x-7}{x^2+5x-6}$

$\dfrac{1}{x-1}$

17. $\dfrac{4y+7}{2y^2+7y-4} - \dfrac{y-5}{2y^2+7y-4}$

$\dfrac{3}{2y-1}$

18. $\dfrac{x+1}{2x^2-5x-12} + \dfrac{x+2}{2x^2-5x-12}$

$\dfrac{1}{x-4}$

19. $\dfrac{2x^2+3x}{x^2-9x+20} + \dfrac{2x^2-3}{x^2-9x+20} - \dfrac{4x^2+2x+1}{x^2-9x+20}$

$\dfrac{1}{x-5}$

20. $\dfrac{2x^2+3x}{x^2-2x-63} - \dfrac{x^2-3x+21}{x^2-2x-63} - \dfrac{x-7}{x^2-2x-63}$

$\dfrac{x-2}{x-9}$

▶ **Objective B**

Simplify.

21. $\dfrac{4}{x} + \dfrac{5}{y}$

$\dfrac{4y + 5x}{xy}$

22. $\dfrac{7}{a} + \dfrac{5}{b}$

$\dfrac{7b + 5a}{ab}$

23. $\dfrac{12}{x} - \dfrac{5}{2x}$

$\dfrac{19}{2x}$

24. $\dfrac{5}{3a} - \dfrac{3}{4a}$

$\dfrac{11}{12a}$

25. $\dfrac{1}{2x} - \dfrac{5}{4x} + \dfrac{7}{6x}$

$\dfrac{5}{12x}$

26. $\dfrac{7}{4y} + \dfrac{11}{6y} - \dfrac{8}{3y}$

$\dfrac{11}{12y}$

27. $\dfrac{5}{3x} - \dfrac{2}{x^2} + \dfrac{3}{2x}$

$\dfrac{19x - 12}{6x^2}$

28. $\dfrac{6}{y^2} + \dfrac{3}{4y} - \dfrac{2}{5y}$

$\dfrac{120 + 7y}{20y^2}$

29. $\dfrac{2}{x} - \dfrac{3}{2y} + \dfrac{3}{5x} - \dfrac{1}{4y}$

$\dfrac{52y - 35x}{20xy}$

30. $\dfrac{5}{2a} + \dfrac{7}{3b} - \dfrac{2}{b} - \dfrac{3}{4a}$

$\dfrac{21b + 4a}{12ab}$

31. $\dfrac{2x + 1}{3x} + \dfrac{x - 1}{5x}$

$\dfrac{13x + 2}{15x}$

32. $\dfrac{4x - 3}{6x} + \dfrac{2x + 3}{4x}$

$\dfrac{14x + 3}{12x}$

33. $\dfrac{x - 3}{6x} + \dfrac{x + 4}{8x}$

$\dfrac{7}{24}$

34. $\dfrac{2x - 3}{2x} + \dfrac{x + 3}{3x}$

$\dfrac{8x - 3}{6x}$

35. $\dfrac{2x + 9}{9x} - \dfrac{x - 5}{5x}$

$\dfrac{x + 90}{45x}$

36. $\dfrac{3y - 2}{12y} - \dfrac{y - 3}{18y}$

$\dfrac{7}{36}$

37. $\dfrac{x + 4}{2x} - \dfrac{x - 1}{x^2}$

$\dfrac{x^2 + 2x + 2}{2x^2}$

38. $\dfrac{x - 2}{3x^2} - \dfrac{x + 4}{x}$

$\dfrac{-3x^2 - 11x - 2}{3x^2}$

39. $\dfrac{x - 10}{4x^2} + \dfrac{x + 1}{2x}$

$\dfrac{2x^2 + 3x - 10}{4x^2}$

40. $\dfrac{x + 5}{3x^2} + \dfrac{2x + 1}{2x}$

$\dfrac{6x^2 + 5x + 10}{6x^2}$

41. $\dfrac{2x + 1}{6x^2} - \dfrac{x - 4}{4x}$

$\dfrac{-3x^2 + 16x + 2}{12x^2}$

Simplify.

42. $\dfrac{x+3}{6x} - \dfrac{x-3}{8x^2}$

$\dfrac{4x^2 + 9x + 9}{24x^2}$

43. $\dfrac{x+2}{xy} - \dfrac{3x-2}{x^2y}$

$\dfrac{x^2 - x + 2}{x^2y}$

44. $\dfrac{3x-1}{xy^2} - \dfrac{2x+3}{xy}$

$\dfrac{3x - 1 - 2xy - 3y}{xy^2}$

45. $\dfrac{4x-3}{3x^2y} + \dfrac{2x+1}{4xy^2}$

$\dfrac{16xy - 12y + 6x^2 + 3x}{12x^2y^2}$

46. $\dfrac{5x+7}{6xy^2} - \dfrac{4x-3}{8x^2y}$

$\dfrac{20x^2 + 28x - 12xy + 9y}{24x^2y^2}$

47. $\dfrac{x-2}{8x^2} - \dfrac{x+7}{12xy}$

$\dfrac{3xy - 6y - 2x^2 - 14x}{24x^2y}$

48. $\dfrac{3x-1}{6y^2} - \dfrac{x+5}{9xy}$

$\dfrac{9x^2 - 2xy - 3x - 10y}{18xy^2}$

49. $\dfrac{4}{x-2} + \dfrac{5}{x+3}$

$\dfrac{9x + 2}{(x-2)(x+3)}$

50. $\dfrac{2}{x-3} + \dfrac{5}{x-4}$

$\dfrac{7x - 23}{(x-3)(x-4)}$

51. $\dfrac{6}{x-7} - \dfrac{4}{x+3}$

$\dfrac{2(x+23)}{(x-7)(x+3)}$

52. $\dfrac{3}{y+6} - \dfrac{4}{y-3}$

$\dfrac{-y - 33}{(y+6)(y-3)}$

53. $\dfrac{2x}{x+1} + \dfrac{1}{x-3}$

$\dfrac{2x^2 - 5x + 1}{(x+1)(x-3)}$

54. $\dfrac{3x}{x-4} + \dfrac{2}{x+6}$

$\dfrac{3x^2 + 20x - 8}{(x-4)(x+6)}$

55. $\dfrac{4x}{2x-1} - \dfrac{5}{x-6}$

$\dfrac{4x^2 - 34x + 5}{(2x-1)(x-6)}$

56. $\dfrac{6x}{x+5} - \dfrac{3}{2x+3}$

$\dfrac{3(4x^2 + 5x - 5)}{(x+5)(2x+3)}$

57. $\dfrac{2a}{a-7} + \dfrac{5}{7-a}$

$\dfrac{2a - 5}{a - 7}$

58. $\dfrac{4x}{6-x} + \dfrac{5}{x-6}$

$\dfrac{-4x + 5}{x - 6}$

59. $\dfrac{x}{x^2-9} + \dfrac{3}{x-3}$

$\dfrac{4x + 9}{(x+3)(x-3)}$

Simplify.

60. $\dfrac{y}{y^2 - 16} + \dfrac{1}{y - 4}$

$\dfrac{2(y + 2)}{(y - 4)(y + 4)}$

61. $\dfrac{2x}{x^2 - x - 6} - \dfrac{3}{x + 2}$

$\dfrac{-x + 9}{(x - 3)(x + 2)}$

62. $\dfrac{5x}{x^2 + 2x - 8} - \dfrac{2}{x + 4}$

$\dfrac{3x + 4}{(x + 4)(x - 2)}$

63. $\dfrac{3x - 1}{x^2 - 10x + 25} - \dfrac{3}{x - 5}$

$\dfrac{14}{(x - 5)(x - 5)}$

64. $\dfrac{2a + 3}{a^2 - 7a + 12} - \dfrac{2}{a - 3}$

$\dfrac{11}{(a - 4)(a - 3)}$

65. $\dfrac{x + 4}{x^2 - x - 42} + \dfrac{3}{7 - x}$

$-\dfrac{2(x + 7)}{(x + 6)(x - 7)}$

66. $\dfrac{x + 3}{x^2 - 3x - 10} + \dfrac{2}{5 - x}$

$\dfrac{-x - 1}{(x - 5)(x + 2)}$

67. $\dfrac{1}{x + 1} + \dfrac{x}{x - 6} - \dfrac{5x - 2}{x^2 - 5x - 6}$

$\dfrac{x - 4}{x - 6}$

68. $\dfrac{x}{x - 4} + \dfrac{5}{x + 5} - \dfrac{11x - 8}{x^2 + x - 20}$

$\dfrac{x + 3}{x + 5}$

69. $\dfrac{3x + 1}{x - 1} - \dfrac{x - 1}{x - 3} + \dfrac{x + 1}{x^2 - 4x + 3}$

$\dfrac{2x + 1}{x - 1}$

70. $\dfrac{4x + 1}{x - 8} - \dfrac{3x + 2}{x + 4} - \dfrac{49x + 4}{x^2 - 4x - 32}$

$\dfrac{x - 2}{x + 4}$

71. $\dfrac{2x + 9}{3 - x} + \dfrac{x + 5}{x + 7} - \dfrac{2x^2 + 3x - 3}{x^2 + 4x - 21}$

$-\dfrac{3(x^2 + 8x + 25)}{(x - 3)(x + 7)}$

72. $\dfrac{3x + 5}{x + 5} - \dfrac{x + 1}{2 - x} - \dfrac{4x^2 - 3x - 1}{x^2 + 3x - 10}$

$\dfrac{4(2x - 1)}{(x + 5)(x - 2)}$

SECTION 7.4   Complex Fractions

Objective A   **To simplify a complex fraction**

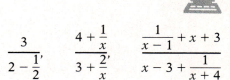

A **complex fraction** is a fraction whose numerator or denominator contains one or more fractions. Examples of complex fractions are shown at the right.

$$\frac{3}{2 - \frac{1}{2}}, \quad \frac{4 + \frac{1}{x}}{3 + \frac{2}{x}}, \quad \frac{\frac{1}{x-1} + x + 3}{x - 3 + \frac{1}{x+4}}$$

Simplify: $\dfrac{1 - \frac{4}{x^2}}{1 + \frac{2}{x}}$

Find the LCM of the denominators of the fractions in the numerator and denominator.

The LCM of x^2 and x is x^2.

Multiply the numerator and denominator of the complex fraction by the LCM. Then simplify.

$$\frac{1 - \frac{4}{x^2}}{1 + \frac{2}{x}} = \frac{1 - \frac{4}{x^2}}{1 + \frac{2}{x}} \cdot \frac{x^2}{x^2} = \frac{1 \cdot x^2 - \frac{4}{x^2} \cdot x^2}{1 \cdot x^2 + \frac{2}{x} \cdot x^2}$$

$$= \frac{x^2 - 4}{x^2 + 2x} = \frac{(x-2)(x+2)}{x(x+2)}$$

$$= \frac{x - 2}{x}$$

Example 1

Simplify: $\dfrac{\frac{1}{x} + \frac{1}{2}}{\frac{1}{x^2} - \frac{1}{4}}$

Solution

The LCM of x, 2, x^2, and 4 is $4x^2$.

$$\frac{\frac{1}{x} + \frac{1}{2}}{\frac{1}{x^2} - \frac{1}{4}} = \frac{\frac{1}{x} + \frac{1}{2}}{\frac{1}{x^2} - \frac{1}{4}} \cdot \frac{4x^2}{4x^2} = \frac{\frac{1}{x} \cdot 4x^2 + \frac{1}{2} \cdot 4x^2}{\frac{1}{x^2} \cdot 4x^2 - \frac{1}{4} \cdot 4x^2}$$

$$= \frac{4x + 2x^2}{4 - x^2}$$

$$= \frac{2x(2 + x)}{(2 - x)(2 + x)}$$

$$= \frac{2x}{2 - x}$$

Example 2

Simplify: $\dfrac{\frac{1}{3} - \frac{1}{x}}{\frac{1}{9} - \frac{1}{x^2}}$

Your solution

$\dfrac{3x}{x + 3}$

Solution on p. A24

Example 3

Simplify: $\dfrac{1 - \dfrac{2}{x} - \dfrac{15}{x^2}}{1 - \dfrac{11}{x} + \dfrac{30}{x^2}}$

Solution

The LCM of x and x^2 is x^2.

$$\dfrac{1 - \dfrac{2}{x} - \dfrac{15}{x^2}}{1 - \dfrac{11}{x} + \dfrac{30}{x^2}} = \dfrac{1 - \dfrac{2}{x} - \dfrac{15}{x^2}}{1 - \dfrac{11}{x} + \dfrac{30}{x^2}} \cdot \dfrac{x^2}{x^2}$$

$$= \dfrac{1 \cdot x^2 - \dfrac{2}{x} \cdot x^2 - \dfrac{15}{x^2} \cdot x^2}{1 \cdot x^2 - \dfrac{11}{x} \cdot x^2 + \dfrac{30}{x^2} \cdot x^2}$$

$$= \dfrac{x^2 - 2x - 15}{x^2 - 11x + 30}$$

$$= \dfrac{\overset{1}{\cancel{(x-5)}}(x+3)}{\underset{1}{\cancel{(x-5)}}(x-6)} = \dfrac{x+3}{x-6}$$

Example 4

Simplify: $\dfrac{1 + \dfrac{4}{x} + \dfrac{3}{x^2}}{1 + \dfrac{10}{x} + \dfrac{21}{x^2}}$

Your solution

$\dfrac{x+1}{x+7}$

Example 5

Simplify: $\dfrac{x - 8 + \dfrac{20}{x+4}}{x - 10 + \dfrac{24}{x+4}}$

Solution

The LCM is $x + 4$.

$$\dfrac{x - 8 + \dfrac{20}{x+4}}{x - 10 + \dfrac{24}{x+4}}$$

$$= \dfrac{x - 8 + \dfrac{20}{x+4}}{x - 10 + \dfrac{24}{x+4}} \cdot \dfrac{x+4}{x+4}$$

$$= \dfrac{(x-8)(x+4) + \dfrac{20}{x+4} \cdot (x+4)}{(x-10)(x+4) + \dfrac{24}{x+4} \cdot (x+4)}$$

$$= \dfrac{x^2 - 4x - 32 + 20}{x^2 - 6x - 40 + 24} = \dfrac{x^2 - 4x - 12}{x^2 - 6x - 16}$$

$$= \dfrac{(x-6)\overset{1}{\cancel{(x+2)}}}{(x-8)\underset{1}{\cancel{(x+2)}}} = \dfrac{x-6}{x-8}$$

Example 6

Simplify: $\dfrac{x + 3 - \dfrac{20}{x-5}}{x + 8 + \dfrac{30}{x-5}}$

Your solution

$\dfrac{x-7}{x-2}$

Solutions on p. A24

| 7.4 | **EXERCISES** |

▶ **Objective A**

Simplify.

1. $\dfrac{1 + \dfrac{3}{x}}{1 - \dfrac{9}{x^2}}$

$\dfrac{x}{x - 3}$

2. $\dfrac{1 + \dfrac{4}{x}}{1 - \dfrac{16}{x^2}}$

$\dfrac{x}{x - 4}$

3. $\dfrac{2 - \dfrac{8}{x + 4}}{3 - \dfrac{12}{x + 4}}$

$\dfrac{2}{3}$

4. $\dfrac{5 - \dfrac{25}{x + 5}}{1 - \dfrac{3}{x + 5}}$

$\dfrac{5x}{x + 2}$

5. $\dfrac{1 + \dfrac{5}{y - 2}}{1 - \dfrac{2}{y - 2}}$

$\dfrac{y + 3}{y - 4}$

6. $\dfrac{2 - \dfrac{11}{2x - 1}}{3 - \dfrac{17}{2x - 1}}$

$\dfrac{4x - 13}{2(3x - 10)}$

7. $\dfrac{4 - \dfrac{2}{x + 7}}{5 + \dfrac{1}{x + 7}}$

$\dfrac{2(2x + 13)}{5x + 36}$

8. $\dfrac{5 + \dfrac{3}{x - 8}}{2 - \dfrac{1}{x - 8}}$

$\dfrac{5x - 37}{2x - 17}$

9. $\dfrac{1 - \dfrac{1}{x} - \dfrac{6}{x^2}}{1 - \dfrac{9}{x^2}}$

$\dfrac{x + 2}{x + 3}$

10. $\dfrac{1 + \dfrac{4}{x} + \dfrac{4}{x^2}}{1 - \dfrac{2}{x} - \dfrac{8}{x^2}}$

$\dfrac{x + 2}{x - 4}$

11. $\dfrac{1 - \dfrac{5}{x} - \dfrac{6}{x^2}}{1 + \dfrac{6}{x} + \dfrac{5}{x^2}}$

$\dfrac{x - 6}{x + 5}$

12. $\dfrac{1 - \dfrac{7}{a} + \dfrac{12}{a^2}}{1 + \dfrac{1}{a} - \dfrac{20}{a^2}}$

$\dfrac{a - 3}{a + 5}$

13. $\dfrac{1 - \dfrac{6}{x} + \dfrac{8}{x^2}}{\dfrac{4}{x^2} + \dfrac{3}{x} - 1}$

$-\dfrac{x - 2}{x + 1}$

14. $\dfrac{1 + \dfrac{3}{x} - \dfrac{18}{x^2}}{\dfrac{21}{x^2} - \dfrac{4}{x} - 1}$

$-\dfrac{x + 6}{x + 7}$

15. $\dfrac{x - \dfrac{4}{x + 3}}{1 + \dfrac{1}{x + 3}}$

$x - 1$

16. $\dfrac{y + \dfrac{1}{y - 2}}{1 + \dfrac{1}{y - 2}}$

$y - 1$

17. $\dfrac{1 - \dfrac{x}{2x + 1}}{x - \dfrac{1}{2x + 1}}$

$\dfrac{1}{2x - 1}$

18. $\dfrac{1 - \dfrac{2x - 2}{3x - 1}}{x - \dfrac{4}{3x - 1}}$

$\dfrac{1}{3x - 4}$

Simplify.

19. $\dfrac{x - 5 + \dfrac{14}{x + 4}}{x + 3 - \dfrac{2}{x + 4}}$

$\dfrac{x - 3}{x + 5}$

20. $\dfrac{a + 4 + \dfrac{5}{a - 2}}{a + 6 + \dfrac{15}{a - 2}}$

$\dfrac{a - 1}{a + 1}$

21. $\dfrac{x + 3 - \dfrac{10}{x - 6}}{x + 2 - \dfrac{20}{x - 6}}$

$\dfrac{x - 7}{x - 8}$

22. $\dfrac{x - 7 + \dfrac{5}{x - 1}}{x - 3 + \dfrac{1}{x - 1}}$

$\dfrac{x - 6}{x - 2}$

23. $\dfrac{y - 6 + \dfrac{22}{2y + 3}}{y - 5 + \dfrac{11}{2y + 3}}$

$\dfrac{2y - 1}{2y + 1}$

24. $\dfrac{x + 2 - \dfrac{12}{2x - 1}}{x + 1 - \dfrac{9}{2x - 1}}$

$\dfrac{2x + 7}{2x + 5}$

25. $\dfrac{x - \dfrac{2}{2x - 3}}{2x - 1 - \dfrac{8}{2x - 3}}$

$\dfrac{x - 2}{2x - 5}$

26. $\dfrac{x + 3 - \dfrac{18}{2x + 1}}{x - \dfrac{6}{2x + 1}}$

$\dfrac{x + 5}{x + 2}$

27. $\dfrac{\dfrac{1}{x} - \dfrac{2}{x - 1}}{\dfrac{3}{x} + \dfrac{1}{x - 1}}$

$-\dfrac{x + 1}{4x - 3}$

28. $\dfrac{\dfrac{3}{n + 1} + \dfrac{1}{n}}{\dfrac{2}{n + 1} + \dfrac{3}{n}}$

$\dfrac{4n + 1}{5n + 3}$

29. $\dfrac{\dfrac{3}{2x - 1} - \dfrac{1}{x}}{\dfrac{4}{x} + \dfrac{2}{2x - 1}}$

$\dfrac{x + 1}{2(5x - 2)}$

30. $\dfrac{\dfrac{4}{3x + 1} + \dfrac{3}{x}}{\dfrac{6}{x} - \dfrac{2}{3x + 1}}$

$\dfrac{13x + 3}{2(8x + 3)}$

31. $\dfrac{\dfrac{3}{b - 4} - \dfrac{2}{b + 1}}{\dfrac{5}{b + 1} - \dfrac{1}{b - 4}}$

$\dfrac{b + 11}{4b - 21}$

32. $\dfrac{\dfrac{5}{x - 5} - \dfrac{3}{x - 1}}{\dfrac{6}{x - 1} + \dfrac{2}{x - 5}}$

$\dfrac{x + 5}{4(x - 4)}$

SECTION 7.5

Solving Equations Containing Fractions

Objective A

To solve an equation containing fractions

To solve an equation containing fractions, **clear denominators** by multiplying each side of the equation by the LCM of the denominators. Then solve for the variable.

Solve: $\frac{3x-1}{4} + \frac{2}{3} = \frac{7}{6}$

The LCM is 12.

$$\frac{3x-1}{4} + \frac{2}{3} = \frac{7}{6}$$

Multiply each side of the equation by the LCM of the denominators.

$$12\left(\frac{3x-1}{4} + \frac{2}{3}\right) = 12 \cdot \frac{7}{6}$$

Simplify using the Distributive Property and the Properties of Fractions.

$$12\left(\frac{3x-1}{4}\right) + 12 \cdot \frac{2}{3} = 12 \cdot \frac{7}{6}$$

$$\frac{\overset{3}{\cancel{12}}}{1}\left(\frac{3x-1}{\cancel{4}}\right) + \frac{\overset{4}{\cancel{12}}}{1} \cdot \frac{2}{\cancel{3}} = \frac{\overset{2}{\cancel{12}}}{1} \cdot \frac{7}{\cancel{6}}$$

Solve for x.

$$9x - 3 + 8 = 14$$
$$9x + 5 = 14$$
$$9x = 9$$
$$x = 1$$

1 checks as a solution.
The solution is 1.

Occasionally, a value of the variable that appears to be a solution will make one of the denominators zero. In this case, the equation has no solution for that value of the variable.

Solve: $\frac{2x}{x-2} = 1 + \frac{4}{x-2}$

The LCM is $x - 2$.

$$\frac{2x}{x-2} = 1 + \frac{4}{x-2}$$

Multiply each side of the equation by the LCM of the denominators.

$$(x-2)\frac{2x}{x-2} = (x-2)\left(1 + \frac{4}{x-2}\right)$$

Simplify using the Distributive Property and Properties of Fractions.

$$(x-2)\left(\frac{2x}{x-2}\right) = (x-2) \cdot 1 + (x-2) \cdot \frac{4}{x-2}$$

$$\frac{\overset{1}{\cancel{x-2}}}{1} \cdot \frac{2x}{\cancel{x-2}} = (x-2) + \frac{\overset{1}{\cancel{x-2}}}{1} \cdot \frac{4}{\cancel{x-2}}$$

Solve for x.

$$2x = x - 2 + 4$$
$$2x = x + 2$$
$$x = 2$$

When x is replaced by 2, the denominators of $\frac{2x}{x-2}$ and $\frac{4}{x-2}$ are zero. Therefore, the equation has no solution

Example 1

Solve: $\dfrac{x}{x+4} = \dfrac{2}{x}$

Solution

The LCM is $x(x+4)$.

$$\frac{x}{x+4} = \frac{2}{x}$$

$$x(x+4)\left(\frac{x}{x+4}\right) = x(x+4)\left(\frac{2}{x}\right)$$

$$\frac{x(\cancel{x+4})}{1} \cdot \frac{x}{\cancel{x+4}} = \frac{\cancel{x}(x+4)}{1} \cdot \frac{2}{\cancel{x}}$$

$$x^2 = (x+4)2$$
$$x^2 = 2x + 8$$

Solve the quadratic equation by factoring.

$$x^2 - 2x - 8 = 0$$
$$(x-4)(x+2) = 0$$
$$x - 4 = 0 \qquad x + 2 = 0$$
$$x = 4 \qquad x = -2$$

Both 4 and −2 check as solutions.
The solutions are 4 and −2.

Example 2

Solve: $\dfrac{x}{x+6} = \dfrac{3}{x}$

Your solution

6 and −3

Example 3

Solve: $\dfrac{3x}{x-4} = 5 + \dfrac{12}{x-4}$

Solution

The LCM is $x - 4$.

$$\frac{3x}{x-4} = 5 + \frac{12}{x-4}$$

$$(x-4)\left(\frac{3x}{x-4}\right) = (x-4)\left(5 + \frac{12}{x-4}\right)$$

$$\frac{(\cancel{x-4})}{1} \cdot \frac{3x}{\cancel{x-4}} = \frac{(x-4)}{1} \cdot 5 + \frac{(\cancel{x-4})}{1} \cdot \frac{12}{\cancel{x-4}}$$

$$3x = (x-4)5 + 12$$
$$3x = 5x - 20 + 12$$
$$3x = 5x - 8$$
$$-2x = -8$$
$$x = 4$$

4 does not check as a solution.
The equation has no solution.

Example 4

Solve: $\dfrac{5x}{x+2} = 3 - \dfrac{10}{x+2}$

Your solution

no solution

Solutions on p. **A25**

Content and Format © 1991 HMCo.

7.5	**EXERCISES**

▶ **Objective A**

Solve.

1. $\dfrac{2x}{3} - \dfrac{5}{2} = -\dfrac{1}{2}$

3

2. $\dfrac{x}{3} - \dfrac{1}{4} = \dfrac{1}{12}$

1

3. $\dfrac{x}{3} - \dfrac{1}{4} = \dfrac{x}{4} - \dfrac{1}{6}$

1

4. $\dfrac{2y}{9} - \dfrac{1}{6} = \dfrac{y}{9} + \dfrac{1}{6}$

3

5. $\dfrac{2x - 5}{8} + \dfrac{1}{4} = \dfrac{x}{8} + \dfrac{3}{4}$

9

6. $\dfrac{3x + 4}{12} - \dfrac{1}{3} = \dfrac{5x + 2}{12} - \dfrac{1}{2}$

2

7. $\dfrac{6}{2a + 1} = 2$

1

8. $\dfrac{12}{3x - 2} = 3$

2

9. $\dfrac{9}{2x - 5} = -2$

$\dfrac{1}{4}$

10. $\dfrac{6}{4 - 3x} = 3$

$\dfrac{2}{3}$

11. $2 + \dfrac{5}{x} = 7$

1

12. $3 + \dfrac{8}{n} = 5$

4

13. $1 - \dfrac{9}{x} = 4$

-3

14. $3 - \dfrac{12}{x} = 7$

-3

15. $\dfrac{2}{y} + 5 = 9$

$\dfrac{1}{2}$

16. $\dfrac{6}{x} + 3 = 11$

$\dfrac{3}{4}$

17. $\dfrac{3}{x - 2} = \dfrac{4}{x}$

8

18. $\dfrac{5}{x + 3} = \dfrac{3}{x - 1}$

7

Solve.

19. $\dfrac{2}{3x - 1} = \dfrac{3}{4x + 1}$

5

20. $\dfrac{5}{3x - 4} = \dfrac{-3}{1 - 2x}$

−7

21. $\dfrac{-3}{2x + 5} = \dfrac{2}{x - 1}$

−1

22. $\dfrac{4}{5y - 1} = \dfrac{2}{2y - 1}$

−1

23. $\dfrac{4x}{x - 4} + 5 = \dfrac{5x}{x - 4}$

5

24. $\dfrac{2x}{x + 2} - 5 = \dfrac{7x}{x + 2}$

−1

25. $2 + \dfrac{3}{a - 3} = \dfrac{a}{a - 3}$

no solution

26. $\dfrac{x}{x + 4} = 3 - \dfrac{4}{x + 4}$

no solution

27. $\dfrac{x}{x - 1} = \dfrac{8}{x + 2}$

2 and 4

28. $\dfrac{x}{x + 12} = \dfrac{1}{x + 5}$

2 and −6

29. $\dfrac{2x}{x + 4} = \dfrac{3}{x - 1}$

$-\dfrac{3}{2}$ and 4

30. $\dfrac{5}{3n - 8} = \dfrac{n}{n + 2}$

$-\dfrac{2}{3}$ and 5

31. $x + \dfrac{6}{x - 2} = \dfrac{3x}{x - 2}$

3

32. $x - \dfrac{6}{x - 3} = \dfrac{2x}{x - 3}$

−1 and 6

33. $\dfrac{8}{y} = \dfrac{2}{y - 2} + 1$

4

34. $\dfrac{8}{r} + \dfrac{3}{r - 1} = 3$

$\dfrac{2}{3}$ and 4

35. $\dfrac{x}{x + 2} + \dfrac{2}{x - 2} = \dfrac{x + 6}{x^2 - 4}$

−1

36. $\dfrac{x}{x + 4} = \dfrac{11}{x^2 - 16} + 2$

−7 and 3

SECTION 7.6   **Ratio and Proportion**

Objective A   **To solve a proportion**

Quantities such as 4 meters, 15 seconds, and 8 gallons are number quantities written with units. In these examples the units are meters, seconds, and gallons.

A **ratio** is the quotient of two quantities that have the same unit.

The length of a living room is 16 ft and the width is 12 ft. The ratio of the length to the width is written

$$\frac{16 \text{ ft}}{12 \text{ ft}} = \frac{16}{12} = \frac{4}{3}$$ A ratio is in simplest form when the two numbers do not have a common factor. Note that the units are not written.

A **rate** is the quotient of two quantities that have different units.

There are 2 lb of salt in 8 gal of water. The salt-to-water rate is

$$\frac{2 \text{ lb}}{8 \text{ gal}} = \frac{1 \text{ lb}}{4 \text{ gal}}$$ A rate is in simplest form when the two numbers do not have a common factor. The units are written as part of the rate.

A **proportion** is an equation that states the equality of two ratios or rates. Examples of proportions are shown at the right.

$$\frac{30 \text{ mi}}{4 \text{ h}} = \frac{15 \text{ mi}}{2 \text{ h}}$$

$$\frac{4}{6} = \frac{8}{12}$$

$$\frac{3}{4} = \frac{x}{8}$$

Solve the proportion $\frac{4}{x} = \frac{2}{3}$.

$$\frac{4}{x} = \frac{2}{3}$$

Multiply each side of the proportion by the LCM of the denominators.
Solve the equation.

$$3x\left(\frac{4}{x}\right) = 3x\left(\frac{2}{3}\right)$$

$$12 = 2x$$

$$6 = x$$

The solution is 6.

Example 1

Solve the proportion $\frac{8}{x+3} = \frac{4}{x}$.

Solution

$$\frac{8}{x+3} = \frac{4}{x}$$

$$x(x+3)\frac{8}{x+3} = x(x+3)\frac{4}{x}$$

$$8x = 4(x+3)$$

$$8x = 4x + 12$$

$$4x = 12$$

$$x = 3$$

The solution is 3.

Example 2

Solve the proportion $\frac{2}{x+3} = \frac{6}{5x+5}$.

Your solution

2

Solution on p. A25

| Objective B | To solve application problems |

Example 3

The monthly loan payment for a car is $28.35 for each $1000 borrowed. At this rate, find the monthly payment for a $6000 car loan.

Strategy

To find the monthly payment, write and solve a proportion, using P to represent the monthly car payment.

Solution

$$\frac{\$28.35}{\$1000} = \frac{P}{\$6000}$$

$$6000\left(\frac{28.35}{1000}\right) = 6000\left(\frac{P}{6000}\right)$$

$$170.10 = P$$

The monthly payment is $170.10.

Example 4

Sixteen ceramic tiles are required to tile a 9-ft² area. At this rate, how many square feet can be tiled using 256 ceramic tiles?

Your strategy

Your solution

144 ft²

Example 5

An investment of $500 earns $60 each year. At the same rate, how much additional money must be invested to earn $90 each year?

Strategy

To find the additional amount of money that must be invested, write and solve a proportion, using x to represent the additional money. Then $500 + x$ is the total amount invested.

Solution

$$\frac{\$60}{\$500} = \frac{\$90}{\$500 + x}$$

$$\frac{3}{25} = \frac{90}{500 + x}$$

$$25(500 + x)\left(\frac{3}{25}\right) = 25(500 + x)\left(\frac{90}{500 + x}\right)$$

$$(500 + x)3 = 25(90)$$

$$1500 + 3x = 2250$$

$$3x = 750$$

$$x = 250$$

An additional $250 must be invested.

Example 6

Three ounces of a medication are required for a 150-lb adult. At the same rate, how many additional ounces of medication are required for a 200-lb adult?

Your strategy

Your solution

1 oz

Solutions on pp. A25–A26

Content and Format © 1991 HMCo.

| 7.6 | **EXERCISES** |

▶ **Objective A**

Solve.

1. $\dfrac{x}{12} = \dfrac{3}{4}$
9

2. $\dfrac{6}{x} = \dfrac{2}{3}$
9

3. $\dfrac{4}{9} = \dfrac{x}{27}$
12

4. $\dfrac{16}{9} = \dfrac{64}{x}$
36

5. $\dfrac{x+3}{12} = \dfrac{5}{6}$
7

6. $\dfrac{3}{5} = \dfrac{x-4}{10}$
10

7. $\dfrac{18}{x+4} = \dfrac{9}{5}$
6

8. $\dfrac{2}{11} = \dfrac{20}{x-3}$
113

9. $\dfrac{2}{x} = \dfrac{4}{x+1}$
1

10. $\dfrac{16}{x-2} = \dfrac{8}{x}$
−2

11. $\dfrac{x+3}{4} = \dfrac{x}{8}$
−6

12. $\dfrac{x-6}{3} = \dfrac{x}{5}$
15

13. $\dfrac{2}{x-1} = \dfrac{6}{2x+1}$
4

14. $\dfrac{9}{x+2} = \dfrac{3}{x-2}$
4

15. $\dfrac{2x}{7} = \dfrac{x-2}{14}$
$-\dfrac{2}{3}$

▶ **Objective B**   *Application Problems*

Solve.

16. Simple syrup used in making some desserts requires 2 cups of sugar for every 2/3 cup of boiling water. At this rate, how many cups of sugar are required for 2 cups of boiling water?
6 cups

17. An exit poll survey showed that 4 out of every 7 voters cast a ballot in favor of an amendment to a city charter. At this rate, how many voters voted in favor of the amendment if 35,000 people voted?
20,000 voters

18. A quality control inspector found 3 defective transistors in a shipment of 500 transistors. At this rate, how many defective transistors are there in a shipment of 2000 transistors?
12 transistors

19. An air conditioning specialist recommends 2 air vents for each 300 ft² of space. At this rate, how many air vents are required for a 21,000-ft² office building?
140 vents

Solve.

20. A company decides to accept a large shipment of 10,000 computer chips if there are 2 or fewer defects in a sample of 100 randomly chosen chips. Assuming that there are 300 defective chips in the shipment and that the rate of defective chips in the sample is the same as the rate in the shipment, will the shipment be accepted?
No

21. A company decides to accept a large shipment of 20,000 precision bearings if there are 3 or fewer defects in a sample of 100 randomly chosen bearings. Assuming that there are 400 defective bearings in the shipment and that the rate of defective bearings in the sample is the same as the rate in the shipment, will the shipment be accepted?
Yes

22. The lighting for some billboards is provided by using solar energy. If 3 small solar energy panels can generate 10 watts of power, how many panels are necessary to provide 600 watts of power?
180 panels

23. A laser printer is rated by the number of pages per minute it can print. An inexpensive laser printer can print 5 pages every 2 minutes. At this rate, how long would it take to print a document 45 pages long?
18 min

24. As part of a conservation effort for a lake, 40 fish are caught, tagged, and then released. Later 80 fish are caught. Four of the 80 fish are found to have tags. Estimate the number of fish in the lake.
800 fish

25. In a wildlife preserve, 10 elk are captured, tagged, and then released. Later 15 elk are captured and 2 are found to have tags. Estimate the number of elk in the preserve.
75 elk

26. A painter estimates that 5 gal of paint will cover 1200 ft^2 of wall space. At this rate, how many additional gallons will be necessary to cover 1680 ft^2?
2 gal

27. A health department estimated that 8 vials of a malaria serum will treat 100 people. At this rate, how many additional vials will be necessary to treat 175 people?
6 vials

28. The engine of a small rocket burns 170,000 lb of fuel in 1 min. At this rate, how many pounds of fuel are burned in 45 s?
127,500 lb

29. To allow as much natural lighting as possible and still conserve energy, an architect suggests that the ratio of the area of a window to the area of the total wall space be 5:12. At this suggested ratio, what should be the area of a window for a wall that measures 8 ft by 12 ft? (Recall that 5:12 can be written as the fraction $\frac{5}{12}$.)
40 ft^2

SECTION 7.7   Literal Equations

Objective A   To solve a literal equation for one of the variables

A **literal equation** is an equation that contains more than one variable. Examples of literal equations are shown at the right.

$$2x + 3y = 6$$
$$4w - 2x + z = 0$$

Formulas are used to express a relationship among physical quantities. A **formula** is a literal equation that states rules about measurements. Examples of formulas are shown at the right.

$$\frac{1}{R_1} + \frac{1}{R_2} = \frac{1}{R}$$ (Physics)

$$s = a + (n - 1)d$$ (Mathematics)

$$A = P + Prt$$ (Business)

The Addition and Multiplication Properties can be used to solve a literal equation for one of the variables. The goal is to rewrite the equation so that the letter being solved for is alone on one side of the equation and all the other numbers and variables are on the other side.

Solve $A = P(1 + i)$ for i.

The goal is to rewrite the equation so that i is on one side of the equation and all other variables are on the other side.

Use the Distributive Property to remove parentheses.

$$A = P(1 + i)$$
$$A = P + Pi$$

Subtract P from each side of the equation.

$$A - P = P - P + Pi$$
$$A - P = Pi$$

Divide each side of the equation by P.

$$\frac{A - P}{P} = \frac{Pi}{P}$$
$$\frac{A - P}{P} = i$$

Example 1

Solve $3x - 4y = 12$ for y.

Solution

$$3x - 4y = 12$$
$$3x - 3x - 4y = -3x + 12$$
$$-4y = -3x + 12$$

$$\frac{-4y}{-4} = \frac{-3x + 12}{-4}$$

$$y = \frac{3}{4}x - 3$$

Example 2

Solve $5x - 2y = 10$ for y.

Your solution

$$y = \frac{5}{2}x - 5$$

Solution on p. A26

Example 3

Solve $I = \dfrac{E}{R + r}$ for R.

Solution

$$I = \frac{E}{R + r}$$

$$(R + r)I = (R + r)\frac{E}{R + r}$$

$$RI + rI = E$$

$$RI + rI - rI = E - rI$$

$$RI = E - rI$$

$$\frac{RI}{I} = \frac{E - rI}{I}$$

$$R = \frac{E - rI}{I}$$

Example 4

Solve $s = \dfrac{A + L}{2}$ for L.

Your solution

$L = 2s - A$

Example 5

Solve $L = a(1 + ct)$ for c.

Solution

$$L = a(1 + ct)$$

$$L = a + act$$

$$L - a = a - a + act$$

$$L - a = act$$

$$\frac{L - a}{at} = \frac{act}{at}$$

$$\frac{L - a}{at} = c$$

Example 6

Solve $S = a + (n - 1)d$ for n.

Your solution

$n = \dfrac{S - a + d}{d}$

Example 7

Solve $S = C - rC$ for C.

Solution

$$S = C - rC$$

$$S = (1 - r)C$$

$$\frac{S}{1 - r} = \frac{(1 - r)C}{1 - r}$$

$$\frac{S}{1 - r} = C$$

Example 8

Solve $S = C + rC$ for C.

Your solution

$C = \dfrac{S}{1 + r}$

Solutions on p. A26

7.7   EXERCISES

▶ **Objective A**

Solve for y.

1. $3x + y = 10$
$y = -3x + 10$

2. $2x + y = 5$
$y = -2x + 5$

3. $4x - y = 3$
$y = 4x - 3$

4. $5x - y = 7$
$y = 5x - 7$

5. $3x + 2y = 6$
$y = -\dfrac{3}{2}x + 3$

6. $2x + 3y = 9$
$y = -\dfrac{2}{3}x + 3$

7. $2x - 5y = 10$
$y = \dfrac{2}{5}x - 2$

8. $5x - 2y = 4$
$y = \dfrac{5}{2}x - 2$

9. $2x + 7y = 14$
$y = -\dfrac{2}{7}x + 2$

10. $6x - 5y = 10$
$y = \dfrac{6}{5}x - 2$

11. $x + 3y = 6$
$y = -\dfrac{1}{3}x + 2$

12. $x + 2y = 8$
$y = -\dfrac{1}{2}x + 4$

13. $x - 4y = 12$
$y = \dfrac{1}{4}x - 3$

14. $x - 3y = 9$
$y = \dfrac{1}{3}x - 3$

15. $7x - 2y - 14 = 0$
$y = \dfrac{7}{2}x - 7$

16. $2x - 9y - 18 = 0$
$y = \dfrac{2}{9}x - 2$

17. $3x - y + 7 = 0$
$y = 3x + 7$

18. $2x - y + 5 = 0$
$y = 2x + 5$

Solve for x.

19. $x + 3y = 6$
$x = -3y + 6$

20. $x + 6y = 10$
$x = -6y + 10$

21. $3x - y = 3$
$x = \dfrac{1}{3}y + 1$

22. $2x - y = 6$
$x = \dfrac{1}{2}y + 3$

23. $2x + 5y = 10$
$x = -\dfrac{5}{2}y + 5$

24. $4x + 3y = 12$
$x = -\dfrac{3}{4}y + 3$

25. $x - 2y + 1 = 0$
$x = 2y - 1$

26. $x - 4y - 3 = 0$
$x = 4y + 3$

27. $5x + 4y + 20 = 0$
$x = -\dfrac{4}{5}y - 4$

28. $3x + 5y + 15 = 0$
$x = -\dfrac{5}{3}y - 5$

29. $3x - 2y - 15 = 0$
$x = \dfrac{2}{3}y + 5$

30. $5x - 8y + 10 = 0$
$x = \dfrac{8}{5}y - 2$

Solve the formula for the given variable.

31. $A = \frac{1}{2}bh$; h (Geometry)

$h = \dfrac{2A}{b}$

32. $P = a + b + c$; b (Geometry)

$b = P - a - c$

33. $d = rt$; t (Physics)

$t = \dfrac{d}{r}$

34. $E = IR$; R (Physics)

$R = \dfrac{E}{I}$

35. $PV = nRT$; T (Chemistry)

$T = \dfrac{PV}{nR}$

36. $A = bh$; h (Geometry)

$h = \dfrac{A}{b}$

37. $P = 2l + 2w$; l (Geometry)

$l = \dfrac{P - 2w}{2}$

38. $F = \frac{9}{5}C + 32$; C (Temperature conversion)

$C = \dfrac{5F - 160}{9}$

39. $A = \frac{1}{2}h(b_1 + b_2)$; b_1 (Geometry)

$b_1 = \dfrac{2A - hb_2}{h}$

40. $C = \frac{5}{9}(F - 32)$; F (Temperature conversion)

$F = \dfrac{9C + 160}{5}$

41. $V = \frac{1}{3}Ah$; h (Geometry)

$h = \dfrac{3V}{A}$

42. $P = R - C$; C (Business)

$C = R - P$

43. $R = \dfrac{C - S}{t}$; S (Business)

$S = C - Rt$

44. $P = \dfrac{R - C}{n}$; R (Business)

$R = Pn + C$

45. $A = P + Prt$; P (Business)

$P = \dfrac{A}{1 + rt}$

46. $T = fm - gm$; m (Engineering)

$m = \dfrac{T}{f - g}$

47. $A = Sw + w$; w (Physics)

$w = \dfrac{A}{S + 1}$

48. $a = S - Sr$; S (Mathematics)

$S = \dfrac{a}{1 - r}$

SECTION 7.8	**Application Problems**

Objective A	To solve work problems

If a painter can paint a room in 4 h, then in 1 h the painter can paint $\frac{1}{4}$ of the room. The painter's rate of work is $\frac{1}{4}$ of the room each hour. The **rate of work** is that part of a task that is completed in one unit of time.

A pipe can fill a tank in 30 min. This pipe can fill $\frac{1}{30}$ of the tank in 1 min. The rate of work is $\frac{1}{30}$ of the tank each minute. If a second pipe can fill the tank in x min, the rate of work for the second pipe is $\frac{1}{x}$ of the tank each minute.

In solving a work problem, the goal is to determine the time it takes to complete a task. The basic equation that is used to solve work problems is

Rate of work × time worked = part of task completed

For example, if a faucet can fill a sink in 6 min, then in 5 min the faucet will fill $\frac{1}{6} \times 5 = \frac{5}{6}$ of the sink. In 5 min the faucet completes $\frac{5}{6}$ of the task.

A painter can paint a wall in 20 min. The painter's apprentice can paint the same wall in 30 min. How long will it take to paint the wall when they work together?

Strategy for Solving a Work Problem

For each person or machine, write a numerical or variable expression for the rate of work, the time worked, and the part of the task completed. The results can be recorded in a table.

Unknown time to paint the wall working together: t

	Rate of work	·	Time worked	=	Part of task completed
Painter	$\frac{1}{20}$	·	t	=	$\frac{t}{20}$
Apprentice	$\frac{1}{30}$	·	t	=	$\frac{t}{30}$

Determine how the parts of the task completed are related. Use the fact that the sum of the parts of the task completed must equal 1; the complete task.

The sum of the part of the task completed by the painter and the part of the task completed by the apprentice is 1.

$$\frac{t}{20} + \frac{t}{30} = 1$$

$$60\left(\frac{t}{20} + \frac{t}{30}\right) = 60 \cdot 1$$

$$3t + 2t = 60$$

$$5t = 60$$

$$t = 12$$

Working together, they will paint the wall in 12 min.

Example 1

A small water pipe takes three times longer to fill a tank than does a large water pipe. With both pipes open it takes 4 h to fill the tank. Find the time it would take the small pipe working alone to fill the tank.

Example 2

Two computer printers that work at the same rate are working together to print the payroll checks for a large corporation. After working together for 2 h, one of the printers quits. The second requires 3 more hours to complete the payroll checks. Find the time it would take one printer working alone to print the payroll.

Strategy

- Time for large pipe to fill the tank: t
  Time for small pipe to fill the tank: $3t$

	Rate	Time	Part
Small pipe	$\frac{1}{3t}$	4	$\frac{4}{3t}$
Large pipe	$\frac{1}{t}$	4	$\frac{4}{t}$

- The sum of the parts of the task completed by each pipe must equal one.

Your strategy

Solution

$$\frac{4}{3t} + \frac{4}{t} = 1$$

$$3t\left(\frac{4}{3t} + \frac{4}{t}\right) = 3t \cdot 1$$

$$4 + 12 = 3t$$
$$16 = 3t$$
$$\frac{16}{3} = t$$

$$3t = 3\left(\frac{16}{3}\right) = 16$$

The small pipe working alone takes 16 h to fill the tank.

Your solution

7 h

Solution on p. A27

Content and Format © 1991 HMCo.

| Objective B | **To solve uniform motion problems** |

A car that travels constantly in a straight line at 30 mph is in uniform motion. **Uniform motion** means that the speed or direction of an object does not change.

The basic equation used to solve uniform motion problems is

$$\text{Distance} = \text{rate} \times \text{time}$$

An alternate form of this equation can be written by solving the equation for time.

$$\frac{\text{Distance}}{\text{Rate}} = \text{time}$$

This form of the equation is useful when the total time of travel for two objects or the time of travel between two points is known.

The speed of a boat in still water is 20 mph. The boat traveled 75 mi down a river in the same amount of time it took to travel 45 mi up the river. Find the rate of the river's current.

Strategy for Solving a Uniform Motion Problem

> For each object, write a numerical or variable expression for the distance, rate, and time. The results can be recorded in a table.

The unknown rate of the river's current: r

	Distance	÷	*Rate*	=	*Time*
Down river	75	÷	$20 + r$	=	$\frac{75}{20 + r}$
Up river	45	÷	$20 - r$	=	$\frac{45}{20 - r}$

> Determine how the times traveled by each object are related. For example, it may be known that the times are equal, or the total time may be known.

The time down the river is equal to the time up the river.

$$\frac{75}{20 + r} = \frac{45}{20 - r}$$

$$(20 + r)(20 - r)\frac{75}{20 + r} = (20 + r)(20 - r)\frac{45}{20 - r}$$

$$(20 - r)75 = (20 + r)45$$

$$1500 - 75r = 900 + 45r$$

$$-120r = -600$$

$$r = 5$$

The rate of the river's current is 5 mph.

Example 3

A cyclist rode the first 20 mi of a trip at a constant rate. For the next 16 mi, the cyclist reduced the speed by 2 mph. The total time for the 36 mi was 4 h. Find the rate of the cyclist for each leg of the trip.

Example 4

The total time for a sailboat to sail back and forth across a lake 6 km wide was 2 h. The rate sailing back was three times the rate sailing across. Find the rate across the lake.

Strategy

■ Rate for the first 20 mi: r

	Distance	*Rate*	*Time*
First 20 mi	20	r	$\dfrac{20}{r}$
Next 16 mi	16	$r - 2$	$\dfrac{16}{r - 2}$

■ The total time for the trip was 4 h.

Your strategy

Solution

$$\frac{20}{r} + \frac{16}{r - 2} = 4$$

$$r(r - 2)\left[\frac{20}{r} + \frac{16}{r - 2}\right] = r(r - 2) \cdot 4$$

$$(r - 2)20 + 16r = 4r^2 - 8r$$
$$20r - 40 + 16r = 4r^2 - 8r$$
$$36r - 40 = 4r^2 - 8r$$

Solve the quadratic equation by factoring.

$$0 = 4r^2 - 44r + 40$$
$$0 = 4(r^2 - 11r + 10)$$
$$0 = 4(r - 10)(r - 1)$$

$$r - 10 = 0 \qquad\quad r - 1 = 0$$
$$r = 10 \qquad\qquad r = 1$$

The solution $r = 1$ mph is not possible, because the rate on the last 16 mi would then be -1 mph.

10 mph was the rate for the first 20 mi.
8 mph was the rate for the next 16 mi.

Your solution

4 km/h

Solution on p. A27

7.8   EXERCISES

▶ **Objective A**   *Application Problems*

Solve.

1. An experienced painter can paint a garage twice as fast as an apprentice. Working together, the painters require 4 h to paint the garage. How long would it take the experienced painter, working alone, to paint the garage?
6 h

2. One grocery clerk can stock a shelf in 20 min, whereas a second clerk requires 30 min to stock the same shelf. How long would it take to stock the shelf if the two clerks worked together?
12 min

3. One person with a skiploader requires 12 h to remove a large quantity of earth. A second, larger skiploader can remove the same amount of earth in 4 h. How long would it take to remove the earth with both skiploaders working together?
3 h

4. One worker can dig the trenches for a sprinkler system in 3 h, whereas a second worker requires 6 h to do the same task. How long would it take to dig the trenches with both people working together?
2 h

5. One computer can solve a complex prime factorization problem in 75 h. A second computer can solve the same problem in 50 h. How long would it take both computers, working together, to solve the problem?
30 h

6. A new machine can make 10,000 aluminum cans three times faster than an older machine. With both machines working, 10,000 cans can be made in 9 h. How long would it take the new machine, working alone, to make the 10,000 cans?
12 h

7. A small air conditioner will cool a room 2° in 15 min. A larger air conditioner will cool the room 2° in 10 min. How long would it take to cool the room 2° with both air conditioners operating?
6 min

8. One printing press can print the first edition of a book in 55 min, whereas a second printing press requires 66 min to print the same number of copies. How long would it take to print the first edition with both presses operating?
30 min

9. Two welders working together can complete a job in 6 h. One of the welders, working alone, can complete the task in 10 h. How long would it take the second welder, working alone, to complete the task?
15 h

10. Two oil pipelines can fill a small tank in 30 min. Using one of the pipelines would require 45 min to fill the tank. How long would it take the second pipeline alone to fill the tank?
90 min

Solve.

11. With two harvesters, a plot of land can be harvested in 1 h. One harvester, working alone, requires 1.5 h to harvest the field. How long would it take the second harvester, working alone, to harvest the field?
    3 h

12. Working together, two dock workers can load a crate in 6 min. One dock worker, working alone, can load the crate in 15 min. How long would it take the second dock worker, working alone, to load the crate?
    10 min

13. A cement mason can build a barbeque in 8 h, while it takes a second mason 12 h to do the same task. After working alone for 4 h, the first mason quits. How long will it take the second mason to complete the task?
    6 h

14. A mechanic requires 2 h to repair a transmission, while an apprentice requires 6 h to make the same repairs. The mechanic worked alone for 1 h and then stopped. How long will it take the apprentice, working alone, to complete the repairs?
    3 h

15. One computer technician can wire a modem in 4 h, while it takes 6 h for a second technician to do the same job. After working alone for 2 h, the first technician quit. How long will it take the second technician to complete the wiring?
    3 h

16. A wallpaper hanger requires 2 h to hang the wallpaper on one wall of a room. A second wallpaper hanger requires 4 h to hang the same amount of paper. The first wallpaper hanger worked alone for one hour and then quit. How long will it take the second wallpaper hanger, working alone, to complete the wall?
    2 h

17. Two welders who work at the same rate are welding the girders of a building. After they work together for 10 h, one of the welders quits. The second welder requires 20 more hours to complete the welds. Find the time it would have taken one of the welders, working alone, to complete the welds.
    40 h

18. A large and a small heating unit are being used to heat the water of a pool. The larger unit, working alone, requires 8 h to heat the pool. After both units have been operating 2 h, the larger unit is turned off. The small unit requires 9 more hours to heat the pool. How long would it take the small unit working alone to heat the pool?
    $14\frac{2}{3}$ h

19. Two machines that fill cereal boxes work at the same rate. After they work together for 7 h, one machine breaks down. The second machine requires 14 more hours to finish filling the boxes. How long would it have taken one of the machines, working alone, to fill the boxes?
    28 h

20. A large and a small drain are opened to drain a pool. The large drain can empty a pool in 6 h. After both drains have been open for one hour, the large drain becomes clogged and is closed. The smaller drain remains open and requires 9 more hours to empty the pool. How long would it have taken the small drain, working alone, to empty the pool?
    12 h

Content and Format © 1991 HMCo.

▶ **Objective B**   *Application Problems*

Solve.

21. A camper drove 90 mi to a recreational area and then hiked 5 mi into the wilderness. The rate of the camper while driving in the car was nine times the rate hiking. The time spent hiking and driving was 3 h. Find the rate at which the camper hiked.
5 mph

22. The president of a company traveled 1800 mi by jet and 300 mi on a prop plane. The rate of the jet is four times the rate of the prop plane. The entire trip took a total of 5 h. Find the rate of the jet plane.
600 mph

23. As part of a conditioning program, a jogger ran 8 mi in the same time a cyclist rode 20 mi. The rate of the cyclist was 12 mph faster than the rate of the jogger. Find the rate of the jogger and that of the cyclist.
jogger 8 mph; cyclist 20 mph

24. An express train travels 600 mi in the same amount of time it takes a freight train to travel 360 mi. The rate of the express train is 20 mph faster than that of the freight train. Find the rate of each train.
freight 30 mph; express 50 mph

25. To assess the damage done by a fire, a forest ranger travels 1080 mi by jet and then an additional 180 mi by helicopter. The rate of the jet is 4 times the rate of the helicopter. The entire trip took a total of 5 h. Find the rate of the jet.
360 mph

26. A twin-engine plane can fly 800 mi in the same time that it takes a single-engine plane to fly 600 mi. The rate of the twin-engine plane is 50 mph faster than that of the single-engine plane. Find the rate of the twin-engine plane.
200 mph

27. Two planes leave an airport and head for another airport 900 mi away. The rate of the first plane is twice that of the second plane. The second plane arrives at the airport 3 h after the first plane. Find the rate of the second plane.
150 mph

28. A car and a bus leave a town at 1 P.M. and head for a town 300 mi away. The rate of the car is twice the rate of the bus. The car arrives 5 h ahead of the bus. Find the rate of the car.
60 mph

29. A car is traveling at a rate that is 36 mph faster than the rate of a cyclist. The car travels 384 mi in the same time it takes the cyclist to travel 96 mi. Find the rate of the car.
48 mph

30. An engineer traveled 165 mi by car and then an additional 660 mi by plane. The rate of the plane was 4 times the rate of the car, and the total trip took 6 h. Find the rate of the car.
55 mph

Solve.

31. A backpacker hiking into a wilderness area walked 9 mi at a constant rate and then reduced this rate by 1 mph. Another 4 mi was hiked at this reduced rate. The time required to hike the 4 mi was 1 h less than the time required to walk the 9 mi. Find the rate at which the hiker walked the first 9 mi.
3 mph

32. A sailboat sailed 15 mi on the first leg of a trip before changing direction and sailing an additional 7 mi. Because of the wind, the change caused the sailboat to increase its speed by 2 mph. The total sailing time was 4 h. Find the rate of sailing for the first leg of the trip.
5 mph

33. A small motor on a fishing boat can move the boat 6 mph in calm water. When trolling in a river, the amount of time it takes to travel 12 mi against the river's current is the same as the time it takes to travel 24 mi with the current. Find the rate of the current.
2 mph

34. A commercial jet can fly 550 mph in calm air. Traveling with the jet stream, the plane flew 2400 mi in the same amount of time it takes to fly 2000 mi against the jet stream. Find the rate of the jet stream.
50 mph

35. A cruise ship can sail at 28 mph in calm water. Sailing with the gulf current, the ship can sail 170 mi in the same amount of time that it can sail 110 mi against the gulf current. Find the rate of the gulf current.
6 mph

36. Paddling in calm water, a canoeist can paddle at a rate of 8 mph. Traveling with the current, the canoeist went 30 mi in the same amount of time it took to travel 18 mi against the current. Find the rate of the current.
2 mph

37. On a recent trip, a trucker traveled 330 mi at a constant rate. Because of road construction, the trucker then had to reduce speed by 25 mph. An additional 30 mi was traveled at the reduced rate. The total time for the entire trip was 7 h. Find the rate of the trucker for the first 330 mi.
55 mph

38. Commuting from work to home, a lab technician traveled 10 mi at a constant rate through congested traffic. Upon reaching the expressway, the technician increased speed by 20 mph. An additional 20 mi was traveled at the increased rate. The total time for the trip was 1 h. Find the rate in the congested traffic.
20 mph

39. Rowing with the current of a river, a rowing team can row 25 mi in the same amount of time it takes to row 15 mi against the current. The rate of the rowing team in calm water is 20 mph. Find the rate of the current.
5 mph

40. A plane can fly 180 mph in calm air. Flying with the wind, the plane can fly 600 mi in the same time it takes to fly 480 mi against the wind. Find the rate of the wind.
20 mph

Calculators and Computers

Simplifying an Algebraic Fraction

The first three sections of Chapter 7 present operations with algebraic fractions. Many students find performing operations on algebraic fractions difficult, especially adding and subtracting those with unlike denominators. The difficulty often lies in the fact that a number of steps must be performed in any one problem, and it is essential that the student understand the function of each step of the process. It is helpful to keep in mind that addition, subtraction, multiplication, and division performed on algebraic fractions involve the same steps as these operations performed on numerical fractions. It is the fractions themselves that make the procedure look more difficult, and the greater amount of time it takes to perform the procedures makes them seem so much more complicated. Therefore, you are encouraged to practice as much as possible.

The program SIMPLIFY AN ALGEBRAIC EXPRESSION on the Math ACE Disk will enable you to practice writing an algebraic fraction in simplest form. The program allows you to choose one of three levels of difficulty. Level one contains the easiest problems, and level three the most difficult.

After you have chosen a level of difficulty, the program will display a problem. Using paper and pencil, simplify the expression. Then press the RETURN key. The correct solution will be displayed. After you have completed a problem, you may continue with problems of the same level of difficulty, return to the menu and change the level, or quit the program.

Simplifying these expressions is an important part of your algebra training. Continued practice of this skill will pay rewards in your future math classes.

Chapter Summary

Key Words

An *algebraic fraction* is a fraction in which the numerator or denominator is a variable expression.

An algebraic fraction is in *simplest form* when the numerator and denominator have no common factors.

The *reciprocal* of a fraction is a fraction with the numerator and denominator interchanged.

The *least common multiple* (LCM) of two or more numbers is the smallest number that contains the prime factorization of each number.

A *complex fraction* is a fraction whose numerator or denominator contains one or more fractions.

A *ratio* is the quotient of two quantities that have the same unit.

A *rate* is the quotient of two quantities that have different units.

A *proportion* is an equation that states the equality of two ratios or rates.

A *literal equation* is an equation that contains more than one variable.

A *formula* is a literal equation that states rules about measurements.

Essential Rules	*To multiply fractions:*	$\dfrac{a}{b} \cdot \dfrac{c}{d} = \dfrac{ac}{bd}$
	To divide fractions:	$\dfrac{a}{b} \div \dfrac{c}{d} = \dfrac{a}{b} \cdot \dfrac{d}{c}$
	To add fractions:	$\dfrac{a}{c} + \dfrac{b}{c} = \dfrac{a + b}{c}$
	To subtract fractions:	$\dfrac{a}{c} - \dfrac{b}{c} = \dfrac{a - b}{c}$
	Equation for Work Problems:	$\begin{array}{c}\text{Rate of} \\ \text{work}\end{array} \times \begin{array}{c}\text{time} \\ \text{worked}\end{array} = \begin{array}{c}\text{part of task} \\ \text{completed}\end{array}$
	Uniform Motion Equation:	Distance = rate × time

Chapter Review

SECTION 7.1

1. Simplify $\dfrac{16x^5y^3}{24xy^{10}}$.

$\dfrac{2x^4}{3y^7}$

2. Simplify $\dfrac{x^2 + x - 30}{15 + 2x - x^2}$.

$-\dfrac{x+6}{x+3}$

3. Simplify $\dfrac{8ab^2}{15x^3y} \cdot \dfrac{5xy^4}{16a^2b}$.

$\dfrac{by^3}{6ax^2}$

4. Simplify $\dfrac{3x^3 + 10x^2}{10x - 2} \cdot \dfrac{20x - 4}{6x^4 + 20x^3}$.

$\dfrac{1}{x}$

5. Simplify $\dfrac{24x^2 - 94x + 15}{12x^2 - 49x + 15} \cdot \dfrac{24x^2 + 7x - 5}{4 - 27x + 18x^2}$.

$\dfrac{8x+5}{3x-4}$

6. Simplify $\dfrac{6a^2b^7}{25x^3y} \div \dfrac{12a^3b^4}{5x^2y^2}$.

$\dfrac{b^3y}{10ax}$

7. Simplify $\dfrac{20x^2 - 45x}{6x^3 + 4x^2} \div \dfrac{40x^3 - 90x^2}{12x^2 + 8x}$.

$\dfrac{1}{x^2}$

8. Simplify $\dfrac{10 - 23y + 12y^2}{6y^2 - y - 5} \div \dfrac{4y^2 - 13y + 10}{18y^2 + 3y - 10}$.

$\dfrac{(3y-2)^2}{(y-1)(y-2)}$

SECTION 7.2

9. Find the LCM of $10x^2 - 11x + 3$ and $20x^2 - 17x + 3$.

$(5x - 3)(2x - 1)(4x - 1)$

10. Write each fraction in terms of the LCM of the denominators.

$$\dfrac{x}{12x^2 + 16x - 3}, \dfrac{4x^2}{6x^2 + 7x - 3}$$

$\dfrac{3x^2 - x}{(2x + 3)(6x - 1)(3x - 1)}, \dfrac{24x^3 - 4x^2}{(2x + 3)(6x - 1)(3x - 1)}$

SECTION 7.3

11. Simplify $\dfrac{5x + 3}{2x^2 + 5x - 3} - \dfrac{3x + 4}{2x^2 + 5x - 3}$.

$\dfrac{1}{x+3}$

12. Simplify $\dfrac{x + 7}{15x} + \dfrac{x - 2}{20x}$.

$\dfrac{7x + 22}{60x}$

13. Simplify $\dfrac{2y}{5y - 7} + \dfrac{3}{7 - 5y}$.

$\dfrac{2y - 3}{5y - 7}$

14. Simplify $\dfrac{x - 1}{x + 2} + \dfrac{3x - 2}{5 - x} + \dfrac{5x^2 + 15x - 11}{x^2 - 3x - 10}$.

$\dfrac{3x - 1}{x - 5}$

SECTION 7.4

15. Simplify $\dfrac{1 - \dfrac{1}{x}}{1 - \dfrac{8x - 7}{x^2}}$.

$\dfrac{x}{x - 7}$

16. Simplify $\dfrac{x + \dfrac{6}{x - 5}}{1 + \dfrac{2}{x - 5}}$.

$x - 2$

17. Simplify $\dfrac{x - \dfrac{16}{5x - 2}}{3x - 4 - \dfrac{88}{5x - 2}}$.

$\dfrac{x - 2}{3x - 10}$

SECTION 7.5

18. Solve $\frac{5}{7} + \frac{x}{2} = 2 - \frac{x}{7}$.

2

19. Solve $\frac{x+8}{x+4} = 1 + \frac{5}{x+4}$.

no solution

20. Solve $\frac{20}{2x+3} = \frac{17x}{2x+3} - 5$.

5

SECTION 7.6

21. Solve $\frac{3}{20} = \frac{x}{80}$.

12

22. Solve $\frac{20}{x+2} = \frac{5}{16}$.

62

23. Solve $\frac{6}{x-7} = \frac{8}{x-6}$.

10

24. A pitcher's ERA, or "earned run average," is the average number of runs allowed in 9 innings of pitching. If a pitcher allows 15 runs in 100 innings, find the pitcher's ERA.

1.35

SECTION 7.7

25. Solve $4x + 9y = 18$ for y.

$y = -\frac{4}{9}x + 2$

26. Solve $i = \frac{100m}{c}$ for c.

$c = \frac{100m}{i}$

27. Solve $T = 2(ab + bc + ca)$ for a.

$a = \frac{T - 2bc}{2b + 2c}$

SECTION 7.8

28. One hose can fill a pool in 15 h. A second hose can fill the pool in 10 h. How long would it take to fill the pool using both hoses?

6 h

29. A car travels 315 mi in the same amount of time that a bus travels 245 mi. The rate of the car is 10 mph faster than the bus. Find the rate of the car.

45 mph

30. The rate of a jet is 400 mph in calm air. Traveling with the wind the jet can fly 2100 mi in the same amount of time as it flies 1900 mi against the wind. Find the rate of the wind.

20 mph

Chapter Test

1. Simplify: $\dfrac{16x^5y}{24x^2y^4}$

   $\dfrac{2x^3}{3y^3}$ [7.1A]

2. Simplify: $\dfrac{x^2 + 4x - 5}{1 - x^2}$

   $-\dfrac{x + 5}{x + 1}$ [7.1A]

3. Simplify: $\dfrac{x^3y^4}{x^2 - 4x + 4} \cdot \dfrac{x^2 - x - 2}{x^6y^4}$

   $\dfrac{x + 1}{x^3(x - 2)}$ [7.1B]

4. Simplify: $\dfrac{x^2 + 2x - 3}{x^2 + 6x + 9} \cdot \dfrac{2x^2 - 11x + 5}{2x^2 + 3x - 5}$

   $\dfrac{(x - 5)(2x - 1)}{(x + 3)(2x + 5)}$ [7.1B]

5. Simplify: $\dfrac{x^2 + 3x + 2}{x^2 + 5x + 4} \div \dfrac{x^2 - x - 6}{x^2 + 2x - 15}$

   $\dfrac{x + 5}{x + 4}$ [7.1C]

6. Find the LCM of $6x - 3$ and $2x^2 + x - 1$.

   $3(2x - 1)(x + 1)$ [7.2A]

7. Write each fraction in terms of the LCM of the denominators.

   $\dfrac{3}{x^2 - 2x}, \dfrac{x}{x^2 - 4}$

   $\dfrac{3x + 6}{x(x - 2)(x + 2)}, \dfrac{x^2}{x(x - 2)(x + 2)}$ [7.2B]

8. Simplify: $\dfrac{2x}{x^2 + 3x - 10} - \dfrac{4}{x^2 + 3x - 10}$

   $\dfrac{2}{x + 5}$ [7.3A]

9. Simplify: $\dfrac{2}{2x - 1} - \dfrac{3}{3x + 1}$

   $\dfrac{5}{(2x - 1)(3x + 1)}$ [7.3B]

10. Simplify: $\dfrac{x}{x + 3} - \dfrac{2x - 5}{x^2 + x - 6}$

    $\dfrac{x^2 - 4x + 5}{(x - 2)(x + 3)}$ [7.3B]

11. Simplify: $\dfrac{1 + \frac{1}{x} - \frac{12}{x^2}}{1 + \frac{2}{x} - \frac{8}{x^2}}$

$\dfrac{x-3}{x-2}$ [7.4A]

12. Solve: $\dfrac{6}{x} - 2 = 1$

2   [7.5A]

13. Solve: $\dfrac{2x}{x+1} - 3 = \dfrac{-2}{x+1}$

no solution   [7.5A]

14. Solve the proportion.
$\dfrac{3}{x+4} = \dfrac{5}{x+6}$

-1 [7.6A]

15. A salt water solution is formed by mixing 4 lb of salt with 10 gal of water. At this rate, how many additional pounds of salt are required for 15 gal of water?
2 lb   [7.6B]

16. A landscape architect uses three sprinklers for each 200 ft² of lawn. At this rate, how many sprinklers are needed for a 3600 ft² lawn?
54 sprinklers   [7.6B]

17. Solve $3x - 8y = 16$ for y.
$y = \frac{3}{8}x - 2$ [7.7A]

18. Solve $d = s + rt$ for t.
$t = \dfrac{d-s}{r}$ [7.7A]

19. A pool can be filled with one pipe in 6 h, while a second pipe requires 12 h to fill the pool. How long would it take to fill the pool with both pipes turned on?
4 h   [7.8A]

20. A small plane can fly at 110 mph in calm air. Flying with the wind the plane can fly 260 mi in the same amount of time as it can fly 180 mi against the wind. Find the rate of the wind.
20 mph   [7.8B]

Cumulative Review

1. Simplify $\left(\frac{2}{3}\right)^2 \div \left(\frac{3}{2} - \frac{2}{3}\right) + \frac{1}{2}$.

$\frac{31}{30}$ [1.5B]

2. Evaluate $-a^2 + (a - b)^2$ when $a = -2$ and $b = 3$. 21 [2.1A]

3. Simplify $-2x - (-3y) + 7x - 5y$.
$5x - 2y$ [2.2A]

4. Simplify $2[3x - 7(x - 3) - 8]$.
$-8x + 26$ [2.2D]

5. Solve: $4 - \frac{2}{3}x = 7$

$-\frac{9}{2}$ [3.2A]

6. Solve: $3[x - 2(x - 3)] = 2(3 - 2x)$
-12 [3.3B]

7. Find $16\frac{2}{3}\%$ of 60.

10 [4.2A]

8. Simplify $(a^2 b^5)(ab^2)$.
$a^3 b^7$ [5.2A]

9. Simplify $(a - 3b)(a + 4b)$.
$a^2 + ab - 12b^2$ [5.3C]

10. Simplify $\frac{15b^4 - 5b^2 + 10b}{5b}$.

$3b^3 - b + 2$ [5.4B]

11. Simplify $(x^3 - 8) \div (x - 2)$.
$x^2 + 2x + 4$ [5.4C]

12. Factor $12x^2 - x - 1$.
$(4x + 1)(3x - 1)$ [6.3A]

13. Factor $y^2 - 7y + 6$.
$(y - 6)(y - 1)$ [6.2A]

14. Factor $2a^3 + 7a^2 - 15a$.
$a(2a - 3)(a + 5)$ [6.3B]

15. Factor $4b^2 - 100$.
$4(b + 5)(b - 5)$ [6.4B]

16. Solve: $(x + 3)(2x - 5) = 0$
-3 and $\frac{5}{2}$ [6.5A]

17. Simplify: $\frac{12x^4 y^2}{18xy^7}$
$\frac{2x^3}{3y^5}$ [7.1A]

18. Simplify: $\frac{x^2 - 7x + 10}{25 - x^2}$
$-\frac{x - 2}{x + 5}$ [7.1A]

19. Simplify: $\dfrac{x^2 - x - 56}{x^2 + 8x + 7} \div \dfrac{x^2 - 13x + 40}{x^2 - 4x - 5}$

   1   [7.1C]

20. Simplify: $\dfrac{2}{2x - 1} - \dfrac{1}{x + 1}$

   $\dfrac{3}{(2x - 1)(x + 1)}$ [7.3B]

21. Simplify: $\dfrac{1 - \dfrac{2}{x} - \dfrac{15}{x^2}}{1 - \dfrac{25}{x^2}}$

   $\dfrac{x + 3}{x + 5}$ [7.4A]

22. Solve: $\dfrac{3x}{x - 3} - 2 = \dfrac{10}{x - 3}$

   4   [7.5A]

23. Solve the proportion.

   $\dfrac{2}{x - 2} = \dfrac{12}{x + 3}$

   3   [7.6A]

24. Solve $f = v + at$ for t.

   $t = \dfrac{f - v}{a}$ [7.7A]

25. Translate "the difference between five times a number and thirteen is the opposite of eight" into an equation and solve.
   $5x - 13 = -8$; $x = 1$ [3.4A]

26. A silversmith mixes 60 g of an alloy which is 40% silver with 120 g of another silver alloy. The resulting alloy is 60% silver. Find the percent of silver in the 120 g alloy.
   70%   [4.5B]

27. The length of the base of a triangle is 2 in. less than twice the height. The area of the triangle is 30 in². Find the base and height of the triangle.
   base 10 in.; height 6 in.   [6.5B]

28. A life insurance policy costs $16 for every $1000 of coverage. At this rate, how much money would a policy of $5000 cost?
   $80   [7.6B]

29. One water pipe can fill a tank in 9 min while a second pipe requires 18 min to fill the tank. How long would it take both pipes working together to fill the tank?
   6 min   [7.8A]

30. The rower of a boat can row at a rate of 5 mph in calm water. Rowing with the current, the boat travels 14 mi in the same amount of time as it travels 6 mi against the current. Find the rate of the current.
   2 mph   [7.8B]

8

Graphs and Linear Equations

OBJECTIVES

▶ To graph points on a rectangular coordinate system
▶ To determine a solution of a linear equation in two variables
▶ To graph a scatter diagram
▶ To graph an equation of the form $y = mx + b$
▶ To graph an equation of the form $Ax + By = C$
▶ To find the x- and y-intercepts of a straight line
▶ To find the slope of a straight line
▶ To graph a line using the slope and the y-intercept
▶ To find the equation of a line using the equation $y = mx + b$
▶ To find the equation of a line using the point-slope formula

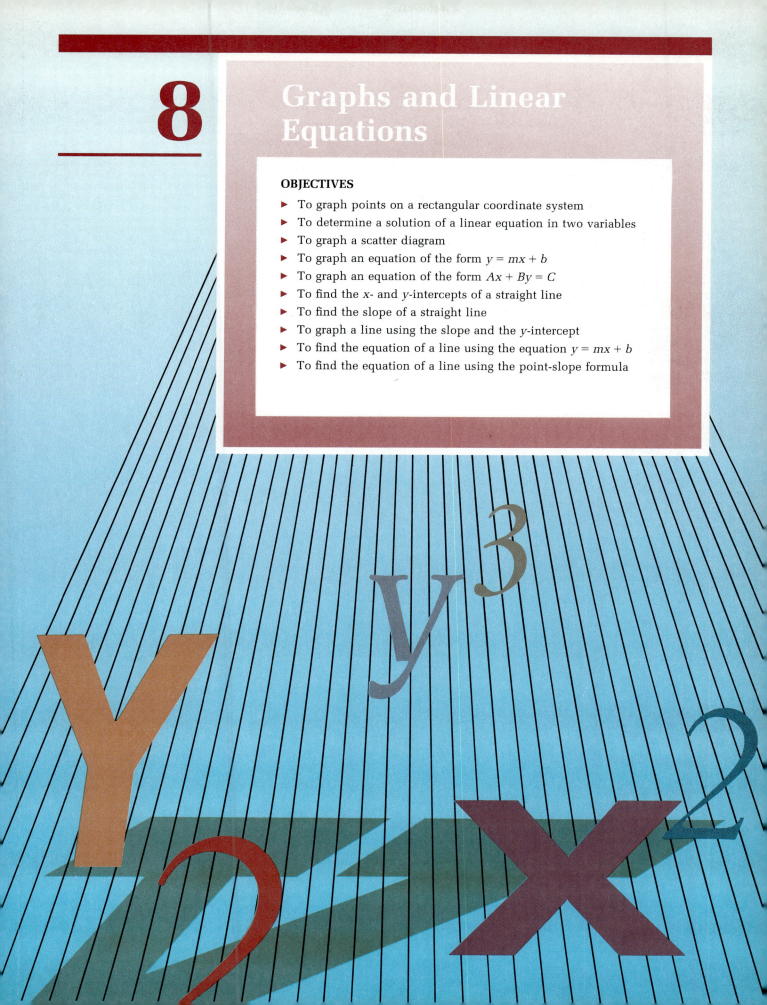

Magic Squares

A magic square is a square array of distinct integers so arranged that the numbers along any row, column, or main diagonal have the same sum. An example of a magic square is shown at the right.

8	3	4
1	5	9
6	7	2

The oldest known example of a magic square comes from China. Estimates are that this magic square is over 4000 years old. It is shown at the left.

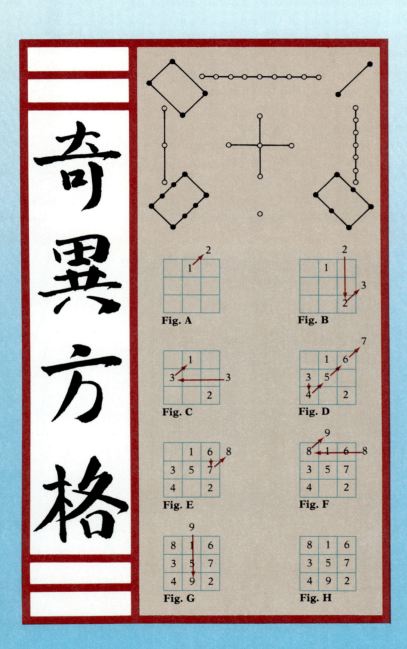

Fig. A

Fig. B

Fig. C

Fig. D

Fig. E

Fig. F

Fig. G

Fig. H

There is a simple way to produce a magic square with an odd number of cells. Start by writing a 1 in the top middle cell. The rule then is to proceed diagonally upward to the right with the successive integers.

When the rule takes you outside the square, write the number by shifting either across the square from right to left or down the square from top to bottom, as the case may be. For example, in Fig. B the second number (2) is outside the square above a column. Because the 2 is above a column, it should be shifted down to the bottom cell in that column. In Fig. C, the 3 is outside the square to the right of a column and should therefore be shifted all the way to the left.

If the rule takes you to a square that is already filled (as shown in Fig. D), then write the number in the cell directly below the last number written. Continue until the entire square is filled.

It is possible to begin a magic square with any integer and proceed by using the above rule and consecutive integers.

For an odd magic square beginning with 1, the sum of a row, column, or diagonal is $\frac{n(n^2+1)}{2}$, where n is the number of rows.

SECTION 8.1	# The Rectangular Coordinate System

| **Objective A** | **To graph points on a rectangular coordinate system** |

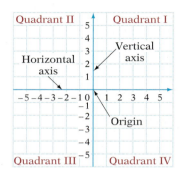

A **rectangular coordinate system** is formed by two number lines, one horizontal and one vertical, that intersect at the zero point of each line. The point of intersection is called the **origin.** The two lines are called the **coordinate axes,** or simply **axes.**

The axes determine a plane and divide the plane into four regions, called **quadrants.** The quadrants are numbered counterclockwise from I to IV.

Each point in the plane can be identified by a pair of numbers called an **ordered pair.** The first number of the pair measures a horizontal distance and is called the **abscissa** or *x*-coordinate. The second number of the pair measures a vertical distance and is called the **ordinate** or *y*-coordinate. The **coordinates** of a point are the numbers in the ordered pair associated with the point.

horizontal distance ⎤ ⎡ vertical distance

ordered pair → (3 , 4)

abscissa ⎦ ⎣ ordinate

The **graph of an ordered pair** is a point in the plane. The graphs of the points (2, 3) and (3, 2) are shown at the right. Note that they are different points. The order in which the numbers in an ordered pair appear *is* important.

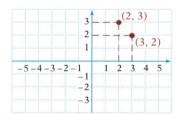

Example 1 Graph the ordered pairs (−2, −3), (3, −2), (1, 3), and (4, 1).

Solution

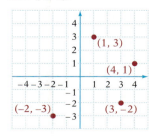

Example 2 Graph the ordered pairs (−1, 3), (1, 4), (−4, 0), and (−2, −1).

Your solution

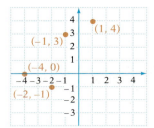

Solution on p. A27

Example 3

Find the coordinates of each of the points.

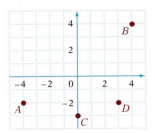

Solution

A (−4, −2)   C (0, −3)
B (4, 4)   D (3, −2)

Example 4

Find the coordinates of each of the points.

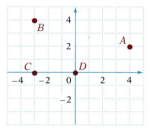

Your solution

A (4, 2)   C (−3, 0)
B (−3, 4)   D (0, 0)

Example 5

a) Name the abscissas of points A and C.
b) Name the ordinates of points B and D.

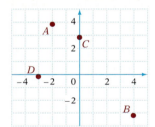

Solution

a) Abscissa of point A: −2
    Abscissa of point C: 0
b) Ordinate of point B: −3
    Ordinate of point D: 0

Example 6

a) Name the abscissas of points A and C.
b) Name the ordinates of points B and D.

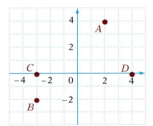

Your solution

a) Abscissa of point A: 2
    Abscissa of point C: −3
b) Ordinate of point B: −2
    Ordinate of point D: 0

Solutions on p. A27

Objective B   **To determine a solution of a linear equation in two variables**

An equation of the form $y = mx + b$, where m and b are constants, is a **linear equation in two variables.** Each exponent of each variable for a linear equation is 1.

Linear Equation in Two Variables
$$\begin{cases} y = 3x + 4 & (m = 3, b = 4) \\ y = 2x - 3 & (m = 2, b = -3) \\ y = -\frac{2}{3}x + 1 & (m = -\frac{2}{3}, b = 1) \\ y = -2x & (m = -2, b = 0) \\ y = x + 2 & (m = 1, b = 2) \end{cases}$$

The equations $y = 2x^2 - 1$ and $y = \frac{1}{x}$ are not linear equations.

A **solution of an equation in two variables** is an ordered pair of numbers (x, y) that makes the equation a true statement.

Is $(1, -2)$ a solution of $y = 3x - 5$?
Replace x with 1, the abscissa.
Replace y with -2, the ordinate.

$$y = 3x - 5$$
$$\begin{array}{c|c} -2 & 3(1) - 5 \\ & 3 - 5 \end{array}$$
$$-2 = -2$$

Compare the results. If the results are equal, the given ordered pair is a solution. If the results are not equal, the given ordered pair is not a solution.

Yes, $(1, -2)$ is a solution of the equation $y = 3x - 5$.

Besides the ordered pair $(1, -2)$, there are many other ordered pair solutions of the equation $y = 3x - 5$. For example, the method used above can be used to show that $(2, 1)$, $(-1, -8)$, $\left(\frac{2}{3}, -3\right)$ and $(0, -5)$ are also solutions.

In general, a linear equation in two variables has an infinite number of solutions. By choosing any value for x and substituting that value into the linear equation, a corresponding value of y can be found.

Find the ordered pair solution of $y = 2x - 5$ corresponding to $x = 1$.

Substitute 1 for x.
Solve for y.

$$y = 2x - 5 = 2 \cdot 1 - 5$$
$$= 2 - 5 = -3$$

The ordered pair solution is $(1, -3)$.

Example 7

Is $(-3, 2)$ a solution of $y = 2x + 2$?

Solution

$$y = 2x + 2$$
$$\begin{array}{c|c} 2 & 2(-3) + 2 \\ & -6 + 2 \\ & -4 \end{array}$$
$$2 \neq -4$$

No, $(-3, 2)$ is not a solution of $y = 2x + 2$.

Example 8

Is $(2, -4)$ a solution of $y = -\frac{1}{2}x - 3$?

Your solution

yes

Example 9

Find the ordered pair solution of $y = \frac{2}{3}x - 1$ corresponding to $x = 3$.

Solution

$$y = \frac{2}{3}x - 1$$
$$= \frac{2}{3}(3) - 1$$
$$= 2 - 1$$
$$= 1$$

The ordered pair solution is $(3, 1)$.

Example 10

Find the ordered pair solution of $y = -\frac{1}{4}x + 1$ corresponding to $x = 4$.

Your solution

$(4, 0)$

Solutions on pp. A27–A28

Objective C

To graph a scatter diagram

8

There are many situations in which the relationship between two variables may be of interest. For example, the price of a car to a dealer (one variable) and the cost of the car to the retail buyer (the second variable). Another example might be the score on an aptitude test and ability to perform a job.

A researcher can investigate these relationships by means of *regression analysis,* a branch of statistics. The study of the relationship between two variables begins with a **scatter diagram,** which is a graph of some of the known data.

For example, the following table gives the record times (rounded to the nearest second) for events of different lengths (in meters) at a track meet.

Length, x	100	200	400	800	1000	1500
Time, y	10	20	50	100	130	210

The scatter diagram for these data is shown in the graph at the right. Each ordered pair represents the length of the race and the record time. For example, the ordered pair (400, 50) indicates that the record for the 400-m race is 50 s.

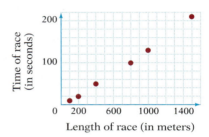

Example 11

To test a heart medicine, a doctor measures the heart rate of 5 patients before and after they take the medicine. The results are recorded in the following table. Graph the scatter diagram for these data.

Before, x	85	80	85	75	90
After, y	75	70	70	80	80

Strategy

Graph the ordered pairs on a rectangular coordinate system with the reading on the horizontal axis representing the heart rate before the medicine and that on the vertical axis representing the heart rate after the medicine.

Solution

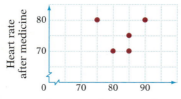

The jags indicate that a portion of the axis has been omitted.

Example 12

The age of five cars in years and the price paid for the cars in hundreds of dollars are recorded in the following table. Graph the scatter diagram for these data.

Age, x	4	5	3	5	2
Price, y	27	22	31	18	32

Your strategy

Your solution

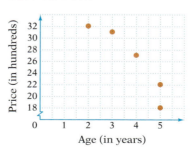

Solution on p. A28

8.1   EXERCISES

▶ **Objective A**

1. Graph the ordered pairs (−2, 1), (3, −5), (−2, 4) and (0, 3).

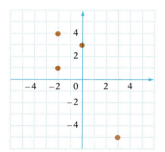

2. Graph the ordered pairs (5, −1), (−3, −3), (−1, 0), and (1, −1).

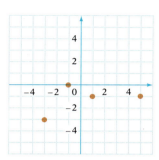

3. Graph the ordered pairs (0, 0), (0, −5), (−3, 0), and (0, 2).

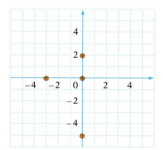

4. Graph the ordered pairs (−4, 5), (−3, 1), (3, −4), and (5, 0).

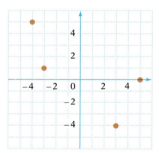

5. Graph the ordered pairs (−1, 4), (−2, −3), (0, 2), and (4, 0).

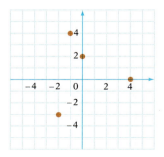

6. Graph the ordered pairs (5, 2), (−4, −1), (0, 0), and (0, 3).

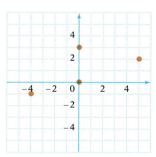

7. Find the coordinates of each of the points.

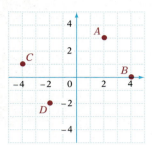

A(2, 3) C(−4, 1)
B(4, 0) D(−2, −2)

8. Find the coordinates of each of the points.

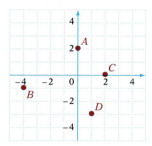

A(0, 2) C(2, 0)
B(−4, −1) D(1, −3)

9. Find the coordinates of each of the points.

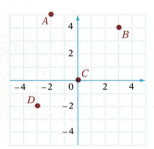

A(−2, 5) C(0, 0)
B(3, 4) D(−3, −2)

10. Find the coordinates of each of the points.

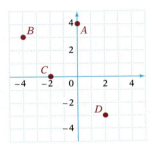

A(0, 4) C(−2, 0)
B(−4, 3) D(2, −3)

11. a) Name the abscissas of points A and C.
    2; −4
   b) Name the ordinates of points B and D.
    1; −3

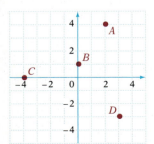

12. a) Name the abscissas of points A and C.
    0; 3
   b) Name the ordinates of points B and D.
    1; −1

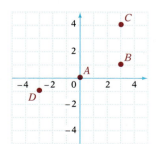

▶ ## Objective B

13. Is $(3, 4)$ a solution of $y = -x + 7$?
yes

14. Is $(2, -3)$ a solution of $y = x + 5$?
no

15. Is $(-1, 2)$ a solution of $y = \frac{1}{2}x - 1$?
no

16. Is $(1, -3)$ a solution of $y = -2x - 1$?
yes

17. Is $(4, 1)$ a solution of $y = \frac{1}{4}x + 1$?
no

18. Is $(-5, 3)$ a solution of $y = -\frac{2}{5}x + 1$?
yes

19. Is $(0, 4)$ a solution of $y = \frac{3}{4}x + 4$?
yes

20. Is $(-2, 0)$ a solution of $y = -\frac{1}{2}x - 1$?
yes

21. Is $(0, 0)$ a solution of $y = 3x + 2$?
no

22. Is $(0, 0)$ a solution of $y = -\frac{3}{4}x$?
yes

23. Find the ordered-pair solution of $y = 3x - 2$ corresponding to $x = 3$.
$(3, 7)$

24. Find the ordered-pair solution of $y = 4x + 1$ corresponding to $x = -1$.
$(-1, -3)$

25. Find the ordered-pair solution of $y = \frac{2}{3}x - 1$ corresponding to $x = 6$.
$(6, 3)$

26. Find the ordered-pair solution of $y = \frac{3}{4}x - 2$ corresponding to $x = 4$.
$(4, 1)$

27. Find the ordered-pair solution of $y = -3x + 1$ corresponding to $x = 0$.
$(0, 1)$

28. Find the ordered-pair solution of $y = \frac{2}{5}x - 5$ corresponding to $x = 0$.
$(0, -5)$

29. Find the ordered-pair solution of $y = \frac{2}{5}x + 2$ corresponding to $x = -5$.
$(-5, 0)$

30. Find the ordered-pair solution of $y = -\frac{1}{6}x - 2$ corresponding to $x = 12$.
$(12, -4)$

▶ **Objective C**

31. The math midterm and final exam scores for six students are given in the following table. Graph the scatter diagram for these data.

Midterm score, x	90	85	75	80	85	70
Final exam score, y	95	75	80	75	90	70

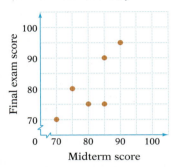

32. The number of aces served by a tennis player during a set and the number of games that player won in the set are given in the following table. Graph the scatter diagram for these data.

Aces, x	6	8	4	10	6
Games won, y	6	5	3	6	4

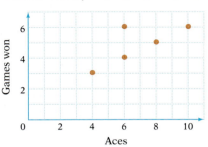

33. The number of concerts given in one year by five musical groups and the number of hit records that group had that year are given in the following table. Graph the scatter diagram for these data.

Hit records, x	11	9	7	12	12
Concerts, y	200	150	200	175	250

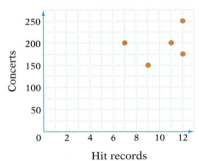

34. The monthly salary (in thousands) of an employee and the number of years of college for that employee are given in the following table. Graph the scatter diagram for these data.

Years of college, x	4	4	5	4	6	8
Monthly salary, y	3	3.5	4	2.5	5	5.5

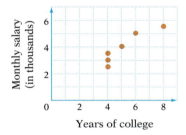

35. The table shows the number of lift tickets sold on a particular day and the number of skiing accidents reported that day. Graph the scatter diagram for these data.

Tickets sold, x	1500	2500	4000	6000
Accidents, y	25	35	50	55

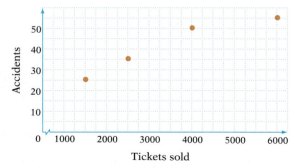

36. The table shows a study of six algebra classes that measured the number of hours the class met per week and the average score of the class on the final exam. Graph the scatter diagram for these data.

Average score, x	72	72	68	75	76	70
Hours in class, y	3	4	3	3	5	4

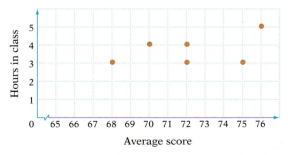

<table>
<tr><td>**SECTION 8.2**</td><td></td></tr>
</table>

Graphs of Straight Lines

Objective A   **To graph an equation of the form $y = mx + b$**

The **graph of an equation in two variables** is a drawing of the ordered pair solutions of the equation. For a linear equation in two variables, the graph is a straight line.

To graph a linear equation, find ordered pair solutions of the equation. Do this by choosing any value of x and finding the corresponding value of y. Repeat this procedure, choosing different values for x, until you have found the number of solutions desired.

Because the graph of a linear equation in two variables is a straight line, and a straight line is determined by two points, it is necessary to find only two solutions. However, it is recommended that at least three points be used to ensure accuracy.

Graph $y = 2x + 1$.

Choose any values of x and then find the corresponding values of y. The numbers 0, 2, and -1 were chosen arbitrarily for x. It is convenient to record these solutions in a table.

x	$y =$	$2x + 1$	y
0		$2 \cdot 0 + 1$	1
2		$2 \cdot 2 + 1$	5
-1		$2(-1) + 1$	-1

The horizontal axis is the x-axis. The vertical axis is the y-axis. Graph the ordered pair solutions $(0,\ 1)$, $(2,\ 5)$, and $(-1,\ -1)$. Draw a line through the ordered pair solutions.

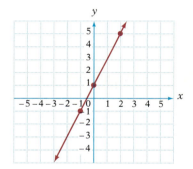

Remember that a graph is a drawing of the ordered pair solutions of the equation. Therefore, every point on the graph is a solution of the equation and every solution of the equation is a point on the graph.

Note that $(-2, -3)$ and $(1, 3)$ are points on the graph and that these points are solutions of the equation $y = 2x + 1$.

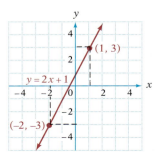

Graph $y = \frac{1}{3}x - 1$.

Find at least three solutions.
When *m is a fraction*, choose values of x that will simplify the evaluation.
Display the ordered pairs in a table.

x	y
0	-1
3	0
-3	-2

Graph the ordered pairs on a rectangular coordinate system and draw a straight line through the points.

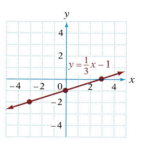

Example 1   Graph $y = 3x - 2$.

Solution

x	y
0	-2
-1	-5
2	4

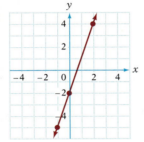

Example 2   Graph $y = 3x + 1$.

Your solution

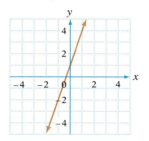

Example 3   Graph $y = 2x$.

Solution

x	y
0	0
2	4
-2	-4

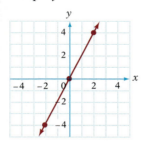

Example 4   Graph $y = -2x$.

Your solution

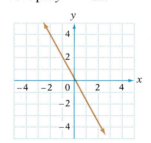

Example 5   Graph $y = \frac{1}{2}x - 1$.

Solution

x	y
0	-1
2	0
-2	-2

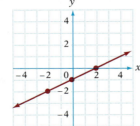

Example 6   Graph $y = \frac{1}{3}x - 3$.

Your solution

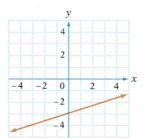

Solutions on p. A28

Objective B   **To graph an equation of the form $Ax + By = C$**

An equation in the form $Ax + By = C$, where A, B, and C are constants, is also a linear equation. Examples of these equations are shown at the right.

$2x + 3y = 6$	($A = 2$, $B = 3$, $C = 6$)
$x - 2y = -4$	($A = 1$, $B = -2$, $C = -4$)
$2x + y = 0$	($A = 2$, $B = 1$, $C = 0$)
$4x - 5y = 2$	($A = 4$, $B = -5$, $C = 2$)

To graph an equation of the form $Ax + By = C$, first solve the equation for y. Then follow the same procedure used for graphing an equation of the form $y = mx + b$.

Graph $3x + 4y = 12$.

Solve the equation for y.

$$3x + 4y = 12$$
$$4y = -3x + 12$$
$$y = -\frac{3}{4}x + 3$$

Find at least three solutions.
Display the ordered pairs in a table.

x	y
0	3
4	0
-4	6

Graph the ordered pairs on a rectangular coordinate system and draw a straight line through the points.

The graph of an equation in which one of the variables is missing is either a horizontal line or a vertical line.

The equation $y = 2$ could be written $0 \cdot x + y = 2$. No matter what value of x is chosen, y is always 2. Some solutions to the equation are $(3, 2)$, $(-1, 2)$, $(0, 2)$, and $(-4, 2)$. The graph is shown at the right.

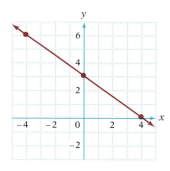

The **graph of $y = b$** is a horizontal line passing through the point $(0, b)$.

The equation $x = -2$ could be written $x + 0 \cdot y = -2$. No matter what value of y is chosen, x is always -2. Some solutions of the equation are $(-2, 3)$, $(-2, -2)$, $(-2, 0)$, and $(-2, 2)$. The graph is shown at the right.

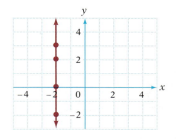

The **graph of $x = a$** is a vertical line passing through the point $(a, 0)$.

Example 7 Graph $2x - 5y = 10$.

Solution

$$2x - 5y = 10$$
$$-5y = -2x + 10$$
$$y = \frac{2}{5}x - 2$$

x	y
0	−2
5	0
−5	−4

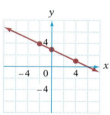

Example 8 Graph $5x - 2y = 10$.

Your solution

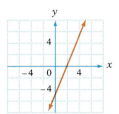

Example 9 Graph $x + 2y = 6$.

Solution

$$x + 2y = 6$$
$$2y = -x + 6$$
$$y = -\frac{1}{2}x + 3$$

x	y
0	3
−2	4
4	1

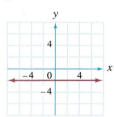

Example 10 Graph $x - 3y = 9$.

Your solution

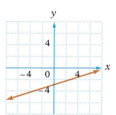

Example 11 Graph $y = -2$.

Solution The graph of an equation of the form $y = b$ is a horizontal line passing through the point $(0, b)$.

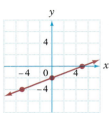

Example 12 Graph $y = 3$.

Your solution

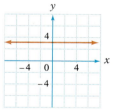

Example 13 Graph $x = 3$.

Solution The graph of an equation of the form $x = a$ is a vertical line passing through the point $(a, 0)$.

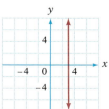

Example 14 Graph $x = -4$.

Your solution

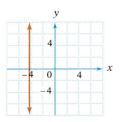

Solutions on pp. A28−A29

8.2   EXERCISES

▶ ## Objective A

Graph.

1. $y = 2x - 3$

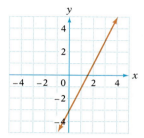

2. $y = -2x + 2$

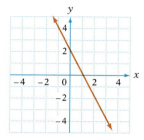

3. $y = \dfrac{1}{3}x$

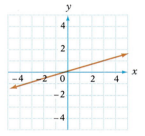

4. $y = -3x$

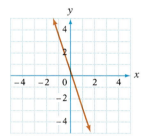

5. $y = \dfrac{2}{3}x - 1$

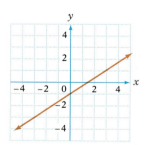

6. $y = \dfrac{3}{4}x + 2$

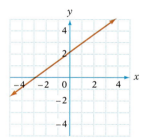

7. $y = -\dfrac{1}{4}x + 2$

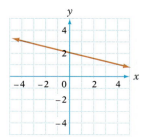

8. $y = -\dfrac{1}{3}x + 1$

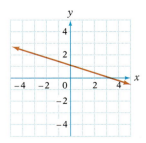

Graph.

9. $y = -\dfrac{2}{5}x + 1$

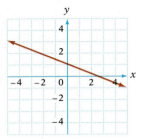

10. $y = -\dfrac{1}{2}x + 3$

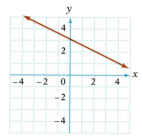

11. $y = 2x - 4$

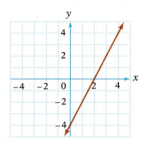

12. $y = 3x - 4$

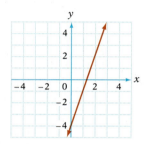

13. $y = -x + 2$

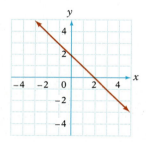

14. $y = -x - 1$

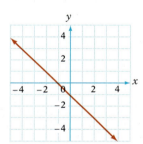

15. $y = -\dfrac{2}{3}x + 1$

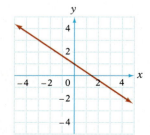

16. $y = 5x - 4$

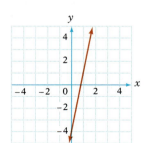

▶ ## Objective B

Graph.

17. $3x + y = 3$

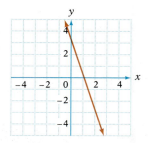

18. $2x + y = 4$

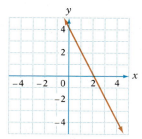

19. $2x + 3y = 6$

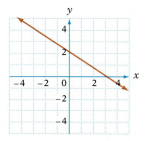

20. $3x + 2y = 4$

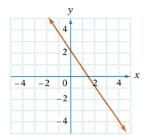

21. $x - 2y = 4$

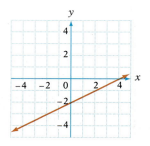

22. $x - 3y = 6$

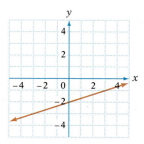

23. $y = 4$

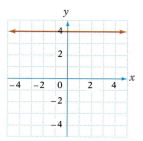

24. $x = -2$

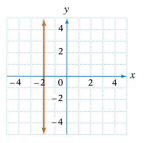

Graph.

25. $2x - 3y = 6$

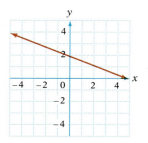

26. $3x - 2y = 8$

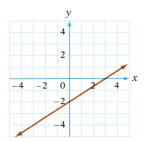

27. $2x + 5y = 10$

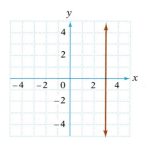

28. $3x + 4y = 12$

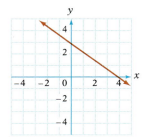

29. $x = 3$

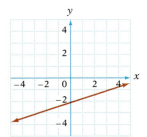

30. $y = -4$

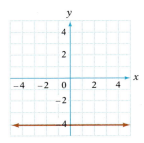

31. $x - 3y = 6$

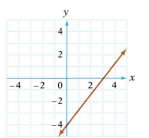

32. $4x - 3y = 12$

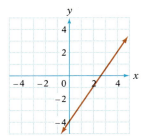

SECTION 8.3   Intercepts and Slopes of Straight Lines

Objective A   **To find the x- and y-intercepts of a straight line**

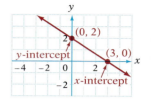

The graph of the equation $2x + 3y = 6$ is shown at the right. The graph crosses the x-axis at the point (3, 0). This point is called the **x-intercept.** The graph also crosses the y-axis at the point (0, 2). This point is called the **y-intercept.**

Find the x-intercept and the y-intercept of the graph of the equation $2x - 3y = 12$.

To find the x-intercept, let $y = 0$.
(Any point on the x-axis has
y-coordinate 0.)

$$2x - 3y = 12$$
$$2x - 3(0) = 12$$
$$2x = 12$$
$$x = 6$$

The x-intercept is (6, 0).

To find the y-intercept, let $x = 0$.
(Any point on the y-axis has
x-coordinate 0.)

$$2x - 3y = 12$$
$$2(0) - 3y = 12$$
$$-3y = 12$$
$$y = -4$$

The y-intercept is (0, −4).

Find the y-intercept of $y = 3x + 4$.

Let $x = 0$.

$$y = 3x + 4 = 3(0) + 4 = 4$$

The y-intercept is (0, 4).

For any equation of the form $y = mx + b$, the y-intercept is (0, b).

A linear equation can be graphed by finding the x- and y-intercepts and then drawing a line through these two points.

Example 1   Find the x- and y-intercepts for $x - 2y = 4$. Graph the line.

Solution x-intercept: y-intercept:
$$x - 2y = 4 \qquad\qquad x - 2y = 4$$
$$x - 2(0) = 4 \qquad\quad 0 - 2y = 4$$
$$x = 4 \qquad\qquad\quad -2y = 4$$
$$(4, 0) \qquad\qquad\qquad y = -2$$
$$(0, -2)$$

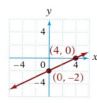

Example 2   Find the x- and y-intercepts for $4x - y = 4$. Graph the line.

Your solution x-intercept (1, 0);
y-intercept (0, −4)

Solution on p. A29

Example 3   Find the x- and y-intercepts for $y = 2x - 4$. Graph the line.

Solution x-intercept: y-intercept:
$$y = 2x - 4 \qquad (0, b)$$
$$0 = 2x - 4 \qquad b = -4$$
$$-2x = -4 \qquad (0, -4)$$
$$x = 2$$
$$(2, 0)$$

Example 4   Find the x- and y-intercepts for $y = 3x - 6$. Graph the line.

Your solution x-intercept (2, 0);
y-intercept (0, −6)

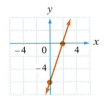

Solution on p. A29

Objective B   To find the slope of a straight line

The graphs of $y = \frac{2}{3}x + 1$ and $y = 2x + 1$ are shown at the right. Each graph crosses the y-axis at the point (0, 1), but the graphs have different slants. The **slope** of a line is a measure of the slant of a line. The symbol for slope is m.

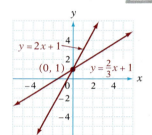

The slope of a line containing two points is the ratio of the change in the y values of the two points to the change in the x values. The line containing the points $(-2, -3)$ and $(6, 1)$ is graphed at the right. The change in the y values is the difference between the two ordinates.

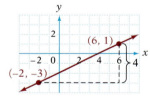

Change in $y = 1 - (-3) = 4$

The change in the x values is the difference between the two abscissas.

Change in $x = 6 - (-2) = 8$

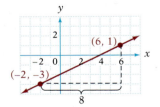

$$\text{Slope} = m = \frac{\text{change in } y}{\text{change in } x} = \frac{4}{8} = \frac{1}{2}$$

Slope Formula

If $P_1(x_1, y_1)$ and $P_2(x_2, y_2)$ are two points on a line and $x_1 \neq x_2$, then $m = \frac{y_2 - y_1}{x_2 - x_1}$.

If $x_1 = x_2$, the slope is undefined.

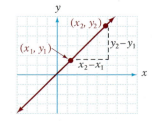

Find the slope of the line containing the points $(-1, 1)$ and $(2, 3)$.

Let P_1 be $(-1, 1)$ and P_2 be $(2, 3)$. Then,
$x_1 = -1, y_1 = 1; x_2 = 2, y_2 = 3.$

$$m = \frac{y_2 - y_1}{x_2 - x_1} = \frac{3 - 1}{2 - (-1)} = \frac{2}{3}$$

The slope is $\frac{2}{3}$.

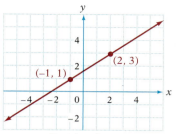
Positive slope

A line that slants upward to the right always has a **positive slope.**

Notice that you get the same result if the points had been named oppositely. If P_1 was $(2, 3)$ and P_2 was $(-1, 1)$, then $x_1 = 2, y_1 = 3; x_2 = -1, y_2 = 1$.

$$m = \frac{y_2 - y_1}{x_2 - x_1} = \frac{1 - 3}{-1 - 2} = \frac{-2}{-3} = \frac{2}{3}$$

The slope is $\frac{2}{3}$.

Therefore, it does not matter which point is named P_1 or P_2, the slope will be the same.

Find the slope of the line containing the points $(-3, 4)$ and $(2, -2)$.

Let P_1 be $(-3, 4)$ and P_2 be $(2, -2)$.

$$m = \frac{y_2 - y_1}{x_2 - x_1} = \frac{-2 - 4}{2 - (-3)} = \frac{-6}{5} = -\frac{6}{5}$$

The slope is $-\frac{6}{5}$.

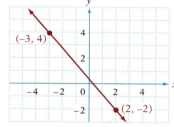
Negative slope

A line that slants downward to the right always has a **negative slope.**

Find the slope of the line containing the points $(-1, 3)$ and $(4, 3)$.

Let P_1 be $(-1, 3)$ and P_2 be $(4, 3)$.

$$m = \frac{y_2 - y_1}{x_2 - x_1} = \frac{3 - 3}{4 - (-1)} = \frac{0}{5} = 0$$

The slope is 0.

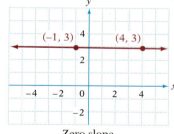
Zero slope

A horizontal line has **zero slope.**

Find the slope of a line containing the points $(2, -2)$ and $(2, 4)$.

Let P_1 be $(2, -2)$ and P_2 be $(2, 4)$.

$$m = \frac{y_2 - y_1}{x_2 - x_1} = \frac{4 - (-2)}{2 - 2} = \frac{6}{0}$$ Not a real number

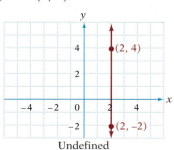
Undefined

The slope of a vertical line is undefined.

Example 5 Find the slope of the line containing the points $(-2, -1)$ and $(3, 4)$.

Solution Let $P_1 = (-2, -1)$ and $P_2 = (3, 4)$.

$$m = \frac{y_2 - y_1}{x_2 - x_1} = \frac{4 - (-1)}{3 - (-2)} = \frac{5}{5} = 1$$

The slope is 1.

Example 6 Find the slope of the line containing the points $(-1, 2)$ and $(1, 3)$.

Your solution $\frac{1}{2}$

Example 7 Find the slope of the line containing the points $(-3, 1)$ and $(2, -2)$.

Solution Let $P_1 = (-3, 1)$ and $P_2 = (2, -2)$.

$$m = \frac{y_2 - y_1}{x_2 - x_1} = \frac{-2 - 1}{2 - (-3)} = \frac{-3}{5}$$

The slope is $-\frac{3}{5}$.

Example 8 Find the slope of the line containing the points $(1, 2)$ and $(4, -5)$.

Your solution $-\frac{7}{3}$

Example 9 Find the slope of the line containing the points $(-1, 4)$ and $(-1, 0)$.

Solution Let $P_1 = (-1, 4)$ and $P_2 = (-1, 0)$.

$$m = \frac{y_2 - y_1}{x_2 - x_1} = \frac{0 - 4}{-1 - (-1)} = \frac{-4}{0}$$

The slope is undefined.

Example 10 Find the slope of the line containing the points $(2, 3)$ and $(2, 7)$.

Your solution undefined

Example 11 Find the slope of the line containing the points $(-1, 2)$ and $(4, 2)$.

Solution Let $P_1 = (-1, 2)$ and $P_2 = (4, 2)$.

$$m = \frac{y_2 - y_1}{x_2 - x_1} = \frac{2 - 2}{4 - (-1)} = \frac{0}{5} = 0$$

The line has zero slope.

Example 12 Find the slope of the line containing the points $(1, -3)$ and $(-5, -3)$.

Your solution zero slope

Solutions on p. A29

| **Objective C** | **To graph a line using the slope and the y-intercept** |

The graph of the equation $y = \frac{2}{3}x + 1$ is shown at the right. The points $(-3, -1)$ and $(3, 3)$ are on the graph. The slope of the line is

$$m = \frac{3 - (-1)}{3 - (-3)} = \frac{4}{6} = \frac{2}{3}$$

Note that the slope of the line has the same value as the coefficent of x.

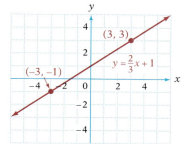

For any equation of the form $y = mx + b$, the slope of the line is m, the coefficient of x. The y-intercept is $(0, b)$. Thus, an equation of the form $y = mx + b$ is called the **slope–intercept form of a straight line.**

Find the slope and the y-intercept of the line $y = -\frac{3}{4}x + 1$.

$$y = \boxed{m}\,x\,\boxed{+ b}$$
$$y = \boxed{-\frac{3}{4}}\,x\,\boxed{+ 1}$$

Slope $= m = -\frac{3}{4}$ ⟶   ⟵ y-intercept $= (0, b) = (0, 1)$

The slope is $-\frac{3}{4}$. The y-intercept is $(0, 1)$.

When the equation of a straight line is in the form $y = mx + b$, the graph can be drawn using the slope and the y-intercept. First locate the y-intercept. Use the slope to find a second point on the line. Then draw a line through the two points.

Graph $y = 2x - 3$.

y-intercept $= (0, b) = (0, -3)$

$$m = 2 = \frac{2}{1} = \frac{\text{change in } y}{\text{change in } x}$$

Beginning at the y-intercept, move right 1 unit (change in x) and then up 2 units (change in y).

$(1, -1)$ is a second point on the graph.

Draw a line through the two points $(0, -3)$ and $(1, -1)$.

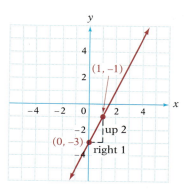

Example 13 Graph $y = -\frac{2}{3}x + 1$ by using the slope and y-intercept.

Solution y-intercept $= (0, b) = (0, 1)$

$$m = -\frac{2}{3} = \frac{-2}{3}$$

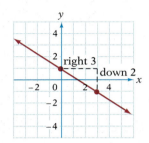

Example 14 Graph $y = -\frac{1}{4}x - 1$ by using the slope and y-intercept.

Your solution

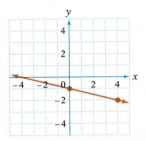

Example 15 Graph $y = -\frac{3}{4}x$ by using the slope and y-intercept.

Solution y-intercept $= (0, b) = (0, 0)$

$$m = -\frac{3}{4} = \frac{-3}{4}$$

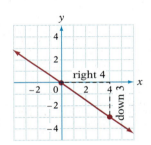

Example 16 Graph $y = -\frac{3}{5}x$ by using the slope and y-intercept.

Your solution

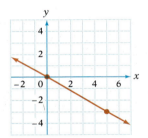

Example 17 Graph $2x - 3y = 6$ by using the slope and y-intercept.

Solution Solve the equation for y.

$$2x - 3y = 6$$
$$-3y = -2x + 6$$
$$y = \frac{2}{3}x - 2$$

y-intercept $= (0, -2)$; $m = \frac{2}{3}$

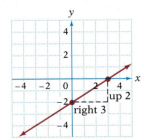

Example 18 Graph $x - 2y = 4$ by using the slope and y-intercept.

Your solution

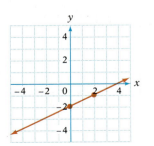

Solutions on pp. A29–A30

Content and Format © 1991 HMCo.

8.3   **EXERCISES**

▶ **Objective A**

Find the *x*- and *y*-intercepts.

1. $x - y = 3$
    (3, 0) and (0, −3)

2. $3x + 4y = 12$
    (4, 0) and (0, 3)

3. $y = 3x - 6$
    (2, 0) and (0, −6)

4. $y = 2x + 10$
    (−5, 0) and (0, 10)

5. $x - 5y = 10$
    (10, 0) and (0, −2)

6. $3x + 2y = 12$
    (4, 0) and (0, 6)

7. $y = 3x + 12$
    (−4, 0) and (0, 12)

8. $y = 5x + 10$
    (−2, 0) and (0, 10)

9. $2x - 3y = 0$
    (0, 0) and (0, 0)

10. $3x + 4y = 0$
    (0, 0) and (0, 0)

11. $y = -\frac{1}{2}x + 3$
    (6, 0) and (0, 3)

12. $y = \frac{2}{3}x - 4$
    (6, 0) and (0, −4)

Find the *x*- and *y*-intercepts and graph.

13. $5x + 2y = 10$

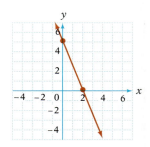

14. $x - 3y = 6$

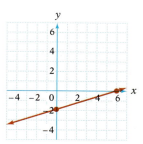

15. $y = \frac{3}{4}x - 3$

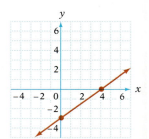

16. $y = \frac{2}{5}x - 2$

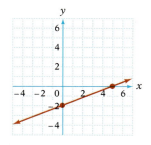

▶ **Objective B**

Find the slope of the line containing the points.

17. $P_1(4, 2)$, $P_2(3, 4)$
-2

18. $P_1(2, 1)$, $P_2(3, 4)$
3

19. $P_1(-1, 3)$, $P_2(2, 4)$
$\dfrac{1}{3}$

20. $P_1(-2, 1)$, $P_2(2, 2)$
$\dfrac{1}{4}$

21. $P_1(2, 4)$, $P_2(4, -1)$
$-\dfrac{5}{2}$

22. $P_1(1, 3)$, $P_2(5, -3)$
$-\dfrac{3}{2}$

23. $P_1(-2, 3)$, $P_2(2, 1)$
$-\dfrac{1}{2}$

24. $P_1(5, -2)$, $P_2(1, 0)$
$-\dfrac{1}{2}$

25. $P_1(8, -3)$, $P_2(4, 1)$
-1

26. $P_1(0, 3)$, $P_2(2, -1)$
-2

27. $P_1(3, -4)$, $P_2(3, 5)$
undefined

28. $P_1(-1, 2)$, $P_2(-1, 3)$
undefined

29. $P_1(4, -2)$, $P_2(3, -2)$
zero slope

30. $P_1(5, 1)$, $P_2(-2, 1)$
zero slope

31. $P_1(0, -1)$, $P_2(3, -2)$
$-\dfrac{1}{3}$

32. $P_1(3, 0)$, $P_2(2, -1)$
1

33. $P_1(-2, 3)$, $P_2(1, 3)$
zero slope

34. $P_1(4, -1)$, $P_2(-3, -1)$
zero slope

35. $P_1(-2, 4)$, $P_2(-1, -1)$
-5

36. $P_1(6, -4)$, $P_2(4, -2)$
-1

37. $P_1(-2, -3)$, $P_2(-2, 1)$
undefined

38. $P_1(5, 1)$, $P_2(5, -2)$
undefined

39. $P_1(-1, 5)$, $P_2(5, 1)$
$-\dfrac{2}{3}$

40. $P_1(-1, 5)$, $P_2(7, 1)$
$-\dfrac{1}{2}$

▶ **Objective C**

Graph by using the slope and *y*-intercept.

41. $y = 3x + 1$

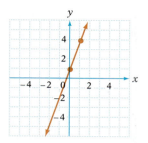

42. $y = -2x - 1$

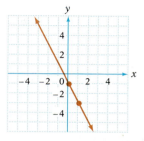

43. $y = \dfrac{2}{5}x - 2$

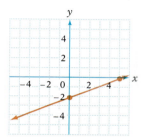

44. $y = \dfrac{3}{4}x + 1$

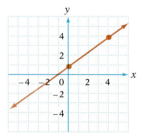

45. $2x + y = 3$

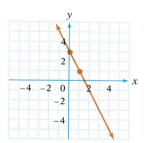

46. $3x - y = 1$

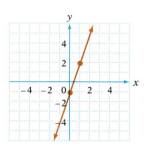

47. $x - 2y = 4$

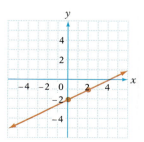

48. $x + 3y = 6$

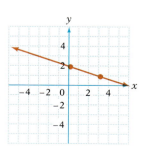

Graph by using the slope and *y*-intercept.

49. $y = \dfrac{2}{3}x$

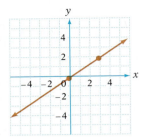

50. $y = \dfrac{1}{2}x$

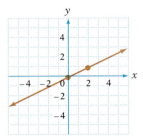

51. $y = -x + 1$

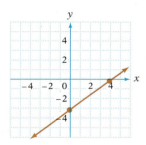

52. $y = -x - 3$

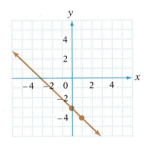

53. $3x - 4y = 12$

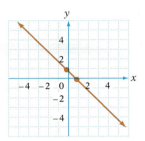

54. $5x - 2y = 10$

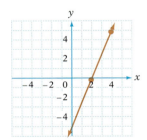

55. $y = -4x + 2$

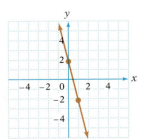

56. $4x - 5y = 20$

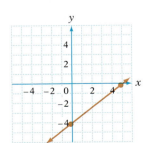

| SECTION 8.4 | **Equations of Straight Lines** |

Objective A   **To find the equation of a line using the equation $y = mx + b$**

When the slope of a line and a point on the line are known, the equation of the line can be written using the slope-intercept form, $y = mx + b$.

Find the equation of the line that has slope 3 and y-intercept (0, 2).

The given slope, 3, is m. Replace m with 3.
$$y = mx + b$$
$$y = 3x + b$$

The given point, (0, 2), is the y-intercept.
Replace b with 2.
$$y = 3x + 2$$

The equation of the line is $y = 3x + 2$.

Find the equation of the line that has slope $\frac{1}{2}$ and contains point (−2, 4).

$$y = mx + b$$

The given slope, $\frac{1}{2}$, is m. Replace m with $\frac{1}{2}$.
$$y = \frac{1}{2}x + b$$

The given point, (−2, 4), is a solution of the equation of the line.
Replace x and y in the equation with the coordinates of the point.
$$4 = \frac{1}{2}(-2) + b$$

Solve for b, the y-intercept.
$$4 = -1 + b$$
$$5 = b$$

Write the equation of the line by replacing m and b in the equation by their values.
$$y = mx + b$$
$$y = \frac{1}{2}x + 5$$

Example 1   Find the equation of the line that contains the point (3, −3) and has slope $\frac{2}{3}$.

Solution
$$y = \frac{2}{3}x + b$$
$$-3 = \frac{2}{3}(3) + b$$
$$-3 = 2 + b$$
$$-5 = b$$

$$y = \frac{2}{3}x - 5$$

Example 2   Find the equation of the line that contains the point (4, −2) and has slope $\frac{3}{2}$.

Your solution $y = \frac{3}{2}x - 8$

Solution on p. A30

| **Objective B** | **To find the equation of a line using the point–slope formula** |

An alternative method for finding the equation of a line, given the slope and a point on the line, involves use of the point–slope formula. The point–slope formula is derived from the formula for slope.

Let (x_1, y_1) be the given point on the line, and let (x, y) be any other point on the line.

Formula for slope
$$\frac{y - y_1}{x - x_1} = m$$

Multiply both sides of the equation by $(x - x_1)$.
$$\frac{y - y_1}{x - x_1}\, x - x_1 = m(x - x_1)$$

Simplify.
$$y - y_1 = m(x - x_1)$$

Point-slope formula

If (x_1, y_1) is a point on a line with slope m, then $y - y_1 = m(x - x_1)$.

In this equation, m is the slope and (x_1, y_1) is the given point.

Find the equation of the line that passes through the point $(-3, 2)$ and has slope $\frac{2}{3}$.

$$(x_1, y_1) = (-3, 2) \qquad m = \frac{2}{3} \qquad y - y_1 = m(x - x_1)$$
$$y - 2 = \frac{2}{3}[x - (-3)]$$
$$y - 2 = \frac{2}{3}(x + 3)$$
$$y - 2 = \frac{2}{3}x + 2$$
$$y = \frac{2}{3}x + 4$$

The equation of the line is $y = \frac{2}{3}x + 4$.

Example 3

Use the point–slope formula to find the equation of the line that passes through the point $(-2, -1)$ and has slope $\frac{3}{2}$.

Solution $\quad m = \frac{3}{2} \qquad (x_1, y_1) = (-2, -1)$

$$y - y_1 = m(x - x_1)$$
$$y - (-1) = \frac{3}{2}[x - (-2)]$$
$$y + 1 = \frac{3}{2}(x + 2)$$
$$y + 1 = \frac{3}{2}x + 3$$
$$y = \frac{3}{2}x + 2$$

The equation of the line is $y = \frac{3}{2}x + 2$.

Example 4

Use the point–slope formula to find the equation of the line that passes through the point $(4, -2)$ and has slope $\frac{3}{4}$.

Your solution $\quad y = \frac{3}{4}x - 5$

Solution on p. A30

8.4   EXECISES

▶ **Objective A**

Use the slope–intercept form.

1. Find the equation of the line that contains the point (0, 2) and has slope 2.
$y = 2x + 2$

2. Find the equation of the line that contains the point (0, −1) and has slope −2.
$y = -2x - 1$

3. Find the equation of the line that contains the point (−1, 2) and has slope −3.
$y = -3x - 1$

4. Find the equation of the line that contains the point (2, −3) and has slope 3.
$y = 3x - 9$

5. Find the equation of the line that contains the point (3, 1) and has slope $\frac{1}{3}$.
$y = \frac{1}{3}x$

6. Find the equation of the line that contains the point (−2, 3) and has slope $\frac{1}{2}$.
$y = \frac{1}{2}x + 4$

7. Find the equation of the line that contains the point (4, −2) and has slope $\frac{3}{4}$.
$y = \frac{3}{4}x - 5$

8. Find the equation of the line that contains the point (2, 3) and has slope $-\frac{1}{2}$.
$y = -\frac{1}{2}x + 4$

9. Find the equation of the line that contains the point (5, −3) and has slope $-\frac{3}{5}$.
$y = -\frac{3}{5}x$

10. Find the equation of the line that contains the point (5, −1) and has slope $\frac{1}{5}$.
$y = \frac{1}{5}x - 2$

11. Find the equation of the line that contains the point (2, 3) and has slope $\frac{1}{4}$.
$y = \frac{1}{4}x + \frac{5}{2}$

12. Find the equation of the line that contains the point (−1, 2) and has slope $-\frac{1}{2}$.
$y = -\frac{1}{2}x + \frac{3}{2}$

▶ **Objective B**

Use the point–slope formula.

13. Find the equation of the line that passes through the point $(1, -1)$ and has slope 2.
$y = 2x - 3$

14. Find the equation of the line that passes through the point $(2, 3)$ and has slope -1.
$y = -x + 5$

15. Find the equation of the line that passes through the point $(-2, 1)$ and has slope -2.
$y = -2x - 3$

16. Find the equation of the line that passes through the point $(-1, -3)$ and has slope -3.
$y = -3x - 6$

17. Find the equation of the line that passes through the point $(0, 0)$ and has slope $\frac{2}{3}$.
$y = \frac{2}{3}x$

18. Find the equation of the line that passes through the point $(0, 0)$ and has slope $-\frac{1}{5}$.
$y = -\frac{1}{5}x$

19. Find the equation of the line that passes through the point $(2, 3)$ and has slope $\frac{1}{2}$.
$y = \frac{1}{2}x + 2$

20. Find the equation of the line that passes through the point $(3, -1)$ and has slope $\frac{2}{3}$.
$y = \frac{2}{3}x - 3$

21. Find the equation of the line that passes through the point $(-4, 1)$ and has slope $-\frac{3}{4}$.
$y = -\frac{3}{4}x - 2$

22. Find the equation of the line that passes through the point $(-5, 0)$ and has slope $-\frac{1}{5}$.
$y = -\frac{1}{5}x - 1$

23. Find the equation of the line that passes through the point $(-2, 1)$ and has slope $\frac{3}{4}$.
$y = \frac{3}{4}x + \frac{5}{2}$

24. Find the equation of the line that passes through the point $(3, -2)$ and has slope $\frac{1}{6}$.
$y = \frac{1}{6}x - \frac{5}{2}$

Calculators and Computers

Graphs of Straight Lines

The program GRAPHS OF STRAIGHT LINES on the Student Disk graphs lines of the form

$$y = mx + b$$

where m is the slope of the line and b is the y-intercept.

This program will enable you to examine the effect on the graph when changes are made in the slope m while the y-intercept remains constant. You can also examine the effect that changes in the y-intercept have on the graph.

After each line is drawn, you will have the option of erasing the line, drawing another line, or quitting. Here is an example to get you started:

Graph $y = x + 1$ $(m = 1,\ \ b = 1)$
           $y = 2x + 1$ $(m = 2,\ \ b = 1)$
           $y = -x + 1$ $(m = -1, b = 1)$
           $y = -2x + 1$ $(m = -2, b = 1)$

For each of these graphs, the y-intercept is 1 and the graph passes through the point (0, 1).

Now erase the graphs and try these.

Graph $y = 2x - 1$ $(m = 2, b = -1)$
           $y = 2x + 2$ $(m = 2, b = 2)$
           $y = 2x - 4$ $(m = 2, b = -4)$

The three graphs are parallel because the slope of each line is 2. The y-intercept changes.

Try this program with your own values. Remember that when entering a fractional slope or y-intercept, you must enter it as a decimal. For example, to see the graph of $y = \frac{3}{4}x + 2$, enter the slope $\frac{3}{4}$ as 0.75.

Chapter Summary

Key Words

A *rectangular coordinate system* is formed by two number lines, one horizontal and one vertical, that intersect at the zero point of each line.

The number lines that make up a rectangular coordinate system are called the *coordinate axes* or simply *axes*.

The *origin* is the point of intersection of the two coordinate axes.

A rectangular coordinate system divides the plane into four regions, called *quadrants*.

An *ordered pair* (a, b) is used to locate a point in a plane.

The first number in an ordered pair is called the *abscissa* or x-coordinate.

The second number in an ordered pair is called the *ordinate* or y-coordinate.

The *coordinates* of a point are the numbers in the ordered pair that is associated with the point.

An equation of the form $y = mx + b$, where m and b are constants, is a *linear equation in two variables*.

The point at which a graph crosses the x-axis is called the *x-intercept*.

The point at which a graph crosses the y-axis is called the *y-intercept*.

The *slope* of a line is the measure of the slant of a line. The symbol for slope is m.

A line that slants upward to the right has a *positive slope*.

A line that slants downward to the right has a *negative slope*.

A horizontal line has *zero slope*.

The slope of a vertical line is *undefined*.

Essential Rules   *Slope of a straight line:* $\text{slope} = m = \dfrac{y_2 - y_1}{x_2 - x_1}$

Slope–intercept form of a straight line: $y = mx + b$

Point–slope form of a straight line: $y - y_1 = m(x - x_1)$

Chapter Test

1. Graph the ordered pairs $(-3, 1)$ and $(0, 2)$.

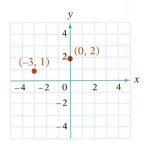

[8.1A]

2. Find the ordered pair solution of $y = -\frac{2}{3}x + 2$ corresponding to $x = 3$.

(3, 0)   [8.1B]

3. Find the ordered pair solution of $3x - 2y = 7$ corresponding to $y = 1$.

(3, 1)   [8.1B]

4. The distance a house is from a fire station and the amount of fire damage that house sustained in a fire are given in the following table. Graph the scatter diagram for these data.

Distance in miles, x	3.5	4.0	5.5	6.0
Damage in thousand of $, y	25	30	40	35

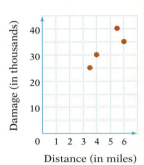

[8.1C]

5. Graph $y = 3x + 1$.

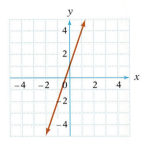

[8.2A]

6. Graph $y = -\frac{3}{4}x + 3$.

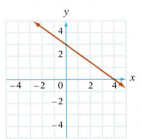

[8.2A]

7. Graph $3x - 2y = 6$.

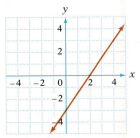

[8.2B]

8. Graph $x + 3 = 0$.

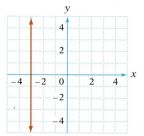

[8.2B]

9. Find the *x*- and *y*-intercepts for
$6x - 4y = 12$.
(2, 0) and (0, −3)   [8.3A]

10. Find the *x*- and *y*-intercepts for $y = \frac{1}{2}x + 1$.
(−2, 0) and (0, 1)   [8.3A]

11. Find the slope of the line containing the points (2, −3) and (4, 1).
2   [8.3B]

12. Find the slope of the line containing the points (3, −4) and (1, −4).
zero slope   [8.3B]

13. Find the slope of the line containing the points (−5, 2) and (−5, 7).
undefined   [8.3B]

14. Find the slope of the line whose equation is $2x + 3y = 6$.

$m = -\dfrac{2}{3}$ [8.3C]

15. Graph the line that has slope $-\frac{2}{3}$ and *y*-intercept (0, 4).

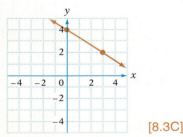

[8.3C]

16. Graph the line that has slope 2 and *y*-intercept −2.

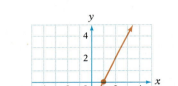

[8.3C]

17. Find the equation of the line that contains the point (0, −1) and has slope 3.
$y = 3x - 1$ [8.4A]

18. Find the equation of the line that contains the point (−3, 1) and has slope $\frac{2}{3}$.

$y = \dfrac{2}{3}x + 3$ [8.4A]

19. Find the equation of the line that contains the point (2, 3) and has slope $\frac{1}{2}$.

$y = \dfrac{1}{2}x + 2$ [8.4B]

20. Find the equation of the line that contains the point (−1, 2) and has slope $-\frac{2}{3}$.

$y = -\dfrac{2}{3}x + \dfrac{4}{3}$ [8.4B]

Cumulative Review

1. Simplify $12 - 18 \div 3 \cdot (-2)^2$.
   -12 [1.5B]

2. Evaluate $\dfrac{a-b}{a^2-c}$ when $a = -2$, $b = 3$, and $c = -4$.
   $-\dfrac{5}{8}$ [2.1A]

3. Simplify $4(2 - 3x) - 5(x - 4)$.
   $-17x + 28$ [2.2D]

4. Solve: $2x - \dfrac{2}{3} = \dfrac{7}{3}$
   $x = \dfrac{3}{2}$ [3.2A]

5. Solve: $3x - 2[x - 3(2 - 3x)] = x - 7$
   $x = \dfrac{19}{18}$ [3.3B]

6. Write $6\frac{2}{3}\%$ as a fraction.
   $\dfrac{1}{15}$ [4.1A]

7. Simplify $(-2x^2y)^3(2xy^2)^2$.
   $-32x^8y^7$ [5.2A]

8. Simplify $\dfrac{-15x^7}{5x^5}$.
   $-3x^2$ [5.4A]

9. Divide: $(x^2 - 4x - 21) \div (x - 7)$
   $x + 3$ [5.4C]

10. Factor $5x^2 + 15x + 10$.
    $5(x + 2)(x + 1)$ [6.2B]

11. Factor $x(a + 2) + y(a + 2)$.
    $(a + 2)(x + y)$ [6.1A]

12. Solve: $x(x - 2) = 8$
    4 and -2 [6.5A]

13. Simplify: $\dfrac{x^5y^3}{x^2 - x - 6} \cdot \dfrac{x^2 - 9}{x^2y^4}$
    $\dfrac{x^3(x + 3)}{y(x + 2)}$ [7.1B]

14. Simplify: $\dfrac{3x}{x^2 + 5x - 24} - \dfrac{9}{x^2 + 5x - 24}$
    $\dfrac{3}{x + 8}$ [7.3A]

15. Solve: $3 - \dfrac{1}{x} = \dfrac{5}{x}$
    $x = 2$ [7.5A]

16. Solve $4x - 5y = 15$ for y.
    $y = \dfrac{4}{5}x - 3$ [7.7A]

17. Find the ordered-pair solution of $y = 2x - 1$ corresponding to $x = -2$.
$(-2, -5)$ [8.1B]

18. Find the slope of the line that contains the points $(2, 3)$ and $(-2, 3)$.
zero slope   [8.3B]

19. Find the equation of the line that contains the point $(2, -1)$ and has slope $\frac{1}{2}$.
$y = \frac{1}{2}x - 2$ [8.4A]

20. Find the equation of the line that contains the point $(0, 2)$ and has slope -3.
$y = -3x + 2$ [8.4A]

21. Find the equation of the line that contains the point $(-1, 0)$ and has slope 2.
$y = 2x + 2$ [8.4B]

22. Find the equation of the line that contains the point $(6, 1)$ and has slope $\frac{2}{3}$.
$y = \frac{2}{3}x - 3$ [8.4B]

23. A suit that regularly sells for $89 is on sale for 30% off the regular price. Find the sale price.
$62.30   [4.3B]

24. A gold coin is 60 years older than a silver coin. Fifteen years ago the gold coin was twice as old as the silver coin was then. Find the present ages of the two coins.
gold coin 135 years; silver coin 75 years [4.8C]

25. The real estate tax for a home that costs $50,000 is $625. At this rate, what is the value of a home for which the real estate tax is $1375?
$110,000   [7.6B]

26. An electrician requires 6 h to wire a garage. An apprentice can do the same job in 10 h. How long would it take to wire the garage if both the electrician and the apprentice were working?
$3\frac{3}{4}$ h   [7.8A]

27. Graph $y = \frac{1}{2}x - 1$.

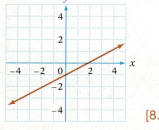

[8.2A]

28. Graph the line that has slope $-\frac{2}{3}$ and y-intercept 2.

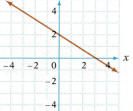

[8.3C]

9

Systems of Linear Equations

OBJECTIVES

▶ To solve a system of linear equations by graphing

▶ To solve a system of linear equations by the substitution method

▶ To solve a system of linear equations by the addition method

▶ To solve rate-of-wind or -current problems

▶ To solve application problems using two variables

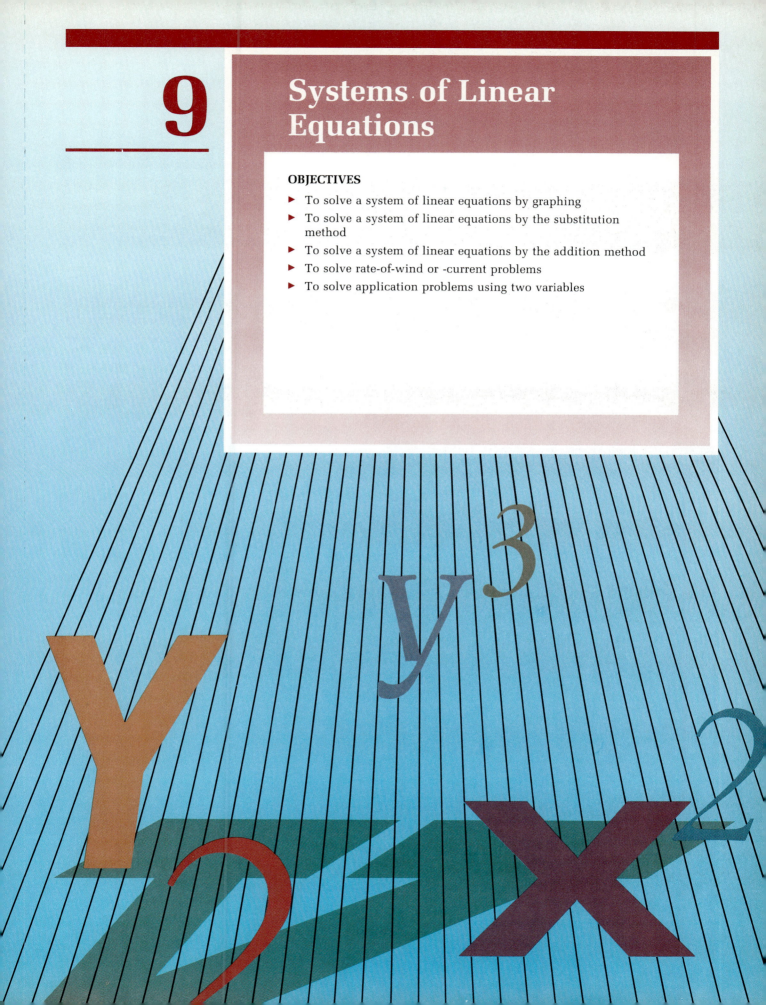

Input–Output Analysis

In 1973, the Nobel Prize in Economics was awarded for applications of mathematics to economics. The technique was to examine various sectors of an economy (steel industry, oil, farms, autos, and many others) and determine how each sector interacted with the others. Over 500 sectors of the economy were studied.

The interaction of each sector with the others was written as a series of equations. This series of equations is called a *system of equations*.

Using a computer, economists searched for a solution to the system of equations that would determine the output levels various sectors would have to meet to satisfy the requests from other sectors. The method is called Input–Output Analysis.

This chapter begins the study of systems of equations.

<table>
<tr><td>SECTION 9.1</td></tr>
</table>

Solving Systems of Linear Equations by Graphing

| Objective A | **To solve a system of linear equations by graphing** | |

Equations considered together are called **systems of equations.** A system of equations is shown at the right.

$$2x + 3y = 2$$
$$3x - 5y = 22$$

A solution of a system of equations is an ordered pair that is a solution of each equation of the system.

Is $(4, -2)$ a solution of the system

$$2x + 3y = 2$$
$$3x - 5y = 22?$$

$2x + 3y = 2$			$3x - 5y = 22$	
$2(4) + 3(-2)$	2		$3(4) - 5(-2)$	22
$8 + (-6)$	2		$12 - (-10)$	22
	$2 = 2$			$22 = 22$

Yes, since $(4, -2)$ is a solution of each equation, it is the solution of the system. However, $(7, -4)$ is not a solution because

$2x + 3y = 2$			$3x - 5y = 22$	
$2(7) + 3(-4)$	2		$3(7) - 5(-4)$	22
$14 + (-12)$	2		$21 - (-20)$	22
	$2 = 2$			$41 \neq 22$

Is $(3, -3)$ a solution of the system

$$2x + y = 3$$
$$x + y = 1?$$

$2x + y = 3$			$x + y = 1$	
$2(3) + (-3)$	3		$3 + (-3)$	1
$6 + (-3)$	3			$0 \neq 1$
	$3 = 3$			

No, $(3, -3)$ is not a solution of the system of equations. It is not a solution of each equation.

Graphing the equations in a system of linear equations is one method of finding a solution of the system of equations. The lines can intersect at one point, intersect at infinitely many points (the graphs are the same line), or the lines can be parallel and do not intersect at all.

These three conditions respectively are called **independent, dependent,** or **inconsistent.**

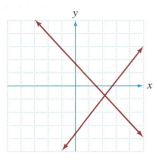
Independent

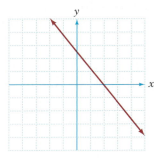
Dependent

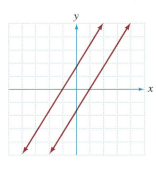
Inconsistent

Solve by graphing: $2x + 3y = 6$
$2x + y = -2$

Graph each line.

Find the point of intersection.

$(-3, 4)$ is a solution of each equation.
The system of equations is *independent*.

The solution is $(-3, 4)$.

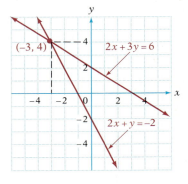

Solve by graphing: $2x - y = 1$
$6x - 3y = 12$

Graph each line.

The lines are parallel and therefore do not intersect. The system of equations is *inconsistent* and has no solution.

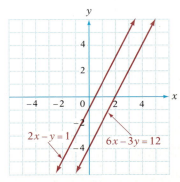

When a system of equations is *dependent*, the graphs of the two equations will be the same line. Therefore, the lines intersect at infinitely many points. The solutions of the system of equations are the ordered pairs that satisfy either one of the two equations of the system of equations.

Solve by graphing: $2x + 3y = 6$
$\qquad\qquad\qquad\quad 6x + 9y = 18$

Graph each line.

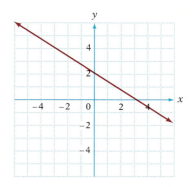

The two equations represent the same line. The system of equations is dependent, and therefore there are an infinite number of solutions. The solutions are the ordered pairs that satisfy both equations. The solutions are stated by using the ordered pairs of one of the equations. Therefore, the solutions are the ordered pairs that satisfy the equation $2x + 3y = 6$.

By choosing values for x, some specific ordered pair solutions can be found. For example, when $x = -3$, 0, and 6, three solutions of the system of equations are $(-3, 4)$, $(0, 2)$ and $(6, -2)$.

Example 1

Is $(1, -3)$ a solution of the system
$3x + 2y = -3$
$x - 3y = 6?$

Solution

$$
\begin{array}{r|l}
3x + 2y = -3 & \\
\hline
3 \cdot 1 + 2(-3) & -3 \\
3 + (-6) & -3 \\
-3 = -3 &
\end{array}
\qquad
\begin{array}{r|l}
x - 3y = 6 & \\
\hline
1 - 3(-3) & 6 \\
1 - (-9) & 6 \\
10 \neq 6 &
\end{array}
$$

No, $(1, -3)$ is not a solution of the system of equations.

Example 2

Is $(-1, -2)$ a solution of the system
$2x - 5y = 8$
$-x + 3y = -5?$

Your solution

yes

Solution on p. A30

Example 3

Solve by graphing:
$x - 2y = 2$
$x + y = 5$

Solution

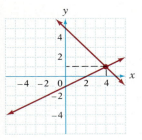

The solution is (4, 1).

Example 4

Solve by graphing:
$x + 3y = 3$
$-x + y = 5$

Your solution

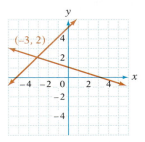

(−3, 2)

Example 5

Solve by graphing:
$4x - 2y = 6$
$y = 2x - 3$

Solution

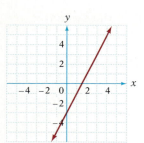

The solutions are the ordered pairs that satisfy the equation $y = 2x - 3$.

Example 6

Solve by graphing:
$y = 3x - 1$
$6x - 2y = -6$

Your solution

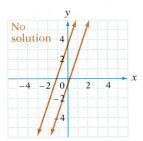

The system of equations is inconsistent and does not have a solution.

Solutions on p. A30

Content and Format © 1991 HMCo.

9.1   EXERCISES

▶ **Objective A**

1. Is (2, 3) a solution of the system
   $3x + 4y = 18$
   $2x - y = 1?$
   yes

2. Is (2, −1) a solution of the system
   $x - 2y = 4$
   $2x + y = 3?$
   yes

3. Is (1, −2) a solution of the system
   $3x - y = 5$
   $2x + 5y = -8?$
   yes

4. Is (−1, −1) a solution of the system
   $x - 4y = 3$
   $3x + y = 2?$
   no

5. Is (4, 3) a solution of the system
   $5x - 2y = 14$
   $x + y = 8?$
   no

6. Is (2, 5) a solution of the system
   $3x + 2y = 16$
   $2x - 3y = 4?$
   no

7. Is (−1, 3) a solution of the system
   $4x - y = -5$
   $2x + 5y = 13?$
   no

8. Is (4, −1) a solution of the system
   $x - 4y = 9$
   $2x - 3y = 11?$
   no

9. Is (0, 0) a solution of the system
   $4x + 3y = 0$
   $2x - y = 1?$
   no

10. Is (2, 0) a solution of the system
    $3x - y = 6$
    $x + 3y = 2?$
    yes

11. Is (2, −3) a solution of the system
    $y = 2x - 7$
    $3x - y = 9?$
    yes

12. Is (−1, −2) a solution of the system
    $3x - 4y = 5$
    $y = x - 1?$
    yes

13. Is (5, 2) a solution of the system
    $y = 2x - 8$
    $y = 3x - 13?$
    yes

14. Is (−4, 3) a solution of the system
    $y = 2x + 11$
    $y = 5x - 19?$
    no

15. Is (−2, −3) a solution of the system
    $3x - 4y = 6$
    $2x - 7y = 17?$
    yes

16. Is (0, 0) a solution of the system
    $y = 2x$
    $3x + 5y = 0?$
    yes

17. Is (0, −3) a solution of the system
    $4x - 3y = 9$
    $2x + 5y = 15?$
    no

18. Is (4, 0) a solution of the system
    $2x + 3y = 8$
    $x - 5y = 4?$
    yes

Solve by graphing.

19. $x - y = 3$
$x + y = 5$

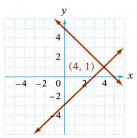

20. $2x - y = 4$
$x + y = 5$

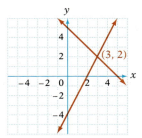

21. $x + 2y = 6$
$x - y = 3$

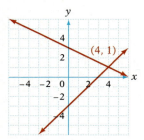

22. $3x - y = 3$
$2x + y = 2$

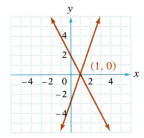

23. $3x - 2y = 6$
$y = 3$

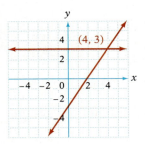

24. $x = 2$
$3x + 2y = 4$

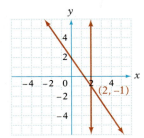

25. $x = 3$
$y = -2$

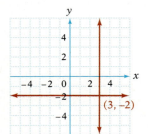

26. $x + 1 = 0$
$y - 3 = 0$

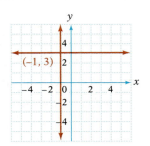

Solve by graphing.

27. $y = 2x - 6$
$x + y = 0$

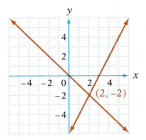

28. $5x - 2y = 11$
$y = 2x - 5$

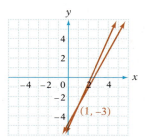

29. $2x + y = -2$
$6x + 3y = 6$

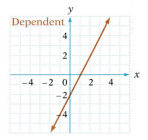

30. $x + y = 5$
$3x + 3y = 6$

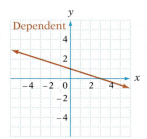

31. $y = 2x - 2$
$4x - 2y = 4$

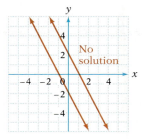

32. $y = -\frac{1}{3}x + 1$
$2x + 6y = 6$

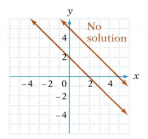

33. $x - y = 5$
$2x - y = 6$

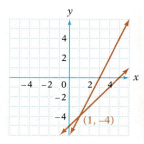

34. $5x - 2y = 10$
$3x + 2y = 6$

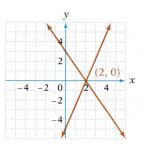

Solve by graphing.

35. $3x + 4y = 0$
$2x - 5y = 0$

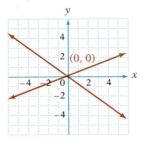

36. $2x - 3y = 0$
$y = -\frac{1}{3}x$

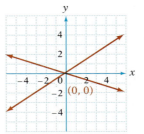

37. $x - 3y = 3$
$2x - 6y = 12$

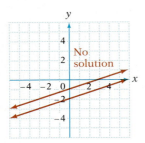

38. $4x + 6y = 12$
$6x + 9y = 18$

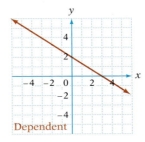

39. $3x + 2y = -4$
$x = 2y + 4$

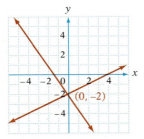

40. $5x + 2y = -14$
$3x - 4y = 2$

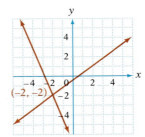

41. $4x - y = 5$
$3x - 2y = 5$

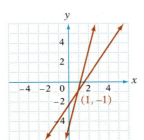

42. $2x - 3y = 9$
$4x + 3y = -9$

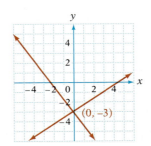

SECTION 9.2

Solving Systems of Linear Equations by the Substitution Method

Objective A

To solve a system of linear equations by the substitution method

A graphical solution of a system of equations may give only an approximate solution of the system. For example, the point $\left(\frac{1}{4}, \frac{1}{2}\right)$ would be difficult to read from the graph. An algebraic method called the **substitution method** can be used to find an exact solution of a system.

In the system of equations at the right, equation (2) states that $y = 3x - 9$. Substitute $3x - 9$ for y in equation (1).

$$(1) \qquad 2x + 5y = -11$$
$$(2) \qquad y = 3x - 9$$

$$2x + 5(3x - 9) = -11$$

Solve for x.

$$2x + 15x - 45 = -11$$
$$17x - 45 = -11$$
$$17x = 34$$
$$x = 2$$

Substitute the value of x into equation (2) and solve for y.

$$(2) \qquad y = 3x - 9$$
$$y = 3 \cdot 2 - 9$$
$$y = 6 - 9$$
$$y = -3$$

The solution is $(2, -3)$.

Solve: $5x + y = 4$
$2x - 3y = 5$

$$(1) \qquad 5x + y = 4$$
$$(2) \qquad 2x - 3y = 5$$

Solve equation (1) for y. Equation (1) is chosen because it is the easier equation to solve for one variable in terms of the other.

$$5x + y = 4$$
$$y = -5x + 4$$

Substitute $-5x + 4$ for y in equation (2).

$$2x - 3(-5x + 4) = 5$$

Solve for x.

$$2x + 15x - 12 = 5$$
$$17x - 12 = 5$$
$$17x = 17$$
$$x = 1$$

Substitute the value of x in equation (1) and solve for y.

$$5x + y = 4$$
$$5(1) + y = 4$$
$$5 + y = 4$$
$$y = -1$$

The solution is $(1, -1)$.

Example 1 Solve by substitution:
(1) $3x + 4y = -2$
(2) $-x + 2y = 4$

Solution Solve equation (2) for x.
$-x + 2y = 4$
$-x = -2y + 4$
$x = 2y - 4$
Substitute in equation (1).
$3(2y - 4) + 4y = -2$
$6y - 12 + 4y = -2$
$10y - 12 = -2$
$10y = 10$
$y = 1$
Substitute in equation (2).
$-x + 2y = 4$
$-x + 2(1) = 4$
$-x + 2 = 4$
$-x = 2$
$x = -2$

The solution is $(-2, 1)$.

Example 2 Solve by substitution:
$7x - y = 4$
$3x + 2y = 9$

Your solution (1, 3)

Example 3 Solve by substitution:
$4x + 2y = 5$
$y = -2x + 1$

Solution
$4x + 2y = 5$
$4x + 2(-2x + 1) = 5$
$4x - 4x + 2 = 5$
$2 = 5$

This is not a true equation. The system of equations is inconsistent and therefore does not have a solution.

Example 4 Solve by substitution:
$3x - y = 4$
$y = 3x + 2$

Your solution no solution

Example 5 Solve by substitution:
$y = 3x - 2$
$6x - 2y = 4$

Solution
$6x - 2y = 4$
$6x - 2(3x - 2) = 4$
$6x - 6x + 4 = 4$
$4 = 4$

This is a true equation. The system of equations is dependent. The solutions are the ordered pairs that satisfy the equation $y = 3x - 2$.

Example 6 Solve by substitution:
$y = -2x + 1$
$6x + 3y = 3$

Your solution The system of equations is dependent. The solutions are the ordered pairs that satisfy the equation $y = -2x + 1$.

Solutions on p. A31

9.2	**EXERCISES**

▶ ## Objective A

Solve by substitution.

1. $2x + 3y = 7$
$x = 2$
(2, 1)

2. $y = 3$
$3x - 2y = 6$
(4, 3)

3. $y = x - 3$
$x + y = 5$
(4, 1)

4. $y = x + 2$
$x + y = 6$
(2, 4)

5. $x = y - 2$
$x + 3y = 2$
(−1, 1)

6. $x = y + 1$
$x + 2y = 7$
(3, 2)

7. $2x + 3y = 9$
$y = x - 2$
(3, 1)

8. $3x + 2y = 11$
$y = x + 3$
(1, 4)

9. $3x - y = 2$
$y = 2x - 1$
(1, 1)

10. $2x - y = -5$
$y = x + 4$
(−1, 3)

11. $x = 2y - 3$
$2x - 3y = -5$
(−1, 1)

12. $x = 3y - 1$
$3x + 4y = 10$
(2, 1)

13. $y = 4 - 3x$
$3x + y = 5$
no solution

14. $y = 2 - 3x$
$6x + 2y = 7$
no solution

15. $x = 3y + 3$
$2x - 6y = 12$
no solution

16. $x = 2 - y$
$3x + 3y = 6$
Dependent. The solutions
satisfy the equation $x = 2 - y$.

17. $3x + 5y = -6$
$x = 5y + 3$
$\left(-\dfrac{3}{4}, -\dfrac{3}{4}\right)$

18. $y = 2x + 3$
$4x - 3y = 1$
(−5, −7)

19. $4x - 3y = -1$
$y = 2x - 3$
(5, 7)

20. $3x - 7y = 28$
$x = 3 - 4y$
(7, −1)

21. $7x + y = 14$
$2x - 5y = -33$
(1, 7)

Solve by substitution.

22. $3x + y = 4$
$4x - 3y = 1$
(1, 1)

23. $x - 4y = 9$
$2x - 3y = 11$
$\left(\dfrac{17}{5}, -\dfrac{7}{5}\right)$

24. $3x - y = 6$
$x + 3y = 2$
(2, 0)

25. $4x - y = -5$
$2x + 5y = 13$
$\left(-\dfrac{6}{11}, \dfrac{31}{11}\right)$

26. $3x - y = 5$
$2x + 5y = -8$
(1, −2)

27. $3x + 4y = 18$
$2x - y = 1$
(2, 3)

28. $4x + 3y = 0$
$2x - y = 0$
(0, 0)

29. $5x + 2y = 0$
$x - 3y = 0$
(0, 0)

30. $2x - y = 2$
$6x - 3y = 6$
Dependent. The solutions
satisfy the equation $2x - y = 2$.

31. $3x + y = 4$
$9x + 3y = 12$
Dependent. The solutions
satisfy the equation $3x + y = 4$.

32. $x - 5y = 6$
$2x - 7y = 9$
(1, −1)

33. $x + 7y = -5$
$2x - 3y = 5$
$\left(\dfrac{20}{17}, -\dfrac{15}{17}\right)$

34. $y = 2x + 11$
$y = 5x - 19$
(10, 31)

35. $y = 2x - 8$
$y = 3x - 13$
(5, 2)

36. $y = -4x + 2$
$y = -3x - 1$
(3, −10)

37. $x = 3y + 7$
$x = 2y - 1$
(−17, −8)

38. $x = 4y - 2$
$x = 6y + 8$
(−22, −5)

39. $x = 3 - 2y$
$x = 5y - 10$
$\left(-\dfrac{5}{7}, \dfrac{13}{7}\right)$

40. $y = 2x - 7$
$y = 4x + 5$
(−6, −19)

41. $3x - y = 11$
$2x + 5y = -4$
(3, −2)

42. $-x + 6y = 8$
$2x + 5y = 1$
(−2, 1)

SECTION 9.3

Solving Systems of Linear Equations by the Addition Method

Objective A

To solve a system of linear equations by the addition method

Another algebraic method for solving a system of equations is called the **addition method.** It is based on the Addition Property of Equations.

Note, for the system of equations at the right, the effect of adding equation (2) to equation (1). Because $2y$ and $-2y$ are opposites, adding the equations results in an equation with only one variable.

(1) $\quad 3x + 2y = 4$
(2) $\quad 4x - 2y = 10$
$\quad\quad 7x + 0y = 14$
$\quad\quad\quad 7x = 14$

The solution of the resulting equation is the first component of the ordered-pair solution of the system.

$\quad\quad 7x = 14$
$\quad\quad\quad x = 2$

The second component is found by substituting the value of x into equation (1) or (2) and then solving for y. Equation (1) is used here.

(1) $\quad 3x + 2y = 4$
$\quad\quad 3 \cdot 2 + 2y = 4$
$\quad\quad 6 + 2y = 4$
$\quad\quad\quad 2y = -2$
$\quad\quad\quad\quad y = -1$

The solution is $(2, -1)$.

Sometimes adding the two equations does not eliminate one of the variables. In this case, use the Multiplication Property of Equations to rewrite one or both of the equations in such a way that when the equations are added, one of the variables is eliminated.

To do this, first choose which variable to eliminate. The coefficients of that variable must be opposites. Multiply each equation by a constant that will produce coefficients that are opposites.

Solve: $\quad 3x + 2y = 7$
$\quad\quad\quad 5x - 4y = 19$

(1) $\quad 3x + 2y = 7$
(2) $\quad 5x - 4y = 19$

Eliminate y. Multiply equation (1) by 2.

$2(3x + 2y) = 2 \cdot 7$
$5y - 4y = 19$

Now the coefficients of the y terms are opposites.

$6x + 4y = 14$
$5x - 4y = 19$

Add the equations.
Solve for x.

$11x + 0y = 33$
$\quad\quad 11x = 33$
$\quad\quad\quad x = 3$

Substitute the value of x into one of the equations and solve for y. Equation (2) is used here.

(2) $\quad 5x - 4y = 19$
$\quad\quad 5 \cdot 3 - 4y = 19$
$\quad\quad 15 - 4y = 19$
$\quad\quad\quad -4y = 4$
$\quad\quad\quad\quad y = -1$

The solution is $(3, -1)$.

Solve: $5x + 6y = 3$
    $2x - 5y = 16$

(1) $5x + 6y = 3$
(2) $2x - 5y = 16$

Eliminate x. Multiply equation (1) by 2 and equation (2) by -5. Note how the constants are selected.

$2 (5x + 6y) = 2 \cdot 3$
$-5 (2x - 5y) = -5 \cdot 16$
↑——The negative is
used so that the
coefficients will
be opposites.

Now the coefficients of the x terms are opposites.

$10x + 12y = 6$
$-10x + 25y = -80$

Add the equations.
Solve for y.

$0x + 37y = -74$
$37y = -74$
$y = -2$

Substitute the value of y into one of the equations and solve for x. Equation (1) is used here.

(1) $5x + 6y = 3$
  $5x + 6(-2) = 3$
   $5x - 12 = 3$
    $5x = 15$
     $x = 3$

The solution is $(3, -2)$.

Solve: $5x = 2y + 19$
    $3x + 4y = 1$

(1) $5x = 2y + 19$
(2) $3x + 4y = 1$

Write equation (1) in the form $Ax + By = C$.

$5x - 2y = 19$
$3x + 4y = 1$

Eliminate y. Multiply equation (1) by 2.

$2(5x - 2y) = 2 \cdot 19$
$3x + 4y = 1$

Now the coefficients of the y terms are opposites.

$10x - 4y = 38$
$3x + 4y = 1$

Add the equations.
Solve for x.

$13x + 0y = 39$
$13x = 39$
$x = 3$

Substitute the value of x into one of the equations and solve for y. Equation (1) is used here.

$5x = 2y + 19$
$5 \cdot 3 = 2y + 19$
$15 = 2y + 19$
$-4 = 2y$
$-2 = y$

The solution is $(3, -2)$.

Solve: $2x + y = 2$ (1) $2x + y = 2$
          $4x + 2y = 5$ (2) $4x + 2y = 5$

Eliminate y. Multiply equation (1) by -2. $-2(2x + y) = -2 \cdot 2$
                                                 $4x + 2y = 5$

                                                 $-4x - 2y = -4$
                                                 $4x + 2y = 5$
Add the equations. $0x + 0y = 1$
                                                 $0 = 1$

This is not a true equation. The system of equations is inconsistent and therefore does not have a solution.

Example 1

Solve by the addition method:
$2x + 4y = 7$
$5x - 3y = -2$

Solution

Eliminate x.

$5(2x + 4y) = 5 \cdot 7$
$-2(5x - 3y) = -2 \cdot (-2)$

$10x + 20y = 35$
$-10x + 6y = 4$

Add the equations.

$26y = 39$

$y = \dfrac{39}{26} = \dfrac{3}{2}$

Replace y in equation (1).

$2x + 4\left(\dfrac{3}{2}\right) = 7$
$2x + 6 = 7$
$2x = 1$

$x = \dfrac{1}{2}$

The solution is $\left(\dfrac{1}{2}, \dfrac{3}{2}\right)$.

Example 2

Solve by the addition method:
$x - 2y = 1$
$2x + 4y = 0$

Your solution

$\left(\dfrac{1}{2}, -\dfrac{1}{4}\right)$

Solution on p. A32

Example 3

Solve by the addition method:
$6x + 9y = 15$
$4x + 6y = 10$

Solution

Eliminate x.

$4(6x + 9y) = 4 \cdot 15$
$-6(4x + 6y) = -6 \cdot 10$

$24x + 36y = 60$
$-24x - 36y = -60$

Add the equations.

$0x + 0y = 0$
$0 = 0$

The system of equations is dependent. The solutions are the ordered pairs that satisfy the equation $6x + 9y = 15$.

Example 4

Solve by the addition method:
$2x - 3y = 4$
$-4x + 6y = -8$

Your solution

The system of equations is dependent. The solutions are the ordered pairs that satisfy the equation $2x - 3y = 4$.

Example 5

Solve by the addition method:
$2x = y + 8$
$3x + 2y = 5$

Solution

Write equation (1) in the form $Ax + By = C$.

$2x = y + 8$
$2x - y = 8$

Eliminate y.

$2(2x - y) = 2 \cdot 8$
$3x + 2y = 5$

$4x - 2y = 16$
$3x + 2y = 5$

Add the equations.

$7x = 21$
$x = 3$

Replace x in equation (1).

$2 \cdot 3 = y + 8$
$6 = y + 8$
$-2 = y$

The solution is $(3, -2)$.

Example 6

Solve by the addition method:
$4x + 5y = 11$
$3y = x + 10$

Your solution

$(-1, 3)$

Solutions on p. A32

9.3   **EXERCISES**

▶ **Objective A**

Solve by the addition method.

1. $x + y = 4$
$x - y = 6$
(5, −1)

2. $2x + y = 3$
$x - y = 3$
(2, −1)

3. $x + y = 4$
$2x + y = 5$
(1, 3)

4. $x - 3y = 2$
$x + 2y = -3$
(−1, −1)

5. $2x - y = 1$
$x + 3y = 4$
(1, 1)

6. $x - 2y = 4$
$3x + 4y = 2$
(2, −1)

7. $4x - 5y = 22$
$x + 2y = -1$
(3, −2)

8. $3x - y = 11$
$2x + 5y = 13$
(4, 1)

9. $2x - y = 1$
$4x - 2y = 2$
Dependent. The solutions satisfy the equation $2x - y = 1$.

10. $x + 3y = 2$
$3x + 9y = 6$
Dependent. The solutions satisfy the equation $x + 3y = 2$.

11. $4x + 3y = 15$
$2x - 5y = 1$
(3, 1)

12. $3x - 7y = 13$
$6x + 5y = 7$
(2, −1)

13. $2x - 3y = 1$
$4x - 6y = 2$
Dependent. The solutions satisfy the equation $2x - 3y = 1$.

14. $2x + 4y = 6$
$3x + 6y = 9$
Dependent. The solutions satisfy the equation $2x + 4y = 6$.

15. $5x - 2y = -1$
$x + 3y = -5$
$\left(-\dfrac{13}{17}, -\dfrac{24}{17}\right)$

16. $4x - 3y = 1$
$8x + 5y = 13$
(1, 1)

17. $5x + 7y = 10$
$3x - 14y = 6$
(2, 0)

18. $7x + 10y = 13$
$4x + 5y = 6$
(−1, 2)

19. $3x - 2y = 0$
$6x + 5y = 0$
(0, 0)

20. $5x + 2y = 0$
$3x + 5y = 0$
(0, 0)

21. $2x - 3y = 16$
$3x + 4y = 7$
(5, −2)

Solve by the addition method.

22. $3x + 4y = 10$
$4x + 3y = 11$
(2, 1)

23. $5x + 3y = 7$
$2x + 5y = 1$
$\left(\dfrac{32}{19}, -\dfrac{9}{19}\right)$

24. $-2x + 7y = 9$
$3x + 2y = -1$
(−1, 1)

25. $7x - 2y = 13$
$5x + 3y = 27$
(3, 4)

26. $12x + 5y = 23$
$2x - 7y = 39$
$\left(\dfrac{178}{47}, -\dfrac{211}{47}\right)$

27. $8x - 3y = 11$
$6x - 5y = 11$
(1, −1)

28. $4x - 8y = 36$
$3x - 6y = 27$
Dependent. The solutions satisfy the equation $4x - 8y = 36$.

29. $5x + 15y = 20$
$2x + 6y = 8$
Dependent. The solutions satisfy the equation $5x + 15y = 20$.

30. $y = 2x - 3$
$3x + 4y = -1$
(1, −1)

31. $3x = 2y + 7$
$5x - 2y = 13$
(3, 1)

32. $2y = 4 - 9x$
$9x - y = 25$
(2, −7)

33. $2x + 9y = 16$
$5x = 1 - 3y$
(−1, 2)

34. $3x - 4 = y + 18$
$4x + 5y = -21$
$\left(\dfrac{89}{19}, -\dfrac{151}{19}\right)$

35. $2x + 3y = 7 - 2x$
$7x + 2y = 9$
(1, 1)

36. $5x - 3y = 3y + 4$
$4x + 3y = 11$
(2, 1)

37. $3x + y = 1$
$5x + y = 2$
$\left(\dfrac{1}{2}, -\dfrac{1}{2}\right)$

38. $2x - y = 1$
$2x - 5y = -1$
$\left(\dfrac{3}{4}, \dfrac{1}{2}\right)$

39. $4x + 3y = 3$
$x + 3y = 1$
$\left(\dfrac{2}{3}, \dfrac{1}{9}\right)$

40. $2x - 5y = 4$
$x + 5y = 1$
$\left(\dfrac{5}{3}, -\dfrac{2}{15}\right)$

41. $3x - 4y = 1$
$4x + 3y = 1$
$\left(\dfrac{7}{25}, -\dfrac{1}{25}\right)$

42. $2x - 7y = -17$
$3x + 5y = 17$
$\left(\dfrac{34}{31}, \dfrac{85}{31}\right)$

<table>
<tr><td>SECTION 9.4</td></tr>
</table>

Application Problems in Two Variables

| Objective A | To solve rate-of-wind or -current problems |

Motion problems that involve an object moving with or against a wind or current normally require two variables to solve.

Flying with the wind, a small plane can fly 600 mi in 3 h. Against the wind, the plane can fly the same distance in 4 h. Find the rate of the plane in calm air and the rate of the wind.

Strategy for Solving Rate-of-wind or -Current Problems

Choose one variable to represent the rate of the object in calm conditions and a second variable to represent the rate of the wind or current. Using these variables, express the rate of the object with and against the wind or current. Use the equation $d = rt$ to write expressions for the distance traveled by the object. The results can be recorded in a table.

Rate of plane in calm air: p
Rate of wind: w

	Rate	·	Time	=	Distance
With the wind	$p + w$	·	3	=	$3(p + w)$
Against the wind	$p - w$	·	4	=	$4(p - w)$

Determine how the expressions for distance are related.

The distance traveled with the wind is 600 mi. $3(p + w) = 600$
The distance traveled against the wind is 600 mi. $4(p - w) = 600$

Solve the system of equations.

$$3(p + w) = 600 \qquad \frac{1}{3} \cdot 3(p + w) = \frac{1}{3} \cdot 600 \qquad p + w = 200$$

$$4(p - w) = 600 \qquad \frac{1}{4} \cdot 4(p + w) = \frac{1}{4} \cdot 600 \qquad p - w = 150$$

$$2p = 350$$
$$p = 175$$

$$p + w = 200$$
$$175 + w = 200$$
$$w = 25$$

The rate of the plane in calm air is 175 mph.
The rate of the wind is 25 mph.

Example 1

A 450-mile trip from one city to another takes 3 h when a plane is flying with the wind. The return trip, against the wind, takes 5 h. Find the rate of the plane in still air and the rate of the wind.

Strategy

- Rate of the plane in still air: p
  Rate of the wind: w

	Rate	Time	Distance
With wind	$p + w$	3	$3(p + w)$
Against wind	$p - w$	5	$5(p - w)$

- The distance traveled with the wind is 450 mi. The distance traveled against the wind is 450 mi.

Solution

$3(p + w) = 450$ $\dfrac{1}{3} \cdot 3(p + w) = \dfrac{1}{3} \cdot 450$

$5(p - w) = 450$ $\dfrac{1}{5} \cdot 5(p - w) = \dfrac{1}{5} \cdot 450$

$$p + w = 150$$
$$p - w = 90$$

$$2p = 240$$
$$p = 120$$

$$p + w = 150$$
$$120 + w = 150$$
$$w = 30$$

The rate of the plane in still air is 120 mph.
The rate of the wind is 30 mph.

Example 2

A canoeist paddling with the current can travel 15 mi in 3 h. Against the current it takes 5 h to travel the same distance. Find the rate of the current and the rate of the canoeist in calm water.

Your strategy

Your solution

rate of current, 1 mph; rate of canoeist in calm water, 4 mph

Solution on p. A33

| **Objective B** | **To solve application problems using two variables** | |

The application problems in this section are varieties of those problems solved earlier in the text. Each of the strategies for the problems in this section will result in a system of equations.

A jeweler purchased 5 oz of a gold alloy and 20 oz of a silver alloy for a total cost of \$540. The next day, at the same prices per ounce, the jeweler purchased 4 oz of the gold alloy and 25 oz of the silver alloy for a total cost of \$450. Find the cost per ounce of the gold and silver alloys.

Strategy for Solving an Application Problem in Two Variables

Choose one variable to represent one of the unknown quantities and a second variable to represent the other unknown quantity. Write numerical or variable expressions for all the remaining quantities. These results can be recorded in two tables, one for each of the conditions.

Cost per ounce of gold: g
Cost per ounce of silver: s

First Day

	Amount	·	Unit Cost	=	Value
Gold	5	·	g	=	$5g$
Silver	20	·	s	=	$20s$

Second Day

	Amount	·	Unit Cost	=	Value
Gold	4	·	g	=	$4g$
Silver	25	·	s	=	$25s$

Determine a system of equations. The strategies presented in Chapter 4 can be used to determine the relationships between the expressions in the tables. Each table will give one equation of the system.

The total value of the purchase on the first day was \$540. $5g + 20s = 540$
The total value of the purchase on the second day was \$450. $4g + 25s = 450$

Solve the system of equations.

$$5g + 20s = 540$$
$$4g + 25s = 450$$

$$4(5g + 20s) = 4 \cdot 540$$
$$-5(4g + 25s) = -5 \cdot 450$$

$$20g + 80s = 2160$$
$$-20g - 125s = -2250$$
$$-45s = -90$$
$$s = 2$$

$$5g + 20s = 540$$
$$5g + 20(2) = 540$$
$$5g + 40 = 540$$
$$5g = 500$$
$$g = 100$$

The cost per ounce of the gold alloy was \$100.
The cost per ounce of the silver alloy was \$2.

Example 3

In five years, an oil painting will be twice as old as a water color painting will be then. Five years ago, the oil painting was three times as old as the water color was then. Find the present age of each painting.

Example 4

Two coin banks contain only dimes and quarters. In the first bank, the total value of the coins is $4.80. In the second bank, there are twice as many quarters as in the first bank and half the number of dimes. The total value of the coins in the second bank is $8.40. Find the number of dimes and the number of quarters in the first bank.

Strategy

■ Present age of oil painting: x
  Present age of water color: y

	Present	*Future*
Oil	x	$x + 5$
Water Color	y	$y + 5$

	Present	*Past*
Oil	x	$x - 5$
Water Color	y	$y - 5$

■ In five years, twice the age of the water color will be the age of the oil painting. Five years ago, three times the age of the water color was the age of the oil painting.

Solution

$$2(y + 5) = x + 5 \qquad 2y + 10 = x + 5$$
$$3(y - 5) = x - 5 \qquad 3y - 15 = x - 5$$

Write both equations in the form $Ax + By = C$.

$$-x + 2y = -5$$
$$-x + 3y = 10$$

Solve by the addition method.

$$(-1)(-x + 2y) = (-1)(-5) \qquad x - 2y = 5$$
$$-x + 3y = 10 \qquad \qquad \underline{-x + 3y = 10}$$
$$\qquad\qquad\qquad\qquad\qquad\qquad y = 15$$

$$x - 2y = 5$$
$$x - 2(15) = 5$$
$$x - 30 = 5$$
$$x = 35$$

The present age of the oil painting is 35 years. The present age of the water color is 15 years.

Your strategy

Your solution

8 dimes; 16 quarters

Solution on p. A33

9.4   EXERCISES

▶ **Objective A**   *Application Problems*

Solve.

1. A plane flying with the jet stream flew from Los Angeles to Chicago, a distance of 2250 mi, in 5 h. Flying against the jet stream, the plane could fly only 1750 mi in the same amount of time. Find the rate of the plane in calm air and the rate of the wind.
   plane 400 mph; wind 50 mph

2. A rowing team rowing with the current traveled 40 km in 2 h. Rowing against the current, the team could only travel 16 km in 2 h. Find the rate of rowing in calm water and the rate of the current.
   rowing 14 km/h; current 6 km/h

3. A motorboat traveling with the current went 35 mi in 3.5 h. Traveling against the current, the boat went 12 mi in 3 h. Find the rate of the boat in calm water and the rate of the current.
   boat 7 mph; current 3 mph

4. A small plane flew 270 mi in 3 h into a headwind. Flying with the wind, the plane traveled 260 mi in 2 h. Find the rate of the plane in calm air and the rate of the wind.
   plane 110 mph; wind 20 mph

5. A plane flying with a tailwind flew 300 mi in 2 h. Against the wind, it took 3 h to travel the same distance. Find the rate of the plane in calm air and the rate of the wind.
   plane 125 mph; wind 25 mph

6. A rowing team rowing with the current traveled 17 mi in 2 h. Against the current, the team rowed 7 mi in the same amount of time. Find the rate of the rowing team in calm water and the rate of the current.
   rowing 6 mph; current 2.5 mph

7. A seaplane pilot flying with the wind flew from an ocean port to a lake, a distance of 240 mi, in 2 h. Flying against the wind, the trip from the lake to the ocean port took 2 h and 40 min. Find the rate of the plane in calm air and the rate of the wind.
   plane 105 mph; wind 15 mph

8. Rowing with the current, a canoeist paddled 14 mi in 2 h. Against the current, the canoeist can paddle only 10 mi in the same amount of time. Find the rate of the canoeist in calm water and the rate of the current.
   canoeist 6 mph; current 1 mph

9. With the wind, a quarterback passes a football 140 ft in 2 s. Against the wind, the same pass would travel 80 ft in 2 s. Find the rate of the pass and the rate of the wind.
   pass 55 ft/s; wind 15 ft/s

10. Traveling with the current, a cruise ship sailed between two islands, a distance of 90 mi, in 3 h. The return trip against the current required 4 h and 30 min. Find the rate of the cruise ship in calm water and the rate of the current.
    cruise ship 25 mph; current 5 mph

▶ **Objective B**   *Application problems*

11. A computer software store received two shipments of software. The value of the first shipment, which contained 12 identical word processing programs and 10 identical spreadsheet programs, was $6190. The second shipment, at the same prices, contained 5 word processing programs and 8 spreadsheet programs. The value of the second shipment was $3825. Find the cost of a word processing program and a spreadsheet program.
word processing $245; spreadsheet $325

12. A baker purchased 12 lb of wheat flour and 15 lb of rye flour for a total cost of $18.30. A second purchase, at the same prices, included 15 lb of wheat flour and 10 lb of rye flour. The cost of the second purchase was $16.75. Find the cost per pound of the wheat flour and the rye flour.
wheat $.65; rye $.70

13. An investor owned 300 shares of an oil company and 200 shares of a movie company. The quarterly dividend from the two stocks was $165. After the investor sold 100 shares of the oil company and bought an additional 100 shares of the movie company, the quarterly dividend became $185. Find the dividend per share for each stock.
oil $.25; movie $.45

14. The charge for 25 min of prime time and 35 min of non-prime time to a customer for using a computerized financial news network was $10.75. A second customer used the system for 30 min of prime time and 45 min of non-prime time for a cost of $13.35. Find the cost per minute for using the financial news network during prime and non-prime time.
prime $.22; non-prime $.15

15. A college football team scored 30 points in one game with only touchdowns and field goals. If the number of touchdowns had been field goals and the number of field goals had been touchdowns, the score would have been 33. Find the number of touchdowns and field goals that were actually scored. Use 6 points for a touchdown and 3 points for a field goal.
touchdowns 3; field goals 4

16. A professional basketball team scored 87 points in two-point baskets and three-point baskets. If the number of two-point baskets had been three-point baskets and the number of three-point baskets had been two-point baskets, the score would have been 93. Find the number of two-point and three-point baskets that were actually scored.
two-point 21; three-point 15

17. The total value of the dimes and quarters in a bank is $7.75. If the dimes were pennies and the quarters were nickels, the value in the bank would be $1.40. Find the number of dimes and quarters in the bank.
dimes 15; quarters 25

18. The total value of the nickels and quarters in a bank is $8.50. If the number of nickels were halved and the number of quarters were doubled, the value in the bank would be $15.50. Find the number of nickels and quarters in the bank.
nickels 20; quarters 30

19. Twenty-five years ago, a coin was twice the age a stamp was then. Twenty-five years from now, the coin will be 25 years older than the stamp is then. Find the present ages of the coin and the stamp.
coin 75; stamp 50

20. Two years ago, the sum of the age of an adult and a child was 40. Two years from now, the adult will be three times the age the child will be then. Find the present ages of the adult and the child.
child 10; adult 34

Calculators and Computers

Systems of Equations

In Chapter 9, three methods of determining the solutions to systems of linear equations are presented: solving by graphing, solving by the substitution method, and solving by the addition method. As stated in the text, solving a system by graphing is not the most efficient method of finding a solution. However, doing so should give you a good visual understanding of the concept of solutions of systems of linear equations.

The substitution method is used most often when one variable is given in terms of the other; for example, in the equation $y = 2x + 5$, y is given in terms of x. Therefore, the value $2x + 5$ is easily substituted for y in another equation.

The addition method is used when neither of the equations is easily solved for one of the variables. Students sometimes find this method difficult at first because of the number of steps involved in finding a solution. Remember that the goal here is the same as it was when finding the solution to one equation: to rewrite the equation in the form *variable = constant*.

By using the addition method, the system of equations $\begin{matrix} ax + by = c \\ dx + ey = f \end{matrix}$ can be solved.

The solution is $x = \frac{ce - bf}{ae - bd}$ and $y = \frac{af - cd}{ae - bd}$, $ae - bd \neq 0$.

Using this solution, a system of equations can be solved by using a calculator. It is helpful to observe that the denominators for each expression are identical. The calculation for the denominator is done first and then stored in the calculator's memory. If the value of the denominator is zero, then the system is dependent or inconsistent, and this calculator method cannot be used.

Solve: $2x - 5y = 9$
            $4x + 3y = 2$

Make a list of the values of a, b, c, d, e, and f.

$$a = 2 \qquad b = -5 \qquad c = 9$$

$$d = 4 \qquad e = 3 \qquad f = 2$$

Calculate the denominator $D = ae - bd$. $D = 2 \cdot 3 - (-5) \cdot 4 = 6 + 20 = 26$

Store the result in memory.   Press $\boxed{M+}$.

Find x. Replace the letters by the given values. $x = \dfrac{ce - bf}{D} = \dfrac{9 \cdot 3 - (-5) \cdot 2}{26}$

Calculate x. $9 \boxed{\times} 3 \boxed{-} \boxed{(} 5 \boxed{+/-} \boxed{\times} 2 \boxed{)} \boxed{\div} \boxed{MR} \boxed{=}$

The result in the display should be 1.423077.

Find y. Replace the letters by the given values.

$$y = \frac{af - cd}{D} = \frac{2 \cdot 2 - 9 \cdot 4}{26}$$

Calculate y.   2 ⊠ 2 ⊟ (9 ⊠ 4) ÷ MR =

The result in the display should be -1.230769.

The solution of the system is $(1.423077, -1.230769)$.

The keys M+ (store in memory) and MR (recall from memory) were used for this illustration. Some calculators use the keys STO (store in memory) and RCL (recall from memory). If your calculator uses these keys, then use these keys in place of the keys shown in the illustration.

Chapter Summary

Key Words   Equations considered together are called a *system of equations*.

A *solution of a system of equations* in two variables is an ordered pair that is a solution of each equation of the system.

An *independent system of equations* has one solution.

A *dependent system of equations* has an infinite number of solutions.

An *inconsistent system of equations* has no solution.

Essential Rules   A system of equations can be solved by *the graphing method, the substitution method,* or *the addition method.*

Chapter Review

SECTION 9.1

1. Is $(-1, -3)$ a solution of the system
$$5x + 4y = -17$$
$$2x - y = 1?$$
yes

2. Is $(-2, 0)$ a solution of the system
$$-x + 9y = 2$$
$$6x - 4y = 12?$$
no

3. Solve by graphing:
$$3x - y = 6$$
$$y = -3$$

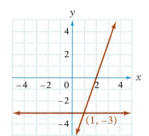

4. Solve by graphing:
$$4x - 2y = 8$$
$$y = 2x - 4$$

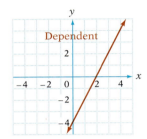

5. Solve by graphing:
$$x + 2y = 3$$
$$y = -\frac{1}{2}x + 1$$

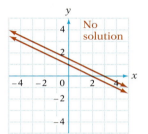

SECTION 9.2

6. Solve by substitution.
$$4x + 7y = 3$$
$$x = y - 2$$
$(-1, 1)$

7. Solve by substitution.
$$8x - y = 2$$
$$y = 5x + 1$$
$(1, 6)$

8. Solve by substitution.
$$6x - y = 0$$
$$7x - y = 1$$
$(1, 6)$

9. Solve by substitution.
$$7x + 3y = -16$$
$$x - 2y = 5$$
$(-1, -3)$

10. Solve by substitution.
$$12x - 9y = 18$$
$$y = \frac{4}{3}x - 3$$
no solution

11. Solve by substitution.
$$4x + 3y = 12$$
$$y = -\frac{4}{3}x + 4$$
Dependent. The solutions satisfy the equation $4x + 3y = 12$.

SECTION 9.3

12. Solve by the addition method.
$$3x + 8y = -1$$
$$x - 2y = -5$$
$(-3, 1)$

13. Solve by the addition method.
$$4x - y = 9$$
$$2x + 3y = -13$$
$(1, -5)$

14. Solve by the addition method.
$$6x + 4y = -3$$
$$12x - 10y = -15$$
$\left(-\dfrac{5}{6}, \dfrac{1}{2}\right)$

15. Solve by the addition method.
$$5x + 7y = 21$$
$$20x + 28y = 63$$
no solution

16. Solve by the addition method.
$$3x + y = -2$$
$$-9x - 3y = 6$$
Dependent. The solutions satisfy the equation $3x + y = -2$.

17. Solve by the addition method.
$$5x + 2y = -9$$
$$12x - 7y = 2$$
$(-1, -2)$

18. Solve by the addition method.
$$6x - 18y = 7$$
$$9x + 24y = 2$$
$\left(\dfrac{2}{3}, -\dfrac{1}{6}\right)$

SECTION 9.4

19. A canoeist traveling with the current traveled the 30 mi between two riverside camp sites in 3 h. The return trip took 5 h. Find the rate of the canoeist in still water and the rate of the current.
rate of canoeist in still water: 8 mph; rate of current: 2 mph

20. A flight crew flew 420 km in 3 h with a tailwind. Flying against the wind, the flight crew flew 440 km in 4 h. Find the rate of the flight crew in calm air and the rate of the wind.
rate of flight crew in calm air: 125 km/h; rate of wind: 15 km/h

21. A small plane flying with the wind flew 360 mi in 3 h. Against a headwind, the plane took 4 h to fly the same distance. Find the rate of the plane in calm air and the rate of the wind.
rate of plane in calm air: 105 mph; rate of wind: 15 mph

22. A sculling team rowing with the current went 24 mi in 2 h. Rowing against the current the sculling team went 18 mi in 3 h. Find the rate of the sculling team in calm water and the rate of the current.
rate of sculling team in calm water: 9 mph; rate of current: 3 mph

23. A small wood carving company mailed 190 advertisements, some requiring 25 cents postage and others 45 cents. If the total cost for mailing is $59.50, find the number of advertisements at each rate.
number requiring 25¢: 130 advertisements
number requiring 45¢: 60 advertisements

24. A silo contains a mixture of lentils and corn. If 50 bushels of lentils were added, there would be twice as many bushels of lentils as corn; if instead, 150 bushels of corn were added, there would be the same amount of corn as lentils. How many bushels of each were originally in the silo?
350 bushels of lentils; 200 bushels of corn

25. An investor bought 1500 shares of stock, some at $6 per share and the rest at $25 per share. If $12,800 worth of stock was purchased, how many shares of each kind did the investor buy?
1300 shares at $6; 200 shares at $25

Chapter Test

1. Is $(-2, 3)$ a solution of the system
$$2x + 5y = 11$$
$$x + 3y = 7?$$
yes   [9.1A]

2. Is $(1, -3)$ a solution of the system
$$3x - 2y = 9$$
$$4x + y = 1?$$
yes   [9.1A]

3. Solve by graphing: $3x + 2y = 6$
$$5x + 2y = 2$$

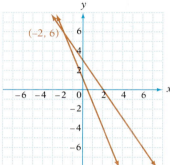

[9.1A]

4. Solve by substitution.
$$4x - y = 11$$
$$y = 2x - 5$$
$(3, 1)$ [9.2A]

5. Solve by substitution.
$$x = 2y + 3$$
$$3x - 2y = 5$$
$(1, -1)$ [9.2A]

6. Solve by substitution.
$$3x + 5y = 1$$
$$2x - y = 5$$
$(2, -1)$ [9.2A]

7. Solve by substitution.
$$3x - 5y = 13$$
$$x + 3y = 1$$
$\left(\dfrac{22}{7}, -\dfrac{5}{7}\right)$ [9.2A]

8. Solve by substitution.
$$2x - 4y = 1$$
$$y = \frac{1}{2}x + 3$$
no solution   [9.2A]

9. Solve by the addition method.
$$4x + 3y = 11$$
$$5x - 3y = 7$$
$(2, 1)$ [9.3A]

10. Solve by the addition method.
$$2x - 5y = 6$$
$$4x + 3y = -1$$
$\left(\dfrac{1}{2}, -1\right)$ [9.3A]

11. Solve by the addition method.

$$x + 2y = 8$$
$$3x + 6y = 24$$

Dependent. The solutions satisfy the equation
$x + 2y = 8$.   [9.3A]

12. Solve by the addition method.

$$7x + 3y = 11$$
$$2x - 5y = 9$$

$(2, -1)$ [9.3A]

13. Solve by the addition method.

$$5x + 6y = -7$$
$$3x + 4y = -5$$

$(1, -2)$ [9.3A]

14. With the wind, a plane flies 240 mi in 2 h. Against the wind, the plane requires 3 h to fly the same distance. Find the rate of the plane in calm air and the rate of the wind.

rate of plane in calm air, 100 mph; rate of wind: 20 mph   [9.4A]

15. For the first performance of a play in a community theater, 50 reserved-seat tickets and 80 general-admission tickets were sold. The total receipts were $980. For the second performance, 60 reserved-seat tickets and 90 general-admission tickets were sold. The total receipts were $1140. Find the price of a reserved-seat ticket and the price of a general-admission ticket.

reserved seat, $10; general admission, $6   [9.4B]

Cumulative Review

1. Evaluate $\frac{a^2 - b^2}{2a}$ when $a = 4$ and $b = -2$.
   $\frac{3}{2}$ [2.1A]

2. Solve: $-\frac{3}{4}x = \frac{9}{8}$
   $x = -\frac{3}{2}$ [3.1C]

3. Solve: $4 - 3(2 - 3x) = 7x - 9$
   $x = -\frac{7}{2}$ [3.3B]

4. Simplify: $(2a^2 - 3a + 1)(2 - 3a)$
   $-6a^3 + 13a^2 - 9a + 2$ [5.3B]

5. Simplify: $\frac{(-2x^2y)^4}{-8x^3y^2}$
   $-2x^5y^2$ [5.4A]

6. Simplify: $(4b^2 - 8b + 4) \div (2b - 3)$
   $2b - 1 + \frac{1}{2b - 3}$ [5.4C]

7. Simplify: $\frac{8x^{-2}y^5}{-2xy^4}$
   $-\frac{4y}{x^3}$ [5.5A]

8. Factor $4x^2y^4 - 64y^2$.
   $4y^2(xy - 4)(xy + 4)$ [6.4B]

9. Solve: $(x - 5)(x + 2) = -6$
   4 and -1 [6.5A]

10. Simplify: $\frac{x^2 - 6x + 8}{2x^3 + 6x^2} \div \frac{2x - 8}{4x^3 + 12x^2}$
    $x - 2$ [7.1C]

11. Simplify: $\frac{x - 1}{x + 2} + \frac{2x + 1}{x^2 + x - 2}$
    $\frac{x^2 + 2}{(x + 2)(x - 1)}$ [7.3B]

12. Simplify: $\dfrac{x + 4 - \frac{7}{x - 2}}{x + 8 + \frac{21}{x - 2}}$
    $\frac{x - 3}{x + 1}$ [7.4A]

13. Solve: $\frac{x}{2x - 3} + 2 = \frac{-7}{2x - 3}$
    $x = -\frac{1}{5}$ [7.5A]

14. Solve $A = P + Prt$ for r.
    $r = \frac{A - P}{Pt}$ [7.7A]

15. Find the x- and y-intercepts for $2x - 3y = 12$.
    x-intercept $(6, 0)$; y-intercept $(0, -4)$ [8.3A]

16. Find the slope of the line containing the points $(2, -3)$ and $(-3, 4)$.
    $-\frac{7}{5}$ [8.3B]

17. Find the equation of the line that contains the point $(-2, 3)$ and has slope $-\frac{3}{2}$.
    $y = -\frac{3}{2}x$ [8.4A]

18. Is $(2, 0)$ a solution of the system
    $5x - 3y = 10$
    $4x + 7y = 8$?
    yes [9.1A]

19. Solve by substitution.
$$3x - 5y = -23$$
$$x + 2y = -4$$
$(-6, 1)$ [9.2A]

20. Solve by the addition method.
$$5x - 3y = 29$$
$$4x + 7y = -5$$
$(4, -3)$ [9.3A]

21. A total of $8750 is invested into two simple interest accounts. On one account, the annual simple interest rate is 9.6%; on the second account, the annual simple interest rate is 7.2%. How much should be invested in each account so that the total interest earned by each account is the same?
at 9.6%, $3750; at 7.2%, $5000   [4.4A]

22. A passenger train leaves a train depot $\frac{1}{2}$ h after a freight train leaves the same depot. The freight train is traveling 8 mph slower than the passenger train. Find the rate of each train if the passenger train overtakes the freight train in 3 h.
freight train, 48 mph; passenger train, 56 mph   [4.6A]

23. The length of each side of a square is extended 4 in. The area of the resulting square is 144 in^2. Find the length of a side of the original square.
8 in.   [6.5B]

24. A plane can travel 160 mph in calm air. Flying with the wind, the plane can fly 570 mi in the same amount of time as it takes to fly 390 mi against the wind. Find the rate of the wind.
30 mph   [7.8B]

25. Graph $2x - 3y = 6$.

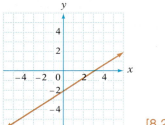

[8.2B]

26. Solve by graphing: $3x + 2y = 6$
$3x - 2y = 6$

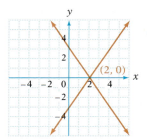

[9.1A]

27. With the current, a motorboat can travel 48 mi in 3 h. Against the current, the boat requires 4 h to travel the same distance. Find the rate of the boat in calm water.
14 mph   [9.4A]

28. Two coin banks contain only dimes and nickels. In the first bank, the total value of the coins is $5.50. In the second bank, there are half as many dimes as in the first bank and 10 fewer nickels. The total value of the coins in the second bank is $3. Find the number of dimes in the first bank.
40 dimes   [9.4B]

10

Inequalities

OBJECTIVES

▶ To write a set using the roster method
▶ To write a set using set builder notation
▶ To graph the solution set of an inequality on the number line
▶ To solve an inequality using the Addition Property of Inequalities
▶ To solve an inequality using the Multiplication Property of Inequalities
▶ To solve application problems
▶ To solve general inequalities
▶ To solve application problems
▶ To graph an inequality in two variables

Calculations of Pi

There are many early references to estimated values for pi. One of the earliest is from the Rhind Papyrus, which was found in Egypt in the 1800s. Scientists have estimated that these tablets were written around 1600 B.C. The Rhind Papyrus contains the estimate 3.1604 for pi.

One of the most famous calculations of pi came around 240 B.C. and was calculated by Archimedes. The calculation was based on finding the perimeter of inscribed and circumscribed six-sided polygons (or hexagons). Once the perimeter for a hexagon figure was calculated, known formulas could be used to calculate the perimeter of polygons with twice that number of sides. Continuing in this way, Archimedes calculated the perimeters for the polygons with 12, 24, 48, and 96 sides.

His calculations resulted in a value of pi between $3\frac{10}{71}$ and $3\frac{1}{7}$. You might recognize $3\frac{1}{7}$, or $\frac{22}{7}$, as an approximation for pi still used today.

After Archimedes' work, calculations to improve the accuracy of pi were continued. One French mathematician, using Archimedes' method, estimated pi by using a polygon of 393,216 sides. A mathematician from the Netherlands estimated pi by using a polygon with over one million sides.

Around the 1650s, new mathematical methods were developed to estimate the value of pi. These methods started yielding estimates of pi that were accurate to over 70 places. By the 1850s an estimate for pi was accurate to 200 places.

Today, using more refined mathematical methods and computers, estimates of the value of pi now exceed one million places.

In 1914, an issue of *Scientific American* contained the following short note:

> "See, I have a rhyme assisting my feeble brain,
> its tasks oftimes resisting."

Can you see what this note has to do with the estimates for the value of pi?

(Each word length represents a digit in the approximation 3.141592653579.)

SECTION 10.1 Sets

Objective A **To write a set using the roster method**

A **set** is a collection of objects. The objects in a set are called the **elements** of the set.

The **roster method** of writing a set encloses a list of the elements in braces.

The set of the last three letters of the alphabet is written {x, y, z}.

The set of the positive integers less than 5 is written {1, 2, 3, 4}.

Use the roster method to write the set of integers between 0 and 10.

A set can be designated by a capital letter. Note that 0 and 10 are not elements of the set.

$A = \{1, 2, 3, 4, 5, 6, 7, 8, 9\}$

Use the roster method to write the set of natural numbers.

The three dots mean that the pattern of numbers continues without end.

$A = \{1, 2, 3, 4,...\}$

The symbol $\in$ means "is an element of."

$2 \in B$ is read "2 is an element of set B."

Given $A = \{3, 5, 9\}$, then $3 \in A$, $5 \in A$, and $9 \in A$.

The **empty set,** or **null set,** is the set that contains no elements. The symbol $\varnothing$ or { } is used to represent the empty set.

The set of people who have run a two-minute mile is the empty set.

The **union** of two sets, written $A \cup B$, is the set that contains the elements of A and the elements of B.

Find $A \cup B$, given $A = \{1, 2, 3, 4\}$ and $B = \{3, 4, 5, 6\}$.

The union of A and B contains all the elements of A and all the elements of B. Any elements that are in both A and B are listed only once.

$A \cup B = \{1, 2, 3, 4, 5, 6\}$

The **intersection** of two sets, written $A \cap B$, is the set that contains the elements that are common to both A and B.

Find $A \cap B$, given $A = \{1, 2, 3, 4\}$ and $B = \{3, 4, 5, 6\}$.

The intersection of A and B contains the elements common to A and B.

$A \cap B = \{3, 4\}$

Example 1

Use the roster method to write the set of the odd positive integers less than 12.

Solution

$A = \{1, 3, 5, 7, 9, 11\}$

Example 2

Use the roster method to write the set of the odd negative integers greater than -10.

Your solution

$A = \{-9, -7, -5, -3, -1\}$

Solution on p. A34

Example 3

Use the roster method to write the set of the even positive integers.

Solution

$A = \{2, 4, 6,...\}$

Example 4

Use the roster method to write the set of the odd positive integers.

Your solution

$A = \{1, 3, 5,...\}$

Example 5

Find $D \cup E$, given
$D = \{6, 8, 10, 12\}$ and
$E = \{-8, -6, 10, 12\}$.

Solution

$D \cup E = \{-8, -6, 6, 8, 10, 12\}$

Example 6

Find $A \cup B$, given
$A = \{-2, -1, 0, 1, 2\}$ and
$B = \{0, 1, 2, 3, 4\}$.

Your solution

$A \cup B = \{-2, -1, 0, 1, 2, 3, 4\}$

Example 7

Find $A \cap B$, given
$A = \{5, 6, 9, 11\}$ and
$B = \{5, 9, 13, 15\}$.

Solution

$A \cap B = \{5, 9\}$

Example 8

Find $C \cap D$, given
$C = \{10, 12, 14, 16\}$ and
$D = \{10, 16, 20, 26\}$.

Your solution

$C \cap D = \{10, 16\}$

Example 9

Find $A \cap B$, given
$A = \{1, 2, 3, 4\}$ and
$B = \{8, 9, 10, 11\}$.

Solution

$A \cap B = \varnothing$

Example 10

Find $A \cap B$, given
$A = \{-5, -4, -3, -2\}$ and
$B = \{2, 3, 4, 5\}$.

Your solution

$A \cap B = \varnothing$

Solutions on p. A34

Objective B   To write a set using set builder notation

Another method of representing sets is called **set builder notation.** Using set builder notation, the set of all positive integers less than 10 is as follows:

$\{x | x < 10, x$ is a positive integer$\}$, which is read "the set of all x such that x is less than 10 and x is a positive integer."

Use set builder notation to write the set of real numbers greater than 4.

"$x \in$ real numbers" is read "x is an element of the real numbers."

$\{x | x > 4, x \in$ real numbers$\}$

Example 11

Use set builder notation to write the set of negative integers greater than −100.

Solution

$\{x \mid x > -100,\ x \text{ is a negative integer}\}$

Example 12

Use set builder notation to write the set of positive even integers less than 59.

$\{x \mid x < 59,\ x \text{ is a positive even integer}\}$

Example 13

Use set builder notation to write the set of real numbers less than 60.

Solution

$\{x \mid x < 60,\ x \in \text{real numbers}\}$

Example 14

Use set builder notation to write the set of real numbers greater than −3.

Your solution

$\{x \mid x > -3,\ x \in \text{real numbers}\}$

Solutions on p. A34

Objective C

To graph the solution set of an inequality on the number line

An expression that contains the symbol $>$, $<$, $\geq$ (is greater than or equal to), or $\leq$ (is less than or equal to) is called an **inequality.** An inequality expresses the relative order of two mathematical expressions. The expressions can be either numerical or variable.

$$4 > 2$$
$$3x \leq 7$$
$$x^2 - 2x > y + 4$$
$\left. \right\}$ Inequalities

The **solution set of an inequality** is a set of real numbers and can be graphed on the number line.

Graph the solution set of $x > 1$.

The solution set is the real numbers greater than 1. The circle on the graph indicates that 1 is not included in the solution set.

Graph the solution set of $x \geq 1$.

The dot at 1 indicates that 1 is included in the solution set.

Graph the solution set of $x < -1$.

The numbers less than −1 are to the left of −1 on the number line.

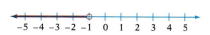

The union of two sets is the set that contains all the elements of each set.

Graph the solution set of $(x > 4) \cup (x < 1)$.

The solution set is the numbers greater than 4 and the numbers less than 1.

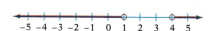

The intersection of two sets is the set that contains the elements common to both sets.

Graph the solution set of $(x > -1) \cap (x < 2)$.

The solution set is the numbers between -1 and 2.

Example 15

Graph the solution set of $x < 5$.

Solution

The solution set is the numbers less than 5.

Example 16

Graph the solution set of $x > -2$.

Your solution

Example 17

Graph the solution set of $(x > -2) \cap (x < 1)$.

Solution

The solution set is the numbers between -2 and 1.

Example 18

Graph the solution set of $(x > -1) \cup (x < -3)$.

Your solution

Example 19

Graph the solution set of $(x \le 5) \cup (x \ge -3)$.

Solution

The solution set is the real numbers.

Example 20

Graph the solution set of $(x < 5) \cup (x \ge -2)$.

Your solution

Example 21

Graph the solution set of $(x > 3) \cup (x < 1)$.

Solution

The solution set is the numbers greater than 3 or the numbers less than 1.

Example 22

Graph the solution set of $(x \le 4) \cap (x \ge -4)$.

Your solution

Solutions on p. A34

Content and Format © 1991 HMCo.

10.1 EXERCISES

▶ Objective A

Use the roster method to write the set.

1. the integers between 15 and 22
$A = \{16, 17, 18, 19, 20, 21\}$

2. the integers between -10 and -4
$A = \{-9, -8, -7, -6, -5\}$

3. the odd integers between 8 and 18
$A = \{9, 11, 13, 15, 17\}$

4. the even integers between -11 and -1
$A = \{-10, -8, -6, -4, -2\}$

5. the letters of the alphabet between a and d
$A = \{b, c\}$

6. the letters of the alphabet between p and v
$A = \{q, r, s, t, u\}$

7. all perfect squares less than 50
$A = \{1, 4, 9, 16, 25, 36, 49\}$

8. positive integers less than 20 that are divisible by 4
$A = \{4, 8, 12, 16\}$

Find $A \cup B$.

9. $A = \{3, 4, 5\}$ $\qquad$ $B = \{4, 5, 6\}$
$A \cup B = \{3, 4, 5, 6\}$

10. $A = \{-3, -2, -1\}$ $\qquad$ $B = \{-2, -1, 0\}$
$A \cup B = \{-3, -2, -1, 0\}$

11. $A = \{-10, -9, -8\}$ $\qquad$ $B = \{8, 9, 10\}$
$A \cup B = \{-10, -9, -8, 8, 9, 10\}$

12. $A = \{a, b, c\}$ $\qquad$ $B = \{x, y, z\}$
$A \cup B = \{a, b, c, x, y, z\}$

13. $A = \{a, b, d, e\}$ $\qquad$ $B = \{c, d, e, f\}$
$A \cup B = \{a, b, c, d, e, f\}$

14. $A = \{m, n, p, q\}$ $\qquad$ $B = \{m, n, o\}$
$A \cup B = \{m, n, o, p, q\}$

15. $A = \{1, 3, 7, 9\}$ $\qquad$ $B = \{7, 9, 11, 13\}$
$A \cup B = \{1, 3, 7, 9, 11, 13\}$

16. $A = \{-3, -2, -1\}$ $\qquad$ $B = \{-1, 1, 2\}$
$A \cup B = \{-3, -2, -1, 1, 2\}$

Find $A \cap B$.

17. $A = \{3, 4, 5\}$ $\qquad$ $B = \{4, 5, 6\}$
$A \cap B = \{4, 5\}$

18. $A = \{-4, -3, -2\}$ $\qquad$ $B = \{-6, -5, -4\}$
$A \cap B = \{-4\}$

19. $A = \{-4, -3, -2\}$ $\qquad$ $B = \{2, 3, 4\}$
$A \cap B = \varnothing$

20. $A = \{1, 2, 3, 4\}$ $\qquad$ $B = \{1, 2, 3, 4\}$
$A \cap B = \{1, 2, 3, 4\}$

21. $A = \{a, b, c, d, e\}$ $\qquad$ $B = \{c, d, e, f, g\}$
$A \cap B = \{c, d, e\}$

22. $A = \{m, n, o, p\}$ $\qquad$ $B = \{k, l, m, n\}$
$A \cap B = \{m, n\}$

23. $A = \{1, 7, 9, 11\}$ $\qquad$ $B = \{7, 11, 17\}$
$A \cap B = \{7, 11\}$

24. $A = \{3, 6, 9, 12\}$ $\qquad$ $B = \{6, 12, 18\}$
$A \cap B = \{6, 12\}$

▶ **Objective B**

Use set builder notation to write the set.

25. the negative integers greater than −5
{x|x > −5, x is a negative integer}

26. the positive integers less than 5
{x|x < 5, x is a positive integer}

27. the integers greater than 30
{x|x > 30, x is an integer}

28. the integers less than −70
{x|x < −70, x is an integer}

29. the even integers greater than 5
{x|x > 5, x is an even integer}

30. the odd integers less than −2
{x|x < −2, x is an odd integer}

31. the real numbers greater than 8
{x|x > 8, x ∈ real numbers}

32. the real numbers less than 57
{x|x < 57, x ∈ real numbers}

33. the real numbers greater than −5
{x|x > −5, x ∈ real numbers}

34. the real numbers less than −63
{x|x < −63, x ∈ real numbers}

▶ **Objective C**

Graph the solution set.

35. $x > 2$

36. $x \geq -1$

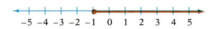

37. $x \leq 0$

38. $x < 4$

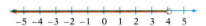

39. $(x > -2) \cup (x < -4)$

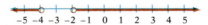

40. $(x > 4) \cup (x < -2)$

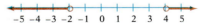

41. $(x > -2) \cap (x < 4)$

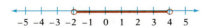

42. $(x > -3) \cap (x < 3)$

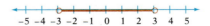

43. $(x \geq -2) \cup (x < 4)$

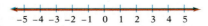

44. $(x > 0) \cup (x \leq 4)$

SECTION 10.2

The Addition and Multiplication Properties of Inequalities

Objective A   **To solve an inequality using the Addition Property of Inequalities**

The **solution set of an inequality** is a set of numbers, each element of which, when substituted for the variable, results in a true inequality.

The inequality at the right is true if the variable is replaced by 7, 9.3, or $\frac{15}{2}$.

$$x + 3 > 8$$
$$7 + 3 > 8$$
$$9.3 + 3 > 8$$
$$\frac{15}{2} + 3 > 8$$
$\Big\}$ True inequalities

The inequality $x + 3 > 8$ is false if the variable is replaced by 4, 1.5, or $-\frac{1}{2}$.

$$4 + 3 > 8$$
$$1.5 + 3 > 8$$
$$-\frac{1}{2} + 3 > 8$$
$\Big\}$ False inequalities

There are many values of the variable x that will make the inequality $x + 5 > 8$ true. The solution set of $x + 5 > 8$ is any number greater than 3.

The graph of the solution set of $x + 5 > 8$

In solving an inequality, the goal is to rewrite the given inequality in the form *variable > constant* or *variable < constant*. The Addition Property of Inequalities is used to rewrite an inequality in this form.

Addition Property of Inequalities

> The same term can be added to each side of an inequality without changing the solution set of the inequality.
>
> $$\text{If } a > b, \text{ then } a + c > b + c.$$
> $$\text{If } a < b, \text{ then } a + c < b + c.$$

The Addition Property of Inequalities also holds true for an inequality containing the symbol $\geq$ or $\leq$.

The Addition Property of Inequalities is used when, in order to rewrite an inequality in the form *variable > constant* or *variable < constant*, a term must be removed from one side of the inequality. Add the same term to each side of the inequality.

To rewrite the inequality at the right, add 4 to each side of the inequality. Then simplify.

$$x - 4 < -3$$
$$x - 4 + 4 < -3 + 4$$
$$x < 1$$

The graph of the solution set of $x + 4 < 5$.

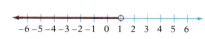

Because subtraction is defined in terms of addition, the Addition Property of Inequalities allows the same term to be subtracted from each side of an inequality.

Solve: $5x - 6 \le 4x - 4$

Subtract $4x$ from each side of the inequality. Simplify.

$$5x - 6 \le 4x - 4$$
$$5x - 4x - 6 \le 4x - 4x - 4$$
$$x - 6 \le -4$$

Add 6 to each side of the inequality. Simplify.

$$x - 6 + 6 \le -4 + 6$$
$$x \le 2$$

Example 1

Solve and graph the solution set of $x + 5 > 3$.

Solution

$$x + 5 > 3$$
$$x + 5 - 5 > 3 - 5$$
$$x > -2$$

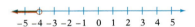

Example 2

Solve and graph the solution set of $x + 2 < -2$.

Your solution

$x < -4$

Example 3

Solve: $7x - 14 \le 6x - 16$

Solution

$$7x - 14 \le 6x - 16$$
$$7x - 6x - 14 \le 6x - 6x - 16$$
$$x - 14 \le -16$$
$$x - 14 + 14 \le -16 + 14$$
$$x \le -2$$

Example 4

Solve: $5x + 3 > 4x + 5$

Your solution

$x > 2$

Solutions on p. A34

Objective B

To solve an inequality using the Multiplication Property of Inequalities

In solving an inequality, the goal is to rewrite the given inequality in the form *variable > constant* or *variable < constant*. The Multiplication Property of Inequalities is used when, in order to rewrite an inequality in this form, a coefficient must be removed from one side of the inequality.

Multiplication Property of Inequalities

Each side of an inequality can be multiplied by the same positive number without changing the solution set of the inequality.

If $a > b$ and $c > 0$, then $ac > bc$.   If $a < b$ and $c > 0$, then $ac < bc$.

$5 > 4$ $6 < 9$
$5(2) > 4(2)$ $6(3) < 9(3)$
$10 > 8$ A true inequality $18 < 27$ A true inequality

If each side of an inequality is multiplied by the same negative number and the inequality symbol is reversed, then the solution set of the inequality is not changed.

If $a > b$ and $c < 0$, then $ac < bc$.   If $a < b$ and $c < 0$, then $ac > bc$.

$5 > 4$ $6 < 9$
$5(-2) < 4(-2)$ $6(-3) > 9(-3)$
$-10 < -8$ A true inequality $-18 > -27$ A true inequality

The Multiplication Property of Inequalities also holds true for an inequality containing the symbol $\geq$ or $\leq$.

Solve $-\frac{3}{2}x \leq 6$ and graph the solution set.

$$-\frac{3}{2}x \leq 6$$

Multiply each side of the inequality by the reciprocal of $-\frac{3}{2}$. Because $-\frac{2}{3}$ is a negative number, the inequality symbol must be reversed.

$$-\frac{2}{3}\left(-\frac{3}{2}x\right) \geq -\frac{2}{3}(6)$$

$$x \geq -4$$

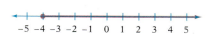

Because division is defined in terms of multiplication, the Multiplication Property of Inequalities allows each side of an inequality to be divided by a non-zero constant.

Solve $-4x > -8$.

$$-4x > 8$$

Divide each side of the inequality by -4. Because each side is divided by a negative number, the inequality symbol is reversed.

$$\frac{-4x}{-4} < \frac{8}{-4}$$

$$x < -2$$

Example 5 Solve and graph the solution set of $-7x > 14$.

Solution $-7x > 14$

$$\frac{-7x}{-7} < \frac{14}{-7}$$

$$x < -2$$

-5 -4 -3 -2 -1 0 1 2 3 4 5

Example 6 Solve and graph the solution set of $-3x > -9$.

Your solution $x < 3$

-5 -4 -3 -2 -1 0 1 2 3 4 5

Solution on p. A34

Example 7

Solve: $-\dfrac{5}{8}x \le \dfrac{5}{12}$

Solution

$$-\dfrac{5}{8}x \le \dfrac{5}{12}$$

$$-\dfrac{8}{5}\left(-\dfrac{5}{8}x\right) \ge -\dfrac{8}{5}\left(\dfrac{5}{12}\right)$$

$$x \ge -\dfrac{2}{3}$$

Example 8

Solve: $-\dfrac{3}{4}x \ge 18$

Your Solution

$x \le -24$

Solution on p. A34

Objective C   To solve application problems

Example 9

A student must have at least 450 points out of 500 points on five tests to receive an A in a course. One student's results on the first four tests were 94, 87, 77, and 95. What scores on the last test will enable this student to receive an A in the course?

Strategy

To find the scores, write and solve an inequality using N to represent the possible scores on the last test.

Solution

Total number of points on the 5 tests	is greater than or equal to	450

$$94 + 87 + 77 + 95 + N \ge 450$$
$$353 + N \ge 450$$
$$353 + (-353) + N \ge 450 + (-353)$$
$$N \ge 97$$

The student's score on the last test must be equal to or greater than 97.

Example 10

An appliance dealer will make a profit on the sale of a television set if the cost of the new set is less than 70% of the selling price. What selling prices will enable the dealer to make a profit on a television set that costs the dealer $314?

Your strategy

Your solution

any price equal to or greater than $448.58

Solution on p. A34

Content and Format © 1991 HMCo.

inequalities

8am – 5pm	Mike 10 am – 3 pm	
Sat 10am–4pm	EJ 10 am – 4 pm	
Sun 1pm–5pm	Mike 1 pm – 5 pm	

Study Skills and Reading Tutors

(May also be cross-listed as math and/or writing tutors)

- ☐ EJ Brown
- ☐ Lara Ferris
- ☐ Jae Hively
- ☐ Aniseh Koro
- ☐ Mike Micallef

10.2   **EXERCISES**

▶ **Objective A**

Solve and graph the solution set.

1. $x + 1 < 3$ $x < 2$

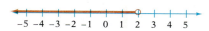

2. $y + 2 < 2$ $y < 0$

3. $x - 5 > -2$ $x > 3$

4. $x - 3 > -2$ $x > 1$

5. $n + 4 \geq 7$ $n \geq 3$

6. $x + 5 \geq 3$ $x \geq -2$

7. $x - 6 \leq -10$ $x \leq -4$

8. $y - 8 \leq -11$ $y \leq -3$

9. $5 + x \geq 4$ $x \geq -1$

10. $-2 + n \geq 0$ $n \geq 2$

Solve.

11. $y - 3 \geq -12$
$y \geq -9$

12. $x + 8 \geq -14$
$x \geq -22$

13. $3x - 5 < 2x + 7$
$x < 12$

14. $5x + 4 < 4x - 10$
$x < -14$

15. $8x - 7 \geq 7x - 2$
$x \geq 5$

16. $3n - 9 \geq 2n - 8$
$n \geq 1$

17. $2x + 4 < x - 7$
$x < -11$

18. $9x + 7 < 8x - 7$
$x < -14$

19. $4x - 8 \leq 2 + 3x$
$x \leq 10$

20. $5b - 9 < 3 + 4b$
$b < 12$

21. $6x + 4 \geq 5x - 2$
$x \geq -6$

22. $7x - 3 \geq 6x - 2$
$x \geq 1$

23. $2x - 12 > x - 10$
$x > 2$

24. $3x + 9 > 2x + 7$
$x > -2$

25. $d + \dfrac{1}{2} < \dfrac{1}{3}$
$d < -\dfrac{1}{6}$

26. $x - \dfrac{3}{8} < \dfrac{5}{6}$
$x < \dfrac{29}{24}$

27. $x + \dfrac{5}{8} \geq -\dfrac{2}{3}$
$x \geq -\dfrac{31}{24}$

28. $y + \dfrac{5}{12} \geq -\dfrac{3}{4}$
$y \geq -\dfrac{7}{6}$

Solve.

29. $x - \dfrac{3}{8} < \dfrac{1}{4}$

$x < \dfrac{5}{8}$

30. $y + \dfrac{5}{9} \leq \dfrac{5}{6}$

$y \leq \dfrac{5}{18}$

31. $2x - \dfrac{1}{2} < x + \dfrac{3}{4}$

$x < \dfrac{5}{4}$

32. $6x - \dfrac{1}{3} \leq 5x - \dfrac{1}{2}$

$x \leq -\dfrac{1}{6}$

33. $3x + \dfrac{5}{8} > 2x + \dfrac{5}{6}$

$x > \dfrac{5}{24}$

34. $4b - \dfrac{7}{12} \geq 3b - \dfrac{9}{16}$

$b \geq \dfrac{1}{48}$

35. $3.8x < 2.8x - 3.8$
$x < -3.8$

36. $1.2x < 0.2x - 7.3$
$x < -7.3$

37. $x + 5.8 \leq 4.6$
$x \leq -1.2$

38. $n - 3.82 \leq 3.95$
$n \leq 7.77$

39. $x - 3.5 < 2.1$
$x < 5.6$

40. $x - 0.23 \leq 0.47$
$x \leq 0.70$

41. $1.33x - 1.62 > 0.33x - 3.1$
$x > -1.48$

42. $2.49x + 1.35 \geq 1.49x - 3.45$
$x \geq -4.80$

▶ **Objective B**

Solve and graph the solution set.

43. $3x < 12$ $x < 4$

44. $8x \leq -24$ $x \leq -3$

45. $5y \geq 15$ $y \geq 3$

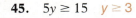

46. $24x > -48$ $x > -2$

47. $16x \leq 16$ $x \leq 1$

48. $3x > 0$ $x > 0$

49. $-8x > 8$ $x < -1$

50. $-2n \leq -8$ $n \geq 4$

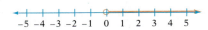

51. $-6b > 24$ $b < -4$

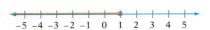

52. $-4x < 8$ $x > -2$

Solve.

53. $-5y \geq 20$
$y \leq -4$

54. $3x < 5$
$x < \dfrac{5}{3}$

55. $7x > 2$
$x > \dfrac{2}{7}$

56. $6x \leq -1$
$x \leq -\dfrac{1}{6}$

57. $2x \leq -5$
$x \leq -\dfrac{5}{2}$

58. $\dfrac{5}{6}n < 15$
$n < 18$

59. $\dfrac{3}{4}x < 12$
$x < 16$

60. $\dfrac{2}{3}y \geq 4$
$y \geq 6$

61. $\dfrac{5}{8}x \geq 10$
$x \geq 16$

62. $-\dfrac{2}{3}x \leq 4$
$x \geq -6$

63. $-\dfrac{3}{7}x \leq 6$
$x \geq -14$

64. $-\dfrac{2}{11}b \geq -6$
$b \leq 33$

65. $-\dfrac{4}{7}x \geq -12$
$x \leq 21$

66. $\dfrac{2}{3}n < \dfrac{1}{2}$
$n < \dfrac{3}{4}$

67. $\dfrac{3}{5}x < \dfrac{7}{10}$
$x < \dfrac{7}{6}$

68. $-\dfrac{2}{3}x \geq \dfrac{4}{7}$
$x \leq -\dfrac{6}{7}$

69. $-\dfrac{3}{8}x \geq \dfrac{9}{14}$
$x \leq -\dfrac{12}{7}$

70. $-\dfrac{3}{5}x < -\dfrac{6}{7}$
$x > \dfrac{10}{7}$

71. $-\dfrac{4}{5}x < -\dfrac{8}{15}$
$x > \dfrac{2}{3}$

72. $-\dfrac{3}{4}y \geq -\dfrac{5}{8}$
$y \leq \dfrac{5}{6}$

73. $-\dfrac{8}{9}x \geq -\dfrac{16}{27}$
$x \leq \dfrac{2}{3}$

74. $1.5x \leq 6.30$
$x \leq 4.2$

75. $2.3x \leq 5.29$
$x \leq 2.3$

76. $-3.5d > 7.35$
$d < -2.1$

77. $-0.24x > 0.768$
$x < -3.2$

78. $4.25m > -34$
$m > -8$

79. $-3.9x \geq -19.5$
$x \leq 5$

80. $0.035x < -0.0735$
$x < -2.1$

81. $0.07x < -0.378$
$x < -5.4$

82. $-11.7x \leq 4.68$
$x \geq -0.4$

83. $0.685y \geq -2.15775$
$y \geq -3.15$

84. $1.38n > -0.9936$
$n > -0.72$

85. $-5.24x < 43.0728$
$x > -8.22$

86. $-0.0663b < 0.19227$
$b > -2.9$

87. $13.58x \leq 95.06$
$x \leq 7$

88. $18.92x < 264.88$
$x < 14$

▶ ## Objective C   *Application Problems*

Solve.

89. The circumference of an official major league baseball is between 9.00 in. and 9.25 in. Find the possible diameters of a major league baseball to the nearest hundredth of an inch. Recall that $C = \pi d$ and use $\pi \approx 3.14$.
2.87 in. $< d <$ 2.95 in.

Solve.

90. To be eligible for a basketball tournament, a basketball team must win at least 60% of its remaining games. If the team has 17 games remaining, how many games must the team win to qualify for the tournament?

≥ 11 games

91. To avoid a tax penalty, at least 90% of a self-employed person's total annual income tax liability must be paid by April 15. What amount of income tax must a person, with an annual income tax liability of $3500, pay?

≥ $3150

92. Computer software engineers are fond of saying that software takes at least twice as long to develop as they think it will. Applying that saying, how many hours will it take to develop a software product that an engineer thinks can be finished in 50 hours?

≥ 100 h

93. A health official recommends a maximum cholesterol level of 220 units. How many units must a patient, with a cholesterol level of 275 units, reduce her cholesterol level to satisfy the recommended maximum level?

≥ 55 units

94. A government agency recommends a minimum daily allowance of vitamin C of 60 mg. How many additional milligrams of vitamin C are needed by a person who drank a glass of orange juice with 10 mg of vitamin C to satisfy the recommended daily allowance?

≥ 50 mg

95. To pass a course with a B grade, a student must have an average of 80 points on five tests. The student's grades on the first 4 tests were 75, 83, 86, and 78. What scores can the student receive on the fifth test to earn a B grade?

≥ 78

96. A professor scores all tests with a maximum of 100 points. To earn an A grade in this course, a student must have an average of 92 on four tests. The student's grades on the first three tests were 89, 86, and 90. Can this student earn an A grade?

no

97. A car sales representative receives a commission that is the greater of $250 or 8% of the selling price of a car. What dollar amounts in the sale price of a car will make the commission offer more attractive?

> $3125

98. A sales representative for a stereo store has the option of a monthly salary of $2000 or a 35% commission on the selling price of each item sold by the representative. What dollar amounts in sales will make the commission more attractive than the monthly salary?

≥ $5715

SECTION 10.3   General Inequalities

| **Objective A** | **To solve general inequalities** |

Solving an inequality frequently requires application of both the Addition and Multiplication Properties of Inequalities.

Solve: $3x - 2 < 5x + 4$

Subtract $5x$ from each side of the inequality. Simplify.

$$3x - 2 < 5x + 4$$
$$3x - 5x - 2 < 5x - 5x + 4$$
$$-2x - 2 < 4$$

Add 2 to each side of the inequality. Simplify.

$$-2x - 2 + 2 < 4 + 2$$
$$-2x < 6$$

Divide each side by the coefficient of x. Because -2 is a negative number, the inequality symbol must be reversed. Simplify.

$$\frac{-2x}{-2} > \frac{6}{-2}$$
$$x > -3$$

When an inequality contains parentheses, one of the steps in solving the inequality requires the use of the Distributive Property.

Solve: $-2(x - 7) > 3 - 4(2x - 3)$

Use the Distributive Property to remove parentheses. Simplify.

$$-2(x - 7) > 3 - 4(2x - 3)$$
$$-2x + 14 > 3 - 8x + 12$$
$$-2x + 14 > 15 - 8x$$

Add $8x$ to each side of the inequality. Simplify.

$$-2x + 8x + 14 > 15 - 8x + 8x$$
$$6x + 14 > 15$$

Subtract 14 from each side of the inequality. Simplify.

$$6x + 14 - 14 > 15 - 14$$
$$6x > 1$$

Divide each side of the inequality by the coefficient of x. Simplify.

$$\frac{6x}{6} > \frac{1}{6}$$
$$x > \frac{1}{6}$$

Example 1   Solve: $7x - 3 \le 3x + 17$

Solution
$$7x - 3 \le 3x + 17$$
$$7x - 3x - 3 \le 3x - 3x + 17$$
$$4x - 3 \le 17$$
$$4x - 3 + 3 \le 17 + 3$$
$$4x \le 20$$
$$\frac{4x}{4} \le \frac{20}{4}$$
$$x \le 5$$

Example 2   Solve: $5 - 4x > 9 - 8x$

Your solution $x > 1$

Solution on p. A35

Example 3   Solve:
$$3(3 - 2x) \geq -5x - 2(3 - x)$$

Solution
$$3(3 - 2x) \geq -5x - 2(3 - x)$$
$$9 - 6x \geq -5x - 6 + 2x$$
$$9 - 6x \geq -3x - 6$$
$$9 - 6x + 3x \geq -3x + 3x - 6$$
$$9 - 3x \geq -6$$
$$9 - 9 - 3x \geq -6 - 9$$
$$-3x \geq -15$$
$$\frac{-3x}{-3} \leq \frac{-15}{-3}$$
$$x \leq 5$$

Example 4   Solve:
$$8 - 4(3x + 5) \leq 6(x - 8)$$

Your solution $x \geq 2$

Solution on p. A35

Objective B   To solve application problems

10

Example 5

A rectangle is 10 ft wide and $(2x + 4)$ ft long. Express as an integer the maximum length of the rectangle when the area is less than 200 ft^2. (The area of a rectangle is equal to its length times its width.)

Strategy

To find the maximum length:

- Replace the variables in the area formula by the given values and solve for x.
- Replace the variable in the expression $2x + 4$ with the value found for x.

Solution

Length times width	is less than	200 ft^2

$$10(2x + 4) < 200$$
$$20x + 40 < 200$$
$$20x + 40 - 40 < 200 - 40$$
$$20x < 160$$
$$\frac{20x}{20} < \frac{160}{20}$$
$$x < 8$$

$$2x + 4 < 20$$

The maximum length is 19 ft.

Example 6

Company A rents cars for $8 a day and 10¢ for every mile driven. Company B rents cars for $10 a day and 8¢ per mile driven. You want to rent a car for one week. What is the maximum number of miles you can drive a Company A car if it is to cost you less than a Company B car?

Your strategy

Your solution

699 mi

Solution on p. A35

10.3 EXERCISES

▶ **Objective A**

Solve.

1. $4x - 8 < 2x$
   $x < 4$

2. $7x - 4 < 3x$
   $x < 1$

3. $2x - 8 > 4x$
   $x < -4$

4. $3y + 2 > 7y$
   $y < \dfrac{1}{2}$

5. $8 - 3x \leq 5x$
   $x \geq 1$

6. $10 - 3x \leq 7x$
   $x \geq 1$

7. $3x + 2 \geq 5x - 8$
   $x \leq 5$

8. $2n - 9 \geq 5n + 4$
   $n \leq -\dfrac{13}{3}$

9. $5x - 2 < 3x - 2$
   $x < 0$

10. $8x - 9 > 3x - 9$
    $x > 0$

11. $0.1(180 + x) > x$
    $x < 20$

12. $x > 0.2(50 + x)$
    $x > 12.5$

13. $0.15x + 55 > 0.10x + 80$
    $x > 500$

14. $-3.6b + 16 < 2.8b + 25.6$
    $b > -1.5$

15. $2(3x - 1) > 3x + 4$
    $x > 2$

16. $5(2x + 7) > -4x - 7$
    $x > -3$

17. $3(2x - 5) \geq 8x - 5$
    $x \leq -5$

18. $5x - 8 \geq 7x - 9$
    $x \leq \dfrac{1}{2}$

19. $2(2y - 5) \leq 3(5 - 2y)$
    $y \leq \dfrac{5}{2}$

20. $2(5x - 8) \leq 7(x - 3)$
    $x \leq -\dfrac{5}{3}$

21. $5(2 - x) > 3(2x - 5)$
    $x < \dfrac{25}{11}$

22. $4(3d - 1) > 3(2 - 5d)$
    $d > \dfrac{10}{27}$

23. $5(x - 2) > 9x - 3(2x - 4)$
    $x > 11$

24. $3x - 2(3x - 5) > 4(2x - 1)$
    $x < \dfrac{14}{11}$

25. $4 - 3(3 - n) \leq 3(2 - 5n)$
    $n \leq \dfrac{11}{18}$

26. $15 - 5(3 - 2x) \leq 4(x - 3)$
    $x \leq -2$

27. $2x - 3(x - 4) \geq 4 - 2(x - 7)$
    $x \geq 6$

28. $4 + 2(3 - 2y) \leq 4(3y - 5) - 6y$
    $y \geq 3$

▶ **Objective B**   *Application Problems*

Solve.

29. The sales agent for a jewelry company is offered a flat monthly salary of $3200 or a salary of $1000 plus an 11% commission on the selling price of each item sold by the agent. If the agent chooses the $3200, what dollar amount does the agent expect to sell in one month?
    ≤ $20,000

30. A baseball player is offered an annual salary of $200,000 or a base salary of $100,000 plus a bonus of $1000 for each hit over 100 hits. How many hits must the baseball player make to earn more than $200,000?
    ≥ 201 hits

31. A computer bulletin board service offers service for a flat fee of $10 per month or $4 per month plus $.10 for each minute the service is used. How many minutes must a person use this service to exceed $10?
    > 60 min

32. A site licensing fee for a computer program is $1500. This fee allows the company to use the program at any computer terminal within the company. Alternatively, the company can choose to pay $200 for each individual computer it has. How many individual computers must a company have for the site license to be more economical for the company?
    < 8 computers

33. For a product to be labeled orange juice, a state agency requires at least 80% of the drink be real orange juice. How many ounces of artificial flavors can be added to 32 ounces of real orange juice and still label the drink orange juice?
    ≤ 8 oz

34. Grade A hamburger can not contain more than 20% fat. How much fat can a butcher mix with 300 lb of lean meat to meet the 20% requirement?
    ≤ 75 lb

35. A shuttle service taking skiers to a ski area charges $8 per person each way. Four skiers are debating whether to take the shuttle bus or rent a car for $45 plus $.25 per mile. Assuming that the skiers will share the cost of the car and that they want the least expensive method of transportation, how far away is the ski area if they choose to take the shuttle service?
    > 76 mi

36. Company A rents a car for $25 per day and $.08 per mile. Company B rents a car for $15 per day and $.14 per mile. Find the maximum number of miles you can drive company B's car before the cost exceeds the cost of company A's car.
    166 mi

37. A maintenance crew requires between 30 and 45 min to prepare an aircraft for its next flight. How many aircraft can this crew prepare for flight in a 6 h period of time?
    between 8 and 12 aircraft

38. To be considered safe, a food product cannot contain more than 1% of a certain preservative. How many ounces of preservative, to the nearest hundredth of an ounce, can be mixed with 28 ounces of a cereal to meet the 1% regulation?
    < 0.28 oz

Content and Format © 1991 HMCo.

SECTION 10.4	**Graphing Linear Inequalities**

Objective A	**To graph an inequality in two variables**	

The graph of the linear equation $y = x - 2$ separates a plane into three sets:

the set of points on the line,

the set of points above the line,

the set of points below the line.

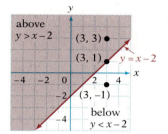

The point (3, 1) is a solution of $y = x - 2$.

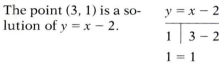

The point (3, 3) is a solution of $y > x - 2$.

$$\begin{array}{c|c} y > x - 2 \\ \hline 3 & 3 - 2 \\ 3 > 1 \end{array}$$

Any point above the line is a solution of $y > x - 2$.

The point (3, −1) is a solution of $y < x - 2$.

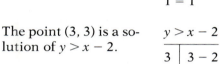

Any point below the line is a solution of $y < x - 2$.

The solution set of $y = x - 2$ is all points on the line. The solution set of $y > x - 2$ is all points above the line. The solution set of $y < x - 2$ is all points below the line. The solution set of an inequality in two variables is a **half-plane.**

The following illustrates the procedure for graphing a linear inequality.

Graph the solution set of $2x + 3y \leq 6$.

Solve the inequality for y.

$$2x + 3y \leq 6$$
$$2x - 2x + 3y \leq -2x + 6$$
$$3y \leq -2x + 6$$
$$\frac{3y}{3} \leq \frac{-2x + 6}{3}$$
$$y \leq -\frac{2}{3}x + 2$$
$$y = -\frac{2}{3}x + 2$$

Change the inequality to an equality and graph the line. If the inequality is **≥ or ≤**, the line is in the solution set and is shown by a **solid line.** If the inequality is **> or <**, the line is not a part of the solution set and is shown by a **dotted line.**

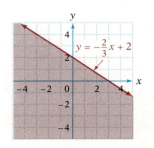

If the inequality is **> or ≥**, shade the **upper half-plane.** If the inequality is **< or ≤**, shade the **lower half-plane.**

Content and Format © 1991 HMCo.

Example 1

Graph the solution set of $3x + y > -2$.

Solution

$$3x + y > -2$$
$$3x - 3x + y > -3x - 2$$
$$y > -3x - 2$$

Graph $y = -3x - 2$ as a dotted line.
Shade the upper half-plane.

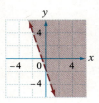

Example 2

Graph the solution set of $x - 3y < 2$.

Your solution

$y > \dfrac{1}{3}x - \dfrac{2}{3}$

Example 3

Graph the solution set of $2x - y \geq 2$.

Solution

$$2x - y \geq 2$$
$$2x - 2x - y \geq -2x + 2$$
$$-y \geq -2x + 2$$
$$-1(-y) \leq -1(-2x + 2)$$
$$y \leq 2x - 2$$

Graph $y = 2x - 2$ as a solid line.
Shade the lower half-plane.

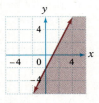

Example 4

Graph the solution set of $2x - 4y \leq 8$.

Your solution

$y \geq \dfrac{1}{2}x - 2$

Example 5

Graph the solution set of $y > 3$.

Solution

$y > 3$

Graph $y = 3$ as a dotted line.
Shade the upper half-plane.

Example 6

Graph the solution set of $x < 3$.

Your solution

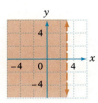

Solutions on pp. A35–A36

10.4   EXERCISES

▶ **Objective A**

Graph the solution set.

1. $x + y > 4$

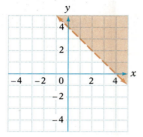

2. $x - y > -3$

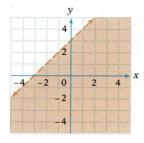

3. $2x - y < -3$

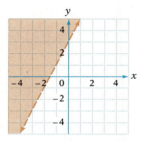

4. $3x - y < 9$

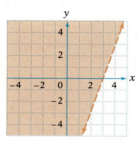

5. $2x + y \geq 4$

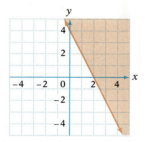

6. $3x + y \geq 6$

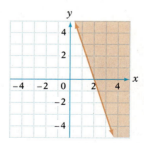

7. $y \leq -2$

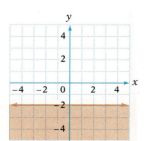

8. $y > 3$

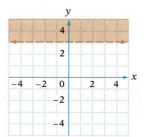

Graph the solution set.

9. $3x - 2y < 8$

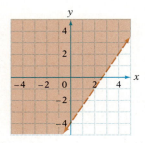

10. $5x + 4y > 4$

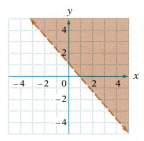

11. $-3x - 4y \geq 4$

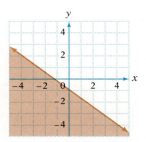

12. $-5x - 2y \geq 8$

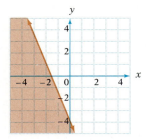

13. $6x + 5y \leq -10$

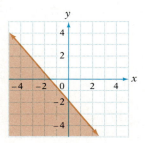

14. $2x + 2y \leq -4$

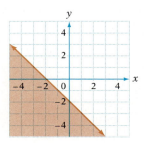

15. $-4x + 3y < -12$

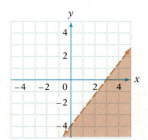

16. $-4x + 5y < 15$

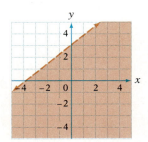

Calculators and Computers

First-Degree Inequalities in One Variable

The program FIRST-DEGREE INEQUALITIES on the Student Disk provides inequalities for you to solve. There are three levels of difficulty; the first level is the easiest, and the third level the most difficult.

Once you choose the level of difficulty, the program will display a problem. Using paper and pencil, solve the inequality. When you are ready, press the RETURN key and compare your solution with the one on the screen. The answers are rounded to the nearest hundredth.

After you complete a problem, you may continue practicing at the same level, return to the menu and select a different level, or quit the program.

Chapter Summary

Key Words

A *set* is a collection of objects. The objects of a set are called the *elements* of the set.

The *roster method* of writing a set encloses a list of the elements in braces.

The *empty set* or *null set,* written $\varnothing$ or { }, is the set that contains no elements.

The *union* of two sets, written $A \cup B$, is the set that contains all the elements of A and all the elements of B (any elements that are in both set A and set B are listed only once).

The *intersection* of two sets, written $A \cap B$, is the set that contains the elements that are common to both A and B.

An *inequality* is an expression that contains one of the symbols $<$, $>$, $\le$, or $\ge$.

The *solution set of an inequality* is a set of numbers each element of which, when substituted for the variable, results in a true inequality. The solution set of an inequality can be graphed on the number line.

The solution set of an inequality in two variables is a *half-plane.*

Essential Rules

Addition Property of Inequalities

The same term can be added to each side of an inequality without changing the solution set of the inequality.

$$\text{If } a > b, \text{ then } a + c > b + c.$$
$$\text{If } a < b, \text{ then } a + c < b + c.$$

The Addition Property of Inequalities also holds true for an inequality containing the symbol $\ge$ or $\le$.

Multiplication Property of Inequalities

Each side of an inequality can be multiplied by the same **positive number** without changing the solution set of the inequality.

If $a > b$ and $c > 0$, then $ac > bc$.
If $a < b$ and $c > 0$, then $ac < bc$.

If each side of an inequality is multiplied by the same **negative number** and the inequality symbol is reversed, then the solution set of the inequality is not changed.

If $a > b$ and $c < 0$, then $ac < bc$.
If $a < b$ and $c < 0$, then $ac > bc$.

The Multiplication Property of Inequalities also holds true for an inequality containing the symbol $\geq$ or $\leq$.

Chapter Review

SECTION 10.1

1. Use the roster method to write the set of odd positive integers less than 8.
A = {1, 3, 5, 7}

2. Find *A* ∪ *B*, given *A* = {6, 8, 10} and *B* = {2, 4, 6}.
A ∪ *B* = {2, 4, 6, 8, 10}

3. Find *A* ∩ *B*, given *A* = {0, 2, 4, 6, 8} and *B* = {−2, −4}.
A ∩ *B* = ∅

4. Find *A* ∩ *B*, given *A* = {1, 5, 9, 13} and *B* = {1, 3, 5, 7, 9}.
A ∩ *B* = {1, 5, 9}

5. Use set builder notation to write the set of odd integers greater than −8.
{*x*|*x* > −8, *x* is an odd integer}

6. Use set builder notation to write the set of real numbers greater than 3.
{*x*|*x* > 3, *x* ∈ real numbers}

7. Graph the solution set of *x* > 3.

8. Graph the solution set of (*x* > −1) ∩ (*x* ≤ 2).

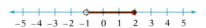

9. Graph the solution set of (*x* < 2) ∪ (*x* > 5).

SECTION 10.2

10. Solve and graph the solution set of *x* − 3 > −1.   *x* > 2

11. Solve 2*x* − 3 > *x* + 15.
x > 18

12. Solve −15*x* ≤ 45.
x ≥ −3

13. Solve $-\frac{3}{4}x > \frac{2}{3}$.
$x < -\frac{8}{9}$

14. Six less than a number is greater than twenty-five. Find the smallest integer that will satisfy the inequality.
32

15. A student's grades on five sociology tests were 68, 82, 90, 73, 95. What is the lowest score the student can receive on the next test and still be able to attain a minimum of 480 points?
72

SECTION 10.3

16. Solve $3x + 4 \geq -8$.
$x \geq -4$

17. Solve $12 - 4(x - 1) \leq 5(x - 4)$.
$x \geq 4$

18. Solve $7x - 2(x + 3) \geq x + 10$.
$x \geq 4$

19. Solve $6x - 9 < 4x + 3(x + 3)$.
$x > -18$

20. Solve $5 - 4(x + 9) > 11(12x - 9)$.
$x < \dfrac{1}{2}$

21. The width of a rectangular garden is 12 ft. The length of the garden is $(3x + 5)$ ft. Express as an integer the minimum length of the garden when the area is greater than 276 ft^2. (The area of a rectangle is equal to its length times its width.)
24 ft

22. Florist A charges a \$3 delivery fee plus \$21 per bouquet delivered. Florist B charges a \$15 delivery fee plus \$18 per bouquet delivered. A church wants to supply each resident of a small nursing home with a bouquet for Grandparent's Day. Find the number of residents of the nursing home if Florist B is more economical than Florist A.
5 or more

SECTION 10.4

23. Graph the solution set of $3x + 2y \leq 12$.

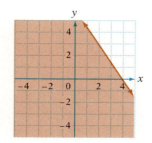

24. Graph the solution set of $2x - 3y < 9$.

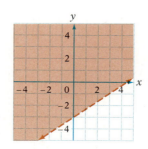

25. Graph the solution set of $5x + 2y < 6$.

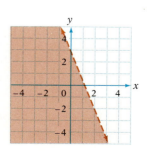

Chapter Test

1. Use the roster method to write the set of the even positive integers between 3 and 9.
$A = \{4, 6, 8\}$ [10.1A]

2. Find $A \cap B$, given $A = \{6, 8, 10, 12\}$ and $B = \{12, 14, 16\}$.
$A \cap B = \{12\}$ [10.1A]

3. Use set builder notation to write the set of the positive integers less than 50.
$\{x|x < 50, x \text{ is a positive integer}\}$ [10.1B]

4. Use set builder notation to write the set of the real numbers greater than -23.
$\{x|x > -23, x \in \text{ real numbers}\}$ [10.1B]

5. Graph the solution set of $x > -2$.
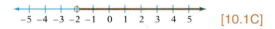
[10.1C]

6. Graph the solution set of $(x < 5) \cap (x > 0)$.

[10.1C]

7. Solve and graph the solution set of $4 + x < 1$.

$x < -3$ [10.2A]

8. Solve: $x + \frac{1}{2} > \frac{5}{8}$
$x > \frac{1}{8}$ [10.2A]

9. Solve and graph the solution set of $\frac{2}{3}x \geq 2$.

$x \geq 3$ [10.2B]

10. Solve: $-\frac{3}{8}x \leq 5$
$x \geq -\frac{40}{3}$ [10.2B]

11. To ride a certain roller coaster at an amusement park, a person must be at least 48 inches tall. How many inches must a child, who is 43 inches tall, grow to be eligible for the roller coaster?
≥ 5 in.   [10.2C]

12. A ball bearing for a rotary engine must have a circumference between 0.1220 inches and 0.1240 inches. What are the allowable diameters for the bearings to the nearest hundredth of an inch. Recall $c = \pi d$ and use $\pi \approx 3.14$.
between 0.0389 inches and 0.0395 inches   [10.2C]

13. Solve: $5 - 3x > 8$
$x < -1$ [10.3A]

14. Solve: $2x - 7 \le 6x + 9$
$x \ge -4$ [10.3A]

15. Solve: $3(2x - 5) \ge 8x - 9$
$x \le -3$ [10.3A]

16. Solve: $6x - 3(2 - 3x) < 4(2x - 7)$
$x < -\dfrac{22}{7}$ [10.3A]

17. A rectangle is 15 ft long and $(2x - 4)$ ft wide. Express, using an integer, how wide the rectangle can be for the area to be less than 180 ft². (The area of a rectangle is equal to its length times its width.)
11 ft   [10.3B]

18. A stock broker receives a monthly salary that is the greater of $2500 or $1000 plus 2% of the total value of all stock transactions the broker processes during the month. What dollar amounts of transactions did the broker process in a month for which the broker's salary was $2500?
< $75,000   [10.3B]

19. Graph the solution set of $3x + y > 4$.

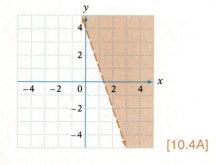

[10.4A]

20. Graph the solution set of $4x - 5y \ge 15$.

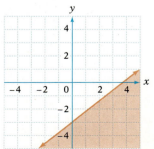

[10.4A]

Cumulative Review

1. Simplify: $2[5a - 3(2 - 5a) - 8]$
   $40a - 28$ [2.2D]

2. Solve: $\frac{5}{8} - 4x = \frac{1}{8}$

   $x = \frac{1}{8}$ [3.2A]

3. Solve: $2x - 3[x - 2(x - 3)] = 2$
   $x = 4$ [3.3B]

4. Simplify: $(-3a)(-2a^3b^2)^2$
   $-12a^7b^4$ [5.2B]

5. Simplify: $\frac{27a^3b^2}{(-3ab^2)^3}$

   $-\frac{1}{b^4}$ [5.4A]

6. Simplify: $(16x^2 - 12x - 2) \div (4x - 1)$

   $4x - 2 - \frac{4}{4x - 1}$ [5.4C]

7. Factor $4x^2 - 21x + 5$.
   $(4x - 1)(x - 5)$ [6.3A]

8. Factor $27a^2x^2 - 3a^2$.
   $3a^2(3x - 1)(3x + 1)$ [6.4D]

9. Simplify: $\frac{x^2 - 2x}{x^2 - 2x - 8} \div \frac{x^3 - 5x^2 + 6x}{x^2 - 7x + 12}$

   $\frac{1}{x + 2}$ [7.1C]

10. Simplify: $\frac{4a}{2a - 3} - \frac{2a}{a + 3}$

    $\frac{18a}{(2a - 3)(a + 3)}$ [7.3B]

11. Solve: $\frac{5y}{6} - \frac{5}{9} = \frac{y}{3} - \frac{5}{6}$

    $y = -\frac{5}{9}$ [7.5A]

12. Solve $R = \frac{C - S}{t}$ for C.

    $C = S + Rt$ [7.7A]

13. Find the slope of the line containing the points $(2, -3)$ and $(-1, 4)$.

    $-\frac{7}{3}$ [8.3B]

14. Find the equation of the line which contains the point $(1, -3)$ and has slope $-\frac{3}{2}$.

    $y = -\frac{3}{2}x - \frac{3}{2}$ [8.4B]

15. Solve by substitution.
    $$x = 3y + 1$$
    $$2x + 5y = 13$$
    $(4, 1)$ [9.2A]

16. Solve by the addition method.
    $$9x - 2y = 17$$
    $$5x + 3y = -7$$
    $(1, -4)$ [9.3A]

17. Find $A \cup B$, given $A = \{0, 1, 2\}$ and
$B = \{-2, -10\}$.
$\{-10, -2, 0, 1, 2\}$ [10.1A]

18. Use set builder notation to write the set of the
real numbers less than 48.
$\{x|x < 48, x \in \text{real numbers}\}$ [10.1B]

19. Graph the solution set of
$(x > 1) \cup (x < -1)$.

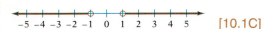
-5 -4 -3 -2 -1 0 1 2 3 4 5   [10.1C]

20. Graph the solution set of $\frac{3}{8}x > -\frac{3}{4}$.

-5 -4 -3 -2 -1 0 1 2 3 4 5   [10.2B]

21. Solve: $-\frac{4}{5}x > 12$
$x < -15$ [10.2B]

22. Solve: $15 - 3(5x - 7) < 2(7 - 2x)$
$x > 2$ [10.3A]

23. Three-fifths of a number is less than negative fifteen. What integers N satisfy this inequality?
$N \le -26$, N is an integer   [10.2C]

24. Company A rents cars for $6 a day and 25¢ for every mile driven. Company
B rents cars for $15 a day and 10¢ per mile. You want to rent a car for 6
days. What is the maximum number of miles you can drive a Company A
car if it is to cost you less than a Company B car?
359 mi   [10.3B]

25. In a lake, 100 fish are caught, tagged, and then released. Later 150 fish are
caught. Three of the 150 fish are found to have tags. Estimate the number of
fish in the lake.
5000 fish   [7.6B]

26. A drawer contains 13¢ stamps and 18¢ stamps. The number of 13¢ stamps
is two less than the number of 18¢ stamps. The total value of all the stamps
is $2.53. How many 13¢ stamps are in the drawer?
7 stamps   [4.8B]

27. Graph $y = 2x - 1$.

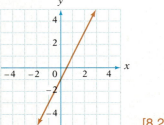

[8.2A]

28. Graph the solution set of $6x - 3y \ge 6$.

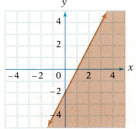

[10.4A]

11

Radical Expressions

OBJECTIVES

▶ To simplify numerical radical expressions
▶ To simplify variable radical expressions
▶ To add and subtract radical expressions
▶ To multiply radical expressions
▶ To divide radical expressions
▶ To solve an equation containing one or more radical expressions
▶ To solve application problems

Y y^3 2 x^2

A Table of Square Roots

The practice of finding the square root of a number has existed for at least two thousand years. Because the process of finding a square root is tedious and time-consuming, it is convenient to have tables of square roots. There is one such table in the back of this book.

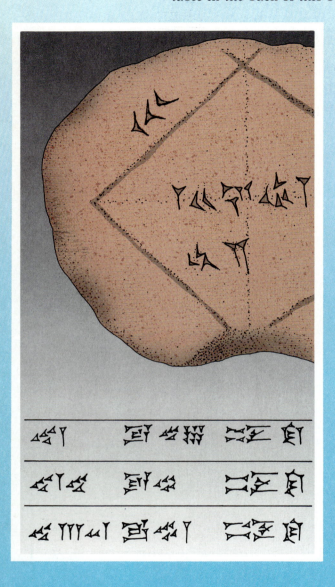

But this table is not the first square root table ever written (nor is it likely to be the last). The table shown below is part of an old Babylonian clay tablet that was written around 350 B.C. It is an incomplete table of square roots written in a style called *cuneiform*.

The number base of the Babylonians was 60 instead of 10, as we use today. The symbol Υ was used for 1, and 10 was written as $\triangleleft$. Some examples of numbers using this system are given below.

$$\text{(cuneiform)} = 9 \qquad \text{(cuneiform)} = 40$$

A translation of the first couple of lines of the table are given next to that line. The number given in parentheses is the equivalent base-10 number that would be used today. You might try to translate the third line. The answer is given at the bottom of this page.

$40 \times 60 + 1$ (= 2401), which is the square of 49

$41 \times 60 + 40$ (= 2500), which is the square of 50

Answer: $43 \times 60 + 21$ (= 2601), which is the square of 51

<table>
<tr><td>**SECTION 11.1**</td></tr>
</table>

Introduction to Radical Expressions

| Objective A | To simplify numerical radical expressions |

A **square root** of a positive number x is a number whose square is x.

A square root of 16 is 4 because $4^2 = 16$.
A square root of 16 is -4 because $(-4)^2 = 16$.

Every positive number has two square roots, one a positive and one a negative number. The symbol "$\sqrt{}$," called a **radical,** is used to indicate the positive or **principal square root** of a number. For example, $\sqrt{16} = 4$ and $\sqrt{25} = 5$. The number under the radical sign is called the **radicand.**

When the negative square root of a number is to be found, a negative sign is placed in front of the radical. For example, $-\sqrt{16} = -4$ and $-\sqrt{25} = -5$.

The square of an integer is a **perfect square.** 49, 81, and 144 are examples of perfect squares.

$7^2 = 49$
$9^2 = 81$
$12^2 = 144$

An integer that is a perfect square can be written as the product of prime factors, each of which has an even exponent when expressed in exponential form.

$49 = 7 \cdot 7 = 7^2$
$81 = 3 \cdot 3 \cdot 3 \cdot 3 = 3^4$
$144 = 2 \cdot 2 \cdot 2 \cdot 2 \cdot 3 \cdot 3 = 2^4 3^2$

To find the square root of a perfect square written in exponential form, remove the radical sign and multiply the exponent by $\frac{1}{2}$.

Simplify $\sqrt{625}$.

Write the prime factorization of the radicand in exponential form.

$\sqrt{625} = \sqrt{5^4}$

Remove the radical sign and multiply the exponent by $\frac{1}{2}$.

$= 5^2$

Simplify.

$= 25$

If a number is not a perfect square, its square root can only be approximated. For example, 2 and 7 are not perfect squares. The square roots of these numbers are **irrational numbers.** Their decimal representations never terminate or repeat.

$$\sqrt{2} \approx 1.4142135 \ldots \qquad \sqrt{7} \approx 2.6457513 \ldots$$

The approximate square roots of the positive integers up to 200 can be found in the Appendix on page A2. The square roots have been rounded to the nearest thousandth.

A radical expression is in simplest form when the radicand contains no factor greater than 1 that is a perfect square. The Product Property of Square Roots is used to simplify radical expressions.

The Product Property of Square Roots

If a and b are positive real numbers, then $\sqrt{ab} = \sqrt{a} \cdot \sqrt{b}$.

Simplify $\sqrt{96}$.

Write the prime factorization of the radicand in exponential form. $\sqrt{96} = \sqrt{2^5 \cdot 3}$

Write the radicand as a product of a perfect square and factors that do not contain a perfect square. Remember that a perfect square has an even exponent. $= \sqrt{2^4(2 \cdot 3)}$

Use the Product Property of Square Roots. $= \sqrt{2^4}\sqrt{2 \cdot 3}$

Simplify. $= 2^2\sqrt{2 \cdot 3}$

$\sqrt{96}$ and $4\sqrt{6}$ represent the same number. $4\sqrt{6}$ is the simplest form of $\sqrt{96}$. $= 4\sqrt{6}$

Simplify $\sqrt{-4}$.

The square root of a negative number is not a real number because the square of a real number is always positive. $\sqrt{-4}$ is not a real number.

Simplify $\sqrt{125}$. Then find the decimal approximation. Round to the nearest thousandth.

Write the prime factorization of the radicand in exponential form. $\sqrt{125} = \sqrt{5^3}$

Write the radicand as a product of a perfect square and factors that do not contain a perfect square. $= \sqrt{5^2 \cdot 5}$

Use the Product Property of Square Roots. $= \sqrt{5^2}\sqrt{5}$

Simplify. $= 5\sqrt{5}$

Replace the radical expression by the decimal approximation found on page A2. $\approx 5(2.236)$

Simplify. ≈ 11.180

Note that in the table on page A2, the decimal approximation of $\sqrt{125}$ is 11.180.

Example 1   Simplify $3\sqrt{90}$.

Solution $3\sqrt{90} = 3\sqrt{2 \cdot 3^2 \cdot 5}$
$= 3\sqrt{3^2(2 \cdot 5)}$
$= 3\sqrt{3^2}\sqrt{2 \cdot 5}$
$= 3 \cdot 3\sqrt{10} = 9\sqrt{10}$

Example 2   Simplify $-5\sqrt{32}$.

Your solution $-20\sqrt{2}$

Example 3   Find the decimal approximation of $\sqrt{252}$. Use the table on page A2.

Solution $\sqrt{252} = \sqrt{2^2 \cdot 3^2 \cdot 7}$
$= \sqrt{2^2 \cdot 3^2}\sqrt{7}$
$= 2 \cdot 3\sqrt{7}$
$= 6\sqrt{7} \approx 6(2.646)$
≈ 15.876

Example 4   Find the decimal approximation of $\sqrt{216}$. Use the table on page A2.

Your solution   14.694

Solutions on p. A36

| **Objective B** | **To simplify variable radical expressions** |

Variable expressions that contain radicals do not always represent real numbers. For example, if $a = -4$, then

$$\sqrt{a^3} = \sqrt{(-4)^3} = \sqrt{-64}$$

and $\sqrt{-64}$ is not a real number.

Now consider the expression $\sqrt{x^2}$ and evaluate this expression for $x = -2$ and $x = 2$.

$$\sqrt{x^2} \qquad\qquad\qquad\qquad \sqrt{x^2}$$
$$\sqrt{(-2)^2} = \sqrt{4} = 2 = |-2| \qquad\qquad \sqrt{2^2} = \sqrt{4} = 2 = |2|$$

This suggests the following:

> For any real number a, $\sqrt{a^2} = |a|$. If $a \geq 0$, then $\sqrt{a^2} = a$.

In order to avoid variable expressions that do not represent real numbers, and so that absolute value signs are not needed for certain expressions, the variables in this chapter will represent *positive* numbers unless otherwise stated.

A variable or a product of variables written in exponential form is a **perfect square** when each exponent is an even number.

To find the square root of a perfect square, remove the radical sign and multiply each exponent by $\frac{1}{2}$.

Simplify $\sqrt{a^6}$.

Remove the radical sign and multiply the exponent by $\frac{1}{2}$. $\sqrt{a^6} = a^3$

A variable radical expression is in simplest form when the radicand contains no factor greater than 1 that is a perfect square.

Simplify $\sqrt{x^7}$.

Write x^7 as the product of a perfect square and x. $\sqrt{x^7} = \sqrt{x^6 \cdot x}$

Use the Product Property of Square Roots. $= \sqrt{x^6}\sqrt{x}$

Simplify the perfect square. $= x^3\sqrt{x}$

Simplify $3x\sqrt{8x^3y^{13}}$.

Write the prime factorization of the coefficient of the radicand in exponential form. $3x\sqrt{8x^3y^{13}} = 3x\sqrt{2^3x^3y^{13}}$

Write the radicand as a product of a perfect square and factors that do not contain a perfect square. $= 3x\sqrt{2^2x^2y^{12}(2xy)}$

Use the Product Property of Square Roots. $= 3x\sqrt{2^2x^2y^{12}}\sqrt{2xy}$

Simplify. $= 3x \cdot 2xy^6\sqrt{2xy}$

$= 6x^2y^6\sqrt{2xy}$

Simplify $\sqrt{25(x+2)^2}$.

Write the prime factorization of the coefficient in exponential form.

$$\sqrt{25(x+2)^2} = \sqrt{5^2(x+2)^2}$$
$$= 5(x+2)$$
$$= 5x + 10$$

Example 5

Simplify $\sqrt{b^{15}}$.

Solution

$\sqrt{b^{15}} = \sqrt{b^{14} \cdot b} = \sqrt{b^{14}} \cdot \sqrt{b} = b^7\sqrt{b}$

Example 6

Simplify $\sqrt{y^{19}}$.

Your solution

$y^9\sqrt{y}$

Example 7

Simplify $\sqrt{24x^5}$.

Solution

$\sqrt{24x^5} = \sqrt{2^3 \cdot 3 \cdot x^5} = \sqrt{2^2x^4(2 \cdot 3x)}$
$\phantom{\sqrt{24x^5}} = \sqrt{2^2 \cdot x^4}\sqrt{2 \cdot 3x}$
$\phantom{\sqrt{24x^5}} = 2x^2\sqrt{6x}$

Example 8

Simplify $\sqrt{45b^7}$.

Your solution

$3b^3\sqrt{5b}$

Example 9

Simplify $2a\sqrt{18a^3b^{10}}$.

Solution

$2a\sqrt{18a^3b^{10}} = 2a\sqrt{2 \cdot 3^2 \cdot a^3b^{10}}$
$\phantom{2a\sqrt{18a^3b^{10}}} = 2a\sqrt{3^2a^2b^{10}(2a)}$
$\phantom{2a\sqrt{18a^3b^{10}}} = 2a\sqrt{3^2a^2b^{10}}\sqrt{2a}$
$\phantom{2a\sqrt{18a^3b^{10}}} = 2a \cdot 3ab^5\sqrt{2a}$
$\phantom{2a\sqrt{18a^3b^{10}}} = 6a^2b^5\sqrt{2a}$

Example 10

Simplify $3a\sqrt{28a^9b^{18}}$.

Your solution

$6a^5b^9\sqrt{7a}$

Example 11

Simplify $\sqrt{16(x+5)^2}$.

Solution

$\sqrt{16(x+5)^2} = \sqrt{2^4(x+5)^2} = 2^2(x+5)$
$\phantom{\sqrt{16(x+5)^2}} = 4(x+5) = 4x + 20$

Example 12

Simplify $\sqrt{25(a+3)^2}$.

Your solution

$5a + 15$

Example 13

Simplify $\sqrt{x^2 + 10x + 25}$.

Solution

$\sqrt{x^2 + 10x + 25} = \sqrt{(x+5)^2} = x + 5$

Example 14

Simplify $\sqrt{x^2 + 14x + 49}$.

Your solution

$x + 7$

Solutions on p. A36

| 11.1 | **EXERCISES** |

▶ **Objective A**

Simplify.

1. $\sqrt{16}$
4

2. $\sqrt{64}$
8

3. $\sqrt{49}$
7

4. $\sqrt{144}$
12

5. $\sqrt{32}$
$4\sqrt{2}$

6. $\sqrt{50}$
$5\sqrt{2}$

7. $\sqrt{8}$
$2\sqrt{2}$

8. $\sqrt{12}$
$2\sqrt{3}$

9. $6\sqrt{18}$
$18\sqrt{2}$

10. $-3\sqrt{48}$
$-12\sqrt{3}$

11. $5\sqrt{40}$
$10\sqrt{10}$

12. $2\sqrt{28}$
$4\sqrt{7}$

13. $\sqrt{15}$
$\sqrt{15}$

14. $\sqrt{21}$
$\sqrt{21}$

15. $\sqrt{29}$
$\sqrt{29}$

16. $\sqrt{13}$
$\sqrt{13}$

17. $-9\sqrt{72}$
$-54\sqrt{2}$

18. $11\sqrt{80}$
$44\sqrt{5}$

19. $\sqrt{45}$
$3\sqrt{5}$

20. $\sqrt{225}$
15

21. $\sqrt{0}$
0

22. $\sqrt{210}$
$\sqrt{210}$

23. $6\sqrt{128}$
$48\sqrt{2}$

24. $9\sqrt{288}$
$108\sqrt{2}$

25. $\sqrt{105}$
$\sqrt{105}$

26. $\sqrt{55}$
$\sqrt{55}$

27. $\sqrt{900}$
30

28. $\sqrt{300}$
$10\sqrt{3}$

29. $5\sqrt{180}$
$30\sqrt{5}$

30. $7\sqrt{98}$
$49\sqrt{2}$

31. $\sqrt{250}$
$5\sqrt{10}$

32. $\sqrt{120}$
$2\sqrt{30}$

33. $\sqrt{96}$
$4\sqrt{6}$

34. $\sqrt{160}$
$4\sqrt{10}$

35. $\sqrt{324}$
18

36. $\sqrt{444}$
$2\sqrt{111}$

Find the decimal approximation. Use the table on page A2.

37. $\sqrt{240}$
15.492

38. $\sqrt{300}$
17.32

39. $\sqrt{288}$
16.968

40. $\sqrt{600}$
24.49

41. $\sqrt{256}$
16

42. $\sqrt{324}$
18

43. $\sqrt{275}$
16.585

44. $\sqrt{450}$
21.21

45. $\sqrt{245}$
15.652

46. $\sqrt{525}$
22.915

47. $\sqrt{352}$
18.76

48. $\sqrt{363}$
19.052

▶ **Objective B**

Simplify.

49. $\sqrt{x^6}$
x^3

50. $\sqrt{x^{12}}$
x^6

51. $\sqrt{y^{15}}$
$y^7\sqrt{y}$

52. $\sqrt{y^{11}}$
$y^5\sqrt{y}$

53. $\sqrt{a^{20}}$
a^{10}

54. $\sqrt{a^{16}}$
a^8

55. $\sqrt{x^4y^4}$
x^2y^2

56. $\sqrt{x^{12}y^8}$
x^6y^4

57. $\sqrt{4x^4}$
$2x^2$

58. $\sqrt{25y^8}$
$5y^4$

59. $\sqrt{24x^2}$
$2x\sqrt{6}$

60. $\sqrt{x^3y^{15}}$
$xy^7\sqrt{xy}$

61. $\sqrt{x^3y^7}$
$xy^3\sqrt{xy}$

62. $\sqrt{a^{15}b^5}$
$a^7b^2\sqrt{ab}$

63. $\sqrt{a^3b^{11}}$
$ab^5\sqrt{ab}$

64. $\sqrt{24y^7}$
$2y^3\sqrt{6y}$

65. $\sqrt{60x^5}$
$2x^2\sqrt{15x}$

66. $\sqrt{72y^7}$
$6y^3\sqrt{2y}$

67. $\sqrt{49a^4b^8}$
$7a^2b^4$

68. $\sqrt{144x^2y^8}$
$12xy^4$

69. $\sqrt{18x^5y^7}$
$3x^2y^3\sqrt{2xy}$

70. $\sqrt{32a^5b^{15}}$
$4a^2b^7\sqrt{2ab}$

71. $\sqrt{40x^{11}y^7}$
$2x^5y^3\sqrt{10xy}$

72. $\sqrt{72x^9y^3}$
$6x^4y\sqrt{2xy}$

73. $\sqrt{80a^9b^{10}}$
$4a^4b^5\sqrt{5a}$

74. $\sqrt{96a^5b^7}$
$4a^2b^3\sqrt{6ab}$

75. $2\sqrt{16a^2b^3}$
$8ab\sqrt{b}$

76. $5\sqrt{25a^4b^7}$
$25a^2b^3\sqrt{b}$

77. $x\sqrt{x^4y^2}$
x^3y

78. $y\sqrt{x^3y^6}$
$xy^4\sqrt{x}$

79. $4\sqrt{20a^4b^7}$
$8a^2b^3\sqrt{5b}$

80. $5\sqrt{12a^3b^4}$
$10ab^2\sqrt{3a}$

81. $3x\sqrt{12x^2y^7}$
$6x^2y^3\sqrt{3y}$

82. $4y\sqrt{18x^5y^4}$
$12x^2y^3\sqrt{2x}$

83. $2x^2\sqrt{8x^2y^3}$
$4x^3y\sqrt{2y}$

84. $3y^2\sqrt{27x^4y^3}$
$9x^2y^3\sqrt{3y}$

85. $\sqrt{25(a+4)^2}$
$5a + 20$

86. $\sqrt{81(x+y)^4}$
$9x^2 + 18xy + 9y^2$

87. $\sqrt{4(x+2)^4}$
$2x^2 + 8x + 8$

88. $\sqrt{9(x+2)^8}$
$3(x+2)^4$

89. $\sqrt{x^2+4x+4}$
$x + 2$

90. $\sqrt{b^2+8b+16}$
$b + 4$

91. $\sqrt{y^2+2y+1}$
$y + 1$

92. $\sqrt{a^2+6a+9}$
$a + 3$

93. $\sqrt{x^2+8x+16}$
$x + 4$

<table>
<tr><td>**SECTION 11.2**</td><td></td></tr>
</table>

Addition and Subtraction of Radical Expressions

Objective A	**To add and subtract radical expressions**

The Distributive Property is used to simplify the sum or difference of radical expressions with like radicands.

$$5\sqrt{2} + 3\sqrt{2} = (5 + 3)\sqrt{2} = 8\sqrt{2}$$
$$6\sqrt{2x} - 4\sqrt{2x} = (6 - 4)\sqrt{2x} = 2\sqrt{2x}$$

Radical expressions that are in simplest form and have unlike radicands cannot be simplified by the Distributive Property.

$2\sqrt{3} + 4\sqrt{2}$ cannot be simplified by the Distributive Property.

Simplify $4\sqrt{8} - 10\sqrt{2}$.

Simplify each term. $4\sqrt{8} - 10\sqrt{2} = 4\sqrt{2^3} - 10\sqrt{2}$

$= 4\sqrt{2^2 \cdot 2} - 10\sqrt{2}$ Do this step mentally.

$= 4\sqrt{2^2}\sqrt{2} - 10\sqrt{2}$
$= 4 \cdot 2\sqrt{2} - 10\sqrt{2}$
$= 8\sqrt{2} - 10\sqrt{2}$

Simplify the expression by using the Distributive Property. $= (8 - 10)\sqrt{2}$ Do this step mentally.

$= -2\sqrt{2}$

Simplify $8\sqrt{18x} - 2\sqrt{32x}$.

Simplify each term. $8\sqrt{18x} - 2\sqrt{32x} = 8\sqrt{2 \cdot 3^2 x} - 2\sqrt{2^5 x}$

$= 8\sqrt{3^2 \cdot 2x} - 2\sqrt{2^4 \cdot 2x}$ Do this step mentally.

$= 8\sqrt{3^2}\sqrt{2x} - 2\sqrt{2^4}\sqrt{2x}$
$= 8 \cdot 3\sqrt{2x} - 2 \cdot 2^2\sqrt{2x}$
$= 24\sqrt{2x} - 8\sqrt{2x}$

Simplify the expression by using the Distributive Property. $= (24 - 8)\sqrt{2x}$ Do this step mentally.

$= 16\sqrt{2x}$

Example 1

Simplify $5\sqrt{2} - 3\sqrt{2} + 12\sqrt{2}$.

Solution

$5\sqrt{2} - 3\sqrt{2} + 12\sqrt{2} = 14\sqrt{2}$

Example 2

Simplify $9\sqrt{3} + 3\sqrt{3} - 18\sqrt{3}$.

Your solution

$-6\sqrt{3}$

Example 3

Simplify $3\sqrt{12} - 5\sqrt{27}$.

Solution

$$3\sqrt{12} - 5\sqrt{27} = 3\sqrt{2^2 \cdot 3} - 5\sqrt{3^3}$$
$$= 3\sqrt{2^2}\sqrt{3} - 5\sqrt{3^2}\sqrt{3}$$
$$= 3 \cdot 2\sqrt{3} - 5 \cdot 3\sqrt{3}$$
$$= 6\sqrt{3} - 15\sqrt{3} = -9\sqrt{3}$$

Example 4

Simplify $2\sqrt{50} - 5\sqrt{32}$.

Your solution

$-10\sqrt{2}$

Example 5

Simplify $3\sqrt{12x^3} - 2x\sqrt{3x}$.

Solution

$$3\sqrt{12x^3} - 2x\sqrt{3x}$$
$$= 3\sqrt{2^2 \cdot 3 \cdot x^3} - 2x\sqrt{3x}$$
$$= 3\sqrt{2^2 \cdot x^2}\sqrt{3x} - 2x\sqrt{3x}$$
$$= 3 \cdot 2 \cdot x\sqrt{3x} - 2x\sqrt{3x}$$
$$= 6x\sqrt{3x} - 2x\sqrt{3x} = 4x\sqrt{3x}$$

Example 6

Simplify $y\sqrt{28y} + 7\sqrt{63y^3}$.

Your solution

$23y\sqrt{7y}$

Example 7

Simplify $2x\sqrt{8y} - 3\sqrt{2x^2y} + 2\sqrt{32x^2y}$.

Solution

$$2x\sqrt{8y} - 3\sqrt{2x^2y} + 2\sqrt{32x^2y}$$
$$= 2x\sqrt{2^3y} - 3\sqrt{2x^2y} + 2\sqrt{2^5x^2y}$$
$$= 2x\sqrt{2^2}\sqrt{2y} - 3\sqrt{x^2}\sqrt{2y} + 2\sqrt{2^4x^2}\sqrt{2y}$$
$$= 2x \cdot 2\sqrt{2y} - 3 \cdot x\sqrt{2y} + 2 \cdot 2^2 \cdot x\sqrt{2y}$$
$$= 4x\sqrt{2y} - 3x\sqrt{2y} + 8x\sqrt{2y} = 9x\sqrt{2y}$$

Example 8

Simplify $2\sqrt{27a^5} - 4a\sqrt{12a^3} + a^2\sqrt{75a}$.

Your solution

$3a^2\sqrt{3a}$

Solutions on p. A36

11.2 EXERCISES

▶ **Objective A**

Simplify.

1. $2\sqrt{2} + \sqrt{2}$
$3\sqrt{2}$

2. $3\sqrt{5} + 8\sqrt{5}$
$11\sqrt{5}$

3. $-3\sqrt{7} + 2\sqrt{7}$
$-\sqrt{7}$

4. $4\sqrt{5} - 10\sqrt{5}$
$-6\sqrt{5}$

5. $-3\sqrt{11} - 8\sqrt{11}$
$-11\sqrt{11}$

6. $-3\sqrt{3} - 5\sqrt{3}$
$-8\sqrt{3}$

7. $2\sqrt{x} + 8\sqrt{x}$
$10\sqrt{x}$

8. $3\sqrt{y} + 2\sqrt{y}$
$5\sqrt{y}$

9. $8\sqrt{y} - 10\sqrt{y}$
$-2\sqrt{y}$

10. $-5\sqrt{2a} + 2\sqrt{2a}$
$-3\sqrt{2a}$

11. $-2\sqrt{3b} - 9\sqrt{3b}$
$-11\sqrt{3b}$

12. $-7\sqrt{5a} - 5\sqrt{5a}$
$-12\sqrt{5a}$

13. $3x\sqrt{2} - x\sqrt{2}$
$2x\sqrt{2}$

14. $2y\sqrt{3} - 9y\sqrt{3}$
$-7y\sqrt{3}$

15. $2a\sqrt{3a} - 5a\sqrt{3a}$
$-3a\sqrt{3a}$

16. $-5b\sqrt{3x} - 2b\sqrt{3x}$
$-7b\sqrt{3x}$

17. $3\sqrt{xy} - 8\sqrt{xy}$
$-5\sqrt{xy}$

18. $-4\sqrt{xy} + 6\sqrt{xy}$
$2\sqrt{xy}$

19. $\sqrt{45} + \sqrt{125}$
$8\sqrt{5}$

20. $\sqrt{32} - \sqrt{98}$
$-3\sqrt{2}$

21. $2\sqrt{2} + 3\sqrt{8}$
$8\sqrt{2}$

22. $4\sqrt{128} - 3\sqrt{32}$
$20\sqrt{2}$

23. $5\sqrt{18} - 2\sqrt{75}$
$15\sqrt{2} - 10\sqrt{3}$

24. $5\sqrt{75} - 2\sqrt{18}$
$25\sqrt{3} - 6\sqrt{2}$

25. $5\sqrt{4x} - 3\sqrt{9x}$
$\sqrt{x}$

26. $-3\sqrt{25y} + 8\sqrt{49y}$
$41\sqrt{y}$

27. $3\sqrt{3x^2} - 5\sqrt{27x^2}$
$-12x\sqrt{3}$

28. $-2\sqrt{8y^2} + 5\sqrt{32y^2}$
$16y\sqrt{2}$

29. $2x\sqrt{xy^2} - 3y\sqrt{x^2y}$
$2xy\sqrt{x} - 3xy\sqrt{y}$

30. $4a\sqrt{b^2a} - 3b\sqrt{a^2b}$
$4ab\sqrt{a} - 3ab\sqrt{b}$

31. $3x\sqrt{12x} - 5\sqrt{27x^3}$
$-9x\sqrt{3x}$

32. $2a\sqrt{50a} + 7\sqrt{32a^3}$
$38a\sqrt{2a}$

33. $4y\sqrt{8y^3} - 7\sqrt{18y^5}$
$-13y^2\sqrt{2y}$

34. $2a\sqrt{8ab^2} - 2b\sqrt{2a^3}$
$2ab\sqrt{2a}$

35. $b^2\sqrt{a^5b} + 3a^2\sqrt{ab^5}$
$4a^2b^2\sqrt{ab}$

36. $y^2\sqrt{x^5y} + x\sqrt{x^3y^5}$
$2x^2y^2\sqrt{xy}$

Simplify.

37. $4\sqrt{2} - 5\sqrt{2} + 8\sqrt{2}$
   $7\sqrt{2}$

38. $3\sqrt{3} + 8\sqrt{3} - 16\sqrt{3}$
   $-5\sqrt{3}$

39. $5\sqrt{x} - 8\sqrt{x} + 9\sqrt{x}$
   $6\sqrt{x}$

40. $\sqrt{x} - 7\sqrt{x} + 6\sqrt{x}$
   0

41. $8\sqrt{2} - 3\sqrt{y} - 8\sqrt{2}$
   $-3\sqrt{y}$

42. $8\sqrt{3} - 5\sqrt{2} - 5\sqrt{3}$
   $3\sqrt{3} - 5\sqrt{2}$

43. $8\sqrt{8} - 4\sqrt{32} - 9\sqrt{50}$
   $-45\sqrt{2}$

44. $2\sqrt{12} - 4\sqrt{27} + \sqrt{75}$
   $-3\sqrt{3}$

45. $-2\sqrt{3} + 5\sqrt{27} - 4\sqrt{45}$
   $13\sqrt{3} - 12\sqrt{5}$

46. $-2\sqrt{8} - 3\sqrt{27} + 3\sqrt{50}$
   $11\sqrt{2} - 9\sqrt{3}$

47. $4\sqrt{75} + 3\sqrt{48} - \sqrt{99}$
   $32\sqrt{3} - 3\sqrt{11}$

48. $2\sqrt{75} - 5\sqrt{20} + 2\sqrt{45}$
   $10\sqrt{3} - 4\sqrt{5}$

49. $\sqrt{25x} - \sqrt{9x} + \sqrt{16x}$
   $6\sqrt{x}$

50. $\sqrt{4x} - \sqrt{100x} - \sqrt{49x}$
   $-15\sqrt{x}$

51. $3\sqrt{3x} + \sqrt{27x} - 8\sqrt{75x}$
   $-34\sqrt{3x}$

52. $5\sqrt{5x} + 2\sqrt{45x} - 3\sqrt{80x}$
   $-\sqrt{5x}$

53. $2a\sqrt{75b} - a\sqrt{20b} + 4a\sqrt{45b}$
   $10a\sqrt{3b} + 10a\sqrt{5b}$

54. $2b\sqrt{75a} - 5b\sqrt{27a} + 2b\sqrt{20a}$
   $-5b\sqrt{3a} + 4b\sqrt{5a}$

55. $x\sqrt{3y^2} - 2y\sqrt{12x^2} + xy\sqrt{3}$
   $-2xy\sqrt{3}$

56. $a\sqrt{27b^2} + 3b\sqrt{147a^2} - ab\sqrt{3}$
   $23ab\sqrt{3}$

57. $3\sqrt{ab^3} + 4a\sqrt{ab} - 5b\sqrt{4ab}$
   $(-7b + 4a)\sqrt{ab}$

58. $5\sqrt{a^3b} + a\sqrt{4ab} - 3\sqrt{49a^3b}$
   $-14a\sqrt{ab}$

59. $3a\sqrt{2ab^2} - \sqrt{a^2b^2} + 4b\sqrt{3a^2b}$
   $3ab\sqrt{2a} - ab + 4ab\sqrt{3b}$

60. $2\sqrt{4a^2b^2} - 3a\sqrt{9ab^2} + 4b\sqrt{a^2b}$
   $4ab - 9ab\sqrt{a} + 4ab\sqrt{b}$

SECTION 11.3

Multiplication and Division of Radical Expressions

Objective A   To multiply radical expressions

The Product Property of Square Roots is used to multiply variable radical expressions.

$$\sqrt{2x}\sqrt{3y} = \sqrt{2x \cdot 3y} = \sqrt{6xy}$$

Simplify $\sqrt{2x^2}\sqrt{32x^5}$.

Use the Product Property of Square Roots.
Multiply the radicands.
Simplify.

$$\sqrt{2x^2}\sqrt{32x^5} \;\; \boxed{= \sqrt{2x^2 \cdot 32x^5}} \quad \text{Do this step}$$
$$\text{mentally.}$$
$$= \sqrt{64x^7}$$
$$= \sqrt{2^6 x^7}$$
$$= \sqrt{2^6 x^6}\sqrt{x}$$
$$= 2^3 x^3 \sqrt{x}$$
$$= 8x^3 \sqrt{x}$$

Simplify $\sqrt{2x}(x + \sqrt{2x})$.

Use the Distributive Property to remove parentheses.
Simplify.

$$\sqrt{2x}(x + \sqrt{2x}) \;\; \boxed{= \sqrt{2x}(x) + \sqrt{2x}\sqrt{2x}} \quad \text{Do this step}$$
$$\text{mentally.}$$
$$= x\sqrt{2x} + \sqrt{4x^2}$$
$$= x\sqrt{2x} + \sqrt{2^2 x^2}$$
$$= x\sqrt{2x} + 2x$$

Simplify $(\sqrt{2} - 3x)(\sqrt{2} + x)$.

Use the FOIL method to remove parentheses.

$$(\sqrt{2} - 3x)(\sqrt{2} + x) = \sqrt{2 \cdot 2} + x\sqrt{2} - 3x\sqrt{2} - 3x^2$$
$$= \sqrt{2^2} + (x - 3x)\sqrt{2} - 3x^2$$
$$= 2 - 2x\sqrt{2} - 3x^2$$

The expressions $a + b$ and $a - b$, which are the sum and difference of two terms, are called **conjugates** of each other.

Simplify $(2 + \sqrt{7})(2 - \sqrt{7})$.

The product of conjugates of the form $(a + b)(a - b) = a^2 - b^2$.

$$(2 + \sqrt{7})(2 - \sqrt{7}) = 2^2 - \sqrt{7}^2$$
$$= 4 - 7$$
$$= -3$$

Simplify $(3 + \sqrt{y})(3 - \sqrt{y})$.

The product of conjugates of the form $(a + b)(a - b) = a^2 - b^2$.

$$(3 + \sqrt{y})(3 - \sqrt{y}) = 3^2 - \sqrt{y}^2$$
$$= 9 - y$$

Example 1

Simplify $\sqrt{3x^4}\sqrt{2x^2y}\sqrt{6xy^2}$.

Solution

$$\sqrt{3x^4}\sqrt{2x^2y}\sqrt{6xy^2} = \sqrt{36x^7y^3}$$
$$= \sqrt{2^23^2x^7y^3}$$
$$= \sqrt{2^23^2x^6y^2}\sqrt{xy}$$
$$= 2 \cdot 3x^3y\sqrt{xy}$$
$$= 6x^3y\sqrt{xy}$$

Example 2

Simplify $\sqrt{5a}\sqrt{15a^3b^4}\sqrt{3b^5}$.

Your solution

$15a^2b^4\sqrt{b}$

Example 3

Simplify $\sqrt{3ab}(\sqrt{3a} + \sqrt{9b})$.

Solution

$$\sqrt{3ab}(\sqrt{3a} + \sqrt{9b})$$
$$= \sqrt{3^2a^2b} + \sqrt{3^3ab^2}$$
$$= \sqrt{3^2a^2}\sqrt{b} + \sqrt{3^2b^2}\sqrt{3a}$$
$$= 3a\sqrt{b} + 3b\sqrt{3a}$$

Example 4

Simplify $\sqrt{5x}(\sqrt{5x} - \sqrt{25y})$.

Your solution

$5x - 5\sqrt{5xy}$

Example 5

Simplify $(\sqrt{a} - \sqrt{b})(\sqrt{a} + \sqrt{b})$.

Solution

$(\sqrt{a} - \sqrt{b})(\sqrt{a} + \sqrt{b}) = \sqrt{a^2} - \sqrt{b^2} = a - b$

Example 6

Simplify $(2\sqrt{x} + 7)(2\sqrt{x} - 7)$.

Your solution

$4x - 49$

Example 7

Simplify $(2\sqrt{x} - \sqrt{y})(5\sqrt{x} - 2\sqrt{y})$.

Solution

$$(2\sqrt{x} - \sqrt{y})(5\sqrt{x} - 2\sqrt{y})$$
$$= 10\sqrt{x^2} - 4\sqrt{xy} - 5\sqrt{xy} + 2\sqrt{y^2}$$
$$= 10x - 9\sqrt{xy} + 2y$$

Example 8

Simplify $(3\sqrt{x} - \sqrt{y})(5\sqrt{x} - 2\sqrt{y})$.

Your solution

$15x - 11\sqrt{xy} + 2y$

Solutions on p. A37

| Objective B | To divide radical expressions |

11

The Quotient Property of Square Roots

If a and b are positive real numbers, then

$$\sqrt{\frac{a}{b}} = \frac{\sqrt{a}}{\sqrt{b}} \text{ and } \frac{\sqrt{a}}{\sqrt{b}} = \sqrt{\frac{a}{b}}.$$

The square root of a quotient is equal to the quotient of the square roots.

Simplify $\sqrt{\dfrac{4x^2}{z^6}}$.

Rewrite the radical expression as the quotient of the square roots.

$$\sqrt{\dfrac{4x^2}{z^6}} \;\left[\; = \dfrac{\sqrt{4x^2}}{\sqrt{z^6}} \;\right]$$ Do this step mentally.

Simplify.

$$= \dfrac{\sqrt{2^2 x^2}}{\sqrt{z^6}} = \dfrac{2x}{z^3}$$

Simplify $\sqrt{\dfrac{24x^3y^7}{3x^7y^2}}$.

Simplify the radicand.

$$\sqrt{\dfrac{24x^3y^7}{3x^7y^2}} = \sqrt{\dfrac{8y^5}{x^4}}$$

Rewrite the radical expression as the quotient of the square roots.

$$\left[\; = \dfrac{\sqrt{8y^5}}{\sqrt{x^4}} \;\right]$$ Do this step mentally.

Simplify.

$$= \dfrac{\sqrt{2^3 y^5}}{\sqrt{x^4}}$$

$$= \dfrac{\sqrt{2^2 y^4}\sqrt{2y}}{\sqrt{x^4}}$$

$$= \dfrac{2y^2\sqrt{2y}}{x^2}$$

Simplify $\dfrac{\sqrt{4x^2y}}{\sqrt{xy}}$.

Use the Quotient Property of Square Roots.

$$\dfrac{\sqrt{4x^2y}}{\sqrt{xy}} = \sqrt{\dfrac{4x^2y}{xy}}$$

Simplify the radicand.

$$= \sqrt{4x}$$

Simplify the radical expression.

$$= \sqrt{2^2}\sqrt{x}$$

$$= 2\sqrt{x}$$

A radical expression is not considered to be in simplest form if a radical remains in the denominator. The procedure used to remove a radical from the denominator is called **rationalizing the denominator.**

Simplify $\dfrac{2}{\sqrt{3}}$.

Multiply the expression by 1 in the form of $\dfrac{\sqrt{3}}{\sqrt{3}}$.

$$\dfrac{2}{\sqrt{3}} = \dfrac{2}{\sqrt{3}} \cdot \dfrac{\sqrt{3}}{\sqrt{3}}$$

The radicand in the denominator is a perfect square.

$$\left[\; = \dfrac{2\sqrt{3}}{\sqrt{3^2}} \;\right]$$ Do this step mentally.

Simplify.

$$= \dfrac{2\sqrt{3}}{3}$$

The radical expression is in simplest form because no radical remains in the denominator and the numerator radical contains no perfect square factors other than 1.

Simplify $\dfrac{\sqrt{2y}}{\sqrt{y}+3}$.

Multiply the numerator and denominator by $\sqrt{y}-3$, the conjugate of $\sqrt{y}+3$.

Simplify.

$$\dfrac{\sqrt{2y}}{\sqrt{y}+3} = \dfrac{\sqrt{2y}}{\sqrt{y}+3} \cdot \dfrac{\sqrt{y}-3}{\sqrt{y}-3}$$

$$= \dfrac{\sqrt{2y^2}-3\sqrt{2y}}{\sqrt{y^2}-3^2}$$

$$= \dfrac{y\sqrt{2}-3\sqrt{2y}}{y-9}$$

Example 9

Simplify $\dfrac{\sqrt{4x^2y^5}}{\sqrt{3x^4y}}$.

Solution

$$\dfrac{\sqrt{4x^2y^5}}{\sqrt{3x^4y}} = \sqrt{\dfrac{2^2x^2y^5}{3x^4y}} = \sqrt{\dfrac{2^2y^4}{3x^2}} = \dfrac{2y^2}{x\sqrt{3}}$$

$$= \dfrac{2y^2}{x\sqrt{3}} \cdot \dfrac{\sqrt{3}}{\sqrt{3}} = \dfrac{2y^2\sqrt{3}}{3x}$$

Example 10

Simplify $\dfrac{\sqrt{15x^6y^7}}{\sqrt{3x^7y^9}}$.

Your solution

$\dfrac{\sqrt{5x}}{xy}$

Example 11

Simplify $\dfrac{\sqrt{2}}{\sqrt{2}-\sqrt{x}}$.

Solution

$$\dfrac{\sqrt{2}}{\sqrt{2}-\sqrt{x}} = \dfrac{\sqrt{2}}{\sqrt{2}-\sqrt{x}} \cdot \dfrac{\sqrt{2}+\sqrt{x}}{\sqrt{2}+\sqrt{x}}$$

$$= \dfrac{2+\sqrt{2x}}{2-x}$$

Example 12

Simplify $\dfrac{\sqrt{y}}{\sqrt{y}+3}$.

Your solution

$\dfrac{y-3\sqrt{y}}{y-9}$

Example 13

Simplify $\dfrac{3-\sqrt{5}}{2+3\sqrt{5}}$.

Solution

$$\dfrac{3-\sqrt{5}}{2+3\sqrt{5}} = \dfrac{3-\sqrt{5}}{2+3\sqrt{5}} \cdot \dfrac{2-3\sqrt{5}}{2-3\sqrt{5}}$$

$$= \dfrac{6-9\sqrt{5}-2\sqrt{5}+3\cdot5}{4-9\cdot5}$$

$$= \dfrac{6-11\sqrt{5}+15}{4-45}$$

$$= \dfrac{21-11\sqrt{5}}{-41} = -\dfrac{21-11\sqrt{5}}{41}$$

Example 14

Simplify $\dfrac{5+\sqrt{y}}{1-2\sqrt{y}}$.

Your solution

$\dfrac{5+11\sqrt{y}+2y}{1-4y}$

Solutions on p. A37

11.3   EXERCISES

▶ **Objective A**

Simplify.

1. $\sqrt{5} \cdot \sqrt{5}$
5

2. $\sqrt{11} \cdot \sqrt{11}$
11

3. $\sqrt{3} \cdot \sqrt{12}$
6

4. $\sqrt{2} \cdot \sqrt{8}$
4

5. $\sqrt{x} \cdot \sqrt{x}$
x

6. $\sqrt{y} \cdot \sqrt{y}$
y

7. $\sqrt{xy^3} \cdot \sqrt{x^5y}$
x^3y^2

8. $\sqrt{a^3b^5} \cdot \sqrt{ab^5}$
a^2b^5

9. $\sqrt{3a^2b^5} \cdot \sqrt{6ab^7}$
$3ab^6\sqrt{2a}$

10. $\sqrt{5x^3y} \cdot \sqrt{10x^2y}$
$5x^2y\sqrt{2x}$

11. $\sqrt{6a^3b^2} \cdot \sqrt{24a^5b}$
$12a^4b\sqrt{b}$

12. $\sqrt{8ab^5} \cdot \sqrt{12a^7b}$
$4a^4b^3\sqrt{6}$

13. $\sqrt{2}(\sqrt{2} - \sqrt{3})$
$2 - \sqrt{6}$

14. $3(\sqrt{12} - \sqrt{3})$
$3\sqrt{3}$

15. $\sqrt{x}(\sqrt{x} - \sqrt{y})$
$x - \sqrt{xy}$

16. $\sqrt{b}(\sqrt{a} - \sqrt{b})$
$\sqrt{ab} - b$

17. $\sqrt{5}(\sqrt{10} - \sqrt{x})$
$5\sqrt{2} - \sqrt{5x}$

18. $\sqrt{6}(\sqrt{y} - \sqrt{18})$
$\sqrt{6y} - 6\sqrt{3}$

19. $\sqrt{8}(\sqrt{2} - \sqrt{5})$
$4 - 2\sqrt{10}$

20. $\sqrt{10}(\sqrt{20} - \sqrt{a})$
$10\sqrt{2} - \sqrt{10a}$

21. $(\sqrt{x} - 3)^2$
$x - 6\sqrt{x} + 9$

22. $(2\sqrt{a} - y)^2$
$4a - 4y\sqrt{a} + y^2$

23. $\sqrt{3a}(\sqrt{3a} - \sqrt{3b})$
$3a - 3\sqrt{ab}$

24. $\sqrt{5x}(\sqrt{10x} - \sqrt{x})$
$5x\sqrt{2} - x\sqrt{5}$

25. $\sqrt{2ac} \cdot \sqrt{5ab} \cdot \sqrt{10cb}$
$10abc$

26. $\sqrt{3xy} \cdot \sqrt{6x^3y} \cdot \sqrt{2y^2}$
$6x^2y^2$

27. $(3\sqrt{x} - 2y)(5\sqrt{x} - 4y)$
$15x - 22y\sqrt{x} + 8y^2$

28. $(5\sqrt{x} + 2\sqrt{y})(3\sqrt{x} - \sqrt{y})$
$15x + \sqrt{xy} - 2y$

29. $(\sqrt{x} - \sqrt{y})(\sqrt{x} + \sqrt{y})$
$x - y$

Simplify.

30. $(\sqrt{3x} + y)(\sqrt{3x} - y)$
$3x - y^2$

31. $(2\sqrt{x} + \sqrt{y})(5\sqrt{x} + 4\sqrt{y})$
$10x + 13\sqrt{xy} + 4y$

32. $(5\sqrt{x} - 2\sqrt{y})(3\sqrt{x} - 4\sqrt{y})$
$15x - 26\sqrt{xy} + 8y$

▶ **Objective B**

Simplify.

33. $\dfrac{\sqrt{32}}{\sqrt{2}}$
4

34. $\dfrac{\sqrt{45}}{\sqrt{5}}$
3

35. $\dfrac{\sqrt{98}}{\sqrt{2}}$
7

36. $\dfrac{\sqrt{48}}{\sqrt{3}}$
4

37. $\dfrac{\sqrt{27a}}{\sqrt{3a}}$
3

38. $\dfrac{\sqrt{72x^5}}{\sqrt{2x}}$
$6x^2$

39. $\dfrac{\sqrt{15x^3y}}{\sqrt{3xy}}$
$x\sqrt{5}$

40. $\dfrac{\sqrt{40x^5y^2}}{\sqrt{5xy}}$
$2x^2\sqrt{2y}$

41. $\dfrac{\sqrt{2a^5b^4}}{\sqrt{98ab^4}}$
$\dfrac{a^2}{7}$

42. $\dfrac{\sqrt{48x^5y^2}}{\sqrt{3x^3y}}$
$4x\sqrt{y}$

43. $\dfrac{1}{\sqrt{3}}$
$\dfrac{\sqrt{3}}{3}$

44. $\dfrac{1}{\sqrt{8}}$
$\dfrac{\sqrt{2}}{4}$

45. $\dfrac{3}{\sqrt{x}}$
$\dfrac{3\sqrt{x}}{x}$

46. $\dfrac{4}{\sqrt{2x}}$
$\dfrac{2\sqrt{2x}}{x}$

47. $\dfrac{\sqrt{8x^2y}}{\sqrt{2x^4y^2}}$
$\dfrac{2\sqrt{y}}{xy}$

48. $\dfrac{\sqrt{9xy^2}}{\sqrt{27x}}$
$\dfrac{y\sqrt{3}}{3}$

49. $\dfrac{\sqrt{4x^2y}}{\sqrt{3xy^3}}$
$\dfrac{2\sqrt{3x}}{3y}$

50. $\dfrac{\sqrt{16x^3y^2}}{\sqrt{8x^3y}}$
$\sqrt{2y}$

51. $\dfrac{1}{\sqrt{2} - 3}$
$-\dfrac{\sqrt{2} + 3}{7}$

52. $\dfrac{5}{\sqrt{7} - 3}$
$-\dfrac{5\sqrt{7} + 15}{2}$

53. $\dfrac{3}{5 + \sqrt{5}}$
$\dfrac{15 - 3\sqrt{5}}{20}$

54. $\dfrac{7}{\sqrt{2} - 7}$
$-\dfrac{7\sqrt{2} + 49}{47}$

55. $\dfrac{3 - \sqrt{6}}{5 - 2\sqrt{6}}$
$3 + \sqrt{6}$

56. $\dfrac{6 - 2\sqrt{3}}{4 + 3\sqrt{3}}$
$-\dfrac{42 - 26\sqrt{3}}{11}$

57. $\dfrac{\sqrt{2} + 2\sqrt{6}}{2\sqrt{2} - 3\sqrt{6}}$
$-\dfrac{20 + 7\sqrt{3}}{23}$

58. $\dfrac{2\sqrt{3} - \sqrt{6}}{5\sqrt{3} + 2\sqrt{6}}$
$\dfrac{14 - 9\sqrt{2}}{17}$

59. $\dfrac{3 + \sqrt{x}}{2 - \sqrt{x}}$
$\dfrac{6 + 5\sqrt{x} + x}{4 - x}$

60. $\dfrac{\sqrt{a} - 4}{2\sqrt{a} + 2}$
$\dfrac{a - 5\sqrt{a} + 4}{2a - 2}$

61. $\dfrac{\sqrt{xy}}{\sqrt{x} - \sqrt{y}}$
$\dfrac{x\sqrt{y} + y\sqrt{x}}{x - y}$

62. $\dfrac{\sqrt{x}}{\sqrt{x} - \sqrt{y}}$
$\dfrac{x + \sqrt{xy}}{x - y}$

SECTION 11.4 Solving Equations Containing Radical Expressions

Objective A

To solve an equation containing one or more radical expressions

An equation that contains a variable expression in a radicand is a **radical equation**.

$$\left.\begin{array}{l} \sqrt{x} = 4 \\ \sqrt{x+2} = \sqrt{x-7} \end{array}\right\} \text{Radical Equations}$$

The following property of equality is used to solve radical equations.

Property of Squaring Both Sides of an Equation

If a and b are real numbers and $a = b$, then $a^2 = b^2$.

If two numbers are equal, then the squares of the numbers are equal.

Solve $\sqrt{x-2} - 7 = 0$.

Rewrite the equation with the radical on one side of the equation and the constant on the other side.

$$\sqrt{x-2} - 7 = 0$$
$$\sqrt{x-2} = 7$$

Square both sides of the equation.

$$(\sqrt{x-2})^2 = 7^2$$

Solve the resulting equation.

$$x - 2 = 49$$
$$x = 51$$

Check the solution.

$$\begin{array}{c|c} \text{Check:} & \sqrt{x-2} - 7 = 0 \\ \hline \sqrt{51-2} - 7 & 0 \\ \sqrt{49} - 7 & 0 \\ \sqrt{7^2} - 7 & 0 \\ 7 - 7 & 0 \\ 0 = 0 & \text{A true equation} \end{array}$$

The solution is 51.

When both sides of an equation are squared, the resulting equation may have a solution that is not a solution of the original equation. Be careful when squaring each side of an equation.

Solve $\sqrt{5+x} + \sqrt{x} = 5$.

Solve for one of the radical expressions.

$$\sqrt{5+x} = 5 - \sqrt{x}$$

Square each side and then simplify. Recall that $(a-b)^2 = a^2 - 2ab + b^2$.

$$(\sqrt{5+x})^2 = (5 - \sqrt{x})^2$$
$$5 + x = 25 - 10\sqrt{x} + x$$
$$2 = \sqrt{x}$$

This is still a radical equation. Square each side.

$$2^2 = (\sqrt{x})^2$$
$$4 = x$$

4 checks as the solution.

Example 1

Solve: $\sqrt{3x} + 2 = 5$

Solution

$$\sqrt{3x} + 2 = 5$$
$$\sqrt{3x} = 3$$
$$(\sqrt{3x})^2 = 3^2$$
$$3x = 9$$
$$x = 3$$

Check:
$$\begin{array}{c|c} \sqrt{3x} + 2 = 5 & \\ \hline \sqrt{3 \cdot 3} + 2 & 5 \\ \sqrt{3^2} + 2 & 5 \\ 3 + 2 & 5 \\ 5 = 5 & \end{array}$$

The solution is 3.

Example 2

Solve: $\sqrt{4x} + 3 = 7$

Your solution

4

Example 3

Solve $\sqrt{x} - \sqrt{x - 5} = 1$.

Solution

$$\sqrt{x} - \sqrt{x - 5} = 1$$
$$\sqrt{x} = 1 + \sqrt{x - 5}$$
$$(\sqrt{x})^2 = (1 + \sqrt{x - 5})^2$$
$$x = 1 + 2\sqrt{x - 5} + (x - 5)$$
$$4 = 2\sqrt{x - 5}$$
$$2 = \sqrt{x - 5}$$
$$2^2 = (\sqrt{x - 5})^2$$
$$4 = x - 5$$
$$9 = x$$

Check:
$$\begin{array}{c|c} \sqrt{9} - \sqrt{9 - 5} = 1 & \\ \hline 3 - 2 & 1 \\ 1 = 1 & \end{array}$$

9 checks as the solution.

Example 4

Solve $\sqrt{x} + \sqrt{x + 9} = 9$.

Your solution

16

Solutions on p. A37

| **Objective B** | **To solve application problems** |

A right triangle contains one 90° angle. The side opposite the 90° angle is called the **hypotenuse.** The other two sides are called legs.

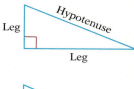

Pythagoras, a Greek mathematician who lived around 550 B.C., is given credit for this theorem that states that the square of the hypotenuse of a right triangle is equal to the sum of the squares of the two legs. Actually, this theorem was known to the Babylonians around 1200 B.C.

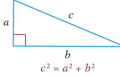

Pythagorean Theorem

If a and b are the lengths of the legs of a right triangle and c is the length of the hypotenuse, then $c^2 = a^2 + b^2$.

Using this theorem, the hypotenuse of a right triangle can be found when the two legs are known. Use the formula

$$\text{Hypotenuse} = \sqrt{(\text{leg})^2 + (\text{leg})^2}$$
$$c = \sqrt{a^2 + b^2}$$
$$= \sqrt{(5)^2 + (12)^2}$$
$$= \sqrt{25 + 144}$$
$$= \sqrt{169}$$
$$= 13$$

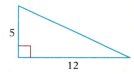

The leg of a right triangle can be found when one leg and the hypotenuse are known. Use the formula

$$\text{Leg} = \sqrt{(\text{hypotenuse})^2 - (\text{leg})^2}$$
$$a = \sqrt{c^2 - b^2}$$
$$= \sqrt{(25)^2 - (20)^2}$$
$$= \sqrt{625 - 400}$$
$$= \sqrt{225}$$
$$= 15$$

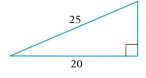

Examples 5 and 6 illustrate the use of the Pythagorean Theorem. Examples 7 and 8 illustrate other applications of radical equations.

Example 5

A guy wire is attached to a point 20 m above the ground on a telephone pole. The wire is anchored to the ground at a point 8 m from the base of the pole. Find the length of the guy wire.

Strategy

To find the length of the guy wire, use the Pythagorean Theorem. One leg is the distance from the bottom of the wire to the base of the telephone pole. The other leg is the distance from the top of the wire to the base of the telephone pole. The guy wire is the hypotenuse. Solve the Pythagorean Theorem for the hypotenuse.

Solution

$$c = \sqrt{a^2 + b^2}$$
$$c = \sqrt{(20)^2 + (8)^2}$$
$$c = \sqrt{400 + 64}$$
$$c = \sqrt{464}$$
$$c = 4\sqrt{29}$$
$$c \approx 21.54$$

The guy wire has a length of 21.54 m.

Example 6

A ladder 8 ft long is resting against a building. How high on the building will the ladder reach when the bottom of the ladder is 3 ft from the building?

Your strategy

Your solution

7.416 ft

Example 7

How far would a submarine periscope have to be above the water to locate a ship 4 mi away? The equation for the distance in miles that the lookout can see is $d = 1.4\sqrt{h}$, where h is the height in feet above the surface of the water. Round to the nearest hundredth.

Strategy

To find the height above the water, replace d in the equation with the given value and solve for h.

Solution

$$1.4\sqrt{h} = d$$
$$1.4\sqrt{h} = 4$$
$$\sqrt{h} = \frac{4}{1.4}$$
$$(\sqrt{h})^2 = \left(\frac{4}{1.4}\right)^2$$
$$h = \frac{16}{1.96}$$
$$h \approx 8.16$$

The periscope must be 8.16 ft above the water.

Example 8

Find the length of a pendulum that makes one swing in 2.5 s. The equation for the time for one swing is $T = 2\pi\sqrt{\frac{L}{32}}$, where T is the time in seconds and L is the length in feet. Use 3.14 for π. Round to the nearest hundredth.

Your strategy

Your solution

5.07 ft

Solutions on p. A38

Content and Format © 1991 HMCo.

11.4   EXERCISES

▶ **Objective A**

Solve and check.

1. $\sqrt{x} = 5$
25

2. $\sqrt{y} = 7$
49

3. $\sqrt{a} = 12$
144

4. $\sqrt{a} = 9$
81

5. $\sqrt{5x} = 5$
5

6. $\sqrt{3x} = 4$
$\dfrac{16}{3}$

7. $\sqrt{4x} = 8$
16

8. $\sqrt{6x} = 3$
$\dfrac{3}{2}$

9. $\sqrt{2x} - 4 = 0$
8

10. $3 - \sqrt{5x} = 0$
$\dfrac{9}{5}$

11. $\sqrt{4x} + 5 = 2$
No solution

12. $\sqrt{3x} + 9 = 4$
No solution

13. $\sqrt{3x - 2} = 4$
6

14. $\sqrt{5x + 6} = 1$
−1

15. $\sqrt{2x + 1} = 7$
24

16. $\sqrt{5x + 4} = 3$
1

17. $0 = 2 - \sqrt{3 - x}$
−1

18. $0 = 5 - \sqrt{10 + x}$
15

19. $\sqrt{5x + 2} = 0$
$-\dfrac{2}{5}$

20. $\sqrt{3x - 7} = 0$
$\dfrac{7}{3}$

21. $\sqrt{3x} - 6 = -4$
$\dfrac{4}{3}$

22. $\sqrt{5x} + 8 = 23$
45

23. $0 = \sqrt{3x - 9} - 6$
15

24. $0 = \sqrt{2x + 7} - 3$
1

25. $\sqrt{x} = \sqrt{x + 3} - 1$
1

26. $\sqrt{x + 5} = \sqrt{x} + 1$
4

27. $\sqrt{2x + 5} = 5 - \sqrt{2x}$
2

28. $\sqrt{2x} + \sqrt{2x + 9} = 9$
8

29. $\sqrt{3x} - \sqrt{3x + 7} = 1$
No solution

30. $\sqrt{x} - \sqrt{x + 9} = 1$
No solution

▶ **Objective B**   *Application Problems*

Solve.

31. The infield of a baseball diamond is a square. The distance between successive bases is 90 ft. The pitcher's mound is on the diagonal between home plate and second base at a distance of 60.5 ft from home plate. Is the pitcher's mound more or less than halfway between home plate and second base?
less than

Solve.

32. The infield of a softball diamond is a square. The distance between successive bases is 60 ft. The pitcher's mound is on the diagonal between home plate and second base at a distance of 46 ft from home plate. Is the pitcher's mound more or less than halfway between home plate and second base?
more than

33. How far would a submarine periscope have to be above the water to locate a ship 5 mi away? The equation for the distance in miles that the lookout can see is $d = 1.4\sqrt{h}$, where h is the height in feet above the surface of the water. Round to the nearest hundredth.
12.76 ft

34. How far would a submarine periscope have to be above the water to locate a ship 6 mi away? The equation for the distance in miles that the lookout can see is $d = 1.4\sqrt{h}$, where h is the height in feet above the surface of the water. Round to the nearest hundredth.
18.37 ft

35. A stone is dropped from a bridge and hits the water 1.5 s later. How high is the bridge? The equation for the distance an object falls in T seconds is $T = \sqrt{\dfrac{d}{16}}$, where d is the distance in feet.
36 ft

36. A stone is dropped into a mine shaft and hits the bottom 3 s later. How deep is the mine shaft? The equation for the distance an object falls in T seconds is $T = \sqrt{\dfrac{d}{16}}$, where d is the distance in feet.
144 ft

37. The measure of a television screen is given by the length of a diagonal across the screen. A 17-in. television has a width of 14.6 in. Find the height of the screen to the nearest tenth of an inch.
8.7 in.

38. The measure of a television screen is given by the length of a diagonal across the screen. A 33-in. big-screen television has a width of 26.4 in. Find the height of the screen to the nearest tenth of an inch.
19.8 in.

39. Find the length of a pendulum that makes one swing in 2 s. The equation for the time of one swing of a pendulum is $T = 2\pi\sqrt{\dfrac{L}{32}}$, where T is the time in seconds and L is the length in feet. Use 3.14 for π. Round to the nearest hundredth.
3.25 ft

40. Find the length of a pendulum that makes one swing in 1.5 s. The equation for the time of one swing of a pendulum is $T = 2\pi\sqrt{\dfrac{L}{32}}$, where T is the time in seconds and L is the length in feet. Use 3.14 for π. Round to the nearest hundredth.
1.83 ft

Calculators and Computers

Simplifying Radical Expressions

Chapter 11 presents simplification of numerical and variable radical expressions and operations with radical expressions (addition, subtraction, multiplication, and division). These concepts are followed by solving equations containing one or more radical expressions and solving application problems which involve radical expressions.

Just as expressions with the same variable part can be added or subtracted, expressions with like radicands can be added or subtracted.

$$2x + 3x = 5x \qquad\qquad 2\sqrt{x} + 3\sqrt{x} = 5\sqrt{x}$$

If the variable parts or the radicands are unlike, the expressions cannot be added or subtracted.

$$2x + 3y = 2x + 3y \qquad\qquad 2\sqrt{x} + 3\sqrt{y} = 2\sqrt{x} + 3\sqrt{y}$$

Expressions with like or unlike variable parts or radicands can be multiplied.

$$(2x)(3x) = 6x^2 \qquad\qquad (2\sqrt{x})(3\sqrt{x}) = 6\sqrt{x^2} = 6x$$
$$(2x)(3y) = 6xy \qquad\qquad (2\sqrt{x})(3\sqrt{y}) = 6\sqrt{xy}$$

A computer program can be written to simplify radical expressions. One such program is on the Math ACE Disk.

The program **RADICAL EXPRESSIONS** on the Math ACE Disk will allow you to practice simplifying radical expressions. The program will display a radical expression. Then, using pencil and paper, simplify the expression. When you are finished, press the RETURN key. The correct solution will be displayed on the screen.

After you complete a problem, you have the opportunity to continue to practice or to quit the program. You will press the letter 'C' to continue or the letter 'Q' to quit.

Chapter Summary

Key Words

A *square root* of a positive number x is a number whose square is x.

The *principal square root* of a number is the positive square root.

The symbol $\sqrt{}$ is called a radical and is used to indicate the principal square root of a number.

The *radicand* is the number under the radical sign.

The square of an integer is a *perfect square*.

If a number is not a perfect square, its square root can only be approximated. Such numbers are *irrational numbers*. Their decimal representations never terminate or repeat.

Conjugates are binomial expressions that differ only in the sign of a term. (The expressions $a + b$ and $a - b$ are conjugates.)

Rationalizing the denominator is the procedure used to remove a radical from the denominator of a fraction.

A *radical equation* is an equation that contains a variable expression in a radicand.

Essential Rules

The Product Property of Square Roots	If a and b are positive real numbers, then $\sqrt{ab} = \sqrt{a}\sqrt{b}.$
The Quotient Property of Square Roots	If a and b are positive real numbers, then $\sqrt{\dfrac{a}{b}} = \dfrac{\sqrt{a}}{\sqrt{b}}$ and $\dfrac{\sqrt{a}}{\sqrt{b}} = \sqrt{\dfrac{a}{b}}.$
Property of Squaring Both Sides of An Equation	If a and b are real numbers and $a = b$, then $a^2 = b^2.$
Pythagorean Theorem	$c^2 = a^2 + b^2$

Chapter Review

SECTION 11.1

1. Simplify $2\sqrt{36}$.
12

2. Simplify $5\sqrt{48}$.
$20\sqrt{3}$

3. Simplify $-3\sqrt{120}$.
$-6\sqrt{30}$

4. Simplify $4\sqrt{250}$.
$20\sqrt{10}$

5. Simplify $3\sqrt{18a^5b}$.
$9a^2\sqrt{2ab}$

6. Simplify $y\sqrt{24y^6}$.
$2y^4\sqrt{6}$

7. Simplify $4y\sqrt{243x^{17}y^9}$.
$36x^8y^5\sqrt{3xy}$

SECTION 11.2

8. Simplify $3\sqrt{12x} + 5\sqrt{48x}$.
$26\sqrt{3x}$

9. Simplify $2x\sqrt{60x^3y^3} + 3x^2y\sqrt{15xy}$.
$7x^2y\sqrt{15xy}$

10. Simplify $6a\sqrt{80b} - \sqrt{180a^2b} + 5a\sqrt{b}$.
$18a\sqrt{5b} + 5a\sqrt{b}$

11. Simplify
$2x^2\sqrt{18x^2y^5} + 6y\sqrt{2x^6y^3} - 9xy^2\sqrt{8x^4y}$.
$-6x^3y^2\sqrt{2y}$

SECTION 11.3

12. Simplify $\sqrt{3}(\sqrt{12} - \sqrt{3})$.
3

13. Simplify $\sqrt{6a}(\sqrt{3a} + \sqrt{2a})$.
$3a\sqrt{2} + 2a\sqrt{3}$

14. Simplify $(4\sqrt{y} - \sqrt{5})(2\sqrt{y} + 3\sqrt{5})$.
$8y + 10\sqrt{5y} - 15$

15. Simplify $\dfrac{\sqrt{98x^7y^9}}{\sqrt{2x^3y}}$.
$7x^2y^4$

16. Simplify $\frac{16}{\sqrt{a}}$.

$\frac{16\sqrt{a}}{a}$

17. Simplify $\frac{8}{\sqrt{x}-3}$.

$\frac{8\sqrt{x}+24}{x-9}$

18. Simplify $\frac{2x}{\sqrt{3}-\sqrt{5}}$.

$-x\sqrt{3}-x\sqrt{5}$

SECTION 11.4

19. Solve $\sqrt{5x}=10$.

20

20. Solve $3-\sqrt{7x}=5$.

No solution

21. Solve $\sqrt{x+1}-\sqrt{x-2}=1$.

3

22. Solve $\sqrt{5x+1}=\sqrt{20x-8}$.

$\frac{3}{5}$

23. The weight of an object is related to the distance the object is above the surface of the earth. An equation for this relationship is $d=4000\sqrt{\frac{W_o}{W_a}}$ where W_o is an object's weight on the surface of the earth and W_a is the object's weight at a distance of d miles above the earth's surface. If a space explorer weighs 36 lb 8000 mi above the surface of the earth, how much does the explorer weigh on the surface of the earth?

144 lb

24. A tsunami is a great sea wave produced by underwater earthquakes or volcanic eruption. The velocity of a tsunami as it approaches land depends on the depth of the water and can be approximated by the equation $v=3\sqrt{d}$, where d is the depth of the water in feet and v is the velocity of the tsunami in feet per second. Find the depth of the water if the velocity is 30 ft per sec.

100 ft

25. A bicycle will overturn if it rounds a corner too sharply or too fast. An equation for the maximum velocity at which a cyclist can turn a corner without tipping over is $v=4\sqrt{r}$, where v is the velocity of the bicycle in miles per hour and r is the radius of the corner in feet. What is the radius of the sharpest corner that a cyclist can safely turn if riding at 20 mph?

25 ft

Chapter Test

1. Simplify $\sqrt{45}$.
$3\sqrt{5}$ [11.1A]

2. Simplify $\sqrt{75}$.
$5\sqrt{3}$ [11.1A]

3. Simplify $\sqrt{121x^8y^2}$.
$11x^4y$ [11.1B]

4. Simplify $\sqrt{72x^7y^2}$.
$6x^3y\sqrt{2x}$ [11.1B]

5. Simplify $\sqrt{32a^5b^{11}}$.
$4a^2b^5\sqrt{2ab}$ [11.1B]

6. Simplify $5\sqrt{8} - 3\sqrt{50}$.
$-5\sqrt{2}$ [11.2A]

7. Simplify $3\sqrt{8y} - 2\sqrt{72x} + 5\sqrt{18y}$.
$21\sqrt{2y} - 12\sqrt{2x}$ [11.2A]

8. Simplify $2x\sqrt{3xy^3} - 2y\sqrt{12x^3y} - 3xy\sqrt{xy}$.
$-2xy\sqrt{3xy} - 3xy\sqrt{xy}$ [11.2A]

9. Simplify $\sqrt{8x^3y}\sqrt{10xy^4}$.
$4x^2y^2\sqrt{5y}$ [11.3A]

10. Simplify $\sqrt{3x^2y}\sqrt{6xy^2}\sqrt{2x}$.
$6x^2y\sqrt{y}$ [11.3A]

11. Simplify $\sqrt{a}(\sqrt{a} - \sqrt{b})$.
   $a - \sqrt{ab}$ [11.3A]

12. Simplify $(\sqrt{y} - 3)(\sqrt{y} + 5)$.
   $y + 2\sqrt{y} - 15$ [11.3A]

13. Simplify $\dfrac{\sqrt{162}}{\sqrt{2}}$.
   9   [11.3B]

14. Simplify $\dfrac{\sqrt{98a^6b^4}}{\sqrt{2a^3b^2}}$.
   $7ab\sqrt{a}$ [11.3B]

15. Simplify $\dfrac{2}{\sqrt{3} - 1}$.
   $\sqrt{3} + 1$ [11.3B]

16. Simplify $\dfrac{2 - \sqrt{5}}{6 + \sqrt{5}}$.
   $\dfrac{17 - 8\sqrt{5}}{31}$ [11.3B]

17. Solve: $\sqrt{9x} + 3 = 18$
   25   [11.4A]

18. Solve: $\sqrt{x} - \sqrt{x + 3} = 1$
   No solution   [11.4A]

19. The square root of the sum of two consecutive odd integers is equal to 10. Find the larger integer.
   51   [11.4B]

20. Find the length of a pendulum that makes one swing in 3 s. The equation for the time of one swing of a pendulum is $T = 2\pi\sqrt{\dfrac{L}{32}}$, where T is the time in seconds and L is the length in feet. Use 3.14 for π. Round to the nearest hundredth.
   7.30 ft   [11.4B]

Cumulative Review

1. Simplify:
   $$\left(\frac{2}{3}\right)^2 \cdot \left(\frac{3}{4} - \frac{3}{2}\right) + \left(\frac{1}{2}\right)^2$$
   $-\dfrac{1}{12}$ [1.5B]

2. Simplify:
   $$-3[x - 2(3 - 2x) - 5x] + 2x$$
   $2x + 18$ [2.2D]

3. Solve:
   $$2x - 4[3x - 2(1 - 3x)] = 2(3 - 4x)$$
   $x = \dfrac{1}{13}$ [3.3B]

4. Simplify $(-3x^2y)(-2x^3y^4)$.
   $6x^5y^5$ [5.2A]

5. Simplify: $\dfrac{12b^4 - 6b^2 + 2}{-6b^2}$
   $-2b^2 + 1 - \dfrac{1}{3b^2}$ [5.4B]

6. Factor $12x^3y^2 - 9x^2y^3$.
   $3x^2y^2(4x - 3y)$ [6.1A]

7. Factor $2a^3 - 16a^2 + 30a$.
   $2a(a - 5)(a - 3)$ [6.3B]

8. Simplify: $\dfrac{3x^3 - 6x^2}{4x^2 + 4x} \cdot \dfrac{3x - 9}{9x^3 - 45x^2 + 54x}$
   $\dfrac{1}{4(x + 1)}$ [7.1B]

9. Simplify: $\dfrac{x + 2}{x - 4} - \dfrac{6}{(x - 4)(x - 3)}$
   $\dfrac{x + 3}{x - 3}$ [7.3B]

10. Solve: $\dfrac{x}{2x - 5} - 2 = \dfrac{3x}{2x - 5}$
    $x = \dfrac{5}{3}$ [7.5A]

11. Find the equation of the line that contains the point $(-2, -3,)$ and has slope $\frac{1}{2}$.
    $y = \dfrac{1}{2}x - 2$ [8.4A]

12. Solve by substitution.
    $$4x - 3y = 1$$
    $$2x + y = 3$$
    $(1, 1)$ [9.2A]

13. Solve by the addition method.
    $$5x + 4y = 7$$
    $$3x - 2y = 13$$
    $(3, -2)$ [9.3A]

14. Solve: $3(x - 7) \geq 5x - 12$
    $x \leq -\dfrac{9}{2}$ [10.3A]

15. Simplify: $\sqrt{108}$
    $6\sqrt{3}$ [11.1A]

16. Simplify: $3\sqrt{32} - 2\sqrt{128}$
    $-4\sqrt{2}$ [11.2A]

17. Simplify: $2a\sqrt{2ab^3} + b\sqrt{8a^3b} - 5ab\sqrt{ab}$
    $4ab\sqrt{2ab} - 5ab\sqrt{ab}$ [11.2A]

18. Simplify: $\sqrt{2a^9b}\sqrt{98ab^3}\sqrt{2a}$
    $14a^5b^2\sqrt{2a}$ [11.3A]

19. Simplify: $\sqrt{3}(\sqrt{6} - \sqrt{x^2})$
    $3\sqrt{2} - x\sqrt{3}$ [11.3A]

20. Simplify: $\dfrac{\sqrt{320}}{\sqrt{5}}$
    8 [11.3B]

21. Simplify: $\dfrac{3}{2 - \sqrt{5}}$
    $-6 - 3\sqrt{5}$ [11.3B]

22. Solve: $\sqrt{3x - 2} - 4 = 0$
    $x = 6$ [11.4A]

23. The selling price for a book is $29.40. The markup rate used by the bookstore is 20%. Find the cost of the book.
    $24.50 [4.3A]

24. How many ounces of pure water must be added to 40 oz of a 12% salt solution to make a salt solution that is 5% salt?
    56 oz [4.5B]

25. The sum of two numbers is twenty-one. The product of the two numbers is one hundred four. Find the two numbers.
    8 and 13 [6.5B]

26. A small water pipe takes twice as long to fill a tank as does a larger water pipe. With both pipes open it takes 16 h to fill the tank. Find the time it would take the small pipe working alone to fill the tank.
    48 h [7.8A]

27. Solve by graphing: $3x - 2y = 8$
    $4x + 5y = 3$

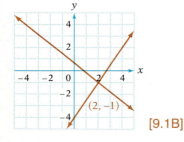

(2, −1)

[9.1B]

28. Graph the solution set of $3x + y \le 2$.

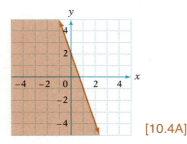

[10.4A]

29. The square root of the sum of two consecutive integers is equal to 9. Find the smaller integer.
    40 [11.4B]

30. A stone is dropped from a building and hits the ground 5 s later. How high is the building? The equation for the distance an object falls in T seconds is $T = \sqrt{\dfrac{d}{16}}$, where d is the distance in feet.
    400 ft [11.4B]

12

Quadratic Equations

OBJECTIVES

▶ To solve a quadratic equation by factoring
▶ To solve a quadratic equation by taking square roots
▶ To solve a quadratic equation by completing the square
▶ To solve a quadratic equation by using the quadratic formula
▶ To graph a quadratic equation of the form $y = ax^2 + bx + c$
▶ To solve application problems

Algebraic Symbolism

The way in which an algebraic expression or equation is written has gone through several stages of development. First there was the *rhetoric*, which was in vogue until the late 13th century. In this method, an expression would be written out in sentences. The word *res* was used to represent an unknown.

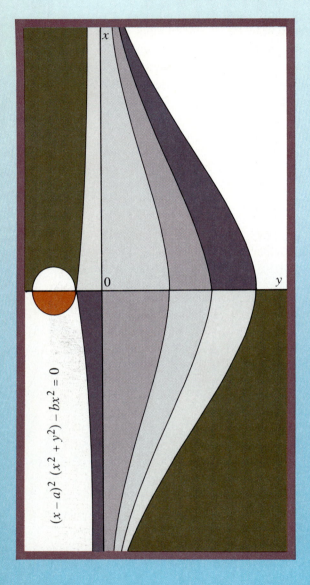

$(x - a)^2 (x^2 + y^2) - bx^2 = 0$

Rhetoric: From the additive *res* in the additive *res* results in a square *res*. From the three in an additive *x* come three additive *res* and from the subtractive four in the additive *res* come subtractive four *res*. From three in subtractive four comes subtractive twelve.

Modern: $(x + 3)(x - 4) = x^2 - x - 12$

The second stage was *syncoptic*, which was a shorthand in which abbreviations were used for words.

Syncoptic: *a* 6 in *b* quad − *c* plano 4 in *b* + *b* cub

Modern: $6ab^2 - 4cb + b^3$

The current modern stage, called the *symbolic* stage, began with the use of exponents rather than words to symbolize exponential expressions. This occurred near the beginning of the 17th century with the publication of the book *La Geometrie* by René Descartes. Modern notation is still evolving as mathematicians continue to search for convenient methods to symbolize concepts.

SECTION 12.1

Solving Quadratic Equations by Factoring or by Taking Square Roots

---

Objective A   **To solve a quadratic equation by factoring**

An equation of the form $ax^2 + bx + c = 0$, where a, b, and c are constants and $a > 0$, is a **quadratic equation.**

$4x^2 - 3x + 1 = 0$, $a = 4$, $b = -3$, $c = 1$

$3x^2 - 4 = 0$, $a = 3$, $b = 0$, $c = -4$

A quadratic equation is also called a **second-degree equation.**

A quadratic equation is in **standard form** when the polynomial is in descending order and equal to zero.

Recall that the Principle of Zero Products states that if the product of two factors is zero, then at least one of the factors must be zero.

If $a \cdot b = 0$, then $a = 0$ or $b = 0$.

The Principle of Zero Products can be used in solving quadratic equations.

Solve by factoring: $2x^2 - x = 1$

$$2x^2 - x = 1$$

Write the equation in standard form.

$$2x^2 - x - 1 = 0$$

Factor.

$$(2x + 1)(x - 1) = 0$$

Let each factor equal zero.

$$2x + 1 = 0 \qquad x - 1 = 0$$

Rewrite each equation in the form *variable = constant.*

$$2x = -1 \qquad x = 1$$
$$x = -\frac{1}{2}$$

Write the solutions.

The solutions are $-\frac{1}{2}$ and 1.

Check:

	$2x^2 - x = 1$		$2x^2 - x = 1$
	$2\left(-\frac{1}{2}\right)^2 - \left(-\frac{1}{2}\right)$ $\Big\vert$ 1		$2(1)^2 - 1$ $\Big\vert$ 1
	$2 \cdot \frac{1}{4} + \frac{1}{2}$ $\Big\vert$ 1		$2 \cdot 1 - 1$ $\Big\vert$ 1
	$\frac{1}{2} + \frac{1}{2}$ $\Big\vert$ 1		$2 - 1$ $\Big\vert$ 1
	$1 = 1$		$1 = 1$

Solve by factoring: $3x^2 - 4x + 8 = (4x + 1)(x - 2)$

$$3x^2 - 4x + 8 = (4x + 1)(x - 2)$$

Simplify the right side of the equation. $\qquad 3x^2 - 4x + 8 = 4x^2 - 7x - 2$

Write the equation in standard form. $\qquad\qquad 0 = x^2 - 3x - 10$

Factor. $\qquad\qquad\qquad\qquad\qquad\qquad 0 = (x - 5)(x + 2)$

Let each factor equal zero. $\qquad\qquad x - 5 = 0 \qquad x + 2 = 0$

Rewrite each equation in the form
variable = constant. $\qquad\qquad\qquad x = 5 \qquad\qquad x = -2$

Write the solutions. $\qquad\qquad$ The solutions are 5 and -2.

Check:

$$
\begin{array}{c|c}
\multicolumn{2}{c}{3x^2 - 4x + 8 = (4x + 1)(x - 2)} \\
\hline
3(5)^2 - 4(5) + 8 & (4[5] + 1)(5 - 2) \\
3(25) - 20 + 8 & (20 + 1)(3) \\
75 - 12 & (21)(3) \\
63 & = 63
\end{array}
\qquad
\begin{array}{c|c}
\multicolumn{2}{c}{3x^2 - 4x + 8 = (4x + 1)(x - 2)} \\
\hline
3(-2)^2 - 4(-2) + 8 & (4[-2] + 1)(-2 - 2) \\
3(4) + 8 + 8 & (-8 + 1)(-4) \\
12 + 16 & (-7)(-4) \\
28 & = 28
\end{array}
$$

Example 1

Solve by factoring:
$x^2 + 10x + 25 = 0$

Solution

$x^2 + 10x + 25 = 0$

$(x + 5)(x + 5) = 0$

$x + 5 = 0 \qquad x + 5 = 0$

$\qquad x = -5 \qquad\qquad x = -5$

-5 is a double root of the equation.

The solution is -5.

Example 2

Solve by factoring:
$2x^2 = (x + 2)(x + 3)$

Your solution

-1 and 6

Solution on p. A38

| **Objective B** | **To solve a quadratic equation by taking square roots** |

Consider a quadratic equation of the form $x^2 = a$. This equation can be solved by factoring.

$$x^2 = 25$$
$$x^2 - 25 = 0$$
$$(x - 5)(x + 5) = 0$$
$$x = 5; \quad x = -5$$

The solutions are 5 and -5. The fact that the solutions are plus or minus the same number is frequently written by using $\pm$; for example, "the solutions are ± 5." Because ± 5 can be written as $\pm\sqrt{25}$, an alternative method of solving this equation is suggested.

Principle of Taking the Square Root of Each Side of an Equation

If $x^2 = a$, then $x = \pm\sqrt{a}$.

Solve by taking square roots: $x^2 = 25$

Take the square root of each side of the equation. Then simplify.

$$x^2 = 25$$
$$\sqrt{x^2} = \sqrt{25}$$
$$x = \pm\sqrt{25} = \pm 5$$

The solutions are 5 and -5.

Solve by taking square roots: $3x^2 = 36$

$$3x^2 = 36$$

Solve for x^2.

$$x^2 = 12$$

Take the square root of each side.

$$\sqrt{x^2} = \sqrt{12}$$

Simplify.

$$x = \pm\sqrt{12} = \pm 2\sqrt{3}$$

The solutions are $2\sqrt{3}$ and $-2\sqrt{3}$.

Solve by taking square roots: $49y^2 - 25 = 0$

$$49y^2 - 25 = 0$$

Write the equation in the form $y^2 = a$.

$$49y^2 = 25$$

Solve for y^2.

$$y^2 = \frac{25}{49}$$

Take the square root of each side.

$$\sqrt{y^2} = \sqrt{\frac{25}{49}}$$

Simplify.

$$y = \pm\frac{5}{7}$$

The solutions are $\frac{5}{7}$ and $-\frac{5}{7}$.

An equation containing the square of a binomial can be solved by taking square roots.

Solve by taking square roots: $2(x - 1)^2 - 36 = 0$

Solve for $(x - 1)^2$.

$$2(x - 1)^2 - 36 = 0$$
$$2(x - 1)^2 = 36$$
$$(x - 1)^2 = 18$$

Take the square root of each side of the equation.

$$\sqrt{(x - 1)^2} = \sqrt{18}$$

Simplify.

$$x - 1 = \pm\sqrt{18} = \pm 3\sqrt{2}$$

Solve for x.

$$x - 1 = 3\sqrt{2} \qquad x - 1 = -3\sqrt{2}$$
$$x = 1 + 3\sqrt{2} \qquad x = 1 - 3\sqrt{2}$$

Write the solutions.

The solutions are $1 + 3\sqrt{2}$ and $1 - 3\sqrt{2}$.

Example 3

Solve by taking square roots:
$x^2 + 16 = 0$

Solution

$$x^2 + 16 = 0$$
$$x^2 = -16$$
$$\sqrt{x^2} = \sqrt{-16}$$

$\sqrt{-16}$ is not a real number.

The equation has no real number solution.

Example 4

Solve by taking square roots:
$x^2 + 81 = 0$

Your solution

No real number solution

Example 5

Solve by taking square roots:
$5(y - 4)^2 = 25$

Solution

$$5(y - 4)^2 = 25$$
$$(y - 4)^2 = 5$$
$$\sqrt{(y - 4)^2} = \sqrt{5}$$
$$y - 4 = \pm\sqrt{5}$$
$$y = 4 \pm \sqrt{5}$$

The solutions are $4 + \sqrt{5}$ and $4 - \sqrt{5}$.

Example 6

Solve by taking square roots:
$7(z + 2)^2 = 21$

Your solution

$-2 + \sqrt{3}$ and $-2 - \sqrt{3}$

Solutions on pp. A38–A39

12.1   EXERCISES

▶ **Objective A**

Solve by factoring.

1. $x^2 + 2x - 15 = 0$
   −5 and 3

2. $t^2 + 3t - 10 = 0$
   −5 and 2

3. $z^2 - 4z + 3 = 0$
   1 and 3

4. $s^2 - 5s + 4 = 0$
   1 and 4

5. $p^2 + 3p + 2 = 0$
   −1 and −2

6. $v^2 + 6v + 5 = 0$
   −1 and −5

7. $x^2 - 6x + 9 = 0$
   3

8. $y^2 - 8y + 16 = 0$
   4

9. $12y^2 + 8y = 0$
   0 and $-\dfrac{2}{3}$

10. $6x^2 - 9x = 0$
    0 and $\dfrac{3}{2}$

11. $r^2 - 10 = 3r$
    −2 and 5

12. $t^2 - 12 = 4t$
    −2 and 6

13. $3v^2 - 5v + 2 = 0$
    $\dfrac{2}{3}$ and 1

14. $2p^2 - 3p - 2 = 0$
    $-\dfrac{1}{2}$ and 2

15. $3s^2 + 8s = 3$
    $\dfrac{1}{3}$ and −3

16. $3x^2 + 5x = 12$
    $\dfrac{4}{3}$ and −3

17. $9z^2 = 12z - 4$
    $\dfrac{2}{3}$

18. $6r^2 = 12 - r$
    $-\dfrac{3}{2}$ and $\dfrac{4}{3}$

19. $4t^2 = 4t + 3$
    $-\dfrac{1}{2}$ and $\dfrac{3}{2}$

20. $5y^2 + 11y = 12$
    −3 and $\dfrac{4}{5}$

21. $4v^2 - 4v + 1 = 0$
    $\dfrac{1}{2}$

22. $9s^2 - 6s + 1 = 0$
    $\dfrac{1}{3}$

23. $x^2 - 9 = 0$
    −3 and 3

24. $t^2 - 16 = 0$
    −4 and 4

25. $4y^2 - 1 = 0$
    $-\dfrac{1}{2}$ and $\dfrac{1}{2}$

26. $9z^2 - 4 = 0$
    $-\dfrac{2}{3}$ and $\dfrac{2}{3}$

27. $x + 15 = x(x - 1)$
    −3 and 5

28. $p + 18 = p(p - 2)$
    −3 and 6

29. $r^2 - r - 2 = (2r - 1)(r - 3)$
    1 and 5

30. $s^2 + 5s - 4 = (2s + 1)(s - 4)$
    0 and 12

31. $x^2 + x + 5 = (3x + 2)(x - 4)$
    −1 and $\dfrac{13}{2}$

▶ **Objective B**

Solve by taking square roots.

32. $x^2 = 36$
−6 and 6

33. $y^2 = 49$
−7 and 7

34. $v^2 - 1 = 0$
−1 and 1

35. $z^2 - 64 = 0$
−8 and 8

36. $4x^2 - 49 = 0$
$-\dfrac{7}{2}$ and $\dfrac{7}{2}$

37. $9w^2 - 64 = 0$
$-\dfrac{8}{3}$ and $\dfrac{8}{3}$

38. $9y^2 = 4$
$-\dfrac{2}{3}$ and $\dfrac{2}{3}$

39. $4z^2 = 25$
$-\dfrac{5}{2}$ and $\dfrac{5}{2}$

40. $16v^2 - 9 = 0$
$-\dfrac{3}{4}$ and $\dfrac{3}{4}$

41. $25x^2 - 64 = 0$
$-\dfrac{8}{5}$ and $\dfrac{8}{5}$

42. $y^2 + 81 = 0$
No real
number solution

43. $z^2 + 49 = 0$
No real
number solution

44. $w^2 - 24 = 0$
$-2\sqrt{6}$ and $2\sqrt{6}$

45. $v^2 - 48 = 0$
$-4\sqrt{3}$ and $4\sqrt{3}$

46. $(x - 1)^2 = 36$
−5 and 7

47. $(y + 2)^2 = 49$
−9 and 5

48. $2(x + 5)^2 = 8$
−3 and −7

49. $4(z - 3)^2 = 100$
−2 and 8

50. $9(x - 1)^2 - 16 = 0$
$-\dfrac{1}{3}$ and $\dfrac{7}{3}$

51. $4(y + 3)^2 - 81 = 0$
$-\dfrac{15}{2}$ and $\dfrac{3}{2}$

52. $49(v + 1)^2 - 25 = 0$
$-\dfrac{2}{7}$ and $-\dfrac{12}{7}$

53. $81(y - 2)^2 - 64 = 0$
$\dfrac{26}{9}$ and $\dfrac{10}{9}$

54. $(x - 4)^2 - 20 = 0$
$4 + 2\sqrt{5}$ and $4 - 2\sqrt{5}$

55. $(y + 5)^2 - 50 = 0$
$-5 + 5\sqrt{2}$ and $-5 - 5\sqrt{2}$

56. $(x + 1)^2 + 36 = 0$
No real number solution

57. $(y - 7)^2 + 49 = 0$
No real number solution

58. $2\left(z - \dfrac{1}{2}\right)^2 = 12$
$\dfrac{1}{2} + \sqrt{6}$ and $\dfrac{1}{2} - \sqrt{6}$

59. $3\left(v + \dfrac{3}{4}\right)^2 = 36$
$-\dfrac{3}{4} + 2\sqrt{3}$ and $-\dfrac{3}{4} - 2\sqrt{3}$

SECTION 12.2

Solving Quadratic Equations by Completing the Square

| **Objective A** | To solve a quadratic equation by completing the square | |

Recall that a perfect square trinomial is the square of a binomial.

Perfect Square Trinomial		Square of a Binomial
$x^2 + 6x + 9$	$=$	$(x + 3)^2$
$x^2 - 10x + 25$	$=$	$(x - 5)^2$
$x^2 + 8x + 16$	$=$	$(x + 4)^2$

For each perfect square trinomial, the square of $\frac{1}{2}$ of the coefficient of x equals the constant term.

$$x^2 + 6x + 9, \quad \left(\frac{1}{2} \cdot 6\right)^2 = 9$$

$$x^2 - 10x + 25, \quad \left[\frac{1}{2}(-10)\right]^2 = 25$$

$$x^2 + 8x + 16, \quad \left(\frac{1}{2} \cdot 8\right)^2 = 16$$

Adding to a binomial the constant term that makes it a perfect square trinomial is called **completing the square.**

Complete the square of $x^2 - 8x$. Write the resulting perfect square trinomial as the square of a binomial.

Find the constant term.

$$\left[\frac{1}{2}(-8)\right]^2 = 16$$

Complete the square of $x^2 - 8x$ by adding the constant term.

$$x^2 - 8x + 16$$

Write the resulting perfect square trinomial as the square of a binomial.

$$x^2 - 8x + 16 = (x - 4)^2$$

Complete the square of $y^2 + 5y$. Write the resulting perfect square trinomial as the square of a binomial.

Find the constant term.

$$\left(\frac{1}{2} \cdot 5\right)^2 = \left(\frac{5}{2}\right)^2 = \frac{25}{4}$$

Complete the square of $y^2 + 5y$ by adding the constant term.

$$y^2 + 5y + \frac{25}{4}$$

Write the resulting perfect square trinomial as the square of a binomial.

$$y^2 + 5y + \frac{25}{4} = \left(y + \frac{5}{2}\right)^2$$

A quadratic equation that cannot be solved by factoring can be solved by completing the square. Add to each side of the equation the term that completes the square. Rewrite the quadratic equation in the form $(x + a)^2 = b$. Take the square root of each side of the equation and then solve for x.

Solve by completing the square: $x^2 - 6x - 3 = 0$

$$x^2 - 6x - 3 = 0$$

Add the opposite of the constant term to each side of the equation.

$$x^2 - 6x = 3$$

Find the constant term that completes the square of $x^2 - 6x$.

$$\left[\frac{1}{2}(-6)\right]^2 = 9 \qquad \text{Do this step mentally.}$$

Add this term to each side of the equation.

$$x^2 - 6x + 9 = 3 + 9$$

Factor the perfect square trinomial.

$$(x - 3)^2 = 12$$

Take the square root of each side of the equation.

$$\sqrt{(x - 3)^2} = \sqrt{12}$$

Simplify.

$$x - 3 = \pm\sqrt{12} = \pm 2\sqrt{3}$$

Solve for x.

$$x - 3 = 2\sqrt{3} \qquad\qquad x - 3 = -2\sqrt{3}$$
$$x = 3 + 2\sqrt{3} \qquad\qquad x = 3 - 2\sqrt{3}$$

Write the solution.

The solutions are $3 + 2\sqrt{3}$ and $3 - 2\sqrt{3}$.

Check:

$$x^2 - 6x - 3 = 0 \qquad\qquad\qquad x^2 - 6x - 3 = 0$$

$$
\begin{array}{c|c}
(3 + 2\sqrt{3})^2 - 6(3 + 2\sqrt{3}) - 3 & 0 \\
9 + 12\sqrt{3} + 12 - 18 - 12\sqrt{3} - 3 & 0 \\
& 0 = 0
\end{array}
\qquad
\begin{array}{c|c}
(3 - 2\sqrt{3})^2 - 6(3 - 2\sqrt{3}) - 3 & 0 \\
9 - 12\sqrt{3} + 12 - 18 + 12\sqrt{3} - 3 & 0 \\
& 0 = 0
\end{array}
$$

Solve by completing the square: $2x^2 - x - 1 = 0$

$$2x^2 - x - 1 = 0$$

Add the opposite of the constant term to each side of the equation.

$$2x^2 - x = 1$$

To complete the square, the coefficient of the x^2 term must be 1. Multiply each term by the reciprocal of the coefficient of x^2.

$$\frac{1}{2}(2x^2 - x) = \frac{1}{2} \cdot 1$$

$$x^2 - \frac{1}{2}x = \frac{1}{2}$$

Find the term that completes the square of $x^2 - \frac{1}{2}x$.

$$\left[\frac{1}{2}\left(-\frac{1}{2}\right)\right]^2 = \left(-\frac{1}{4}\right)^2 = \frac{1}{16} \qquad \text{Do this step mentally.}$$

Add this term to each side of the equation.

$$x^2 - \frac{1}{2}x + \frac{1}{16} = \frac{1}{2} + \frac{1}{16}$$

Factor the perfect square trinomial.

$$\left(x - \frac{1}{4}\right)^2 = \frac{9}{16}$$

Take the square root of each side of the equation.

$$\sqrt{\left(x - \frac{1}{4}\right)^2} = \sqrt{\frac{9}{16}}$$

Simplify.

$$x - \frac{1}{4} = \pm\frac{3}{4}$$

Solve for x.

$$x - \frac{1}{4} = \frac{3}{4} \qquad x - \frac{1}{4} = -\frac{3}{4}$$

$$x = 1 \qquad x = -\frac{1}{2}$$

The solutions are 1 and $-\frac{1}{2}$.

Example 1

Solve by completing the square:
$2x^2 - 4x - 1 = 0$

Solution

$$2x^2 - 4x - 1 = 0$$

$$2x^2 - 4x = 1$$

$$\frac{1}{2}(2x^2 - 4x) = \frac{1}{2} \cdot 1$$

$$x^2 - 2x = \frac{1}{2}$$

Complete the square.

$$x^2 - 2x + 1 = \frac{1}{2} + 1$$

$$(x - 1)^2 = \frac{3}{2}$$

$$\sqrt{(x - 1)^2} = \sqrt{\frac{3}{2}}$$

$$x - 1 = \pm\sqrt{\frac{3}{2}} = \pm\frac{\sqrt{6}}{2}$$

$$x - 1 = \frac{\sqrt{6}}{2} \qquad x - 1 = -\frac{\sqrt{6}}{2}$$

$$x = 1 + \frac{\sqrt{6}}{2} = \frac{2 + \sqrt{6}}{2}$$

$$x = 1 - \frac{\sqrt{6}}{2} = \frac{2 - \sqrt{6}}{2}$$

The solutions are $\frac{2 + \sqrt{6}}{2}$ and $\frac{2 - \sqrt{6}}{2}$.

Example 2

Solve by completing the square:
$3x^2 - 6x - 2 = 0$

Your solution

$$\frac{3 + \sqrt{15}}{3} \text{ and } \frac{3 - \sqrt{15}}{3}$$

Solution on p. A39

Example 3

Solve by completing the square:
$x^2 + 4x + 5 = 0$

Solution

$x^2 + 4x + 5 = 0$
$\quad x^2 + 4x = -5$

Complete the square.

$x^2 + 4x + 4 = -5 + 4$
$\quad\quad (x + 2)^2 = -1$
$\quad \sqrt{(x + 2)^2} = \sqrt{-1}$

$\sqrt{-1}$ is not a real number.

The quadratic equation has no real number solution.

Example 4

Solve by completing the square:
$x^2 + 6x + 12 = 0$

Your solution

No real number solution

Example 5

Solve by completing the square of
$x^2 + 6x + 4 = 0$. Approximate the
solutions. Use a calculator or the Table
of Square Roots on page A2.

Solution

$x^2 + 6x + 4 = 0$
$\quad x^2 + 6x = -4$

Complete the square.

$x^2 + 6x + 9 = -4 + 9$
$\quad\quad (x + 3)^2 = 5$
$\quad \sqrt{(x + 3)^2} = \sqrt{5}$
$\quad\quad\quad x + 3 = \pm\sqrt{5}$

$x + 3 = \sqrt{5} \quad\quad\quad x + 3 = -\sqrt{5}$
$\quad x = -3 + \sqrt{5} \quad\quad x = -3 - \sqrt{5}$
$\quad\quad \approx -3 + 2.236 \quad\quad \approx -3 - 2.236$
$\quad\quad \approx -0.764 \quad\quad\quad \approx -5.236$

The solutions are approximately -0.764
and -5.236.

Example 6

Solve by completing the square of
$x^2 + 8x + 8 = 0$. Approximate the
solutions. Use a calculator or the Table
of Square Roots on page A2.

Your solution

-1.172 and -6.828

Solutions on p. A39

12.2 EXERCISES

▶ **Objective A**

Solve by completing the square.

1. $x^2 + 2x - 3 = 0$
    −3 and 1

2. $y^2 + 4y - 5 = 0$
    −5 and 1

3. $z^2 - 6z - 16 = 0$
    −2 and 8

4. $w^2 + 8w - 9 = 0$
    −9 and 1

5. $x^2 = 4x - 4$
    2

6. $z^2 = 8z - 16$
    4

7. $v^2 - 6v + 13 = 0$
    No real number
    solution

8. $x^2 + 4x + 13 = 0$
    No real number
    solution

9. $y^2 + 5y + 4 = 0$
    −1 and −4

10. $v^2 - 5v - 6 = 0$
    −1 and 6

11. $w^2 + 7w = 8$
    −8 and 1

12. $y^2 + 5y = -4$
    −1 and −4

13. $v^2 + 4v + 1 = 0$
    $-2 + \sqrt{3}$ and $-2 - \sqrt{3}$

14. $y^2 - 2y - 5 = 0$
    $1 + \sqrt{6}$ and $1 - \sqrt{6}$

15. $x^2 + 6x = 5$
    $-3 + \sqrt{14}$ and
    $-3 - \sqrt{14}$

16. $w^2 - 8w = 3$
    $4 + \sqrt{19}$ and $4 - \sqrt{19}$

17. $z^2 = 2z + 1$
    $1 + \sqrt{2}$ and $1 - \sqrt{2}$

18. $y^2 = 10y - 20$
    $5 + \sqrt{5}$ and $5 - \sqrt{5}$

19. $p^2 + 3p = 1$
    $\dfrac{-3 + \sqrt{13}}{2}$ and $\dfrac{-3 - \sqrt{13}}{2}$

20. $r^2 + 5r = 2$
    $\dfrac{-5 + \sqrt{33}}{2}$ and $\dfrac{-5 - \sqrt{33}}{2}$

21. $t^2 - 3t = -2$
    1 and 2

22. $z^2 - 5z = -3$
    $\dfrac{5 + \sqrt{13}}{2}$ and $\dfrac{5 - \sqrt{13}}{2}$

23. $v^2 + v - 3 = 0$
    $\dfrac{-1 + \sqrt{13}}{2}$ and $\dfrac{-1 - \sqrt{13}}{2}$

24. $x^2 - x = 1$
    $\dfrac{1 + \sqrt{5}}{2}$ and $\dfrac{1 - \sqrt{5}}{2}$

25. $y^2 = 7 - 10y$
    $-5 + 4\sqrt{2}$ and $-5 - 4\sqrt{2}$

26. $v^2 = 14 + 16v$
    $8 + \sqrt{78}$ and $8 - \sqrt{78}$

27. $r^2 - 3r = 5$
    $\dfrac{3 + \sqrt{29}}{2}$ and $\dfrac{3 - \sqrt{29}}{2}$

28. $s^2 + 3s = -1$
    $\dfrac{-3 + \sqrt{5}}{2}$ and $\dfrac{-3 - \sqrt{5}}{2}$

29. $t^2 - t = 4$
    $\dfrac{1 + \sqrt{17}}{2}$ and $\dfrac{1 - \sqrt{17}}{2}$

30. $y^2 + y - 4 = 0$
    $\dfrac{-1 + \sqrt{17}}{2}$ and $\dfrac{-1 - \sqrt{17}}{2}$

31. $x^2 - 3x + 5 = 0$
    No real number solution

32. $z^2 + 5z + 7 = 0$
    No real number solution

33. $2t^2 - 3t + 1 = 0$
    1 and $\dfrac{1}{2}$

Solve by completing the square.

34. $2x^2 - 7x + 3 = 0$
3 and $\frac{1}{2}$

35. $2r^2 + 5r = 3$
-3 and $\frac{1}{2}$

36. $2y^2 - 3y = 9$
$-\frac{3}{2}$ and 3

37. $2s^2 = 7s - 6$
2 and $\frac{3}{2}$

38. $2x^2 = 3x + 20$
$-\frac{5}{2}$ and 4

39. $2v^2 = v + 1$
1 and $-\frac{1}{2}$

40. $2z^2 = z + 3$
-1 and $\frac{3}{2}$

41. $3r^2 + 5r = 2$
-2 and $\frac{1}{3}$

42. $3t^2 - 8t = 3$
$-\frac{1}{3}$ and 3

43. $3y^2 + 8y + 4 = 0$
-2 and $-\frac{2}{3}$

44. $3z^2 - 10z - 8 = 0$
4 and $-\frac{2}{3}$

45. $4x^2 + 4x - 3 = 0$
$\frac{1}{2}$ and $-\frac{3}{2}$

46. $4v^2 + 4v - 15 = 0$
$\frac{3}{2}$ and $-\frac{5}{2}$

47. $6s^2 + 7s = 3$
$\frac{1}{3}$ and $-\frac{3}{2}$

48. $6z^2 = z + 2$
$\frac{2}{3}$ and $-\frac{1}{2}$

49. $6p^2 = 5p + 4$
$-\frac{1}{2}$ and $\frac{4}{3}$

50. $6t^2 = t - 2$
No real number solution

51. $4v^2 - 4v - 1 = 0$
$\frac{1 + \sqrt{2}}{2}$ and $\frac{1 - \sqrt{2}}{2}$

52. $2s^2 - 4s - 1 = 0$
$\frac{2 + \sqrt{6}}{2}$ and $\frac{2 - \sqrt{6}}{2}$

53. $4z^2 - 8z = 1$
$\frac{2 + \sqrt{5}}{2}$ and $\frac{2 - \sqrt{5}}{2}$

54. $3r^2 - 2r = 2$
$\frac{1 + \sqrt{7}}{3}$ and $\frac{1 - \sqrt{7}}{3}$

55. $3y - 6 = (y - 1)(y - 2)$
2 and 4

56. $7s + 55 = (s + 5)(s + 4)$
-7 and 5

57. $4p + 2 = (p - 1)(p + 3)$
$1 + \sqrt{6}$ and $1 - \sqrt{6}$

58. $v - 10 = (v + 3)(v - 4)$
$1 + \sqrt{3}$ and $1 - \sqrt{3}$

Solve by completing the square. Approximate the solutions to the nearest thousandth. Use a calculator or the Table of Square Roots on page A2.

59. $y^2 + 3y = 5$
-4.193 and 1.193

60. $w^2 + 5w = 2$
-5.373 and 0.373

61. $2z^2 - 3z = 7$
2.766 and -1.266

62. $2x^2 + 3x = 11$
1.712 and -3.212

63. $4x^2 + 6x - 1 = 0$
-1.652 and 0.152

64. $4x^2 + 2x - 3 = 0$
-1.152 and 0.652

SECTION 12.3   Solving Quadratic Equations by Using the Quadratic Formula

Objective A

To solve a quadratic equation by using the quadratic formula

Any quadratic equation can be solved by completing the square. Applying this method to the standard form of a quadratic equation produces a formula that can be used to solve any quadratic equation.

Solve $ax^2 + bx + c = 0$ by completing the square.

$$ax^2 + bx + c = 0$$

Add the opposite of the constant term to each side of the equation.

$$ax^2 + bx + c + (-c) = 0 + (-c)$$
$$ax^2 + bx = -c$$

Multiply each side of the equation by the reciprocal of a, the coefficient of x^2.

$$\frac{1}{a}(ax^2 + bx) = \frac{1}{a}(-c)$$
$$x^2 + \frac{b}{a}x = -\frac{c}{a}$$

Complete the square by adding $\left(\frac{1}{2} \cdot \frac{b}{a}\right)^2$ to each side of the equation.

$$x^2 + \frac{b}{a}x + \left(\frac{1}{2} \cdot \frac{b}{a}\right)^2 = \left(\frac{1}{2} \cdot \frac{b}{a}\right)^2 - \frac{c}{a}$$
$$x^2 + \frac{b}{a}x + \frac{b^2}{4a^2} = \frac{b^2}{4a^2} - \frac{c}{a}$$

Simplify the right side of the equation.

$$x^2 + \frac{b}{a}x + \frac{b^2}{4a^2} = \frac{b^2}{4a^2} - \left(\frac{c}{a} \cdot \frac{4a}{4a}\right)$$
$$x^2 + \frac{b}{a}x + \frac{b^2}{4a^2} = \frac{b^2}{4a^2} - \frac{4ac}{4a^2}$$
$$x^2 + \frac{b}{a}x + \frac{b^2}{4a^2} = \frac{b^2 - 4ac}{4a^2}$$

Factor the perfect square trinomial on the left side of the equation.

$$\left(x + \frac{b}{2a}\right)^2 = \frac{b^2 - 4ac}{4a^2}$$

Take the square root of each side of the equation.

$$\sqrt{\left(x + \frac{b}{2a}\right)^2} = \sqrt{\frac{b^2 - 4ac}{4a^2}}$$
$$\left(x + \frac{b}{2a}\right) = \pm\frac{\sqrt{b^2 - 4ac}}{2a}$$

Solve for x.

$$x + \frac{b}{2a} = \frac{\sqrt{b^2 - 4ac}}{2a} \qquad x + \frac{b}{2a} = -\frac{\sqrt{b^2 - 4ac}}{2a}$$
$$x = -\frac{b}{2a} + \frac{\sqrt{b^2 - 4ac}}{2a} \qquad x = -\frac{b}{2a} - \frac{\sqrt{b^2 - 4ac}}{2a}$$
$$= \frac{-b + \sqrt{b^2 - 4ac}}{2a} \qquad\qquad = \frac{-b - \sqrt{b^2 - 4ac}}{2a}$$

The Quadratic Formula

The solution of $ax^2 + bx + c = 0$, $a \neq 0$, is

$$\frac{-b \pm \sqrt{b^2 - 4ac}}{2a}.$$

Solve by using the quadratic formula: $2x^2 = 4x - 1$

Write the equation in standard form.
$a = 2, b = -4,$ and $c = 1.$

$$2x^2 - 4x + 1 = 0$$

Replace $a, b,$ and c in the quadratic formula by their values.

$$x = \frac{-b \pm \sqrt{b^2 - 4ac}}{2a}$$

$$= \frac{-(-4) \pm \sqrt{(-4)^2 - 4 \cdot 2 \cdot 1}}{2 \cdot 2}$$

Simplify.

$$= \frac{4 \pm \sqrt{16 - 8}}{4} = \frac{4 \pm \sqrt{8}}{4}$$

$$= \frac{4 \pm 2\sqrt{2}}{4} = \frac{2 \pm \sqrt{2}}{2}$$

Write the solutions.

The solutions are $\frac{2 + \sqrt{2}}{2}$ and $\frac{2 - \sqrt{2}}{2}$.

Example 1

Solve by using the quadratic formula:
$2x^2 - 3x + 1 = 0$

Solution

$2x^2 - 3x + 1 = 0$
$a = 2, b = -3, c = 1$

$$x = \frac{-(-3) \pm \sqrt{(-3)^2 - 4(2)(1)}}{2 \cdot 2}$$

$$= \frac{3 \pm \sqrt{9 - 8}}{4} = \frac{3 \pm \sqrt{1}}{4} = \frac{3 \pm 1}{4}$$

$$x = \frac{3 + 1}{4} \qquad x = \frac{3 - 1}{4}$$

$$= \frac{4}{4} = 1 \qquad = \frac{2}{4} = \frac{1}{2}$$

The solutions are 1 and $\frac{1}{2}$.

Example 2

Solve by using the quadratic formula:
$3x^2 + 4x - 4 = 0$

Your solution

-2 and $\frac{2}{3}$

Example 3

Solve by using the quadratic formula:
$2x^2 = 8x - 5$

Solution

$$2x^2 = 8x - 5$$
$$2x^2 - 8x + 5 = 0$$
$$a = 2, b = -8, c = 5$$

$$x = \frac{-(-8) \pm \sqrt{(-8)^2 - 4(2)(5)}}{2 \cdot 2}$$

$$= \frac{8 \pm \sqrt{64 - 40}}{4} = \frac{8 \pm \sqrt{24}}{4}$$

$$= \frac{8 \pm 2\sqrt{6}}{4} = \frac{4 \pm \sqrt{6}}{2}$$

The solutions are $\frac{4 + \sqrt{6}}{2}$ and $\frac{4 - \sqrt{6}}{2}$.

Example 4

Solve by using the quadratic formula:
$x^2 + 2x = 1$

Your solution

$-1 + \sqrt{2}$ and $-1 - \sqrt{2}$

Solutions on p. A40

12.3 EXERCISES

▶ **Objective A**

Solve by using the quadratic formula.

1. $x^2 - 4x - 5 = 0$
−1 and 5

2. $y^2 + 3y + 2 = 0$
−1 and −2

3. $z^2 - 2z - 15 = 0$
−3 and 5

4. $v^2 + 5v + 4 = 0$
−1 and −4

5. $z^2 + 6z - 7 = 0$
−7 and 1

6. $s^2 + 3s - 10 = 0$
−5 and 2

7. $t^2 + t - 6 = 0$
−3 and 2

8. $x^2 - x - 2 = 0$
−1 and 2

9. $y^2 = 2y + 3$
−1 and 3

10. $w^2 = 3w + 18$
−3 and 6

11. $r^2 = 5 - 4r$
−5 and 1

12. $z^2 = 3 - 2z$
−3 and 1

13. $2y^2 - y - 1 = 0$
$-\dfrac{1}{2}$ and 1

14. $2t^2 - 5t + 3 = 0$
$\dfrac{3}{2}$ and 1

15. $w^2 + 3w + 5 = 0$
No real number solution

16. $x^2 - 2x + 6 = 0$
No real number solution

17. $p^2 - p = 0$
0 and 1

18. $2v^2 + v = 0$
$-\dfrac{1}{2}$ and 0

19. $4t^2 - 9 = 0$
$-\dfrac{3}{2}$ and $\dfrac{3}{2}$

20. $4s^2 - 25 = 0$
$-\dfrac{5}{2}$ and $\dfrac{5}{2}$

21. $4y^2 + 4y = 15$
$-\dfrac{5}{2}$ and $\dfrac{3}{2}$

22. $4r^2 + 4r = 3$
$-\dfrac{3}{2}$ and $\dfrac{1}{2}$

23. $3t^2 = 7t + 6$
$-\dfrac{2}{3}$ and 3

24. $3x^2 = 10x + 8$
$-\dfrac{2}{3}$ and 4

25. $5z^2 + 11z = 12$
$\dfrac{4}{5}$ and −3

26. $4v^2 = v + 3$
$-\dfrac{3}{4}$ and 1

27. $6s^2 - s - 2 = 0$
$-\dfrac{1}{2}$ and $\dfrac{2}{3}$

28. $6y^2 + 5y - 4 = 0$
$-\dfrac{4}{3}$ and $\dfrac{1}{2}$

29. $2x^2 + x + 1 = 0$
No real number solution

30. $3r^2 - r + 2 = 0$
No real number solution

31. $t^2 - 2t = 5$
$1 + \sqrt{6}$ and $1 - \sqrt{6}$

32. $y^2 - 4y = 6$
$2 + \sqrt{10}$ and $2 - \sqrt{10}$

33. $t^2 + 6t - 1 = 0$
$-3 + \sqrt{10}$ and $-3 - \sqrt{10}$

Solve by using the quadratic formula.

34. $z^2 + 4z + 1 = 0$
$-2 + \sqrt{3}$ and $-2 - \sqrt{3}$

35. $w^2 = 4w + 9$
$2 + \sqrt{13}$ and $2 - \sqrt{13}$

36. $y^2 = 8y + 3$
$4 + \sqrt{19}$ and $4 - \sqrt{19}$

37. $4t^2 - 4t - 1 = 0$
$\dfrac{1 + \sqrt{2}}{2}$ and $\dfrac{1 - \sqrt{2}}{2}$

38. $4x^2 - 8x - 1 = 0$
$\dfrac{2 + \sqrt{5}}{2}$ and $\dfrac{2 - \sqrt{5}}{2}$

39. $v^2 + 6v + 1 = 0$
$-3 + 2\sqrt{2}$ and $-3 - 2\sqrt{2}$

40. $s^2 + 4s - 8 = 0$
$-2 + 2\sqrt{3}$ and
$-2 - 2\sqrt{3}$

41. $4t^2 - 12t - 15 = 0$
$\dfrac{3 + 2\sqrt{6}}{2}$ and $\dfrac{3 - 2\sqrt{6}}{2}$

42. $4w^2 - 20w + 5 = 0$
$\dfrac{5 + 2\sqrt{5}}{2}$ and $\dfrac{5 - 2\sqrt{5}}{2}$

43. $9y^2 + 6y - 1 = 0$
$\dfrac{-1 + \sqrt{2}}{3}$ and $\dfrac{-1 - \sqrt{2}}{3}$

44. $9s^2 - 6s - 2 = 0$
$\dfrac{1 + \sqrt{3}}{3}$ and $\dfrac{1 - \sqrt{3}}{3}$

45. $4p^2 + 4p + 1 = 0$
$-\dfrac{1}{2}$

46. $9z^2 + 12z + 4 = 0$
$-\dfrac{2}{3}$

47. $2x^2 = 4x - 5$
No real number solution

48. $3r^2 = 5r - 6$
No real number solution

49. $4p^2 + 16p = -11$
$\dfrac{-4 + \sqrt{5}}{2}$ and $\dfrac{-4 - \sqrt{5}}{2}$

50. $4y^2 - 12y = -1$
$\dfrac{3 + 2\sqrt{2}}{2}$ and $\dfrac{3 - 2\sqrt{2}}{2}$

51. $4x^2 = 4x + 11$
$\dfrac{1 + 2\sqrt{3}}{2}$ and $\dfrac{1 - 2\sqrt{3}}{2}$

52. $4s^2 + 12s = 3$
$\dfrac{-3 + 2\sqrt{3}}{2}$ and $\dfrac{-3 - 2\sqrt{3}}{2}$

53. $9v^2 = -30v - 23$
$\dfrac{-5 + \sqrt{2}}{3}$ and $\dfrac{-5 - \sqrt{2}}{3}$

54. $9t^2 = 30t + 17$
$\dfrac{5 + \sqrt{42}}{3}$ and $\dfrac{5 - \sqrt{42}}{3}$

Solve by using the quadratic formula. Approximate the solutions to the nearest thousandth. Use a calculator or the Table of Square Roots on page A2.

55. $x^2 - 2x - 21 = 0$
5.690 and -3.690

56. $y^2 + 4y - 11 = 0$
1.873 and -5.873

57. $s^2 - 6s - 13 = 0$
7.690 and -1.690

58. $w^2 + 8w - 15 = 0$
1.568 and -9.568

59. $2p^2 - 7p - 10 = 0$
4.590 and -1.090

60. $3t^2 - 8t - 1 = 0$
2.786 and -0.120

61. $4z^2 + 8z - 1 = 0$
0.118 and -2.118

62. $4x^2 + 7x + 1 = 0$
-0.157 and -1.593

63. $5v^2 - v - 5 = 0$
1.105 and -0.905

SECTION 12.4 Graphing Quadratic Equations in Two Variables

Objective A

To graph a quadratic equation of the form $y = ax^2 + bx + c$

An equation of the form $y = ax^2 + bx + c$, $a \neq 0$, is a **quadratic equation in two variables.** Examples of quadratic equations in two variables are shown at the right.

$$y = 3x^2 - x + 1$$
$$y = -x^2 - 3$$
$$y = 2x^2 - 5x$$

The graph of a quadratic equation in two variables is a **parabola.** The graph is "cup shaped" and opens either up or down. The graphs of two parabolas are shown below.

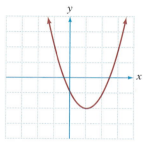

Parabola that opens up

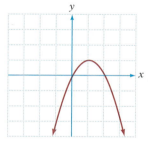

Parabola that opens down

Graph $y = x^2 - 2x - 3$.

Find several solutions of the equation. Because the graph is not a straight line, several solutions must be found in order to determine the cup shape. Display the ordered pair solutions in a table.

x	y
0	-3
1	-4
-1	0
2	-3
3	0

Graph the ordered-pair solutions on a rectangular coordinate system. Draw a parabola through the points.

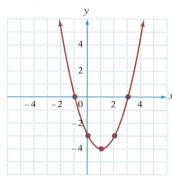

Graph $y = -2x^2 + 1$.

Find enough solutions of the equation to determine the cup shape. Display the ordered pair solutions in a table.

x	y
0	1
1	-1
-1	-1
2	-7
-2	-7

Graph the ordered pair solutions on a rectangular coordinate system. Draw a parabola through the points.

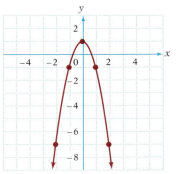

Note in the example on page 459 that the coefficient of x^2 is **positive** and the graph **opens up.** In the second example, the coefficient of x^2 is **negative** and the graph **opens down.**

Example 1   Graph $y = x^2 - 2x$.

Solution

x	y
0	0
1	-1
-1	3
2	0
3	3

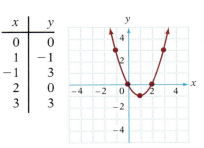

Example 2   Graph $y = x^2 + 2$.

Your solution

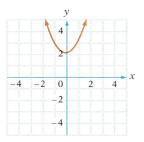

Example 3   Graph $y = -x^2 + 4x - 4$.

Solution

x	y
0	-4
1	-1
2	0
3	-1
4	-4

Example 4   Graph $y = -x^2 - 2x - 1$.

Your solution

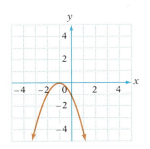

Solutions on p. A40

12.4   EXERCISES

▶ **Objective A**

Graph.

1. $y = x^2$

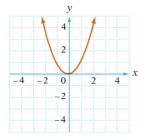

2. $y = -x^2$

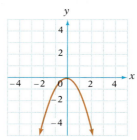

3. $y = -x^2 + 1$

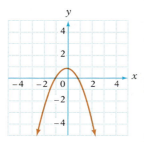

4. $y = x^2 - 1$

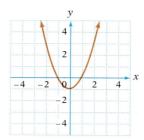

5. $y = 2x^2$

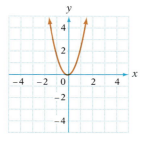

6. $y = \frac{1}{2}x^2$

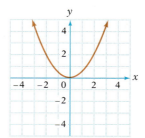

7. $y = -\frac{1}{2}x^2 + 1$

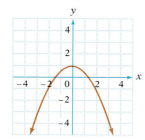

8. $y = 2x^2 - 1$

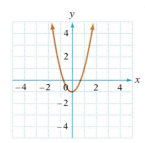

Graph.

9. $y = x^2 - 4x$

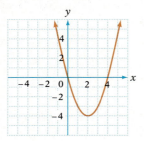

10. $y = x^2 + 4x$

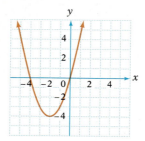

11. $y = x^2 - 2x + 3$

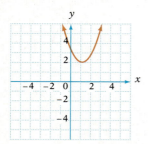

12. $y = x^2 - 4x + 2$

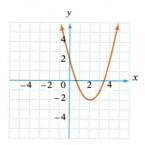

13. $y = -x^2 + 2x + 3$

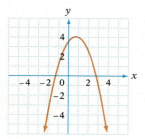

14. $y = -x^2 - 2x + 3$

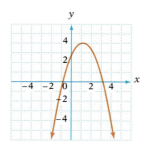

15. $y = -x^2 + 4x - 4$

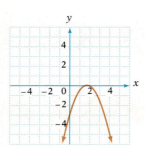

16. $y = -x^2 + 6x - 9$

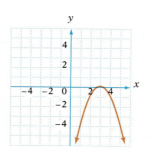

SECTION 12.5   Application Problems

Objective A   **To solve application problems**

The application problems in this section are varieties of those problems solved earlier in the text. Each of the strategies for the problems in this section will result in a quadratic equation.

In 5 h, two campers rowed 12 mi down a stream and then rowed back to their campsite. The rate of the stream's current was 1 mph. Find the rate at which the campers rowed.

Strategy For Solving An Application Problem

> Determine the type of problem. For example, is it a distance-rate problem, a geometry problem, a work problem, or an age problem?

The problem is a distance-rate problem.

> Choose a variable to represent the unknown quantity. Write numerical or variable expressions for all the remaining quantities. These results can be recorded in a table.

The unknown rate of the campers: r

	Distance	$\div$	*Rate*	$=$	*Time*
Downstream	12	$\div$	$r + 1$	$=$	$\dfrac{12}{r+1}$
Upstream	12	$\div$	$r - 1$	$=$	$\dfrac{12}{r-1}$

> Determine how the quantities are related. If necessary, review the strategies presented in Chapter 4.

The total time of the trip was 5 h.

$$\frac{12}{r+1} + \frac{12}{r-1} = 5$$

$$(r+1)(r-1)\left(\frac{12}{r+1} + \frac{12}{r-1}\right) = (r+1)(r-1)5$$

$$(r-1)12 + (r+1)12 = (r^2 - 1)5$$
$$12r - 12 + 12r + 12 = 5r^2 - 5$$
$$24r = 5r^2 - 5$$
$$0 = 5r^2 - 24r - 5$$
$$0 = (5r + 1)(r - 5)$$

$$5r + 1 = 0 \qquad\qquad r - 5 = 0$$
$$5r = -1 \qquad\qquad r = 5$$
$$r = -\frac{1}{5}$$

The rowing rate was 5 mph.

The solution $r = -\frac{1}{5}$ is not possible, because the rate cannot be a negative number.

Example 1

A painter and the painter's apprentice working together can paint a room in 2 h. The apprentice working alone requires 3 more hours to paint the room than the painter requires working alone. How long does it take the painter working alone to paint the room?

Strategy

- This is a work problem.
- Time for the painter to paint the room: t
  Time for the apprentice to paint the room: $t + 3$

	Rate	Time	Part
Painter	$\frac{1}{t}$	2	$\frac{2}{t}$
Apprentice	$\frac{1}{t+3}$	2	$\frac{2}{t+3}$

- The sum of the parts of the task completed must equal 1.

Solution

$$\frac{2}{t} + \frac{2}{t+3} = 1$$

$$t(t+3)\left(\frac{2}{t} + \frac{2}{t+3}\right) = t(t+3) \cdot 1$$

$$(t+3)2 + t(2) = t(t+3)$$
$$2t + 6 + 2t = t^2 + 3t$$
$$0 = t^2 - t - 6$$
$$0 = (t-3)(t+2)$$

$$t - 3 = 0 \qquad t + 2 = 0$$
$$t = 3 \qquad\quad t = -2$$

The time is 3 h.

The solution $t = -2$ is not possible.

Example 2

The length of a rectangle is 2 m more than the width. The area is 15 m². Find the width.

Your strategy

Your solution

3 m

Solution on p. A40

Content and Format © 1991 HMCo.

12.5   **EXERCISES**

▶ **Objective A**   *Application Problems*

Solve.

1. The area of the batter's box on a major league baseball field is 24 ft². The length of the batter's box is 2 ft more than the width. Find the length and width of the batter's box. (Area = lw)
length 6 ft; width 4 ft

2. The length of the batter's box on a softball field is 1 ft less than twice the width. The area of the batter's box is 15 ft². Find the length and width of the batter's box.
length 5 ft; width 3 ft

3. The length of a swimming pool is twice the width. The area of the pool is 5000 ft². Find the length and width of the pool. (Area = lw)
length 100 ft; width 50 ft

4. The length of the singles tennis court is 24 ft more than twice the width. The area of the tennis court is 2106 ft². Find the length and width of the court. (Area = lw)
length 78 ft; width 27 ft

5. The sum of the squares of two positive odd integers is 130. Find the two integers.
7 and 9

6. The sum of the squares of two consecutive positive even integers is 164. Find the two integers.
8 and 10

7. The sum of two integers is 12. The product of the two integers is 35. Find the two integers.
5 and 7

8. The difference between two integers is 4. The product of the two integers is 60. Find the integers.
6 and 10 or −10 and −6

9. Twice an integer equals the square of the integer. Find the integer.
0 or 2

10. The square of an integer equals the integer. Find the integer.
0 or 1

11. A silver coin is twice the age of a gold coin. Three years ago the product of the sum of their ages and the difference between their ages was 45. Find the present ages of the coins.
silver coin, 10 years; gold coin, 5 years

12. An oil painting is twice the age of a watercolor. One year ago the product of their ages was 10. Find the present age of the paintings.
oil, 6 years; watercolor, 3 years

13. One coin is two years older than a second coin. Two years ago the product of their ages was 24. Find the present ages of the coins.
first coin, 8 years; second coin, 6 years

Solve.

14. One car is three times the age of a second car. Eight years ago the product of their ages was 19. Find the present ages of the cars.
first car, 27 years; second car, 9 years

15. One computer takes 21 min longer to calculate the value of a complex equation than a second computer. Working together, these computers complete the calculation in 10 min. How long would it take each computer, working seperately, to calculate the value?
first computer, 35 min; second computer, 14 min

16. A tank has two drains. One drain takes 16 minutes longer to empty the tank than does a second drain. With both drains open, the tank is emptied in 6 minutes. How long would it take each drain, working alone, to empty the tank?
first drain, 24 min; second drain, 8 min

17. Using one engine of a ferryboat, it takes 6 h longer to cross a channel than it does using a second engine alone. Using both engines, the ferryboat can make the crossing in 4 h. How long would it take each engine, working alone, to power the ferryboat across the channel?
first engine, 12 h; second engine, 6 h

18. An apprentice mason takes 8 h longer to build a small fireplace than an experienced mason. Working together, they can build the fireplace in 3 h. How long would it take each mason, working alone, to complete the fireplace?
apprentice mason, 12 h, experienced mason, 4 h

19. It took a small plane 2 h more to fly 375 mi against the wind than it took to fly the same distance with the wind. The rate of the wind was 25 mph. Find the rate of the plane in calm air.
100 mph

20. It took a motorboat 1 h more to travel 36 mi against the current than it took to go 36 mi with the current. The rate of the current was 3 mph. Find the rate of the boat in calm water.
15 mph

21. A cruise ship sailed through a 20-mi inland passageway at a constant rate before increasing its speed by 15 mph. Another 75 mi was traveled at the increased rate. The total time for the 95-mi trip was 5 h. Find the rate during the last 75 mi.
25 mph

22. A motorist traveled 150 mi at a constant rate before decreasing speed by 15 mph. Another 35 mi was driven at the decreased speed. The total time for the 185-mi trip was 4 h. Find the rate during the first 150 mi.
50 mph

Content and Format © 1991 HMCo.

Calculators and Computers

Checking Solutions to Quadratic Equations

A calculator can be used to check solutions to quadratic equations. Here are some examples:

Solve and check: $x^2 - 6x - 16 = 0$

Solve by factoring. The solutions are 8 and -2.

To check the solutions, the $\boxed{x^2}$ key will be used. The method of evaluating polynomials given in Chapter 5 could also be used.

To check the solutions, replace x by 8.

$$8^2 - 6 \cdot 8 - 16 \overset{?}{=} 0$$

Enter the following keystrokes:

8 $\boxed{x^2}$ $\boxed{-}$ $\boxed{(}$ 6 $\boxed{\times}$ 8 $\boxed{)}$ $\boxed{-}$ 16 $\boxed{=}$

The result in the display should be zero. The solution is correct. The solution -2 can be checked in a similar manner.

One note about the calculation—the parentheses keys are used to ensure that multiplication is completed before subtraction. If your calculator follows the Order of Operations Agreement, the parentheses keys are not necessary for this calculation.

Solve and check: $2x^2 - x - 9 = 0$

Use the quadratic formula. The solutions are $\frac{1 + \sqrt{73}}{4}$ and $\frac{1 - \sqrt{73}}{4}$.

To check the solutions, first evaluate the expression $\frac{1 + \sqrt{73}}{4}$ and store the result in the calculator's memory.

1 $\boxed{+}$ 73 $\boxed{\sqrt{}}$ $\boxed{=}$ $\boxed{\div}$ 4 $\boxed{=}$ $\boxed{M+}$

Now replace x in the equation by the solution and determine whether the left and right sides of the equation are equal. The $\boxed{MR}$ key recalls the solution from memory.

2 $\boxed{\times}$ $\boxed{MR}$ $\boxed{x^2}$ $\boxed{-}$ $\boxed{MR}$ $\boxed{-}$ 9 $\boxed{=}$

Is the result in the display zero? Probably not! Nonetheless, the answer is very close to zero. The result in our display was $9.000000 - 10$. The -10 at the end of the display means that the decimal point should be moved ten places to the left. That makes the number 0.0000000009, which is indeed close to zero.

The reason why the answer was not exactly zero is that $\sqrt{73}$ is an irrational number and therefore has an infinitely long decimal representation. The calculator, on the other hand, can store only 8 or 9 places past the decimal point. Thus the calculator is using only an approximation of $\sqrt{73}$, and when we check the solution, the result is not exactly zero.

The solution $\frac{1 - \sqrt{73}}{4}$ can be checked in a similar manner.

Chapter Summary

Key Words   A *quadratic equation* is an equation that can be written in the form $ax^2 + bx + c = 0$, where a, b, and c are constants and $a \neq 0$. A quadratic equation is also called a *second-degree equation.*

A quadratic equation is in *standard form* when the polynomial is in descending order and equal to zero.

The graph of an equation of the form $y = ax^2 + bx + c$, $a \neq 0$, is a *parabola.*

Essential Rules   ***The Quadratic Formula*** $x = \dfrac{-b \pm \sqrt{b^2 - 4ac}}{2a}$

Chapter Review

SECTION 12.1

1. Solve by factoring: $6x^2 + 13x - 28 = 0$
   $\dfrac{4}{3}$ and $-\dfrac{7}{2}$

2. Solve by factoring: $12x^2 + 10 = 29x$
   2 and $\dfrac{5}{12}$

3. Solve by factoring: $(x + 9)^2 = x + 11$
   -7 and -10

4. Solve by factoring: $6x(x + 1) = x - 1$
   $-\dfrac{1}{3}$ and $-\dfrac{1}{2}$

5. Solve by taking square roots:
   $49x^2 = 25$
   $-\dfrac{5}{7}$ and $\dfrac{5}{7}$

6. Solve by taking square roots:
   $4y^2 + 9 = 0$
   No real number solution

7. Solve by taking square roots:
   $(y + 4)^2 - 25 = 0$
   -9 and 1

8. Solve by taking square roots:
   $(x + 2)^2 - 24 = 0$
   $-2 - 2\sqrt{6}$ and $-2 + 2\sqrt{6}$

9. Solve by taking square roots:
   $(x - \frac{1}{2})^2 = \frac{9}{4}$
   -1 and 2

SECTION 12.2

10. Solve by completing the square:
    $x^2 + 2x - 24 = 0$
    -6 and 4

11. Solve by completing the square:
    $2x^2 + 5x = 12$
    -4 and $\dfrac{3}{2}$

12. Solve by completing the square:
    $x^2 - 4x + 1 = 0$
    $2 - \sqrt{3}$ and $2 + \sqrt{3}$

13. Solve by completing the square:
    $x^2 + 6x + 12 = 0$
    No real number solution

14. Solve by completing the square:
    $4x^2 + 16x = 7$
    $\dfrac{-4 - \sqrt{23}}{2}$ and $\dfrac{-4 + \sqrt{23}}{2}$

SECTION 12.3

15. Solve by using the quadratic formula:
    $x^2 + 5x - 6 = 0$
    -6 and 1

16. Solve by using the quadratic formula:
    $2x^2 + 3 = 5x$
    1 and $\dfrac{3}{2}$

17. Solve by using the quadratic formula:
$x^2 - 4x + 8 = 0$
No real number solution

18. Solve by using the quadratic formula:
$x^2 - 3x - 5 = 0$
$\dfrac{3 - \sqrt{29}}{2}$ and $\dfrac{3 + \sqrt{29}}{2}$

19. Solve by using the quadratic formula:
$2x^2 + 5x + 2 = 0$
-2 and $-\dfrac{1}{2}$

SECTION 12.4

20. Graph $y = -3x^2$.

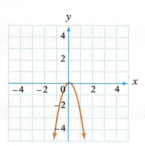

21. Graph $y = -\frac{1}{4}x^2$.

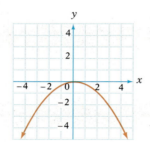

22. Graph $y = 2x^2 + 1$.

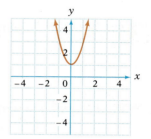

23. Graph $y = x^2 - 4x + 3$.

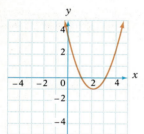

24. Graph $y = -x^2 + 4x - 5$.

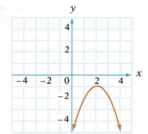

SECTION 12.5

25. It took a hawk half an hour more to fly 70 mi against the wind than it did to go 40 mi with the wind. The rate of the wind was 5 mph. Find the rate of the hawk in calm air.
75 mph

Chapter Test

1. Solve by factoring: $x^2 - 5x - 6 = 0$
   6 and -1 [12.1A]

2. Solve by factoring: $3x^2 + 7x = 20$
   -4 and $\dfrac{5}{3}$ [12.1A]

3. Solve by taking square roots:
   $2(x - 5)^2 - 50 = 0$
   0 and 10   [12.1B]

4. Solve by taking square roots:
   $3(x + 4)^2 - 60 = 0$
   $-4 + 2\sqrt{5}$ and $-4 - 2\sqrt{5}$ [12.1B]

5. Solve by completing the square:
   $x^2 + 4x - 16 = 0$
   $-2 + 2\sqrt{5}$ and $-2 - 2\sqrt{5}$ [12.2A]

6. Solve by completing the square:
   $x^2 + 3x = 8$
   $\dfrac{-3 + \sqrt{41}}{2}$ and $\dfrac{-3 - \sqrt{41}}{2}$ [12.2A]

7. Solve by completing the square:
   $2x^2 - 6x + 1 = 0$
   $\dfrac{3 + \sqrt{7}}{2}$ and $\dfrac{3 - \sqrt{7}}{2}$ [12.2A]

8. Solve by completing the square:
   $2x^2 + 8x = 3$
   $\dfrac{-4 + \sqrt{22}}{2}$ and $\dfrac{-4 - \sqrt{22}}{2}$ [12.2A]

9. Solve by using the quadratic formula:
   $x^2 + 4x + 2 = 0$
   $-2 + \sqrt{2}$ and $-2 - \sqrt{2}$ [12.3A]

10. Solve by using the quadratic formula:
    $x^2 - 3x = 6$
    $\dfrac{3 + \sqrt{33}}{2}$ and $\dfrac{3 - \sqrt{33}}{2}$ [12.3A]

11. Solve by using the quadratic formula:
$2x^2 - 5x - 3 = 0$

$-\dfrac{1}{2}$ and 3   [12.3A]

12. Solve by using the quadratic formula:
$3x^2 - x = 1$

$\dfrac{1 + \sqrt{13}}{6}$ and $\dfrac{1 - \sqrt{13}}{6}$ [12.3A]

13. Graph $y = x^2 + 2x - 4$.

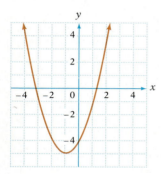

[12.4A]

14. The length of a rectangle is 2 ft less than twice the width. The area of the rectangle is 40 ft². Find the length and width of the rectangle.
length 8 ft; width 5 ft   [12.5A]

15. It took a motorboat 1 h more to travel 60 mi against a current than it did to travel 60 mi with the current. The rate of the current was 1 mph. Find the rate of the boat in calm water.
11 mph   [12.5A]

Cumulative Review

1. Simplify: $2x - 3[2x - 4(3 - 2x) + 2] - 3$
   $-28x + 27$ [2.2D]

2. Solve: $-\frac{3}{5}x = -\frac{9}{10}$
   $x = \frac{3}{2}$ [3.1C]

3. Solve: $2x - 3(4x - 5) = -3x - 6$
   $x = 3$ [3.3B]

4. Simplify: $(2a^2b)^2(-3a^4b^2)$
   $-12a^8b^4$ [5.2B]

5. Simplify: $(x^2 - 8) \div (x - 2)$
   $x + 2 - \dfrac{4}{x - 2}$ [5.4C]

6. Factor $3x^3 + 2x^2 - 8x$.
   $x(3x - 4)(x + 2)$ [6.3B]

7. Simplify: $\dfrac{3x^2 - 6x}{4x - 6} \div \dfrac{2x^2 + x - 6}{6x^3 - 24x}$
   $\dfrac{9x^2(x - 2)^2}{(2x - 3)^2}$ [7.1C]

8. Simplify: $\dfrac{x}{2(x - 1)} - \dfrac{1}{(x - 1)(x + 1)}$
   $\dfrac{x + 2}{2(x + 1)}$ [7.3B]

9. Simplify: $\dfrac{1 - \dfrac{7}{x} + \dfrac{12}{x^2}}{2 - \dfrac{1}{x} - \dfrac{15}{x^2}}$
   $\dfrac{x - 4}{2x + 5}$ [7.4A]

10. Find the x- and y-intercepts for the graph of the line $4x - 3y = 12$.
    x-intercept (3,0); y-intercept: (0,-4)   [8.3A]

11. Find the equation of the line that contains the point $(-3,2)$ and has slope $-\frac{4}{3}$.
    $y = -\frac{4}{3}x - 2$ [8.4B]

12. Solve by substitution:
    $3x - y = 5$
    $y = 2x - 3$
    (2,1)   [9.2A]

13. Solve by the addition method:
    $3x + 2y = 2$
    $5x - 2y = 14$
    (2,-2)   [9.3A]

14. Solve: $2x - 3(2 - 3x) > 2x - 5$
    $x > \dfrac{1}{9}$ [10.3A]

15. Simplify: $(\sqrt{a} - \sqrt{2})(\sqrt{a} + \sqrt{2})$
    $a - 2$ [11.3A]

16. Simplify: $\dfrac{\sqrt{108a^7b^3}}{\sqrt{3a^4b}}$
    $6ab\sqrt{a}$ [11.3B]

17. Simplify: $\dfrac{\sqrt{3}}{5 + 2\sqrt{3}}$

$\dfrac{-6 + 5\sqrt{3}}{13}$ [11.3B]

18. Solve: $3 = 8 - \sqrt{5x}$
$x = 5$ [11.4A]

19. Solve by factoring: $6x^2 - 17x = -5$
$\dfrac{5}{2}$ and $\dfrac{1}{3}$ [12.1A]

20. Solve by taking square roots:
$2(x - 5)^2 = 36$
$5 + 3\sqrt{2}$ and $5 - 3\sqrt{2}$ [12.1B]

21. Solve by completing the square:
$3x^2 + 7x = -3$
$\dfrac{-7 + \sqrt{13}}{6}$ and $\dfrac{-7 - \sqrt{13}}{6}$ [12.2A]

22. Solve by using the quadratic formula:
$2x^2 - 3x - 2 = 0$
2 and $-\dfrac{1}{2}$ [12.3A]

23. Find the selling price per pound of a mixture made from 20 lb of cashews that cost $3.50 per pound and 50 lb of peanuts that cost $1.75 per pound.
$2.25 per pound   [4.5A]

24. A stock investment of 100 shares paid a dividend of $215. At this rate, how many additional shares are required to earn a dividend of $752.50?
250 shares   [7.6B]

25. A 720-mi trip from one city to another takes 3 h when a plane is flying with the wind. The return trip, against the wind, takes 4.5 h. Find the rate of the plane in still air and the rate of the wind.
plane 200 mph; wind 40 mph   [9.4A]

26. A student received a 70, a 91, an 85, and a 77 on four tests in a math class. What scores on the last test will enable the student to receive a minimum of 400 points?
≥ 77 [10.2C]

27. The sum of the squares of three consecutive odd integers is 83. Find the middle odd integer.
-5 or 5 [12.5A]

28. A jogger ran 7 mi at a constant rate and then reduced the rate by 3 mph. An additional 8 mi was run at the reduced rate. The total time spent jogging the 15 mi was 3 h. Find the rate for the last 8 mi.
4 mph   [12.5A]

29. Graph $2x - 3y > 6$.

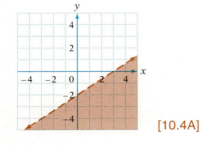

[10.4A]

30. Graph $y = x^2 - 2x - 3$.

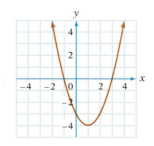

[12.4A]

Final Exam

1. Evaluate $-|-3|$.
   -3 [1.1B]

2. Subtract: $-15 - (-12) - 3$
   -6 [1.2B]

3. Simplify $-2^4 \cdot (-2)^4$.
   -256 [1.5A]

4. Simplify $-7 - \frac{12 - 15}{2 - (-1)} \cdot (-4)$.
   -11 [1.5B]

5. Evaluate $\frac{a^2 - 3b}{2a - 2b^2}$ when
   $a = 3$ and $b = -2$.
   $-\frac{15}{2}$ [2.1A]

6. Simplify $6x - (-4y) - (-3x) + 2y$.
   $9x + 6y$ [2.2A]

7. Simplify $(-15z)(-\frac{2}{5})$.
   $6z$ [2.2B]

8. Simplify $-2[5 - 3(2x - 7) - 2x]$.
   $16x - 52$ [2.2D]

9. Solve: $20 = -\frac{2}{5}x$
   $x = -50$ [3.1C]

10. Solve: $4 - 2(3x + 1) = 3(2 - x) + 5$
    $x = -3$ [3.3B]

11. Write $\frac{1}{8}$ as a percent.
    12.5% [4.1B]

12. Find 19% of 80.
    15.2 [4.2A]

13. Simplify $(2x^2 - 5x + 1) - (5x^2 - 2x - 7)$.
    $-3x^2 - 3x + 8$ [5.1B]

14. Simplify $(-3xy^3)^4$.
    $81x^4y^{12}$ [5.2B]

15. Simplify $(3x^2 - x - 2)(2x + 3)$.
    $6x^3 + 7x^2 - 7x - 6$ [5.3B]

16. Simplify $\frac{(-2x^2y^3)^3}{(-4xy^4)^2}$.
    $-\frac{x^4y}{2}$ [5.4A]

17. Simplify: $\frac{12x^2y - 16x^3y^2 - 20y^2}{4xy^2}$
    $\frac{3x}{y} - 4x^2 - \frac{5}{x}$ [5.4B]

18. Simplify: $(5x^2 - 2x - 1) \div (x + 2)$
    $5x - 12 + \frac{23}{x + 2}$ [5.4C]

19. Simplify: $(4x^{-2}y)^2(2xy^{-2})^{-2}$
    $\frac{4y^6}{x^6}$ [5.4A]

20. Factor $2x^2 - x - 3$.
    $(2x - 3)(x + 1)$ [6.3A]

21. Factor $x^2 - 5x - 6$.
    $(x - 6)(x + 1)$ [6.2A]

22. Factor $6x^2 - 5x - 6$.
    $(3x + 2)(2x - 3)$ [6.3A]

23. Factor $8x^3 - 28x^2 + 12x$.
$4x(2x - 1)(x - 3)$ [6.3B]

24. Factor $25x^2 - 16$.
$(5x - 4)(5x + 4)$ [6.4A]

25. Factor $2a(4 - x) - 6(x - 4)$.
$2(a + 3)(4 - x)$ [6.1B]

26. Factor $75y - 12x^2y$.
$3y(5 - 2x)(5 + 2x)$ [6.4B]

27. Solve: $2x^2 = 7x - 3$
$\dfrac{1}{2}$ and 3   [6.5A]

28. Simplify: $\dfrac{2x^2 - 3x + 1}{4x^2 - 2x} \cdot \dfrac{4x^2 + 4x}{x^2 - 2x + 1}$
$\dfrac{2(x + 1)}{x - 1}$ [7.1B]

29. Simplify: $\dfrac{5}{x + 3} - \dfrac{3x}{2x - 5}$
$\dfrac{-3x^2 + x - 25}{(2x - 5)(x + 3)}$ [7.3B]

30. Simplify: $x - \dfrac{1}{1 - \frac{1}{x}}$
$\dfrac{x^2 - 2x}{x - 1}$ [7.4A]

31. Solve: $\dfrac{5x}{3x - 5} - 3 = \dfrac{7}{3x - 5}$
$x = 2$ [7.5A]

32. Solve $a = 3a - 2b$ for a.
$a = b$ [7.7A]

33. Find the slope of the line that contains the points $(-1, -3)$ and $(2, -1)$.
$\dfrac{2}{3}$ [8.3B]

34. Find the equation of the line that contains the point $(3, -4)$ and has slope $-\frac{2}{3}$.
$y = -\dfrac{2}{3}x - 2$ [8.4A]

35. Solve by substitution.
$y = 4x - 7$
$y = 2x + 5$
$(6, 17)$ [9.2A]

36. Solve by the addition method.
$4x - 3y = 11$
$2x + 5y = -1$
$(2, -1)$ [9.3A]

37. Solve: $4 - x \geq 7$
$x \leq -3$ [10.2A]

38. Solve: $2 - 2(y - 1) \leq 2y - 6$
$y \geq \dfrac{5}{2}$ [10.3A]

39. Simplify $\sqrt{49x^6}$.
$7x^3$ [11.1B]

40. Simplify $2\sqrt{27a} + 8\sqrt{48a}$.
$38\sqrt{3a}$ [11.2A]

41. Simplify $\dfrac{\sqrt{3}}{\sqrt{5} - 2}$.
$\sqrt{15} + 2\sqrt{3}$ [11.3B]

42. Solve: $\sqrt{x + 4} - \sqrt{x - 1} = 1$
$x = 5$ [11.4A]

43. Solve by factoring:
$3x^2 - x = 4$
-1 and $\dfrac{4}{3}$ [12.1A]

44. Solve by using the quadratic formula:
$4x^2 - 2x - 1 = 0$
$\dfrac{1 + \sqrt{5}}{4}$ and $\dfrac{1 - \sqrt{5}}{4}$ [12.3A]

Final Exam

45. Translate and simplify "the sum of twice a number and three times the difference of the number and two."
$2x + 3(x - 2)$; $5x - 6$ [2.3C]

46. Because of depreciation, the value of an office machine is now $2400. This is 80% of its original value. Find the original value.
$3000   [4.2B]

47. The manufacturer's cost for a laser printer is $900. The manufacturer then sells the printer for $1485. What is the markup rate?
65%   [4.3A]

48. An investment of $3000 is made at an annual simple interest rate of 8%. How much additional money must be invested at 11% so that the total interest earned is 10% of the total investment?
$6000   [4.4A]

49. A grocer mixes 4 lb of peanuts that cost $2 per pound with 2 lb of walnuts that cost $5 per pound. What is the cost per pound of the resulting mixture?
$3 per pound   [4.5A]

50. A pharmacist mixes together 20 L of a solution that is 60% acid and 30 L of a solution that is 20% acid. What is the percent concentration of the acid in the mixture?
36%   [4.5B]

51. At 2:00 P.M. a small plane had been flying 1 h when a change of wind direction doubled its average ground speed. The pilot completed the 860-km trip in 2.5 h. How far did the plane travel in the first hour?
215 km   [4.6A]

52. The angles of a triangle are such that the second angle is 10° more than the first angle, and the third angle is 10° more than the second angle. Find the measure of each of the three angles.
50°, 60°, 70°   [4.7B]

53. A coin bank contains quarters and dimes. There are three times the number of dimes as quarters. The total value of the coins in the bank is $11. Find the number of dimes in the bank.
60 dimes   [4.8B]

54. The length of a rectangle is 5 m more than the width. The area of the rectangle is 50 m². Find the dimensions of the rectangle.
width 5 m; length 10 m   [6.5B]

55. A paint formula requires 2 oz of dye for every 15 oz of base paint. How many ounces of dye are required for 120 oz of base paint?
16 oz   [7.6B]

56. It takes a chef 1 h to prepare a dinner. The chef's apprentice can prepare the dinner in 1.5 h. How long would it take the chef and the apprentice, working together, to prepare the dinner?
0.6 h   [7.8A]

57. With the current, a motorboat travels 50 mi in 2.5 h. Against the current, it takes twice as long to travel 50 mi. Find the rate of the boat in calm water and the rate of the current.
boat 15 mph; current 5 mph   [9.4A]

58. Flying against the wind, it took a pilot $\frac{1}{2}$ h longer to travel 500 miles than it took flying with the wind. The rate of the plane in calm air is 225 mph. Find the rate of the wind.
25 mph   [12.5A]

59. Graph the line with slope $-\frac{1}{2}$ and y-intercept $(0, -3)$.

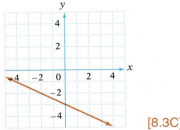

[8.3C]

60. Graph $y = x^2 - 4x + 3$.

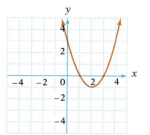

[12.4A]

APPENDIX

Table of Square Roots

Decimal approximations have been rounded to the nearest thousandth.

Number	Square Root	Number	Square Root	Number	Square Root	Number	Square Root
1	1	51	7.141	101	10.050	151	12.288
2	1.414	52	7.211	102	10.100	152	12.329
3	1.732	53	7.280	103	10.149	153	12.369
4	2	54	7.348	104	10.198	154	12.410
5	2.236	55	7.416	105	10.247	155	12.450
6	2.449	56	7.483	106	10.296	156	12.490
7	2.646	57	7.550	107	10.344	157	12.530
8	2.828	58	7.616	108	10.392	158	12.570
9	3	59	7.681	109	10.440	159	12.610
10	3.162	60	7.746	110	10.488	160	12.649
11	3.317	61	7.810	111	10.536	161	12.689
12	3.464	62	7.874	112	10.583	162	12.728
13	3.606	63	7.937	113	10.630	163	12.767
14	3.742	64	8	114	10.677	164	12.806
15	3.873	65	8.062	115	10.724	165	12.845
16	4	66	8.124	116	10.770	166	12.884
17	4.123	67	8.185	117	10.817	167	12.923
18	4.243	68	8.246	118	10.863	168	12.961
19	4.359	69	8.307	119	10.909	169	13
20	4.472	70	8.367	120	10.954	170	13.038
21	4.583	71	8.426	121	11	171	13.077
22	4.690	72	8.485	122	11.045	172	13.115
23	4.796	73	8.544	123	11.091	173	13.153
24	4.899	74	8.602	124	11.136	174	13.191
25	5	75	8.660	125	11.180	175	13.229
26	5.099	76	8.718	126	11.225	176	13.267
27	5.196	77	8.775	127	11.269	177	13.304
28	5.292	78	8.832	128	11.314	178	13.342
29	5.385	79	8.888	129	11.358	179	13.379
30	5.477	80	8.944	130	11.402	180	13.416
31	5.568	81	9	131	11.446	181	13.454
32	5.657	82	9.055	132	11.489	182	13.491
33	5.745	83	9.110	133	11.533	183	13.528
34	5.831	84	9.165	134	11.576	184	13.565
35	5.916	85	9.220	135	11.619	185	13.601
36	6	86	9.274	136	11.662	186	13.638
37	6.083	87	9.327	137	11.705	187	13.675
38	6.164	88	9.381	138	11.747	188	13.711
39	6.245	89	9.434	139	11.790	189	13.748
40	6.325	90	9.487	140	11.832	190	13.784
41	6.403	91	9.539	141	11.874	191	13.820
42	6.481	92	9.592	142	11.916	192	13.856
43	6.557	93	9.644	143	11.958	193	13.892
44	6.633	94	9.695	144	12	194	13.928
45	6.708	95	9.747	145	12.042	195	13.964
46	6.782	96	9.798	146	12.083	196	14
47	6.856	97	9.849	147	12.124	197	14.036
48	6.928	98	9.899	148	12.166	198	14.071
49	7	99	9.950	149	12.207	199	14.107
50	7.071	100	10	150	12.247	200	14.142

Table of Properties

Properties of Real Numbers

Associative Property of Addition

If a, b, and c are real numbers, then $(a + b) + c = a + (b + c)$.

Associative Property of Multiplication

If a, b, and c are real numbers, then $(a \cdot b) \cdot c = a \cdot (b \cdot c)$.

Commutative Property of Addition

If a and b are real numbers, then $a + b = b + a$.

Commutative Property of Multiplication

If a and b are real numbers, then $a \cdot b = b \cdot a$.

Addition Property of Zero

If a is a real number, then $a + 0 = 0 + a = a$.

Multiplication Property of One

If a is a real number, then $a \cdot 1 = 1 \cdot a = a$.

Inverse Property of Addition

If a is a real number, then $a + (-a) = (-a) + a = 0$.

Inverse Property of Multiplication

If a is a real number and $a \neq 0$, then $a \cdot \dfrac{1}{a} = \dfrac{1}{a} \cdot a = 1$.

Distributive Property

If a, b, and c are real numbers, then $a(b + c) = ab + ac$.

Properties of Equations

Addition Property of Equations

If $a = b$, then $a + c = b + c$.

Multiplication Property of Equations

If $a = b$ and $c \neq 0$, then $a \cdot c = b \cdot c$.

Properties of Exponents

If m and n are integers, then $x^m \cdot x^n = x^{m+n}$.

If m and n are integers and $x \neq 0$, then $\dfrac{x^m}{x^n} = x^{m-n}$.

If m and n are integers, then $(x^m)^n = x^{m \cdot n}$.

If m, n, and p are integers, then $(x^m \cdot y^n)^p = x^{m \cdot p} y^{n \cdot p}$.

If x is a real number and $x \neq 0$, then $x^0 = 1$.

If n is a positive integer and $x \neq 0$, then $x^{-n} = \dfrac{1}{x^n}$.

Property of Zero Products

If $a \cdot b = 0$, then $a = 0$ or $b = 0$.

Properties of Radical Expressions

If a and b are positive real numbers, then $\sqrt{ab} = \sqrt{a} \cdot \sqrt{b}$.

If a and b are positive real numbers, then $\sqrt{\dfrac{a}{b}} = \dfrac{\sqrt{a}}{\sqrt{b}}$.

Properties of Inequalities

Addition Property of Inequalities

If $a > b$, then $a + c > b + c$.
If $a < b$, then $a + c < b + c$.

Multiplication Property of Inequalities

If $a > b$ and $c > 0$, then $ac > bc$.
If $a < b$ and $c > 0$, then $ac < bc$.
If $a > b$ and $c < 0$, then $ac < bc$.
If $a < b$ and $c < 0$, then $ac > bc$.

Property of Squaring Both Sides of an Equation

If a and b are real numbers and $a = b$, then $a^2 = b^2$.

Table of Symbols

$+$	add		$<$	is less than
$-$	subtract		$\leq$	is less than or equal to
$\cdot,\ \times,\ (a)(b)$	multiply		$>$	is greater than
$\frac{a}{b},\ \div\ a\overline{)b}$	divide		$\geq$	is greater than or equal to
$(\ )$	parentheses, a grouping symbol		(a, b)	an ordered pair whose first component is a and whose second component is b
$[\ ]$	brackets, a grouping symbol			
π	pi, a number approximately equal to $\frac{22}{7}$ or 3.14		$^\circ$	degree (for angles)
			$\sqrt{a}$	the principal square root of a
$-a$	the opposite, or additive inverse, of a		$\varnothing,\ \{\ \}$	the empty set
$\frac{1}{a}$	the reciprocal, or multiplicative inverse, of a		$|a|$	the absolute value of a
			$\cup$	union of two sets
$=$	is equal to		$\cap$	intersection of two sets
$\approx$	is approximately equal to		$\in$	is an element of (for sets)
$\neq$	is not equal to			

Table of Measurement Abbreviations

U.S. Customary System

Length		**Capacity**		**Weight**		**Area**	
in.	inches	oz	ounces	oz	ounces	in^2	square inches
ft	feet	c	cups	lb	pounds	ft^2	square feet
yd	yards	qt	quarts				
mi	miles	gal	gallons				

Metric System

Length		**Capacity**		**Weight/Mass**		**Area**	
mm	millimeter (0.001 m)	ml	milliliter (0.001 L)	mg	milligram (0.001 g)	cm^2	square centimeters
cm	centimeter (0.01 m)	cl	centiliter (0.01 L)	cg	centigram (0.01 g)	m^2	square meters
dm	decimeter (0.1 m)	dl	deciliter (0.1 L)	dg	decigram (0.1 g)		
m	meter	L	liter	g	gram		
dam	decameter (10 m)	dal	decaliter (10 L)	dag	decagram (10 g)		
hm	hectometer (100 m)	hl	hectoliter (100 L)	hg	hectogram (100 g)		
km	kilometer (1000 m)	kl	kiloliter (1000 L)	kg	kilogram (1000 g)		

Time

h	hours	min	minutes	s	seconds

SOLUTIONS to Chapter 1 Examples

SECTION 1.1   *pages 3–4*

Example 2   a. $5 > -13$ b. $-8 > -22$ **Example 4**   a. 9 b. -62

Example 6 $|-5| = 5$ $-|-9| = -9$

SECTION 1.2   *pages 7–10*

Example 2
$-154 + (-37)$
-191

Example 4
$-36 + 17 + (-21)$
$-19 + (-21)$
-40

Example 6
$-5 + (-2) + 9 + (-3)$
$-7 + 9 + (-3)$
$2 + (-3)$
-1

Example 8
$-8 - 14$
$-8 + (-14)$
-22

Example 10
$3 - (-4) - 15$
$3 + 4 + (-15)$
$7 + (-15)$
-8

Example 12
$4 - (-3) - 12 - (-7) - 20$
$4 + 3 + (-12) + 7 + (-20)$
$7 + (-12) + 7 + (-20)$
$-5 + 7 + (-20)$
$2 + (-20)$
-18

Example 14
$17 - 10 - 2 - (-6) - 9$
$17 + (-10) + (-2) + 6 + (-9)$
$7 + (-2) + 6 + (-9)$
$5 + 6 + (-9)$
$11 + (-9)$
2

SECTION 1.3   *pages 13–16*

Example 2
$(-3) \cdot 4 \cdot (-5)$
$(-12) \cdot (-5)$
60

Example 4
$-38 \cdot 51$
-1938

Example 6
$-6 \cdot 8 \cdot (-11) \cdot 3$
$-48 \cdot (-11) \cdot 3$
$528 \cdot 3$
1584

Example 8
$-7(-8)(9)(-2)$
$56(9)(-2)$
$504(-2)$
-1008

Example 10
$(-135) \div (-9)$
15

Example 12
$-72 \div 4$
-18

Example 14
$84 \div (-6)$
-14

Example 16

Strategy   To find the difference, subtract the average temperature throughout the earth's stratosphere ($-70°$) from the average temperature on earth's surface ($57°$).

Solution $57 - (-70)$
$57 + 70$
127

The difference is $127°$F.

Example 18

Strategy   To find the average daily low temperature:
- Add the seven temperature readings.
- Divide by 7.

Solution $-6 + (-7) + 1 + 0 + (-5) + (-10) + (-1)$
$-13 + 1 + 0 + (-5) + (-10) + (-1)$
$-12 + 0 + (-5) + (-10) + (-1)$
$-12 + (-5) + (-10) + (-1)$
$-17 + (-10) + (-1)$
$-27 + (-1)$
-28

$-28 \div 7 = -4$

The average daily low temperature was $-4°$ C.

SECTION 1.4   *pages 21–26*

Example 2

$$\begin{array}{r} 0.16 \\ 25\overline{)4.00} \\ \underline{-2\ 5} \\ 1\ 50 \\ \underline{-1\ 50} \\ 0 \end{array}$$

$\dfrac{4}{25} = 0.16$

Example 4

$$\begin{array}{r} 0.444 \\ 9\overline{)4.000} \\ \underline{-3\ 6} \\ 40 \\ \underline{-36} \\ 40 \\ \underline{-36} \\ 4 \end{array}$$

$\dfrac{4}{9} = 0.\overline{4}$

Example 6   The LCM of 9 and 12 is 36.

$$\frac{5}{9} - \frac{11}{12} = \frac{20}{36} - \frac{33}{36} = \frac{20}{36} + \frac{-33}{36}$$
$$= \frac{20 + (-33)}{36} = \frac{-13}{36} = -\frac{13}{36}$$

Example 8   The LCM of 8, 6 and 2 is 24.

$$-\frac{7}{8} - \frac{5}{6} + \frac{1}{2} = \frac{-21}{24} - \frac{20}{24} + \frac{12}{14}$$
$$= \frac{-21}{24} + \frac{-20}{24} + \frac{12}{24}$$
$$= \frac{-21 - 20 + 12}{24}$$
$$= \frac{-29}{24} = -\frac{29}{24}$$

Example 10

$$\begin{array}{r} 3.907 \\ 4.9 \\ +\ 6.63 \\ \hline 15.437 \end{array}$$

Example 12

$$\begin{array}{r} 67.910 \\ -16.127 \\ \hline 51.783 \end{array}$$

$16.127 - 67.91 = -51.783$

Example 14 $2.7 + (-9.44) + 6.2$
$-6.74 + 6.2$
-0.54

Example 16   The product is negative.

$$-\frac{7}{12} \times \frac{9}{14} = -\frac{7 \cdot 9}{12 \cdot 14}$$
$$= -\frac{\overset{1}{\cancel{7}} \cdot \overset{1}{\cancel{3}} \cdot 3}{2 \cdot 2 \cdot \underset{1}{\cancel{3}} \cdot 2 \cdot \underset{1}{\cancel{7}}}$$
$$= -\frac{3}{8}$$

Example 18 The quotient is positive.

$$-\frac{3}{8} \div \left(-\frac{5}{12}\right) = \frac{3}{8} \times \frac{12}{5}$$

$$= \frac{3 \cdot 12}{8 \cdot 5} = \frac{3 \cdot \overset{1}{\cancel{2}} \cdot \overset{1}{\cancel{2}} \cdot 3}{\underset{1}{\cancel{2}} \cdot \underset{1}{\cancel{2}} \cdot 2 \cdot 5} = \frac{9}{10}$$

Example 20

$$\begin{array}{r} 5.44 \\ \times\ \ 3.8 \\ \hline 4352 \\ 1632\ \ \\ \hline 20.672 \end{array}$$

$$-5.44 \times 3.8 = -20.672$$

Example 22 $3.44 \times (-1.7) \times 0.6$
$(-5.848) \times 0.6$
-3.5088

Example 24

$$\begin{array}{r} 0.231 \\ 1.7{\overline{\smash{\big)}\,0{.}3{.}940}} \\ \underline{-3\ 4} \\ 54 \\ \underline{-51} \\ 30 \\ \underline{-17} \\ 13 \end{array}$$

$$-0.394 \div 1.7 \approx -0.23$$

SECTION 1.5 *pages 31–34*

Example 2 $-6^3 = -(6 \cdot 6 \cdot 6) = -216$

Example 4 $(-3)^4 = (-3)(-3)(-3)(-3) = 81$

Example 6 $(3^3)(-2)^3 = (3)(3)(3) \cdot (-2)(-2)(-2)$
$= 27(-8) = -216$

Example 8 $\left(-\frac{2}{5}\right)^2 = \left(-\frac{2}{5}\right)\left(-\frac{2}{5}\right) = \frac{4}{25}$

Example 10 $-3(0.3)^3 = -3(0.3)(0.3)(0.3)$
$= -0.9(0.3)(0.3)$
$= -0.27(0.3) = -0.081$

Example 12 $18 - 5[8 - 2(2 - 5)] \div 10$
$18 - 5[8 - 2(-3)] \div 10$
$18 - 5[8 + 6] \div 10$
$18 - 5[14] \div 10$
$18 - 70 \div 10$
$18 - 7$
11

Example 14 $36 \div (8 - 5)^2 - (-3)^2 \cdot 2$
$36 \div (3)^2 - (-3)^2 \cdot 2$
$36 \div 9 - 9 \cdot 2$
$4 - 9 \cdot 2$
$4 - 18$
-14

Example 16 $(6.97 - 4.72)^2 \times 4.5 \div 0.05$
$(2.25)^2 \times 4.5 \div 0.05$
$5.0625 \times 4.5 \div 0.05$
$22.78125 \div 0.05$
455.625

Example 18 $\dfrac{5}{8} \div \left(\dfrac{1}{3} - \dfrac{3}{4}\right) + \dfrac{7}{12}$

$\dfrac{5}{8} \div \left(\dfrac{-5}{12}\right) + \dfrac{7}{12}$

$\dfrac{5}{8} \cdot \left(-\dfrac{12}{5}\right) + \dfrac{7}{12}$

$-\dfrac{3}{2} + \dfrac{7}{12}$

$-\dfrac{18}{12} + \dfrac{7}{12}$

$-\dfrac{11}{12}$

SOLUTIONS to Chapter 2 Examples

SECTION 2.1   *pages 45–46*

Example 2 -4 is the constant term.

Example 4
$2xy + y^2$
$2(-4)(2) + (2)^2$
$2(-4)(2) + 4$
$(-8)(2) + 4$
$(-16) + 4$
-12

Example 6
$\dfrac{a^2 + b^2}{a + b}$

$\dfrac{5^2 + (-3)^2}{5 + (-3)}$

$\dfrac{25 + 9}{5 + (-3)}$

$\dfrac{34}{2}$

17

Example 8
$x^3 - 2(x + y) + z^2$
$(2)^3 - 2[2 + (-4)] + (-3)^2$
$8 - 2(-2) + 9$
$8 + 4 + 9$
$12 + 9$
21

SECTION 2.2   *pages 49–54*

Example 2
$3a - 2b - 5a + 6b$
$-2a + 4b$

Example 4
$-3y^2 + 7 + 8y^2 - 14$
$5y^2 - 7$

Example 6
$-5(4y^2)$
$-20y^2$

Example 8
$-7(-2a)$
$14a$

Example 10
$(-5x)(-2)$
$10x$

Example 12
$5(3 + 7b)$
$15 + 35b$

Example 14
$(3a - 1)5$
$15a - 5$

Example 16
$-8(-2a + 7b)$
$16a - 56b$

Example 18
$-(5x - 12)$
$-5x + 12$

Example 20
$3(-a^2 - 6a + 7)$
$-3a^2 - 18a + 21$

Example 22
$3y - 2(y - 7x)$
$3y - 2y + 14x$
$14x + y$

Example 24
$-2(x - 2y) + 4(x - 3y)$
$-2x + 4y + 4x - 12y$
$2x - 8y$

Example 26
$-5(-2y - 3x) + 4y$
$10y + 15x + 4y$
$15x + 14y$

Example 28
$3y - 2[x - 4(2 - 3y)]$
$3y - 2[x - 8 + 12y]$
$3y - 2x + 16 - 24y$
$-2x - 21y + 16$

SECTION 2.3   *pages 59–62*

Example 2   the <u>difference between</u> <u>twice</u> *n* and one third <u>of</u> *n*

$$2n - \frac{1}{3}n$$

Example 4   the <u>quotient of</u> 7 <u>less than</u> *b* and 15

$$\frac{b-7}{15}$$

Example 6   an unknown number: *n*
the cube of the number: n^3
the total of ten and the cube of the number: $10 + n^3$

$$-4(10 + n^3)$$

Example 8   the first integer: *x*
the second integer: *x* + 1
the third integer: *x* + 2
$x + (x + 1) + (x + 2)$; $3x + 3$

Example 10   the unknown number: *x*
the difference between the number and sixty: *x* − 60

$$5(x - 60); \ 5x - 300$$

Example 12   the speed of the older model: *s*
the new laptop is twice the speed of the older model: 2*s*

Example 14   the length of the longer piece: *y*
the length of the shorter piece: 6 − *y*

SOLUTIONS to Chapter 3 Examples

SECTION 3.1   *pages 77–80*

Example 2
$$5 - 4x = 8x + 2$$

$$
\begin{array}{c|c}
5 - 4\left(\frac{1}{4}\right) & 8\left(\frac{1}{4}\right) + 2 \\
5 - 1 & 2 + 2 \\
4 & = 4
\end{array}
$$

Yes, $\frac{1}{4}$ is a solution.

Example 4
$$10x - x^2 = 3x - 10$$

$$
\begin{array}{c|c}
10(5) - (5)^2 & 3(5) - 10 \\
50 - 25 & 15 - 10 \\
25 & \neq 5
\end{array}
$$

No, 5 is not a solution.

Example 6
$$\frac{1}{2} = x - \frac{2}{3}$$
$$\frac{1}{2} + \frac{2}{3} = x - \frac{2}{3} + \frac{2}{3}$$
$$\frac{7}{6} = x$$

The solution is $\frac{7}{6}$.

Example 8
$$-\frac{2}{5}x = 6$$
$$\left(-\frac{5}{2}\right)\left(-\frac{2}{5}x\right) = \left(-\frac{5}{2}\right)(6)$$
$$x = -15$$

The solution is −15.

Example 10
$$4x - 8x = 16$$
$$-4x = 16$$
$$\frac{-4x}{-4} = \frac{16}{-4}$$
$$x = -4$$

The solution is −4.

SECTION 3.2   *pages 85–86*

Example 2

$$5x + 7 = 10$$
$$5x + 7 - 7 = 10 - 7$$
$$5x = 3$$
$$\frac{5x}{5} = \frac{3}{5}$$
$$x = \frac{3}{5}$$

The solution is $\frac{3}{5}$.

Example 4

$$2 = 11 + 3x$$
$$2 - 11 = 11 - 11 + 3x$$
$$-9 = 3x$$
$$\frac{-9}{3} = \frac{3x}{3}$$
$$-3 = x$$

The solution is -3.

Example 6

$$x - 5 + 4x = 25$$
$$5x - 5 = 25$$
$$5x - 5 + 5 = 25 + 5$$
$$5x = 30$$
$$\frac{5x}{5} = \frac{30}{5}$$
$$x = 6$$

The solution is 6.

Example 8

Strategy   To find the depth, replace P with the given value and solve for D.

Solution

$$P = 15 + \frac{1}{2}D$$
$$45 = 15 + \frac{1}{2}D$$
$$45 - 15 = 15 - 15 + \frac{1}{2}D$$
$$30 = \frac{1}{2}D$$
$$2(30) = 2 \cdot \frac{1}{2}D$$
$$60 = D$$

The depth is 60 ft.

SECTION 3.3   *pages 91–94*

Example 2

$$5x + 4 = 6 + 10x$$
$$5x - 10x + 4 = 6 + 10x - 10x$$
$$-5x + 4 = 6$$
$$-5x + 4 - 4 = 6 - 4$$
$$-5x = 2$$
$$\frac{-5x}{-5} = \frac{2}{-5}$$
$$x = -\frac{2}{5}$$

The solution is $-\frac{2}{5}$.

Example 4

$$5x - 10 - 3x = 6 - 4x$$
$$2x - 10 = 6 - 4x$$
$$2x + 4x - 10 = 6 - 4x + 4x$$
$$6x - 10 = 6$$
$$6x - 10 + 10 = 6 + 10$$
$$6x = 16$$
$$\frac{6x}{6} = \frac{16}{6}$$
$$x = \frac{8}{3}$$

The solution is $\frac{8}{3}$.

Example 6

$5x - 4(3 - 2x) = 2(3x - 2) + 6$
$5x - 12 + 8x = 6x - 4 + 6$
$13x - 12 = 6x + 2$
$13x - 6x - 12 = 6x - 6x + 2$
$7x - 12 = 2$
$7x - 12 + 12 = 2 + 12$
$7x = 14$
$\dfrac{7x}{7} = \dfrac{14}{7}$
$x = 2$

The solution is 2.

Example 8

$-2[3x - 5(2x - 3)] = 3x - 8$
$-2[3x - 10x + 15] = 3x - 8$
$-2[-7x + 15] = 3x - 8$
$14x - 30 = 3x - 8$
$14x - 3x - 30 = 3x - 3x - 8$
$11x - 30 = -8$
$11x - 30 + 30 = -8 + 30$
$11x = 22$
$\dfrac{11x}{11} = \dfrac{22}{11}$
$x = 2$

The solution is 2.

Example 10

Strategy

To find the location of the fulcrum when the system balances, replace the variables F_1, F_2, and d in the lever system equation by the given values and solve for x.

Solution

$F_1 \cdot x = F_2 \cdot (d - x)$
$45x = 80(25 - x)$
$45x = 2000 - 80x$
$45x + 80x = 2000 - 80x + 80x$
$125x = 2000$
$\dfrac{125x}{125} = \dfrac{2000}{125}$
$x = 16$

The fulcrum is 16 ft from the 45-pound force.

SECTION 3.4 *pages 99–103*

Example 2

The unknown number: n

four less than one third of a number	equals	five minus two thirds of the number

$\dfrac{1}{3}n - 4 = 5 - \dfrac{2}{3}n$

$\dfrac{1}{3}n + \dfrac{2}{3}n - 4 = 5 - \dfrac{2}{3}n + \dfrac{2}{3}n$

$n - 4 = 5$
$n - 4 + 4 = 5 + 4$
$n = 9$

The number is 9.

Example 4

The unknown number: n

two times the difference between a number and eight	is equal to	the sum of six times the number and eight

$2(n - 8) = 6n + 8$
$2n - 16 = 6n + 8$
$2n - 6n - 16 = 6n - 6n + 8$
$-4n - 16 = 8$
$-4n - 16 + 16 = 8 + 16$
$-4n = 24$
$\dfrac{-4n}{-4} = \dfrac{24}{-4}$
$n = -6$

The number is -6.

Example 6

The smaller number: n
The larger number: $12 - n$

the total of three times the smaller and six	amounts to	seven less than the product of four and the larger

$$3n + 6 = 4(12 - n) - 7$$
$$3n + 6 = 48 - 4n - 7$$
$$3n + 6 = 41 - 4n$$
$$3n + 4n + 6 = 41 - 4n + 4n$$
$$7n + 6 = 41$$
$$7n + 6 - 6 = 41 - 6$$
$$7n = 35$$
$$\frac{7n}{7} = \frac{35}{7}$$
$$n = 5$$

$$12 - n = 12 - 5 = 7$$

The smaller number is 5.
The larger number is 7.

Example 10

Strategy

To find the Fahrenheit temperature, write and solve an equation using F to represent the Fahrenheit temperature.

Solution

20	is	$\frac{5}{9}$ of the difference between the Fahrenheit temperature and 32

$$20 = \frac{5}{9}(F - 32)$$
$$20 = \frac{5}{9}F - \frac{160}{9}$$
$$20 + \frac{160}{9} = \frac{5}{9}F - \frac{160}{9} + \frac{160}{9}$$
$$\frac{340}{9} = \frac{5}{9}F$$
$$\frac{9}{5} \cdot \frac{340}{9} = \frac{9}{5} \cdot \frac{5}{9}F$$
$$68 = F$$

The Fahrenheit temperature is 68°.

Example 8

Strategy

To find the number of carbon atoms in a butane molecule, write and solve an equation using c to represent the number of carbon atoms in a butane molecule.

Solution

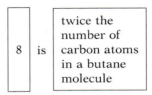

8	is	twice the number of carbon atoms in a butane molecule

$$8 = 2c$$
$$\frac{8}{2} = \frac{2c}{2}$$
$$4 = c$$

There are 4 carbon atoms in a butane molecule.

Example 12

Strategy

To find the length of the wire which produces an A note, write and solve an equation using A to represent the length of the wire.

Solution

10	is	6 in. less than $\frac{1}{2}$ the length of the wire

$$10 = \frac{1}{2}A - 6$$
$$10 + 6 = \frac{1}{2}A - 6 + 6$$
$$16 = \frac{1}{2}A$$
$$2 \cdot 16 = 2 \cdot \frac{1}{2}A$$
$$32 = A$$

The length of the wire which produces an A note is 32 in.

Example 14

Strategy

To find the number of color TV's made each day, write and solve an equation, using x to represent the number of color TV's and $140 - x$ to represent the number of black and white TV's.

Solution

three times the number of black and white TV's	equals	20 less than the number of color TV's

$$3(140 - x) = x - 20$$
$$420 - 3x = x - 20$$
$$420 - 3x - x = x - x - 20$$
$$420 - 4x = -20$$
$$420 - 420 - 4x = -20 - 420$$
$$-4x = -440$$
$$\frac{-4x}{-4} = \frac{-440}{-4}$$
$$x = 110$$

There are 110 color TV's made each day.

SOLUTIONS to Chapter 4 Examples

SECTION 4.1   *pages 117–118*

Example 2 $125\% = 125\left(\frac{1}{100}\right) = \frac{125}{100} = 1\frac{1}{4}$
$125\% = 125(0.01) = 1.25$

Example 4 $16\frac{2}{3}\% = 16\frac{2}{3}\left(\frac{1}{100}\right) = \frac{50}{3}\left(\frac{1}{100}\right) = \frac{1}{6}$

Example 6 $0.5\% = 0.5(0.01) = 0.005$

Example 8 $0.043 = 0.043(100\%) = 4.3\%$

Example 10 $2.57 = 2.57(100\%) = 257\%$

Example 12 $\frac{5}{9} = \frac{5}{9}(100\%) = \frac{500}{9}\% \approx 55.6\%$

Example 14 $\frac{9}{16} = \frac{9}{16}(100\%) = \frac{900}{16}\% = 56\frac{1}{4}\%$

SECTION 4.2   *pages 121–122*

Example 2

$P \times B = A$
$P(60) = 27$
$\frac{P(60)}{60} = \frac{27}{60}$
$P = 0.45$

The percent is 45%.

Example 4

Strategy

To find the percent of defective wheel bearings, solve the basic percent equation using $A = 6$ and $B = 200$. The percent is unknown.

Solution

$$P \times B = A$$
$$P(200) = 6$$
$$\frac{P(200)}{200} = \frac{6}{200}$$
$$P = 0.03$$

3% of the wheel bearings were defective.

Example 6

Strategy

To find the percent decrease, solve the basic percent equation using $B = 2500$ and $A = 500$. The percent is unknown.

Solution

$$P \times B = A$$
$$P(2500) = 500$$
$$\frac{P(2500)}{2500} = \frac{500}{2500}$$
$$P = 0.2$$

The percent decrease is 20%.

SECTION 4.3   *pages 125–126*

Example 2

Strategy

Given: $C = \$40$
         $S = \$60$
Unknown: r
Use the equation $S = C + rC$.

Solution

$$S = C + rC$$
$$60 = 40 + 40r$$
$$20 = 40r$$
$$0.5 = r$$

The markup rate is 50%.

Example 4

Strategy

Given: $R = \$27.60$
         $S = \$20.70$
Unknown: r
Use the equation $S = R - rR$.

Solution

$$S = R - rR$$
$$20.70 = 27.60 - 27.60r$$
$$-6.90 = -27.60r$$
$$0.25 = r$$

The discount rate is 25%.

SECTION 4.4   *pages 129–130*

Example 2

Strategy

■ Additional amount: x

Principal	Rate	Interest
5000	0.08	0.08(5000)
x	0.14	0.14x
5000 + x	0.11	0.11(5000 + x)

■ The sum of the interest earned by the two investments equals the interest earned on the total investment.

Solution

$$0.08(5000) + 0.14x = 0.11(5000 + x)$$
$$400 + 0.14x = 550 + 0.11x$$
$$400 + 0.03x = 550$$
$$0.03x = 150$$
$$x = 5000$$

$5000 more must be invested at 14%.

SECTION 4.5   *pages 133–136*

Example 2

Strategy

- Pounds of $.55 fertilizer: x

	Amount	Cost	Value
$.80 fertilizer	20	$.80	0.80(20)
$.55 fertilizer	x	$.55	0.55x
$.75 fertilizer	20 + x	$.75	0.75(20 + x)

- The sum of the values before mixing equals the value after mixing.

Solution

$0.80(20) + 0.55x = 0.75(20 + x)$
$16 + 0.55x = 15 + 0.75x$
$16 - 0.20x = 15$
$-0.20x = -1$
$x = 5$

5 lb of the $.55 fertilizer must be added.

Example 4

Strategy

- Liters of water: x

	Amount	Percent	Quantity
Water	x	0	0x
12%	5	0.12	5(0.12)
8%	$x + 5$	0.08	0.08($x + 5$)

- The sum of the quantities before mixing is equal to the quantity after mixing.

Solution

$0x + 5(0.12) = 0.08(x + 5)$
$0.60 = 0.08x + 0.40$
$0.20 = 0.08x$
$2.5 = x$

The pharmacist adds 2.5 L of water to the 12% solution to get an 8% solution.

SECTION 4.6   *pages 141–142*

Example 2

Strategy

- Rate of the first train: r
  Rate of the second train: $2r$

	Rate	Time	Distance
1st train	r	3	3r
2nd train	$2r$	3	3(2r)

- The sum of the distances traveled by each train equals 288 mi.

Solution

$3r + 3(2r) = 288$
$3r + 6r = 288$
$9r = 288$
$r = 32$

$2r = 2(32) = 64$

The first train is traveling at 32 mph.
The second train is traveling at 64 mph.

Example 4

Strategy

- Time spent flying out: t
  Time spent flying back: $5 - t$

	Rate	Time	Distance
Out	150	t	150t
Back	100	$5 - t$	100(5 − t)

- The distance out equals the distance back.

Solution

$150t = 100(5 - t)$
$150t = 500 - 100t$
$250t = 500$
$t = 2$ (The time out was 2 h.)

The distance = $150t = 150(2) = 300$ mi.

The parcel of land was 300 mi away.

SECTION 4.7   *pages 145–146*

Example 2

Strategy

- Width of the rectangle: w
  Length of the rectangle: $w + 3$
- Use the equation for the perimeter of a rectangle.

Solution

$$2l + 2w = P$$
$$2(w + 3) + 2w = 34$$
$$2w + 6 + 2w = 34$$
$$4w + 6 = 34$$
$$4w = 28$$
$$w = 7$$

The width of the rectangle is 7 m.

Example 4

Strategy

- Measure of the first angle: $2x$
  Measure of the second angle: x
  Measure of the third angle: $x - 4$
- Use the equation $A + B + C = 180°$.

Solution

$$A + B + C = 180$$
$$2x + x + (x - 4) = 180$$
$$4x - 4 = 180$$
$$4x = 184$$
$$x = 46$$

$$2x = 2(46) = 92$$

$$x - 4 = 46 - 4 = 42$$

The measure of the first angle is 92°.
The measure of the second angle is 46°.
The measure of the third angle is 42°.

SECTION 4.8   *pages 149–152*

Example 2

Strategy

- First consecutive integer: n
  Second consecutive integer: $n + 1$
  Third consecutive integer: $n + 2$
- The sum of the three integers is -6.

Solution

$$n + (n + 1) + (n + 2) = -6$$
$$3n + 3 = -6$$
$$3n = -9$$
$$n = -3$$

$$n + 1 = -3 + 1 = -2$$

$$n + 2 = -3 + 2 = -1$$

The three consecutive integers are -3, -2, and -1.

Example 4

Strategy

- Number of dimes: x
  Number of nickels: $4x$
  Number of quarters: $x + 5$

Coin	Number	Value	Total Value
Dime	x	10	$10x$
Nickel	$4x$	5	$5(4x)$
Quarter	$x + 5$	25	$25(x + 5)$

- The sum of the total values of each denomination of coin equals the total value of all the coins (675 cents).

Solution

$$10x + 5(4x) + 25(x + 5) = 675$$
$$10x + 20x + 25x + 125 = 675$$
$$55x + 125 = 675$$
$$55x = 550$$
$$x = 10$$

$$4x = 4(10) = 40$$

$$x + 5 = 10 + 5 = 15$$

The bank contains 10 dimes, 40 nickels, and 15 quarters.

Example 6

Strategy

▪ The number of years ago: x

	Present age	*Past age*
Half dollar	25	$25 - x$
Dime	15	$15 - x$

▪ At a past age, the half dollar was twice as old as the dime.

Solution

$25 - x = 2(15 - x)$
$25 - x = 30 - 2x$
$25 + x = 30$
$\qquad x = 5$

Five years ago the half dollar was twice as old as the dime.

SOLUTIONS to Chapter 5 Examples

SECTION 5.1   *pages 167–168*

Example 2

$2x^2 + 4x - 3$
$\underline{5x^2 - 6x \qquad}$
$7x^2 - 2x - 3$

Example 6

$8y^2 - 4xy + \ x^2$
$\underline{-(2y^2 - \ xy + 5x^2)}$

$8y^2 - 4xy + \ x^2$
$\underline{-2y^2 + \ xy - 5x^2}$
$6y^2 - 3xy - 4x^2$

Example 4

$(-3x^2 + 2y^2) + (-8x^2 + 9xy)$
$-11x^2 + 9xy + 2y^2$

Example 8

$(-3a^2 - 4a + 2) - (5a^3 + 2a - 6)$
$(-3a^2 - 4a + 2) + (-5a^3 - 2a + 6)$
$-5a^3 - 3a^2 - 6a + 8$

SECTION 5.2   *pages 171–172*

Example 2

$(3x^2)(6x^3) = (3 \cdot 6)(x^2 \cdot x^3) = 18x^5$

Example 6

$(3x)(2x^2y)^3 = (3x)(2^3x^6y^3)$
$\qquad = (3x)(8x^6y^3)$
$\qquad = (3 \cdot 8)(x \cdot x^6)y^3 = 24x^7y^3$

Example 4

$(-3xy^2)(-4x^2y^3)$
$\qquad = [(-3)(-4)](x \cdot x^2)(y^2 \cdot y^3)$
$\qquad = 12x^3y^5$

Example 8

$(3x^2)^2(-2xy^3)^3 = (3^2x^4)[(-2)^3x^3y^6]$
$\qquad = (9x^4)(-8x^3y^6) = [9(-8)](x^4 \cdot x^3)y^6$
$\qquad = -72x^7y^6$

SECTION 5.3   *pages 175–178*

Example 2

$(-2y + 3)(-4y) = 8y^2 - 12y$

Example 4

$-a^2(3a^2 + 2a - 7)$
$\qquad = -3a^4 - 2a^3 + 7a^2$

Example 6

$$2y^3 + 2y^2 \quad\; - 3$$
$$\underline{\qquad\qquad 3y \;\; - 1 \qquad}$$
$$\underline{-2y^3 - 2y^2 \qquad + 3 \qquad}$$
$$\underline{6y^4 + 6y^3 \qquad - 9y \qquad}$$
$$6y^4 + 4y^3 - 2y^2 - 9y + 3$$

Example 8

$$(4y - 5)(2y - 3)$$
$$= 8y^2 - 12y - 10y + 15$$
$$= 8y^2 - 22y + 15$$

Example 10

$$(3b + 2)(3b - 5)$$
$$= 9b^2 - 15b + 6b - 10$$
$$= 9b^2 - 9b - 10$$

Example 12 $(2a + 5c)(2a - 5c) = 4a^2 - 25c^2$

Example 14 $(3x + 2y)^2 = 9x^2 + 12xy + 4y^2$

Example 16

Strategy   To find the area, replace the variable r in the equation $A = \pi r^2$ by the given value and solve for A.

Solution $A = \pi r^2$
$A = 3.14(x - 4)^2$
$A = 3.14(x^2 - 8x + 16)$
$A = 3.14x^2 - 25.12x + 50.24$

The area is $3.14x^2 - 25.12x + 50.24$.

SECTION 5.4   *pages 183–188*

Example 2 $\dfrac{2^2}{2^{-3}} = 2^{2-(-3)} = 2^5 = 32$

Example 4 $(-2x^2)(x^{-3}y^{-4})^{-2} = (-2x^2)(x^6 y^8)$
$= -2x^8 y^8$

Example 6 $\dfrac{18y^3}{-27y^7} = \dfrac{9 \cdot 2y^{3-7}}{-9 \cdot 3} = -\dfrac{2y^{-4}}{3} = -\dfrac{2}{3y^4}$

Example 8 $\dfrac{12x^{-8}y^4}{-16xz^{-2}} = \dfrac{4 \cdot 3x^{-8-1}y^4 z^2}{-4 \cdot 4} = -\dfrac{3y^4 z^2}{4x^9}$

Example 10 $\dfrac{(6a^{-2}b^3)^{-1}}{(4a^3 b^{-2})^{-2}} = \dfrac{6^{-1}a^2 b^{-3}}{4^{-2}a^{-6}b^4}$
$= 4^2 \cdot 6^{-1}a^8 b^{-7}$
$= \dfrac{16a^8}{6b^7} = \dfrac{8a^8}{3b^7}$

Example 12 $\left[\dfrac{6r^3 s^{-3}}{9r^3 s^{-1}}\right]^{-2} = \left[\dfrac{2r^0 s^{-2}}{3}\right]^{-2}$
$= \dfrac{2^{-2}s^4}{3^{-2}} = \dfrac{9s^4}{4}$

Example 14 $\dfrac{4x^3 y + 8x^2 y^2 - 4xy^3}{2xy}$
$= \dfrac{4x^3 y}{2xy} + \dfrac{8x^2 y^2}{2xy} - \dfrac{4xy^3}{2xy}$
$= 2x^2 + 4xy - 2y^2$

Example 16 $\dfrac{24x^2 y^2 - 18xy + 6y}{6xy}$
$= \dfrac{24x^2 y^2}{6xy} - \dfrac{18xy}{6xy} + \dfrac{6y}{6xy}$
$= 4xy - 3 + \dfrac{1}{x}$

Example 18

$$\begin{array}{r}
x^2 + 2x \;\; - 1 \\
2x - 3 \overline{\smash{)}\, 2x^3 + \;\; x^2 - 8x - 3} \\
\underline{2x^3 - 3x^2 } \\
4x^2 - 8x \\
\underline{4x^2 - 6x } \\
-2x - 3 \\
\underline{-2x + 3} \\
-6
\end{array}$$

$(2x^3 + x^2 - 8x - 3) \div (2x - 3)$
$= x^2 + 2x - 1 - \dfrac{6}{2x - 3}$

SOLUTIONS to Chapter 6 Examples

SECTION 6.1 *pages 203–206*

Example 2

The GCF is $7a^2$.

$$14a^2 - 21a^4b = 7a^2(2) + 7a^2(-3a^2b)$$
$$= 7a^2(2 - 3a^2b)$$

Example 4

The GCF is 9.

$$27b^2 + 18b + 9$$
$$= 9(3b^2) + 9(2b) + 9(1)$$
$$= 9(3b^2 + 2b + 1)$$

Example 6

The GCF is $3x^2y^2$.

$$6x^4y^2 - 9x^3y^2 + 12x^2y^4$$
$$= 3x^2y^2(2x^2) + 3x^2y^2(-3x) + 3x^2y^2(4y^2)$$
$$= 3x^2y^2(2x^2 - 3x + 4y^2)$$

Example 8

$$2y(5x - 2) - 3(2 - 5x) = 2y(5x - 2) + 3(5x - 2)$$
$$= (5x - 2)(2y + 3)$$

Example 10

$$a^2 - 3a + 2ab - 6b = (a^2 - 3a) + (2ab - 6b)$$
$$= a(a - 3) + 2b(a - 3)$$
$$= (a - 3)(a + 2b)$$

Example 12

$$2mn^2 - n + 8mn - 4 = (2mn^2 - n) + (8mn - 4)$$
$$= n(2mn - 1) + 4(2mn - 1)$$
$$= (2mn - 1)(n + 4)$$

SECTION 6.2 *pages 209–212*

Example 2

Find the positive factors of 20 whose sum is 9.

Factors	Sums
1, 20	21
2, 10	12
4, 5	9

$$x^2 + 9x + 20 = (x + 4)(x + 5)$$

Example 4

Find the factors of -18 whose sum is 7.

Factors	Sums
+1, −18	−17
−1, +18	17
+2, −9	−7
−2, +9	7
+3, −6	−3
−3, +6	3

$$x^2 + 7x - 18 = (x + 9)(x - 2)$$

Example 6

The GCF is $3b$.

$$3a^2b - 18ab - 81b = 3b(a^2 - 6a - 27)$$

Factor the trinomial.

Find the factors of -27 whose sum is -6.

Factors	Sums
+1, −27	−26
−1, +27	26
+3, −9	−6
−3, +9	6

$$3a^2b - 18ab - 81b = 3b(a + 3)(a - 9)$$

Example 8

The GCF is 3.

$$3x^2 - 9xy - 12y^2 = 3(x^2 - 3xy - 4y^2)$$

Factor the trinomial.

Find the factors of -4 whose sum is -3.

Factors	Sums
+1, −4	−3
−1, +4	3
+2, −2	0

$$3x^2 - 9xy - 12y^2 = 3(x + y)(x - 4y)$$

SECTION 6.3   *pages 217–220*

Example 2

Factors of 2: 1, 2   Factors of −3: +1, −3
                                      −1, +3

Trial Factors	Middle Term
$(1x + 1)(2x - 3)$	$-3x + 2x = -x$
$(1x - 3)(2x + 1)$	$x - 6x = -5x$
$(1x - 1)(2x + 3)$	$3x - 2x = x$
$(1x + 3)(2x - 1)$	$-x + 6x = 5x$

$2x^2 - x - 3 = (x + 1)(2x - 3)$

Example 4

The GCF is $3y$.

$12y + 12y^2 - 45y^3 = 3y(4 + 4y - 15y^2)$

Factor the trinomial.

Factors of 4: 1, 4   Factors of −15: +1, −15
                      2, 2   −1, +15
                                               +3, −5
                                               −3, +5

Trial Factors	Middle Term
$(1 + 1y)(4 - 15y)$	$-15y + 4y = -11y$
$(1 - 15y)(4 + 1y)$	$y - 60y = -59y$
$(1 - 1y)(4 + 15y)$	$15y - 4y = 11y$
$(1 + 15y)(4 - 1y)$	$-y + 60y = 59y$
$(1 + 3y)(4 - 5y)$	$-5y + 12y = 7y$
$(1 - 5y)(4 + 3y)$	$3y - 20y = -17y$
$(1 - 3y)(4 + 5y)$	$5y - 12y = -7y$
$(1 + 5y)(4 - 3y)$	$-3y + 20y = 17y$
$(2 + 1y)(2 - 15y)$	$-30y + 2y = -28y$
$(2 - 1y)(2 + 15y)$	$30y - 2y = 28y$
$(2 + 3y)(2 - 5y)$	$-10y + 6y = -4y$
$(2 - 3y)(2 + 5y)$	$10y - 6y = 4y$

$12y + 12y^2 - 45y^3 = 3y(2 - 3y)(2 + 5y)$

Example 6

Factors of −14 [2(−7)]	Sum
−1, +14	13

$$2a^2 + 13a - 7 = 2a^2 - a + 14a - 7$$
$$= (2a^2 - a) + (14a - 7)$$
$$= a(2a - 1) + 7(2a - 1)$$
$$= (2a - 1)(a + 7)$$

Example 8

The GCF is $5x$.

$15x^3 + 40x^2 - 80x = 5x(3x^2 + 8x - 16)$

Factors of −48 [3(−16)]	Sum
−1, +48	47
+1, −48	−47
−2, +24	22
+2, −24	−22
−3, +16	13
+3, −16	−13
−4, +12	8

$$3x^2 + 8x - 16 = 3x^2 - 4x + 12x - 16$$
$$= (3x^2 - 4x) + (12x - 16)$$
$$= x(3x - 4) + 4(3x - 4)$$
$$= (3x - 4)(x + 4)$$

$$15x^3 + 40x^2 - 80x = 5x(3x^2 + 8x - 16)$$
$$= 5x(3x - 4)(x + 4)$$

SECTION 6.4   *pages 225–228*

Example 2

$25a^2 - b^2 = (5a)^2 - b^2 = (5a + b)(5a - b)$

Example 4

$n^4 - 81 = (n^2)^2 - 9^2 = (n^2 + 9)(n^2 - 9)$
$$= (n^2 + 9)(n + 3)(n - 3)$$

Example 6

Because $16y^2 = (4y)^2$, $1 = 1^2$, and $8y = 2(4y)(1)$ the trinomial is a perfect square trinomial.

$16y^2 + 8y + 1 = (4y + 1)^2$

Example 8

Because $x^2 = (x)^2$, $36 = 6^2$, and $15x \neq 2(x)(6)$, the trinomial is not a perfect square trinomial. Try to factor the trinomial by another method.

$x^2 + 15x + 36 = (x + 3)(x + 12)$

Example 10

$(x^2 - 6x + 9) - y^2 = (x - 3)^2 - y^2$
$$= (x - 3 - y)(x - 3 + y)$$

Example 12

The GCF is $3x$.

$12x^3 - 75x = 3x(4x^2 - 25)$
$$= 3x(2x + 5)(2x - 5)$$

Example 14

The common binomial factor is $b - 7$.

$a^2(b - 7) + (7 - b) = a^2(b - 7) - (b - 7)$
$$= (b - 7)(a^2 - 1)$$
$$= (b - 7)(a + 1)(a - 1)$$

Example 16

The GCF is $4x$.

$4x^3 + 28x^2 - 120x = 4x(x^2 + 7x - 30)$
$$= 4x(x + 10)(x - 3)$$

SECTION 6.5   *pages 233–236*

Example 2

$2x(x + 7) = 0$

$2x = 0 \qquad x + 7 = 0$
$\ x = 0 \qquad\quad x = -7$

The solutions are 0 and -7.

Example 4

$4x^2 - 9 = 0$
$(2x - 3)(2x + 3) = 0$

$2x - 3 = 0 \qquad 2x + 3 = 0$
$\quad 2x = 3 \qquad\quad 2x = -3$
$\quad\ x = \dfrac{3}{2} \qquad\quad x = -\dfrac{3}{2}$

The solutions are $\dfrac{3}{2}$ and $-\dfrac{3}{2}$.

Example 6

$(x + 2)(x - 7) = 52$
$x^2 - 5x - 14 = 52$
$x^2 - 5x - 66 = 0$
$(x + 6)(x - 11) = 0$

$x + 6 = 0 \qquad x - 11 = 0$
$\quad x = -6 \qquad\quad x = 11$

The solutions are -6 and 11.

Example 8

Strategy

First positive consecutive integer: n
Second positive consecutive integer: $n + 1$

The sum of the squares of two positive consecutive integers is 61.

Solution

$n^2 + (n + 1)^2 = 61$
$n^2 + n^2 + 2n + 1 = 61$
$2n^2 + 2n + 1 = 61$
$2n^2 + 2n - 60 = 0$
$2(n^2 + n - 30) = 0$
$2(n - 5)(n + 6) = 0$

$n - 5 = 0 \qquad n + 6 = 0$
$\quad n = 5 \qquad\qquad n = -6$

Since -6 is not a positive integer, it is not a solution.

$n = 5$
$n + 1 = 5 + 1 = 6$

The two integers are 5 and 6.

Example 10

Strategy

Width $= x$
Length $= 2x + 4$

The area of a rectangle is 96 in.^{2}.
Use the equation $A = l \cdot w$.

Solution

$A = l \cdot w$
$96 = (2x + 4)x$
$96 = 2x^2 + 4x$
$0 = 2x^2 + 4x - 96$
$0 = 2(x^2 + 2x - 48)$
$0 = 2(x + 8)(x - 6)$

$x + 8 = 0 \qquad x - 6 = 0$
$\quad x = -8 \qquad\quad x = 6$

Since the width cannot be a negative number, -8 is not a solution.

$x = 6$
$2x + 4 = 2(6) + 4 = 12 + 4 = 16$

The width is 6 in.
The length is 16 in.

SOLUTIONS to Chapter 7 Examples

SECTION 7.1   *pages 251–254*

Example 2

$$\frac{6x^5y}{12x^2y^3} = \frac{\overset{1}{\cancel{2}} \cdot \overset{1}{\cancel{3}} \cdot x^5y}{\underset{1}{\cancel{2}} \cdot 2 \cdot \underset{1}{\cancel{3}} \cdot x^2y^3} = \frac{x^3}{2y^2}$$

Example 4

$$\frac{x^2 + 2x - 24}{16 - x^2} = \frac{\overset{-1}{\cancel{(x - 4)}}(x + 6)}{\underset{1}{\cancel{(4 - x)}}(4 + x)} = -\frac{x + 6}{x + 4}$$

Example 6

$$\frac{x^2 + 4x - 12}{x^2 - 3x + 2} = \frac{(x - 2)(x + 6)}{(x - 1)(x - 2)} = \frac{x + 6}{x - 1}$$

Example 8

$$\frac{12x^2 + 3x}{10x - 15} \cdot \frac{8x - 12}{9x + 18} = \frac{3x(4x + 1)}{5(2x - 3)} \cdot \frac{4(2x - 3)}{9(x + 2)}$$

$$= \frac{\overset{1}{\cancel{3}}x(4x + 1) \cdot 2 \cdot 2\overset{1}{\cancel{(2x - 3)}}}{5\underset{1}{\cancel{(2x - 3)}} \cdot \underset{1}{\cancel{3}} \cdot 3(x + 2)}$$

$$= \frac{4x(4x + 1)}{15(x + 2)}$$

Example 10

$$\frac{x^2 + 2x - 15}{9 - x^2} \cdot \frac{x^2 - 3x - 18}{x^2 - 7x + 6}$$

$$= \frac{(x - 3)(x + 5)}{(3 - x)(3 + x)} \cdot \frac{(x + 3)(x - 6)}{(x - 1)(x - 6)}$$

$$= \frac{\overset{-1}{\cancel{(x - 3)}}(x + 5) \cdot \overset{1}{\cancel{(x + 3)}}\overset{1}{\cancel{(x - 6)}}}{\underset{1}{\cancel{(3 - x)}}\underset{1}{\cancel{(3 + x)}} \cdot (x - 1)\underset{1}{\cancel{(x - 6)}}} = -\frac{x + 5}{x - 1}$$

Example 12

$$\frac{a^2}{4bc^2 - 2b^2c} \div \frac{a}{6bc - 3b^2} = \frac{a^2}{4bc^2 - 2b^2c} \cdot \frac{6bc - 3b^2}{a}$$

$$= \frac{a^2 \cdot 3\overset{1}{\cancel{b}}\overset{1}{\cancel{(2c - b)}}}{2\underset{1}{\cancel{b}}c\underset{1}{\cancel{(2c - b)}} \cdot a} = \frac{3a}{2c}$$

Example 14

$$\frac{3x^2 + 26x + 16}{3x^2 - 7x - 6} \div \frac{2x^2 + 9x - 5}{x^2 + 2x - 15}$$

$$= \frac{3x^2 + 26x + 16}{3x^2 - 7x - 6} \cdot \frac{x^2 + 2x - 15}{2x^2 + 9x - 5}$$

$$= \frac{\overset{1}{\cancel{(3x + 2)}}(x + 8)}{\underset{1}{\cancel{(3x + 2)}}\overset{1}{\cancel{(x - 3)}}} \cdot \frac{\overset{1}{\cancel{(x + 5)}}\overset{1}{\cancel{(x - 3)}}}{(2x - 1)\underset{1}{\cancel{(x + 5)}}} = \frac{x + 8}{2x - 1}$$

SECTION 7.2   *pages 259–260*

Example 2

$8uv^2 = 2 \cdot 2 \cdot 2 \cdot u \cdot v \cdot v$
$12uw = 2 \cdot 2 \cdot 3 \cdot u \cdot w$

$LCM = 2 \cdot 2 \cdot 2 \cdot 3 \cdot u \cdot v \cdot v \cdot w = 24uv^2w$

Example 4

$m^2 - 6m + 9 = (m - 3)(m - 3)$
$m^2 - 2m - 3 = (m + 1)(m - 3)$

$LCM = (m - 3)(m - 3)(m + 1)$

Example 6

The LCM is $36xy^2z$.

$$\frac{x - 3}{4xy^2} = \frac{x - 3}{4xy^2} \cdot \frac{9z}{9z} = \frac{9xz - 27z}{36xy^2z}$$

$$\frac{2x + 1}{9y^2z} = \frac{2x + 1}{9y^2z} \cdot \frac{4x}{4x} = \frac{8x^2 + 4x}{36xy^2z}$$

Example 8

The LCM is $(x + 2)(x - 5)(x + 5)$.

$$\frac{x + 4}{x^2 - 3x - 10} = \frac{x + 4}{(x + 2)(x - 5)} \cdot \frac{x + 5}{x + 5}$$
$$= \frac{x^2 + 9x + 20}{(x + 2)(x - 5)(x + 5)}$$

$$\frac{2x}{25 - x^2} = \frac{2x}{-(x^2 - 25)} = -\frac{2x}{(x - 5)(x + 5)} \cdot \frac{x + 2}{x + 2}$$
$$= -\frac{2x^2 + 4x}{(x + 2)(x - 5)(x + 5)}$$

SECTION 7.3   *pages 263–266*

Example 2

$$\frac{3}{xy} + \frac{12}{xy} = \frac{3 + 12}{xy} = \frac{15}{xy}$$

Example 4

$$\frac{2x^2}{x^2 - x - 12} - \frac{7x + 4}{x^2 - x - 12}$$
$$= \frac{2x^2 - (7x + 4)}{x^2 - x - 12} = \frac{2x^2 - 7x - 4}{x^2 - x - 12}$$
$$= \frac{(2x + 1)(x - 4)}{(x + 3)(x - 4)} = \frac{2x + 1}{x + 3}$$

Example 6

$$\frac{x^2 - 1}{x^2 - 8x + 12} - \frac{2x + 1}{x^2 - 8x + 12} + \frac{x}{x^2 - 8x + 12}$$
$$= \frac{(x^2 - 1) - (2x + 1) + x}{x^2 - 8x + 12} = \frac{x^2 - 1 - 2x - 1 + x}{x^2 - 8x + 12}$$
$$= \frac{x^2 - x - 2}{x^2 - 8x + 12} = \frac{(x + 1)(x - 2)}{(x - 2)(x - 6)} = \frac{x + 1}{x - 6}$$

Example 8

The LCM of the denominators is $24y$.

$$\frac{z}{8y} - \frac{4z}{3y} + \frac{5z}{4y} = \frac{z}{8y} \cdot \frac{3}{3} - \frac{4z}{3y} \cdot \frac{8}{8} + \frac{5z}{4y} \cdot \frac{6}{6}$$
$$= \frac{3z}{24y} - \frac{32z}{24y} + \frac{30z}{24y}$$
$$= \frac{3z - 32z + 30z}{24y} = \frac{z}{24y}$$

Example 10

$2 - x = -(x - 2)$

Therefore, $\dfrac{3}{2 - x} = \dfrac{-3}{x - 2}$.

The LCM is $x - 2$.

$$\frac{5x}{x - 2} - \frac{3}{2 - x} = \frac{5x}{x - 2} - \frac{-3}{x - 2}$$
$$= \frac{5x - (-3)}{x - 2} = \frac{5x + 3}{x - 2}$$

Example 12

The LCM is $(3x - 1)(x + 4)$.

$$\frac{4x}{3x - 1} - \frac{9}{x + 4} = \frac{4x}{3x - 1} \cdot \frac{x + 4}{x + 4} - \frac{9}{x + 4} \cdot \frac{3x - 1}{3x - 1}$$
$$= \frac{4x^2 + 16x}{(3x - 1)(x + 4)} - \frac{27x - 9}{(3x - 1)(x + 4)}$$
$$= \frac{(4x^2 + 16x) - (27x - 9)}{(3x - 1)(x + 4)}$$
$$= \frac{4x^2 + 16x - 27x + 9}{(3x - 1)(x + 4)}$$
$$= \frac{4x^2 - 11x + 9}{(3x - 1)(x + 4)}$$

Example 14

The LCM is $(x + 5)(x - 5)$.

$$\frac{2}{5 - x} = \frac{-2}{x - 5}$$

$$\frac{2x - 1}{x^2 - 25} + \frac{2}{5 - x} = \frac{2x - 1}{(x + 5)(x - 5)} + \frac{-2}{x - 5}$$

$$= \frac{2x - 1}{(x + 5)(x - 5)} + \frac{-2}{x - 5} \cdot \frac{x + 5}{x + 5}$$

$$= \frac{2x - 1}{(x + 5)(x - 5)} + \frac{-2(x + 5)}{(x + 5)(x - 5)}$$

$$= \frac{2x - 1 + (-2)(x + 5)}{(x + 5)(x - 5)}$$

$$= \frac{2x - 1 - 2x - 10}{(x + 5)(x - 5)}$$

$$= \frac{-11}{(x + 5)(x - 5)}$$

$$= -\frac{11}{(x + 5)(x - 5)}$$

Example 16

The LCM is $(3x + 2)(x - 1)$.

$$\frac{2x - 3}{3x^2 - x - 2} + \frac{5}{3x + 2} - \frac{1}{x - 1}$$

$$= \frac{2x - 3}{(3x + 2)(x - 1)} + \frac{5}{3x + 2} \cdot \frac{x - 1}{x - 1} - \frac{1}{x - 1} \cdot \frac{3x + 2}{3x + 2}$$

$$= \frac{2x - 3}{(3x + 2)(x - 1)} + \frac{5x - 5}{(3x + 2)(x - 1)} - \frac{3x + 2}{(3x + 2)(x - 1)}$$

$$= \frac{(2x - 3) + (5x - 5) - (3x + 2)}{(3x + 2)(x - 1)}$$

$$= \frac{2x - 3 + 5x - 5 - 3x - 2}{(3x + 2)(x - 1)}$$

$$= \frac{4x - 10}{(3x + 2)(x - 1)} = \frac{2(2x - 5)}{(3x + 2)(x - 1)}$$

SECTION 7.4   *pages 271–272*

Example 2

The LCM of 3, x, 9, and x^2 is $9x^2$.

$$\frac{\frac{1}{3} - \frac{1}{x}}{\frac{1}{9} - \frac{1}{x^2}} = \frac{\frac{1}{3} - \frac{1}{x}}{\frac{1}{9} - \frac{1}{x^2}} \cdot \frac{9x^2}{9x^2} = \frac{\frac{1}{3} \cdot 9x^2 - \frac{1}{x} \cdot 9x^2}{\frac{1}{9} \cdot 9x^2 - \frac{1}{x^2} \cdot 9x^2} =$$

$$\frac{3x^2 - 9x}{x^2 - 9} = \frac{3x(x - 3)}{(x - 3)(x + 3)} = \frac{3x}{x + 3}$$

Example 4

The LCM of x and x^2 is x^2.

$$\frac{1 + \frac{4}{x} + \frac{3}{x^2}}{1 + \frac{10}{x} + \frac{21}{x^2}} = \frac{1 + \frac{4}{x} + \frac{3}{x^2}}{1 + \frac{10}{x} + \frac{21}{x^2}} \cdot \frac{x^2}{x^2}$$

$$= \frac{1 \cdot x^2 + \frac{4}{x} \cdot x^2 + \frac{3}{x^2} \cdot x^2}{1 \cdot x^2 + \frac{10}{x} \cdot x^2 + \frac{21}{x^2} \cdot x^2}$$

$$= \frac{x^2 + 4x + 3}{x^2 + 10x + 21} = \frac{(x + 1)(x + 3)}{(x + 3)(x + 7)} = \frac{x + 1}{x + 7}$$

Example 6

The LCM is $x - 5$.

$$\frac{x + 3 - \frac{20}{x - 5}}{x + 8 + \frac{30}{x - 5}} = \frac{x + 3 - \frac{20}{x - 5}}{x + 8 + \frac{30}{x - 5}} \cdot \frac{x - 5}{x - 5}$$

$$= \frac{(x + 3)(x - 5) - \frac{20}{x - 5} \cdot (x - 5)}{(x + 8)(x - 5) + \frac{30}{x - 5} \cdot (x - 5)}$$

$$= \frac{x^2 - 2x - 15 - 20}{x^2 + 3x - 40 + 30} = \frac{x^2 - 2x - 35}{x^2 + 3x - 10}$$

$$= \frac{(x + 5)(x - 7)}{(x - 2)(x + 5)} = \frac{x - 7}{x - 2}$$

SECTION 7.5   *pages 275–276*

Example 2

$$\frac{x}{x+6} = \frac{3}{x} \quad \text{The LCM is } x(x+6).$$

$$\frac{x(x+6)}{1} \cdot \frac{x}{x+6} = \frac{x(x+6)}{1} \cdot \frac{3}{x}$$

$$x^2 = (x+6)3$$
$$x^2 = 3x + 18$$
$$x^2 - 3x - 18 = 0$$
$$(x+3)(x-6) = 0$$

$$x + 3 = 0 \qquad x - 6 = 0$$
$$x = -3 \qquad x = 6$$

Both -3 and 6 check as solutions.
The solutions are -3 and 6.

Example 4

$$\frac{5x}{x+2} = 3 - \frac{10}{x+2} \quad \text{The LCM is } x+2.$$

$$\frac{(x+2)}{1} \cdot \frac{5x}{x+2} = \frac{(x+2)}{1}\left(3 - \frac{10}{x+2}\right)$$

$$\frac{x+2}{1} \cdot \frac{5x}{x+2} = \frac{x+2}{1} \cdot 3 - \frac{x+2}{1} \cdot \frac{10}{x+2}$$

$$5x = (x+2)3 - 10$$
$$5x = 3x + 6 - 10$$
$$5x = 3x - 4$$
$$2x = -4$$
$$x = -2$$

-2 does not check as a solution.
The equation has no solution.

SECTION 7.6   *pages 279–280*

Example 2

$$\frac{2}{x+3} = \frac{6}{5x+5}$$

$$\frac{(x+3)(5x+5)}{1} \cdot \frac{2}{x+3} = \frac{(x+3)(5x+5)}{1} \cdot \frac{6}{5x+5}$$

$$\frac{(x+3)(5x+5)}{1} \cdot \frac{2}{x+3} = \frac{(x+3)(5x+5)}{1} \cdot \frac{6}{5x+5}$$

$$(5x+5)2 = (x+3)6$$
$$10x + 10 = 6x + 18$$
$$4x + 10 = 18$$
$$4x = 8$$
$$x = 2$$

The solution is 2.

Example 4

Strategy

To find the total area that 256 ceramic tiles will cover, write and solve a proportion using x to represent the number of square feet that 256 tiles will cover.

Solution

$$\frac{9}{16} = \frac{x}{256}$$

$$256\left(\frac{9}{16}\right) = 256\left(\frac{x}{256}\right)$$

$$144 = x$$

A 144-square-foot area can be tiled using 256 ceramic tiles.

Example 6

Strategy

To find the additional amount of medication required for a 200-pound adult, write and solve a proportion using x to represent the additional medication. Then $3 + x$ is the total amount required for a 200-pound adult.

Solution

$$\frac{150}{3} = \frac{200}{3 + x}$$

$$\frac{50}{1} = \frac{200}{3 + x}$$

$$(3 + x) \cdot 50 = (3 + x) \cdot \frac{200}{3 + x}$$

$$(3 + x) \cdot 50 = 200$$

$$150 + 50x = 200$$

$$50x = 50$$

$$x = 1$$

One additional ounce is required for a 200-pound adult.

SECTION 7.7   *pages 283–284*

Example 2

$$5x - 2y = 10$$
$$5x - 5x - 2y = -5x + 10$$
$$-2y = -5x + 10$$
$$\frac{-2y}{-2} = \frac{-5x + 10}{-2}$$
$$y = \frac{5}{2}x - 5$$

Example 6

$$S = a + (n - 1)d$$
$$S = a + nd - d$$
$$S - a = a - a + nd - d$$
$$S - a = nd - d$$
$$S - a + d = nd - d + d$$
$$S - a + d = nd$$
$$\frac{S - a + d}{d} = \frac{nd}{d}$$
$$\frac{S - a + d}{d} = n$$

Example 4

$$s = \frac{A + L}{2}$$

$$2 \cdot s = 2\left(\frac{A + L}{2}\right)$$

$$2s = A + L$$

$$2s - A = A - A + L$$

$$2s - A = L$$

Example 8

$$S = C + rC$$
$$S = (1 + r)C$$
$$\frac{S}{1 + r} = \frac{(1 + r)C}{1 + r}$$
$$\frac{S}{1 + r} = C$$

SECTION 7.8   *pages 287–290*

Example 2

Strategy

- Time for one printer to complete the job: t

	Rate	*Time*	*Part*
1st printer	$\dfrac{1}{t}$	2	$\dfrac{2}{t}$
2nd printer	$\dfrac{1}{t}$	5	$\dfrac{5}{t}$

- The sum of the parts of the task completed must equal 1.

Solution

$$\frac{2}{t} + \frac{5}{t} = 1$$

$$t\left(\frac{2}{t} + \frac{5}{t}\right) = t \cdot 1$$

$$2 + 5 = t$$

$$7 = t$$

Working alone, one printer takes 7 h to print the payroll.

Example 4

Strategy

- Rate sailing across the lake: r
  Rate sailing back: $3r$

	Distance	*Rate*	*Time*
Across	6	r	$\dfrac{6}{r}$
Back	6	$3r$	$\dfrac{6}{3r}$

- The total time for the trip was 2 h.

Solution

$$\frac{6}{r} + \frac{6}{3r} = 2$$

$$3r\left(\frac{6}{r} + \frac{6}{3r}\right) = 3r(2)$$

$$3r \cdot \frac{6}{r} + 3r \cdot \frac{6}{3r} = 6r$$

$$18 + 6 = 6r$$

$$24 = 6r$$

$$4 = r$$

The rate across the lake was 4 km/h.

SOLUTIONS to Chapter 8 Examples

SECTION 8.1   *pages 305–308*

Example 2

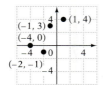

Example 6

a) Abscissa of point A: 2
   Abscissa of point C: -3
b) Ordinate of point B: -2
   Ordinate of point D: 0

Example 4

$A(4, 2)$
$B(-3, 4)$
$C(-3, 0)$
$D(0, 0)$

Example 8

$$y = -\frac{1}{2}x - 3$$

-4	$-\dfrac{1}{2}(2) - 3$
	$-1 - 3$
	-4

$$-4 = -4$$

Yes, $(2, -4)$ is a solution of
$y = -\frac{1}{2}x - 3$.

Example 10 $y = -\frac{1}{4}x + 1$

$\qquad = -\frac{1}{4}(4) + 1$

$\qquad = -1 + 1$

$\qquad = 0$

The ordered pair solution is (4, 0).

Example 12

Strategy   Graph the ordered pairs on a rectangular coordinate system. The horizontal axis represents the age of the car, and the vertical axis represents the price of the car.

Solution

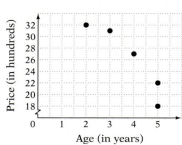

SECTION 8.2   *pages 313–316*

Example 2

Example 4

Example 6

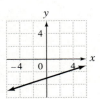

Example 8 $5x - 2y = 10$

$\qquad -2y = -5x + 10$

$\qquad y = \frac{5}{2}x - 5$

Example 10 $x - 3y = 9$

$\qquad -3y = -x + 9$

$\qquad y = \frac{1}{3}x - 3$

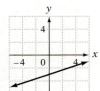

Example 12

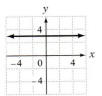

Example 14

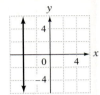

SECTION 8.3 *pages 321–326*

Example 2

x-intercept:	y-intercept:
$4x - y = 4$	$4x - y = 4$
$4x - 0 = 4$	$4(0) - y = 4$
$4x = 4$	$-y = 4$
$x = 1$	$y = -4$
$(1, 0)$	$(0, -4)$

Example 4

x-intercept:	y-intercept:
$y = 3x - 6$	$y = 3x - 6$
$0 = 3x - 6$	$b = -6$
$-3x = -6$	$(0, -6)$
$x = 2$	
$(2, 0)$	

Example 6 Let $P_1 = (-1, 2)$ and $P_2 = (1, 3)$.

$$m = \frac{y_2 - y_1}{x_2 - x_1} = \frac{3 - 2}{1 - (-1)} = \frac{1}{2}$$

The slope is $\frac{1}{2}$.

Example 8 Let $P_1 = (1, 2)$ and $P_2 = (4, -5)$.

$$m = \frac{y_2 - y_1}{x_2 - x_1} = \frac{-5 - 2}{4 - 1} = \frac{-7}{3}$$

The slope is $-\frac{7}{3}$.

Example 10 Let $P_1 = (2, 3)$ and $P_2 = (2, 7)$.

$$m = \frac{y_2 - y_1}{x_2 - x_1} = \frac{7 - 3}{2 - 2} = \frac{4}{0}$$

The slope of the line is undefined.

Example 12 Let $P_1 = (1, -3)$ and $P_2 = (-5, -3)$.

$$m = \frac{y_2 - y_1}{x_2 - x_1} = \frac{-3 - (-3)}{-5 - 1} = \frac{0}{-6} = 0$$

The line has zero slope.

Example 14 y-intercept $= (0, b) = (0, -1)$

$$m = -\frac{1}{4}$$

Example 16 y-intercept $= (0, b) = (0, 0)$

$$m = -\frac{3}{5}$$

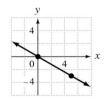

Example 18   Solve the equation for y.

$x - 2y = 4$

$-2y = -x + 4$

$y = \dfrac{1}{2}x - 2$

y-intercept $= (0, b) = (0, -2)$

$m = \dfrac{1}{2}$

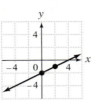

SECTION 8.4   *pages 331–332*

Example 2

$y = \dfrac{3}{2}x + b$

$-2 = \dfrac{3}{2}(4) + b$

$-2 = 6 + b$

$-8 = b$

$y = \dfrac{3}{2}x - 8$

Example 4 $m = \dfrac{3}{4}$ $(x_1, y_1) = (4, -2)$

$y - y_1 = m(x - x_1)$

$y - (-2) = \dfrac{3}{4}(x - 4)$

$y + 2 = \dfrac{3}{4}x - 3$

$y = \dfrac{3}{4}x - 5$

The equation of the line is
$y = \dfrac{3}{4}x - 5$.

SOLUTIONS to Chapter 9 Examples

SECTION 9.1   *pages 345–348*

Example 2

$2x - 5y = 8$		$-x + 3y = -5$	
$2(-1) - 5(-2)$	8	$-(-1) + 3(-2)$	-5
$-2 + 10$	8	$1 + (-6)$	-5
$8 = 8$		$-5 = -5$	

Yes, $(-1, -2)$ is a solution of the system of equations.

Example 4

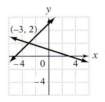

The solution is $(-3, 2)$.

Example 6

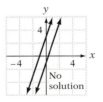

The lines are parallel. The system of equations is inconsistent and therefore does not have a solution.

SECTION 9.2 *pages 353–354*

Example 2

(1) $7x - y = 4$
(2) $3x + 2y = 9$

Solve equation (1) for y.

$$7x - y = 4$$
$$-y = -7x + 4$$
$$y = 7x - 4$$

Substitute in equation (2).

$$3x + 2y = 9$$
$$3x + 2(7x - 4) = 9$$
$$3x + 14x - 8 = 9$$
$$17x - 8 = 9$$
$$17x = 17$$
$$x = 1$$

Substitute in equation (1).

$$7x - y = 4$$
$$7(1) - y = 4$$
$$7 - y = 4$$
$$-y = -3$$
$$y = 3$$

The solution is $(1, 3)$.

Example 6

(1) $y = -2x + 1$
(2) $6x + 3y = 3$

$$6x + 3y = 3$$
$$6x + 3(-2x + 1) = 3$$
$$6x - 6x + 3 = 3$$
$$3 = 3$$

The system of equations is dependent. The solutions are the ordered pairs that satisfy the equation $y = -2x + 1$.

Example 4

(1) $3x - y = 4$
(2) $y = 3x + 2$

$$3x - y = 4$$
$$3x - (3x + 2) = 4$$
$$3x - 3x - 2 = 4$$
$$-2 = 4$$

This is not a true equation. The system of equations is inconsistent and therefore does not have a solution.

SECTION 9.3   *pages 357–360*

Example 2   (1) $x - 2y = 1$
(2) $2x + 4y = 0$

Eliminate y.
$$2(x - 2y) = 2 \cdot 1$$
$$2x + 4y = 0$$

$$2x - 4y = 2$$
$$2x + 4y = 0$$

Add the equations.
$$4x = 2$$
$$x = \frac{2}{4} = \frac{1}{2}$$

Replace x in equation (2).
$$2\left(\frac{1}{2}\right) + 4y = 0$$
$$1 + 4y = 0$$
$$4y = -1$$
$$y = -\frac{1}{4}$$

The solution is $\left(\frac{1}{2}, -\frac{1}{4}\right)$.

Example 6   (1) $4x + 5y = 11$
(2) $3y = x + 10$

Write equation (2) in the form
$Ax + By = C$.
$$3y = x + 10$$
$$-x + 3y = 10$$

Eliminate x.
$$4x + 5y = 11$$
$$4(-x + 3y) = 4 \cdot 10$$

$$4x + 5y = 11$$
$$-4x + 12y = 40$$

Add the equations.
$$17y = 51$$
$$y = 3$$

Replace y in equation (1).
$$4x + 5y = 11$$
$$4x + 5 \cdot 3 = 11$$
$$4x + 15 = 11$$
$$4x = -4$$
$$x = -1$$

The solution is $(-1, 3)$.

Example 4   (1) $2x - 3y = 4$
(2) $-4x + 6y = -8$

Eliminate y.
$$2(2x - 3y) = 2 \cdot 4$$
$$-4x + 6y = -8$$

$$4x - 6y = 8$$
$$-4x + 6y = -8$$

Add the equations.
$$0 + 0 = 0$$
$$0 = 0$$

The system of equations is dependent. The solutions are the ordered pairs that satisfy the equation $2x - 3y = 4$.

SECTION 9.4   *pages 363–366*

Example 2

Strategy

- Rate of the current: c
  Rate of the canoeist in calm water: r

	Rate	Time	Distance
With current	$r + c$	3	$3(r + c)$
Against current	$r - c$	5	$5(r - c)$

- The distance traveled with the current is 15 mi. The distance traveled against the current is 15 mi.

Solution

$$3(r + c) = 15 \qquad \frac{1}{3} \cdot 3(r + c) = \frac{1}{3} \cdot 15$$

$$5(r - c) = 15 \qquad \frac{1}{5} \cdot 5(r - c) = \frac{1}{5} \cdot 15$$

$$r + c = 5$$
$$r - c = 3$$
$$2r = 8$$
$$r = 4$$

$$r + c = 5$$
$$4 + c = 5$$
$$c = 1$$

The rate of the current is 1 mph.
The rate of the canoeist in calm water is 4 mph.

Example 4

Strategy

- The number of dimes in the first bank: d
  The number of quarters in the first bank: q

First bank:

	Number	Value	Total Value
Dimes	d	10	$10d$
Quarters	q	25	$25q$

Second bank:

	Number	Value	Total Value
Dimes	$\frac{1}{2}d$	10	$5d$
Quarters	$2q$	25	$50q$

- The total value of the coins in the first bank is $4.80.
  The total value of the coins in the second bank is $8.40.

Solution

$$10d + 25q = 480 \qquad\qquad 10d + 25q = 480$$
$$5d + 50q = 840 \qquad -2(5d + 50q) = -2 \cdot 840$$

$$10d + 25q = 480$$
$$-10d - 100q = -1680$$
$$-75q = -1200$$
$$q = 16$$

$$10d + 25q = 480$$
$$10d + 25(16) = 480$$
$$10d + 400 = 480$$
$$10d = 80$$
$$d = 8$$

There are 8 dimes and 16 quarters in the first bank.

SOLUTIONS to Chapter 10 Examples

SECTION 10.1   *pages 379–382*

Example 2 $A = \{-9, -7, -5, -3, -1\}$

Example 4 $A = \{1, 3, 5, \ldots\}$

Example 6 $A \cup B = \{-2, -1, 0, 1, 2, 3, 4\}$

Example 8 $C \cap D = \{10, 16\}$

Example 10 $A \cap B = \varnothing$

Example 12 $\{x \mid x < 59, x \text{ is a positive even integer}\}$

Example 14 $\{x \mid x > -3, x \in \text{real numbers}\}$

Example 16   The solution set is the numbers greater than -2.

Example 18   The solution set is the numbers greater than -1 and the numbers less than -3.

Example 20   The solution set is the real numbers.

Example 22   The solution set is the numbers which are less than or equal to 4 and greater than or equal to -4.

SECTION 10.2   *pages 385–388*

Example 2
$$x + 2 < -2$$
$$x + 2 - 2 < -2 - 2$$
$$x < -4$$

Example 4
$$5x + 3 > 4x + 5$$
$$5x - 4x + 3 > 4x - 4x + 5$$
$$x + 3 > 5$$
$$x + 3 - 3 > 5 - 3$$
$$x > 2$$

Example 6
$$-3x > -9$$
$$\frac{-3x}{-3} < \frac{-9}{-3}$$
$$x < 3$$

Example 8
$$-\frac{3}{4}x \geq 18$$
$$-\frac{4}{3}\left(-\frac{3}{4}x\right) \leq -\frac{4}{3}(18)$$
$$x \leq -24$$

Example 10   To find the selling prices, write and solve an inequality using p to represent the possible selling prices.

Solution $0.70p \geq 314$
$$p \geq 448.571$$

The dealer will make a profit if the selling price is greater than or equal to $448.58

SECTION 10.3   *pages 393–394*

Example 2

$$5 - 4x > 9 - 8x$$
$$5 - 4x + 8x > 9 - 8x + 8x$$
$$5 + 4x > 9$$
$$5 - 5 + 4x > 9 - 5$$
$$4x > 4$$
$$\frac{4x}{4} > \frac{4}{4}$$
$$x > 1$$

Example 4

$$8 - 4(3x + 5) \leq 6(x - 8)$$
$$8 - 12x - 20 \leq 6x - 48$$
$$-12 - 12x \leq 6x - 48$$
$$-12 - 12x - 6x \leq 6x - 6x - 48$$
$$-12 - 18x \leq -48$$
$$-12 + 12 - 18x \leq -48 + 12$$
$$-18x \leq -36$$
$$\frac{-18x}{-18} \geq \frac{-36}{-18}$$
$$x \geq 2$$

Example 6

Strategy   To find the maximum number of miles:
- Write an expression for the cost of each car, using x to represent the number of miles driven during the week.
- Write and solve an inequality.

Solution

Cost of a Company A car	is less than	Cost of a Company B car

$$8(7) + 0.10x < 10(7) + 0.08x$$
$$56 + 0.10x < 70 + 0.08x$$
$$56 + 0.10x - 0.08x < 70 + 0.08x - 0.08x$$
$$56 + 0.02x < 70$$
$$56 - 56 + 0.02x < 70 - 56$$
$$0.02x < 14$$
$$\frac{0.02x}{0.02} < \frac{14}{0.02}$$
$$x < 700$$

The maximum number of miles is 699.

SECTION 10.4   *pages 397–398*

Example 2

$$x - 3y < 2$$
$$x - x - 3y < -x + 2$$
$$-3y < -x + 2$$
$$\frac{-3y}{-3} > \frac{-x + 2}{-3}$$
$$y > \frac{1}{3}x - \frac{2}{3}$$

Example 4

$$2x - 4y \leq 8$$
$$2x - 2x - 4y \leq -2x + 8$$
$$-4y \leq -2x + 8$$
$$\frac{-4y}{-4} \geq \frac{-2x + 8}{-4}$$
$$y \geq \frac{1}{2}x - 2$$

Example 6 $x < 3$

SOLUTIONS to Chapter 11 Examples

SECTION 11.1   *pages 411–414*

Example 2

$-5\sqrt{32} = -5\sqrt{2^5} = -5\sqrt{2^4 \cdot 2} = -5\sqrt{2^4}\sqrt{2}$
$= -5 \cdot 2^2\sqrt{2} = -20\sqrt{2}$

Example 6

$\sqrt{y^{19}} = \sqrt{y^{18} \cdot y} = \sqrt{y^{18}}\sqrt{y} = y^9\sqrt{y}$

Example 10

$3a\sqrt{28a^9b^{18}} = 3a\sqrt{2^2 \cdot 7 \cdot a^9b^{18}}$
$= 3a\sqrt{2^2a^8b^{18}(7a)} = 3a\sqrt{2^2a^8b^{18}}\sqrt{7a}$
$= 3a \cdot 2 \cdot a^4b^9\sqrt{7a} = 6a^5b^9\sqrt{7a}$

Example 14

$\sqrt{x^2 + 14x + 49} = \sqrt{(x + 7)^2} = x + 7$

Example 4

$\sqrt{216} = \sqrt{2^3 \cdot 3^3} = \sqrt{2^2 \cdot 3^2(2 \cdot 3)} = \sqrt{2^2 \cdot 3^2}\sqrt{2 \cdot 3}$
$= 2 \cdot 3\sqrt{2 \cdot 3} = 6\sqrt{6} \approx 6(2.449) \approx 14.694$

Example 8

$\sqrt{45b^7} = \sqrt{3^2 \cdot 5 \cdot b^7} = \sqrt{3^2b^6(5 \cdot b)} = \sqrt{3^2b^6}\sqrt{5b}$
$= 3b^3\sqrt{5b}$

Example 12

$\sqrt{25(a + 3)^2} = \sqrt{5^2(a + 3)^2} = 5(a + 3) = 5a + 15$

SECTION 11.2   *pages 417–418*

Example 2

$9\sqrt{3} + 3\sqrt{3} - 18\sqrt{3} = -6\sqrt{3}$

Example 6

$y\sqrt{28y} + 7\sqrt{63y^3} = y\sqrt{2^2 \cdot 7y} + 7\sqrt{3^2 \cdot 7 \cdot y^3}$
$= y\sqrt{2^2}\sqrt{7y} + 7\sqrt{3^2 \cdot y^2}\sqrt{7y}$
$= y \cdot 2\sqrt{7y} + 7 \cdot 3 \cdot y\sqrt{7y}$
$= 2y\sqrt{7y} + 21y\sqrt{7y} = 23y\sqrt{7y}$

Example 4

$2\sqrt{50} - 5\sqrt{32} = 2\sqrt{2 \cdot 5^2} - 5\sqrt{2^5}$
$= 2\sqrt{5^2}\sqrt{2} - 5\sqrt{2^4}\sqrt{2}$
$= 2 \cdot 5\sqrt{2} - 5 \cdot 2^2\sqrt{2} = 10\sqrt{2} - 20\sqrt{2}$
$= -10\sqrt{2}$

Example 8

$2\sqrt{27a^5} - 4a\sqrt{12a^3} + a^2\sqrt{75a}$
$= 2\sqrt{3^3 \cdot a^5} - 4a\sqrt{2^2 \cdot 3 \cdot a^3} + a^2\sqrt{3 \cdot 5^2 \cdot a}$
$= 2\sqrt{3^2 \cdot a^4}\sqrt{3a} - 4a\sqrt{2^2 \cdot a^2}\sqrt{3a} + a^2\sqrt{5^2}\sqrt{3a}$
$= 2 \cdot 3 \cdot a^2\sqrt{3a} - 4a \cdot 2 \cdot a\sqrt{3a} + a^2 \cdot 5\sqrt{3a}$
$= 6a^2\sqrt{3a} - 8a^2\sqrt{3a} + 5a^2\sqrt{3a} = 3a^2\sqrt{3a}$

SECTION 11.3   *pages 421–424*

Example 2

$$\sqrt{5a}\sqrt{15a^3b^4}\sqrt{3b^5} = \sqrt{225a^4b^9} = \sqrt{3^25^2a^4b^9}$$
$$= \sqrt{3^25^2a^4b^8}\sqrt{b} = 3 \cdot 5a^2b^4\sqrt{b}$$
$$= 15a^3b^4\sqrt{b}$$

Example 4

$$\sqrt{5x}(\sqrt{5x} - \sqrt{25y})$$
$$= \sqrt{5^2x^2} - \sqrt{5^3xy}$$
$$= \sqrt{5^2x^2} - \sqrt{5^2}\sqrt{5xy} = 5x - 5\sqrt{5xy}$$

Example 6

$$(2\sqrt{x} + 7)(2\sqrt{x} - 7) = 4\sqrt{x^2} - 7^2 = 4x - 49$$

Example 8

$$(3\sqrt{x} - \sqrt{y})(5\sqrt{x} - 2\sqrt{y})$$
$$= 15\sqrt{x^2} - 6\sqrt{xy} - 5\sqrt{xy} + 2\sqrt{y^2}$$
$$= 15\sqrt{x^2} - 11\sqrt{xy} + 2\sqrt{y^2}$$
$$= 15x - 11\sqrt{xy} + 2y$$

Example 10

$$\frac{\sqrt{15x^6y^7}}{\sqrt{3x^7y^9}} = \sqrt{\frac{15x^6y^7}{3x^7y^9}} = \sqrt{\frac{5}{xy^2}}$$

$$= \frac{\sqrt{5}}{y\sqrt{x}} = \frac{\sqrt{5}}{y\sqrt{x}} \cdot \frac{\sqrt{x}}{\sqrt{x}} = \frac{\sqrt{5x}}{xy}$$

Example 12

$$\frac{\sqrt{y}}{\sqrt{y} + 3} = \frac{\sqrt{y}}{\sqrt{y} + 3} \cdot \frac{\sqrt{y} - 3}{\sqrt{y} - 3} = \frac{y - 3\sqrt{y}}{y - 9}$$

Example 14

$$\frac{5 + \sqrt{y}}{1 - 2\sqrt{y}} = \frac{5 + \sqrt{y}}{1 - 2\sqrt{y}} \cdot \frac{1 + 2\sqrt{y}}{1 + 2\sqrt{y}} = \frac{5 + 11\sqrt{y} + 2y}{1 - 4y}$$

SECTION 11.4   *pages 427–430*

Example 2

$$\sqrt{4x} + 3 = 7$$
$$\sqrt{4x} = 4$$
$$(\sqrt{4x})^2 = 4^2$$
$$4x = 16$$
$$x = 4$$

Check: $\sqrt{4x} + 3 = 7$
$$\sqrt{4 \cdot 4} + 3 = 7$$
$$\sqrt{4^2} + 3 = 7$$
$$4 + 3 = 7$$
$$7 = 7$$

The solution is 4.

Example 4

$$\sqrt{x} + \sqrt{x + 9} = 9$$
$$\sqrt{x} = 9 - \sqrt{x + 9}$$
$$(\sqrt{x})^2 = (9 - \sqrt{x + 9})^2$$
$$x = 81 - 18\sqrt{x + 9} + (x + 9)$$
$$-90 = -18\sqrt{x + 9}$$
$$5 = \sqrt{x + 9}$$
$$5^2 = (\sqrt{x + 9})^2$$
$$25 = x + 9$$
$$16 = x$$

Check:
$$\sqrt{16} + \sqrt{16 + 9} = 9$$
$$4 + 5 = 9$$
$$9 = 9$$

The solution is 16.

Example 6

Strategy

To find the distance, use the Pythagorean Theorem. The hypotenuse is the length of the ladder. One leg is the distance from the bottom of the ladder to the base of the building. The distance along the building from the ground to the top of the ladder is the unknown leg.

Solution

$a^2 = \sqrt{c^2 - b^2}$
$a^2 = \sqrt{(8)^2 - (3)^2}$
$a^2 = \sqrt{64 - 9}$
$a^2 = \sqrt{55}$
$a^2 \approx 7.416$

The distance is 7.416 ft.

Example 8

Strategy

To find the length of the pendulum, replace T in the equation with the given value and solve for L.

Solution

$$T = 2\pi\sqrt{\frac{L}{32}}$$

$$2.5 = 2(3.14)\sqrt{\frac{L}{32}}$$

$$2.5 = 6.28\sqrt{\frac{L}{32}}$$

$$\frac{2.5}{6.28} = \sqrt{\frac{L}{32}}$$

$$\left(\frac{2.5}{6.28}\right)^2 = \left(\sqrt{\frac{L}{32}}\right)^2$$

$$\frac{6.25}{39.4384} = \frac{L}{32}$$

$$(32)\left(\frac{6.25}{39.4384}\right) = (32)\left(\frac{L}{32}\right)$$

$$\frac{200}{39.4384} = L$$

$$5.07 \approx L$$

The length of the pendulum is 5.07 ft.

SOLUTIONS to Chapter 12 Examples

SECTION 12.1   *pages 443–446*

Example 2
$$2x^2 = (x + 2)(x + 3)$$
$$2x^2 = x^2 + 5x + 6$$
$$x^2 - 5x - 6 = 0$$
$$(x + 1)(x - 6) = 0$$

$x + 1 = 0 \qquad x - 6 = 0$
$\qquad x = -1 \qquad\qquad x = 6$

The solutions are -1 and 6.

Example 4
$$x^2 + 81 = 0$$
$$x^2 = -81$$
$$\sqrt{x^2} = \sqrt{-81}$$

$\sqrt{-81}$ is not a real number.

The equation has no real number solution.

Example 6 $7(z + 2)^2 = 21$
$(z + 2)^2 = 3$
$\sqrt{(z + 2)^2} = \sqrt{3}$
$z + 2 = \pm\sqrt{3}$
$z = -2 \pm \sqrt{3}$

The solutions are $-2 + \sqrt{3}$
and $-2 - \sqrt{3}$.

SECTION 12.2   *pages 449–452*

Example 2 $3x^2 - 6x - 2 = 0$
$3x^2 - 6x = 2$
$\frac{1}{3}(3x^2 - 6x) = \frac{1}{3} \cdot 2$
$x^2 - 2x = \frac{2}{3}$

Complete the square.
$x^2 - 2x + 1 = \frac{2}{3} + 1$
$(x - 1)^2 = \frac{5}{3}$
$\sqrt{(x - 1)^2} = \sqrt{\frac{5}{3}}$

$x - 1 = \pm\sqrt{\frac{5}{3}} = \pm\frac{\sqrt{15}}{3}$

$x - 1 = \frac{\sqrt{15}}{3}$ $x - 1 = -\frac{\sqrt{15}}{3}$

$x = 1 + \frac{\sqrt{15}}{3}$ $x = 1 - \frac{\sqrt{15}}{3}$

$= \frac{3 + \sqrt{15}}{3}$ $= \frac{3 - \sqrt{15}}{3}$

The solutions are $\frac{3 + \sqrt{15}}{3}$ and $\frac{3 - \sqrt{15}}{3}$.

Example 4 $x^2 + 6x + 12 = 0$
$x^2 + 6x = -12$
$x^2 + 6x + 9 = -12 + 9$
$(x + 3)^2 = -3$
$\sqrt{(x + 3)^2} = \sqrt{-3}$

$\sqrt{-3}$ is not a real number.

The quadratic equation has no real number solution.

Example 6 $x^2 + 8x + 8 = 0$
$x^2 + 8x = -8$
$x^2 + 8x + 16 = -8 + 16$
$(x + 4)^2 = 8$
$\sqrt{(x + 4)^2} = \sqrt{8}$
$x + 4 = \pm\sqrt{8} = \pm 2\sqrt{2}$
$x + 4 = 2\sqrt{2}$ $x + 4 = -2\sqrt{2}$
$x = -4 + 2\sqrt{2}$ $x = -4 - 2\sqrt{2}$
$\approx -4 + 2(1.414)$ $\approx -4 - 2(1.414)$
$\approx -4 + 2.828$ $\approx -4 - 2.828$
≈ -1.172 ≈ -6.828

The solutions are approximately -1.172 and -6.828.

SECTION 12.3   *pages 455–456*

Example 2 $3x^2 + 4x - 4 = 0$
$a = 3, b = 4, c = -4$

$$x = \frac{-(4) \pm \sqrt{(4)^2 - 4(3)(-4)}}{2 \cdot 3}$$

$$= \frac{-4 \pm \sqrt{16 + 48}}{6}$$

$$= \frac{-4 \pm \sqrt{64}}{6} = \frac{-4 \pm 8}{6}$$

$$x = \frac{-4 + 8}{6} \qquad x = \frac{-4 - 8}{6}$$

$$= \frac{4}{6} = \frac{2}{3} \qquad = \frac{-12}{6} = -2$$

The solutions are $\frac{2}{3}$ and -2.

Example 4 $x^2 + 2x = 1$
$x^2 + 2x - 1 = 0$
$a = 1, b = 2, c = -1$

$$x = \frac{-(2) \pm \sqrt{(2)^2 - 4(1)(-1)}}{2 \cdot 1}$$

$$= \frac{-2 \pm \sqrt{4 + 4}}{2} = \frac{-2 \pm \sqrt{8}}{2}$$

$$= \frac{-2 \pm 2\sqrt{2}}{2} = -1 \pm \sqrt{2}$$

The solutions are $-1 + \sqrt{2}$ and $-1 - \sqrt{2}$.

SECTION 12.4   *pages 459–460*

Example 2

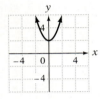

Example 4

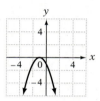

SECTION 12.5   *pages 463–464*

Example 2

Strategy
- This is a geometry problem.
- Width of the rectangle: w
  Length of the rectangle: $w + 2$
- Use the equation $A = l \cdot w$.

Solution

$$A = l \cdot w$$
$$15 = (w + 2)w$$
$$15 = w^2 + 2w$$
$$0 = w^2 + 2w - 15$$
$$0 = (w + 5)(w - 3)$$

$$w + 5 = 0 \qquad w - 3 = 0$$
$$w = -5 \qquad\quad w = 3$$

The solution -5 is not possible.
The width is 3 m.

ANSWERS to Chapter 1 Odd-numbered Exercises

SECTION 1.1   *pages 5–6*

1. $3 < 5$ **3.** $-2 > -5$ **5.** $-16 < 1$ **7.** $3 > -7$ **9.** $-11 < -8$ **11.** $-1 > -6$ **13.** $0 > -3$
15. $6 > -8$ **17.** $-14 < 16$ **19.** $35 > 28$ **21.** $-42 < 27$ **23.** $21 > -34$ **25.** $-27 > -39$
27. $-87 < 63$ **29.** $68 > -79$ **31.** $-62 > -84$ **33.** $94 > 83$ **35.** $59 > -67$ **37.** $-93 < -55$
39. $-88 < 57$ **41.** $0 < 129$ **43.** $-131 < 101$ **45.** $-194 < -180$ **47.** -16 **49.** 3 **51.** -45
53. 88 **55.** 2 **57.** 6 **59.** 5 **61.** 1 **63.** -8 **65.** 0 **67.** 19 **69.** 22 **71.** -20
73. -18 **75.** 23 **77.** -27 **79.** -41 **81.** 25 **83.** 30 **85.** -34 **87.** -45 **89.** 36
91. 61 **93.** -52 **95.** -93 **97.** 119

SECTION 1.2   *pages 11–12*

1. -2 **3.** 20 **5.** -11 **7.** -9 **9.** -3 **11.** 1 **13.** -5 **15.** -30 **17.** 9 **19.** 1
21. -10 **23.** -28 **25.** -41 **27.** -392 **29.** 8 **31.** -7 **33.** -9 **35.** 9 **37.** -3
39. 18 **41.** -9 **43.** 11 **45.** 0 **47.** 11 **49.** 2 **51.** -138 **53.** -8 **55.** -337

SECTION 1.3   *pages 17–20*

1. 42 **3.** -24 **5.** 6 **7.** 18 **9.** -20 **11.** -16 **13.** 25 **15.** 0 **17.** -72 **19.** -102
21. 140 **23.** -228 **25.** -320 **27.** -156 **29.** -70 **31.** 162 **33.** 120 **35.** 36
37. 192 **39.** -108 **41.** -2100 **43.** 0 **45.** $-251{,}636$ **47.** -2 **49.** 8 **51.** 0 **53.** -9
55. -9 **57.** 9 **59.** -24 **61.** -12 **63.** 31 **65.** 17 **67.** 15 **69.** -13 **71.** -18
73. 19 **75.** 13 **77.** -19 **79.** 17 **81.** 26 **83.** 23 **85.** 25 **87.** -34 **89.** 11
91. -13 **93.** 13 **95.** 12 **97.** -14 **99.** 290 **101.** The difference in elevation is 20,602 ft.
103. The difference in elevation is 30,314 ft.   **105.** The average daily low temperature is $-3°$C.   **107.** The difference in temperature is 99°F.   **109.** The difference in balance is $65,065.

SECTION 1.4   *pages 27–30*

1. $0.\overline{3}$ **3.** 0.25 **5.** 0.4 **7.** $0.1\overline{6}$ **9.** 0.125 **11.** $0.\overline{2}$ **13.** $0.\overline{45}$ **15.** $0.58\overline{3}$ **17.** $0.2\overline{6}$
19. 0.5625 **21.** $0.3\overline{8}$ **23.** 0.05 **25.** 0.24 **27.** $0.2\overline{3}$ **29.** 0.225 **31.** $0.36\overline{1}$ **33.** $0.68\overline{1}$
35. $0.458\overline{3}$ **37.** $0.\overline{15}$ **39.** $0.\overline{081}$ **41.** $\frac{13}{12}$ **43.** $-\frac{5}{24}$ **45.** $-\frac{19}{24}$ **47.** $\frac{5}{26}$ **49.** $\frac{7}{24}$ **51.** 0

53. $-\frac{7}{16}$ **55.** $\frac{11}{24}$ **57.** 1 **59.** $\frac{11}{8}$ **61.** $\frac{169}{315}$ **63.** -38.8 **65.** -6.192 **67.** 13.355 **69.** 4.676

71. -10.03 **73.** -37.19 **75.** -17.5 **77.** 853.2594 **79.** $-\frac{3}{8}$ **81.** $\frac{1}{10}$ **83.** $-\frac{4}{9}$ **85.** $-\frac{7}{30}$

87. $\frac{2}{27}$ **89.** $-\frac{10}{9}$ **91.** $-\frac{147}{32}$ **93.** $\frac{25}{8}$ **95.** $\frac{2}{3}$ **97.** 4.164 **99.** 4.347

101. -4.028 **103.** -2.22 **105.** 12.26448 **107.** -274.44 **109.** -2.59 **111.** -5.11
113. -2060.55 **115.** 4781.29

SECTION 1.5   *pages 35–36*

1. 36 **3.** -49 **5.** 9 **7.** 81 **9.** $\frac{1}{4}$ **11.** 0.09 **13.** 12 **15.** 0.216 **17.** -12 **19.** 16

21. -864 **23.** -1008 **25.** 3 **27.** $-77{,}760$ **29.** 9 **31.** 12 **33.** 1 **35.** 13 **37.** -36

39. 13 **41.** 4 **43.** 15 **45.** -1 **47.** 4 **49.** 0.51 **51.** 1.7 **53.** $\frac{1}{4}$ **55.** $\frac{17}{48}$

CHAPTER REVIEW   *pages 39–40*

1. $14 < 29$ **2.** $-4 > -6$ **3.** 17 **4.** -6 **5.** 16 **6.** -7 **7.** -6 **8.** -13 **9.** -4
10. 1 **11.** -42 **12.** 81 **13.** 8 **14.** -20 **15.** The difference in temperature is 396°C.
16. The difference in temperature is 714°C.   **17.** 0.28 **18.** $0.1\overline{3}$ **19.** $\frac{7}{12}$ **20.** -1.068 **21.** $-\frac{8}{15}$
22. -4.6224 **23.** -8 **24.** -25 **25.** $\frac{16}{81}$ **26.** -45 **27.** 10 **28.** 60 **29.** -1 **30.** 31

31. 18 **32.** 1

CHAPTER TEST   *pages 41–42*

1. $-2 > -40$ (1.1A)   **2.** $-1 < 0$ (1.1A)   **3.** 4 (1.1B)   **4.** -4 (1.1B)   **5.** -3 (1.2A)   **6.** -16 (1.2A)
7. -14 (1.2B)   **8.** 4 (1.2B)   **9.** -48 (1.3A)   **10.** 90 (1.3A)   **11.** -9 (1.3B)   **12.** 17 (1.3B)
13. 7°C (1.3C)   **14.** The average low temperature is -2°F. (1.3C)   **15.** 0.85 (1.4A)   **16.** $0.\overline{7}$ (1.4A)
17. $\frac{1}{15}$ (1.4B)   **18.** -6.881 (1.4B)   **19.** $-\frac{1}{2}$ (1.4C)   **20.** -5.3578 (1.4C)   **21.** -108 (1.5A)
22. 12 (1.5A)   **23.** 9 (1.5B)   **24.** 8(1.5B)   **25.** 17 (1.5B)

Note: The numbers in parentheses following the answers in the Chapter Test are a reference to the objective which corresponds with that problem. For example, the reference (1.2A) stands for Section 1.2, Objective A. This notation will be used for all Chapter Tests and Cumulative Reviews throughout the text.

ANSWERS to Chapter 2 Odd-Numbered Exercises

SECTION 2.1   *pages 47–48*

1. $2x^2, 5x, \underline{-8}$ **3.** $\underline{6}, -a^4$ **5.** $7x^2y, 6xy^2$ **7.** 1, -9 **9.** 1, $-4, -1$ **11.** 10   **13.** 32   **15.** 21
17. 16   **19.** -9 **21.** 3   **23.** -7 **25.** 13   **27.** -15 **29.** 41   **31.** 1   **33.** 5   **35.** 1
37. 57   **39.** 5   **41.** 12   **43.** 6   **45.** 10   **47.** 8   **49.** -3 **51.** -2 **53.** -22 **55.** 4
57. 20   **59.** 24   **61.** 4.96   **63.** -5.68

SECTION 2.2   *pages 55–58*

1. $14x$ **3.** $5a$ **5.** $-6y$ **7.** $-3b - 7$ **9.** $5a$ **11.** $-2ab$ **13.** $5xy$ **15.** 0   **17.** $-\frac{5}{6}x$

19. $-\frac{1}{24}x^2$ **21.** $11x$ **23.** $7a$ **25.** $-14x^2$ **27.** $-x + 3y$ **29.** $17x - 3y$ **31.** $-2a - 6b$

33. $-3x - 8y$ **35.** $-4x^2 - 2x$ **37.** $12x$ **39.** $-21a$ **41.** $6y$ **43.** $8x$ **45.** $-6a$ **47.** $12b$
49. $-15x^2$ **51.** x^2 **53.** a **55.** x **57.** n **59.** x **61.** y **63.** $3x$ **65.** $-2x$ **67.** $-8a^2$
69. $8y$ **71.** $4y$ **73.** $-2x$ **75.** $6a$ **77.** $-x - 7$ **79.** $10x - 35$ **81.** $-5a - 80$ **83.** $-15y + 35$
85. $20 - 14b$ **87.** $-35 + 50x$ **89.** $18x^2 + 12x$ **91.** $10x - 35$ **93.** $-14x + 49$ **95.** $-30x^2 - 15$
97. $-24y^2 + 96$ **99.** $5x^2 + 5y^2$ **101.** $-4x^2 + 20y^2$ **103.** $3x^2 + 6x - 18$ **105.** $-2y^2 + 4y - 8$
107. $-2a^2 - 4a + 6$ **109.** $10x^2 + 15x - 35$ **111.** $6x^2 + 3xy - 9y^2$ **113.** $-3a^2 - 5a + 4$ **115.** $-2x - 16$
117. $-12y - 9$ **119.** $7n - 7$ **121.** $-2x + 41$ **123.** $3y - 3$ **125.** $-a - 7b$ **127.** $-4x + 24$
129. $-2x - 16$ **131.** $-3x + 21$ **133.** $-7x + 24$ **135.** $-x + 50$ **137.** $20x - 41y$

SECTION 2.3   *pages 63–66*

1. $8 + y$ **3.** $t + 10$ **5.** $z + 14$ **7.** $x^2 - 20$ **9.** $\frac{3}{4}n + 12$ **11.** $8 + \frac{n}{4}$ **13.** $3(y + 7)$ **15.** $t(t + 16)$

17. $\frac{1}{2}x^2 + 15$ **19.** $5n^3 + n^2$ **21.** $r - \frac{r}{3}$ **23.** $x^2 - (x + 17)$ **25.** $9(z + 4)$ **27.** $12 - x$ **29.** $\frac{2}{3}x$

31. $\frac{2x}{9}$ **33.** $11x - 8$ **35.** $(x + 2) - 9; x - 7$ **37.** $\frac{7}{5 + x}$ **39.** $5 + \frac{1}{2}(x + 3); \frac{1}{2}x + \frac{13}{2}$

41. $x + (2x - 4); 3x - 4$ **43.** $(x - 5)7; 7x - 35$ **45.** $\frac{2x + 5}{x}$ **47.** $x - (3x - 8); -2x + 8$ **49.** $x + 3x; 4x$

51. $(x + 6) + 5; x + 11$ **53.** $x - (x + 10); -10$ **55.** $\frac{1}{6}x + \frac{4}{9}x; \frac{11}{18}x$ **57.** $\frac{x}{3} + x; \frac{4x}{3}$ **59.** $14\left(\frac{1}{7}x\right); 2x$

61. $10x - 2x; 8x$ **63.** $(x - 6) + 16; x + 10$ **65.** $4x - x; 3x$ **67.** $15x - 5x; 10x$ **69.** rate of propeller

plane: $\frac{1}{2}j$ **71.** diameter of baseball: $\frac{1}{4}x$ **73.** highest income tax bracket: $2t + 3$ **75.** amount of crude oil

poured into small container: $20 - L$ **77.** speed of newer model: $\frac{1}{2}x + 7$

CHAPTER REVIEW   *pages 69–70*

1. -7 **2.** 29 **3.** -4 **4.** 79 **5.** $11x$ **6.** $-5y$ **7.** $8a - 4b$ **8.** $-4x^2 + 6x$ **9.** $9c - 5d$
10. $-8r + 8s$ **11.** $20x$ **12.** $36y$ **13.** $-6a$ **14.** $-42x^2$ **15.** $-5n$ **16.** $10x - 35$
17. $12y^2 + 8y - 10$ **18.** $-63 - 36x$ **19.** $3x^2 - 24x - 21$ **20.** $28a^2 - 8a + 12$ **21.** $-4x + 20$
22. $7x + 46$ **23.** $24y + 30$ **24.** $-90x + 25$ **25.** $4x$ **26.** $x - 6$ **27.** $\frac{2}{3}(x + 10)$ **28.** $x + 2x$; $3x$
29. $3x + 5(x - 1)$; $8x - 5$ **30.** $2x - \frac{1}{2}x$; $\frac{3}{2}x$ **31.** number of calories in a candy bar: $2a + 8$ **32.** number
of five-dollar bills: $35 - T$ **33.** diameter of #8 copper wire: $13d - 2$ **34.** number of National League
cards: $5A$

CHAPTER TEST   *pages 71–72*

1. 22 (2.1A)   **2.** 3 (2.1A)   **3.** $5x$ (2.2A)   **4.** y^2 (2.2A)   **5.** $-9x - 7y$ (2.2A)   **6.** $-2x - 5y$
(2.2A)   **7.** $2x$ (2.2B)   **8.** $3x$ (2.2B)   **9.** $36y$ (2.2B)   **10.** $-10a$ (2.2B)   **11.** $15 - 35b$ (2.2C)
12. $-4x + 8$ (2.2C)   **13.** $-6x^2 + 21y^2$ (2.2C)   **14.** $-10x^2 + 15x - 30$ (2.2C)   **15.** $-x + 6$ (2.2D)
16. $7x + 38$ (2.2D)   **17.** $-7x + 33$ (2.2D)   **18.** $2x + y$ (2.2D)   **19.** $a^2 - b^2$ (2.3A)   **20.** $x + 2x^2$
(2.3A)   **21.** $\frac{6}{x} - 3$ (2.3A)   **22.** $b - 7b$; $-6b$ (2.3B)   **23.** $10(x - 3)$; $10x - 30$ (2.3B)   **24.** speed of
fastball: $2s$ (2.3C)   **25.** longer piece: $4x - 3$ (2.3C)

CUMULATIVE REVIEW   *pages 73–74*

1. -7 (1.2A)   **2.** 5 (1.2B)   **3.** 24 (1.3A)   **4.** -5 (1.3B)   **5.** 1.25 (1.4A)   **6.** $\frac{11}{48}$ (1.4B)

7. $\frac{1}{6}$ (1.4C)   **8.** $\frac{1}{4}$ (1.4C)   **9.** $\frac{8}{3}$ (1.5A)   **10.** -5 (1.5B)   **11.** $\frac{53}{48}$ (1.5B)   **12.** 16 (2.1A)

13. $5x^2$ (2.2A)   **14.** $-7a - 10b$ (2.2A)   **15.** $6a$ (2.2B)   **16.** $30b$ (2.2B)   **17.** $24 - 6x$ (2.2C)
18. $6y - 18$ (2.2C)   **19.** $-8x^2 + 12y^2$ (2.2C)   **20.** $-9y^2 + 9y + 21$ (2.2C)   **21.** $-7x + 14$ (2.2D)
22. $5x - 43$ (2.2D)   **23.** $17x - 24$ (2.2D)   **24.** $-3x + 21y$ (2.2D)   **25.** $\frac{1}{2}b + b$ (2.3A)   **26.** $\frac{10}{y - 2}$
(2.3A)   **27.** $8 - \frac{x}{12}$ (2.3B)   **28.** $x + (x + 2)$; $2x + 2$ (2.3B)   **29.** $4[x + (x + 1)]$; $8x + 4$ (2.3B)
30. $(3 + x)5 + 12$; $27 + 5x$ (2.3B)

ANSWERS to Chapter 3 Odd-numbered Exercises

SECTION 3.1   *pages 81–84*

1. Yes   **3.** No   **5.** No   **7.** Yes   **9.** Yes   **11.** Yes   **13.** No   **15.** No   **17.** Yes   **19.** No
21. Yes   **23.** No   **25.** No   **27.** Yes   **29.** Yes   **31.** $x = 2$ **33.** $b = 15$ **35.** $a = 6$
37. $m = -6$ **39.** $n = 3$ **41.** $b = 0$ **43.** $a = -2$ **45.** $z = -7$ **47.** $m = -7$ **49.** $x = -12$
51. $b = 2$ **53.** $x = 6$ **55.** $x = -5$ **57.** $m = 15$ **59.** $w = 9$ **61.** $b = 14$ **63.** $a = 19$

65. $c = -1$ **67.** $x = 1$ **69.** $a = -\frac{2}{3}$ **71.** $b = -\frac{1}{2}$ **73.** $n = \frac{4}{15}$ **75.** $c = \frac{5}{12}$ **77.** $w = 1.869$

79. $t = 0.884$ **81.** $x = 7.251$ **83.** $y = 7$ **85.** $a = -7$ **87.** $m = -4$ **89.** $n = 5$ **91.** $t = 9$
93. $x = -8$ **95.** $x = 0$ **97.** $x = -7$ **99.** $y = 8$ **101.** $t = -7$ **103.** $x = 12$ **105.** $b = -18$

107. $x = 15$ **109.** $m = -20$ **111.** $x = -12$ **113.** $x = 15$ **115.** $x = 25$ **117.** $x = \frac{8}{3}$ **119.** $y = \frac{1}{3}$

121. $m = \frac{15}{7}$ **123.** $n = 4$ **125.** $y = 3$ **127.** $x = 4.745$ **129.** $a = 2.06$ **131.** $x = -2.13$

SECTION 3.2   *pages 87–90*

1. $x = 3$ **3.** $a = 6$ **5.** $x = -1$ **7.** $x = -3$ **9.** $d = 2$ **11.** $c = -2$ **13.** $w = 2$ **15.** $t = 2$
17. $a = 5$ **19.** $b = -3$ **21.** $x = 6$ **23.** $x = 3$ **25.** $a = 1$ **27.** $b = 6$ **29.** $m = -7$ **31.** $y = 0$

33. $c = 2$ **35.** $x = \frac{6}{7}$ **37.** $a = \frac{2}{3}$ **39.** $n = -1$ **41.** $x = \frac{3}{4}$ **43.** $x = \frac{4}{9}$ **45.** $x = \frac{1}{3}$ **47.** $w = -\frac{1}{2}$

49. $b = -\frac{3}{4}$ **51.** $a = \frac{1}{3}$ **53.** $x = -\frac{1}{6}$ **55.** $a = 1$ **57.** $x = 1$ **59.** $x = 0$ **61.** $x = \frac{13}{10}$ **63.** $a = \frac{2}{5}$

65. $x = -\frac{4}{3}$ **67.** $x = -\frac{3}{2}$ **69.** $m = 18$ **71.** $n = 8$ **73.** $b = -16$ **75.** $y = 25$ **77.** $c = 21$

79. $w = 15$ **81.** $x = -16$ **83.** $x = -21$ **85.** $x = \frac{15}{2}$ **87.** $y = -\frac{18}{5}$ **89.** $y = 2$ **91.** $z = 3$

93. $b = 1$ **95.** $m = -2$ **97.** $y = -0.74$ **99.** $x = 0.15$ **101.** 19   **103.** -1 **105.** -11
107. The initial velocity is 8 ft/s.   **109.** The depreciated value will be \$38,000 after 2 years.   **111.** The approximate length is 31.8 in. to the nearest tenth.   **113.** The estimated population is 51,000 people to the nearest thousand.   **115.** The approximate year was 1952.

SECTION 3.3   *pages 95–98*

1. $x = 2$ **3.** $m = 3$ **5.** $x = 3$ **7.** $y = -1$ **9.** $x = -1$ **11.** $x = 2$ **13.** $x = -2$ **15.** $b = -3$
17. $x = -8$ **19.** $y = 0$ **21.** $x = -1$ **23.** $x = -3$ **25.** $x = -1$ **27.** $m = 4$ **29.** $x = -2$

31. $b = \frac{2}{3}$ **33.** $x = \frac{5}{6}$ **35.** $n = -\frac{2}{3}$ **37.** $y = 3.5$ **39.** $x = 2.45$ **41.** -17 **43.** 41   **45.** 8

47. $y = 1$ **49.** $x = 4$ **51.** $m = -1$ **53.** $b = 1$ **55.** $x = -1$ **57.** $y = 2$ **59.** $x = \frac{5}{6}$

61. $a = -\frac{7}{10}$ **63.** $x = \frac{1}{4}$ **65.** $x = 5$ **67.** $x = \frac{20}{3}$ **69.** $x = 2$ **71.** $y = 2.5$ **73.** 26   **75.** -2

77. The force on the lip of the can is 1770 lb.   **79.** The child must sit 10 ft from the fulcrum.   **81.** The see-saw is not balanced.   **83.** There must be 190 cellular telephones sold.   **85.** There must be 3000 softball bats sold.

SECTION 3.4   *pages 103–106*

1. $x - 15 = 7$; $x = 22$ **3.** $7x = -21$; $x = -3$ **5.** $3x - 4 = 5$; $x = 3$ **7.** $4(2x + 3) = 12$; $x = 0$
9. $12 = 6(x - 3)$; $x = 5$ **11.** $4x + 7 = 3$; $x = -1$ **13.** $22 = 6x - 2$; $x = 4$ **15.** $3x = 2x + 4$; $x = 4$
17. $4x + 7 = 2x + 3$; $x = -2$ **19.** $5x - 8 = 8x + 4$; $x = -4$ **21.** $2(x - 25) = 3x$; $x = -50$
23. $3x = 2(20 - x)$; 8 and 12   **25.** $3x - 4 = 2(24 - x) - 12$; 8 and 16   **27.** $3x + 2(18 - x) = 44$; 8 and 10
29. The cost of the compact disc player is \$142.   **31.** The speed of the newer computer is 32 megahertz.
33. There were 4 field goals scored.   **35.** There are 150 research assistants employed.   **37.** There are 3 lb iron, 6 lb potassium, and 15 lb mulch in the soil supplement.   **39.** The monthly payment to the nearest cent is \$117.75.   **41.** The hours of labor required is 5 h.   **43.** The amounts deposited are \$2000 and \$3000.
45. The width is 50 ft, and the length is 80 ft.   **47.** There are 200 vertical pixels.   **49.** The perimeter of the larger square is 8 ft.   **51.** The width of the largest door frame is 3 ft.

CHAPTER REVIEW   *pages 109–110*

1. No   **2.** Yes   **3.** Yes   **4.** No   **5.** $x = 21$ **6.** $a = \frac{5}{6}$ **7.** $x = 2.5$ **8.** $a = 20$ **9.** $x = 7$

10. $x = -3$ **11.** $x = 6$ **12.** $x = -\frac{6}{7}$ **13.** In 15 years, the interest will be \$1800.   **14.** The cost of the cookbook is \$28.00.   **15.** The temperature is 37°C.   **16.** The discount is \$79.25.   **17.** $x = 4$ **18.** $y = \frac{1}{3}$
19. $x = 4$ **20.** $x = -5$ **21.** $x = -2$ **22.** $x = 3$ **23.** $x = 10$ **24.** $x = -1$ **25.** A 24 lb force must be applied to the other end of the lever.   **26.** The fulcrum is 3 ft from the 25 lb force.
27. $5n - 4 = 16$; $n = 4$ **28.** $6(n + 3) = 2n - 10$; $n = -7$ **29.** $3x = 2(21 - x) - 2$; 8 and 13   **30.** The two lengths of wire are 14 in. and 21 in.   **31.** The consulting fee consisted of 8 h of consulting.   **32.** The Eiffel Tower is 993 ft tall.

CHAPTER TEST   *pages 111–112*

1. No (3.1A)   **2.** Yes (3.1A)   **3.** $x = -5$ (3.1B)   **4.** $x = -12$ (3.1C)   **5.** $x = -3$ (3.2A)

6. $x = 5$ (3.2A)   **7.** 200 calculators were produced. (3.2B)   **8.** $x = -5$ (3.3A)   **9.** $x = -\frac{1}{2}$ (3.3A)

10. $x = 2$ (3.3A)   **11.** $x = 2$ (3.3B)   **12.** $x = -\frac{1}{3}$ (3.3B)   **13.** $x = \frac{12}{11}$ (3.3B)   **14.** The final temperature is 60°C. (3.3C)   **15.** $3x - 15 = 27$; $x = 14$ (3.4A)   **16.** $5x + 6 = 3(x + 12)$; $x = 15$ (3.4A)
17. 8 and 10 (3.4A)   **18.** The time is 7 h. (3.4B)   **19.** The pieces measured 6 ft and 12 ft. (3.4B)
20. The new disk controller requires 15 milliseconds to access the data. (3.4B)

CUMULATIVE REVIEW   *pages 113–114*

1. 6 (1.2B)   **2.** −48 (1.3A)   **3.** $-\frac{19}{48}$ (1.4B)   **4.** −2 (1.4C)   **5.** 54 (1.5A)   **6.** 24 (1.5B)
7. 6 (2.1A)   **8.** −17x (2.2A)   **9.** −5a − 2b (2.2A)   **10.** 2x (2.2B)   **11.** 36y (2.2B)
12. $2x^2 + 6x - 4$ (2.2C)   **13.** −4x + 14 (2.2D)   **14.** 6x − 34 (2.2D)   **15.** Yes (3.1A)
16. No (3.1A)   **17.** x = −5 (3.1B)   **18.** x = −25 (3.1C)   **19.** x = −3 (3.2A)   **20.** x = 3 (3.2A)
21. x = 13 (3.3B)   **22.** x = 2 (3.3B)   **23.** x = −3 (3.3A)   **24.** $x = \frac{1}{2}$ (3.3A)   **25.** 250 cameras were
produced. (3.2B)   **26.** The final temperature is 60°C. (3.3C)   **27.** 12 − 5x = −18; x = 6 (3.4A)
28. 6x + 13 = 3x − 5; x = −6 (3.4A)   **29.** The area of the garage is 600 ft². (3.4B)   **30.** The pieces
measure 5 ft and 11 ft. (3.4B)

ANSWERS to Chapter 4 Odd-numbered Exercises

SECTION 4.1   *pages 119–120*

1. $\frac{3}{4}$; 0.75   **3.** $\frac{1}{2}$; 0.50   **5.** $\frac{16}{25}$; 0.64   **7.** $1\frac{1}{4}$; 1.25   **9.** $\frac{19}{100}$; 0.19   **11.** $\frac{1}{20}$; 0.05   **13.** $4\frac{1}{2}$; 4.50
15. $\frac{2}{25}$; 0.08   **17.** $\frac{1}{9}$ **19.** $\frac{1}{8}$ **21.** $\frac{5}{16}$ **23.** $\frac{1}{400}$ **25.** $\frac{23}{400}$ **27.** $\frac{1}{16}$ **29.** 0.073   **31.** 0.158
33. 0.003   **35.** 0.0915   **37.** 0.1823   **39.** 0.0015   **41.** 15%   **43.** 5%   **45.** 17.5%   **47.** 115%
49. 62%   **51.** 316.5%   **53.** 0.8%   **55.** 6.5%   **57.** 54%   **59.** 33.3%   **61.** 45.5%   **63.** 87.5%
65. 166.7%   **67.** 128.6%   **69.** 34%   **71.** $37\frac{1}{2}$%   **73.** $35\frac{5}{7}$%   **75.** $18\frac{3}{4}$%   **77.** 125%   **79.** $155\frac{5}{9}$%

SECTION 4.2   *pages 123–124*

1. 24%   **3.** 7.2   **5.** 400   **7.** 9   **9.** 25%   **11.** 5   **13.** 200%   **15.** 400   **17.** 7.7   **19.** 200
21. 400   **23.** 20   **25.** 80.34%   **27.** 19% of the students are in the fine arts college.   **29.** This
presents a 40% increase.   **31.** The discount is $127.50.   **33.** The minimum number of votes is 67.
35. 177 tickets would be sold.

SECTION 4.3   *pages 127–128*

1. The selling price is $35.   **3.** The markup rate is 87.5%.   **5.** The selling price is $196.   **7.** The
markup rate is 40%.   **9.** The selling price is $69.75.   **11.** The markup rate is 23.3%.   **13.** The sale price
is $41.25.   **15.** The discount rate is 25%.   **17.** The price of the used book is $24.50.   **19.** The discount
rate is 40%.   **21.** The sale price per shirt is $14.45.   **23.** The discount rate to the nearest percent is 26%.

SECTION 4.4   *pages 131–132*

1. There must be an additional $2500 added.   **3.** There was $9000 invested at 7% and $6000 at 6.5%.
5. There was $1500 deposited in the mutual fund.   **7.** The university deposited $200,000 at 10% and
$100,000 at 8.5%.   **9.** The mechanic invests $3000 in additional bonds.   **11.** The charity deposited $40,500
at 8% and $13,500 at 12%.   **13.** The total amount invested was $650,000.   **15.** The total amount invested
was $500,000.

SECTION 4.5   *pages 137–140*

1. The mixture contains 2 lb of diet supplement and 3 lb of vitamin supplement.   **3.** The selling price is
$6.98 per pound.   **5.** The combination contained 56 oz at $4.30 and 144 oz at $1.80.   **7.** The selling price is
$2.90 per lb.   **9.** There must be 10 kg of hard candy.   **11.** The mixture contains 30 lb at $2.20 and 20 lb at
$4.20.   **13.** There must be 8 kg of soil supplement.   **15.** The mixture contains 63 lb of walnuts and 37 lb of
almonds.   **17.** The cereal costs $0.70 per pound.   **19.** There must be 9.6 lb of lima beans.   **21.** The
solution must contain 20 ml of 13% acid and 30 ml of 18% acid.   **23.** There is a 50% concentration of silver.
25. There was 30 lb of 60% mixture used.   **27.** There is 0.74% of hydrocortizone.   **29.** The hair dye
contains 100 ml of 7% and 200 ml of 4%.   **31.** There must be 25 oz of pure water added.   **33.** The percent
concentration is 27%.   **35.** There must be 10 oz of pure bran flakes added.   **37.** The mixture contains 3 lb
of 20% and 2 lb of 15%.   **39.** There should be 1.2 oz of dried apricots added.

SECTION 4.6   *pages 143–144*

1. The plane flew 2 h at 105 mph and 3 h at 115 mph.   **3.** The sailboat traveled 36 mi.   **5.** The rate of the passenger train is 50 mph, and the rate of the freight train is 30 mph.   **7.** The rate of the cyclist is 16 mph. **9.** The rate of the first plane is 95 mph, and the rate of the second plane is 120 mph.   **11.** They will meet after 1 hour.   **13.** The second runner will overtake the first runner after 3 hours.   **15.** It took 20 min downstream.   **17.** The distance between the two airports is 360 mi.   **19.** The jet overtakes the propeller plane 600 mi from the starting point.

SECTION 4.7   *pages 147–148*

1. The lengths are 50 ft, 50 ft, and 25 ft.   **3.** The lengths are 9 m, 9 m, and 3 m.   **5.** The length is 40 ft, and the width is 20 ft.   **7.** The lengths are 11 ft, 10 ft, and 12 ft.   **9.** The length is 130 ft, and the width is 39 ft.   **11.** The lengths are 3.5 m, 6.9 m, and 8 m.   **13.** The angles are 37°, 37°, and 106°.   **15.** The angles are 40°, 20°, and 120°.   **17.** The angles are 63°, 21°, and 96°.   **19.** The angles are 36°, 36°, and 108°. **21.** The angles are 31°, 59°, and 90°.   **23.** The angles are 60°, 55°, and 65°.

SECTION 4.8   *pages 153–156*

1. The integers are 6 and 8.   **3.** The integers are −1, 1, 3.   **5.** The integers are 19, 20, 21.   **7.** The integers are 5 and 7.   **9.** The integers are 12, 14, 16.   **11.** The integers are −8, −6, −4.   **13.** The integers are 23, 25, 27.   **15.** The integers are 12, 14, 16.   **17.** There are four 15¢ stamps in the drawer. **19.** The bank contains 14 dimes and 16 quarters.   **21.** There are fourteen 3¢ stamps, twenty 7¢ stamps, and ten 12¢ stamps in the collection.   **23.** There were six 15¢ stamps and thirty-four 20¢ stamps.   **25.** There are fifteen 6¢ stamps, five 8¢ stamps, and twenty-two 15¢ stamps in the collection.   **27.** There are 25 one-dollar bills and 5 five-dollar bills in the cash box.   **29.** There are 12 pennies in the bank.   **31.** The silver coin is 62 years old, and the gold coin is 146 years old.   **33.** Seven years ago the oval rug was five times as old as the circular rug.   **35.** The younger car is 3 years old, and the older car is 5 years old.   **37.** Twelve years ago the 2¢ stamp was twice as old as the 3¢ stamp.   **39.** The younger child is 4 years old, and the older child is 14 years old.   **41.** The present age of the lithograph is 15 years, and the present age of the painting is 25 years.   **43.** In 18 years, the ruby ring will be twice the age of the diamond ring.

CHAPTER REVIEW   *pages 159–160*

1. $\frac{159}{200}$ **2.** 0.062   **3.** 62.5%   **4.** $54\frac{2}{7}\%$ **5.** 15   **6.** 67.5%   **7.** It represents a 400% increase. **8.** The astronaut would weigh 30 lb on the moon.   **9.** The cost of the curio cabinet is $671.25.   **10.** The sale price of the shoes is $41.25.   **11.** The regular price of the carpet sweeper is $32.   **12.** The discount rate is $33\frac{1}{3}\%$.   **13.** The amount invested at 6.75% is $1400. The amount invested at 9.45% is $1000.   **14.** An additional $5600 must be deposited.   **15.** The mixture contains 7 qt of cranberry juice and 3 qt of apple juice. **16.** The selling price of the mixture is $1.84 per pound.   **17.** The mixture is 14% butterfat.   **18.** One liter of pure water should be added.   **19.** The rate of the motorcyclist is 45 mph.   **20.** The swimmers will meet 4 min after they begin.   **21.** The sides measure 8 in., 12 in., and 15 in.   **22.** The length is 80 ft. The width is 20 ft.   **23.** The measures of the angles are 16°, 82°, and 82°.   **24.** The measure of the angles are 75°, 60°, and 45°.   **25.** The integers are 9 and 10.   **26.** There are 30 one-dollar bills and 23 five-dollar bills. **27.** Ten years ago the typewriter was twice the age of the telephone.   **28.** The present age of the grandparent is 48 years. The present age of the child is 6 years.

CHAPTER TEST   *pages 161–162*

1. $\frac{9}{20}$, 0.45 (4.1A)   **2.** $\frac{3}{8}$ (4.1A)   **3.** 102.5% (4.1B)   **4.** $83\frac{1}{3}\%$ (4.1B)   **5.** 25.2 (4.2A) **6.** 150% (4.2A)   **7.** The value was $44,000. (4.2B)   **8.** The percent of students was 60%. (4.2B) **9.** The cost was $350. (4.3A)   **10.** The discount rate was 20%. (4.3B)   **11.** The markup rate was $33\frac{1}{3}\%$. (4.3A)   **12.** They should invest $3500 at 6% and $1500 at 9%. (4.4A)   **13.** The baker should use 10 lb at $0.70 and 5 lb at $0.40. (4.5A)   **14.** They should mix 1.25 gal of water with the salt solution. (4.5B) **15.** The rate of the snowmobile is 6 mph. (4.6A)   **16.** The length is 6 ft, and the width is 4 ft. (4.7A) **17.** The measure is 70°. (4.7B)   **18.** The integers are 10, 12, 14. (4.8A)   **19.** There are 20 quarters. (4.8B)   **20.** The stamp is 3 years old, and the coin is 21 years old. (4.8C)

CUMULATIVE REVIEW   *pages 163–164*

1. -36 (1.5B)   **2.** $-\frac{1}{16}$ (1.5A)   **3.** $\frac{49}{40}$ (1.5B)   **4.** 9 (2.1A)   **5.** $8a - 9b$ (2.2A)

6. $-9x^2 - 12x + 21$ (2.2C)   **7.** $10x - 4$ (2.2D)   **8.** Yes (3.1A)   **9.** $x = -2$ (3.1B)

10. $x = -12$ (3.1C)   **11.** $x = 2$ (3.2A)   **12.** $x = 2$ (3.3B)   **13.** $\frac{11}{20}$ (4.1A)   **14.** $\frac{2}{3}$ (4.1A)

15. 4 and 6 (3.4A)   **16.** 1.5 h (3.4B)   **17.** 103% (4.1B)   **18.** 45% (4.1B)   **19.** 15 (4.2A)

20. 120 (4.2A)   **21.** The value of the investment was \$4000. (4.2B)   **22.** An additional \$1200 must be deposited. (4.4A)   **23.** The markup rate is 40%. (4.3A)   **24.** The sale price is \$37.20. (4.3B)

25. 20 lb of oat flour must be used. (4.5A)   **26.** 25 g of pure gold must be used. (4.5B)   **27.** The length is 12 ft, and the width is 10 ft. (4.7A)   **28.** The measure of one angle is 60°. (4.7B)   **29.** The integers are $-1, 0, 1$. (4.8A)   **30.** There are 44 dimes in the bank. (4.8B)

ANSWERS to Chapter 5 Odd-numbered Exercises

SECTION 5.1   *pages 169–170*

1. $-2x^2 + 3x$ **3.** $y^2 - 8$ **5.** $5x^2 + 7x + 20$ **7.** $x^3 + 2x^2 - 6x - 6$ **9.** $2a^3 - 3a^2 - 11a + 2$
11. $5x^2 + 8x$ **13.** $7x^2 + xy - 4y^2$ **15.** $3a^2 - 3a + 17$ **17.** $5x^3 + 10x^2 - x - 4$ **19.** $3r^3 + 2r^2 - 11r + 7$
21. $-2x^3 + 3x^2 + 10x + 11$ **23.** $4x$ **25.** $3y^2 - 4y - 2$ **27.** $-7x - 7$ **29.** $4x^3 + 3x^2 + 3x + 1$
31. $y^3 - y^2 + 6y - 6$ **33.** $-y^2 - 13xy$ **35.** $2x^2 - 3x - 1$ **37.** $-2x^3 + x^2 + 2$ **39.** $3a^3 - 2$
41. $4y^3 - 2y^2 + 2y - 4$

SECTION 5.2   *pages 173–174*

1. $2x^2$ **3.** $12x^2$ **5.** $6a^7$ **7.** x^3y^5 **9.** $-10x^9y$ **11.** x^7y^8 **13.** $-6x^3y^5$ **15.** x^4y^5z
17. $a^3b^5c^4$ **19.** $-a^5b^8$ **21.** $-6a^5b$ **23.** $40y^{10}z^6$ **25.** $-20a^2b^3$ **27.** $x^3y^5z^3$ **29.** $-12a^{10}b^7$
31. $-36a^3b^2c^3$ **33.** 81   **35.** -27 **37.** -512 **39.** y^8 **41.** y^{15} **43.** $-x^6$ **45.** $27y^3$
47. $9y^6$ **49.** $x^{15}y^{20}$ **51.** $16a^4b^{12}$ **53.** $16a^{12}b^2$ **55.** $-54y^{13}$ **57.** a^6b^4 **59.** $x^{13}y^5$
61. $192x^6y^{10}$ **63.** $24x^6y^4$ **65.** $9a^4b^{10}$ **67.** $-24a^3b^8$ **69.** $729a^9b^6$

SECTION 5.3   *pages 179–182*

1. $x^2 - 2x$ **3.** $-x^2 - 7x$ **5.** $3a^3 - 6a^2$ **7.** $-5x^4 + 5x^3$ **9.** $-3x^5 + 7x^3$ **11.** $12x^3 - 6x^2$
13. $6x^2 - 12x$ **15.** $3x^2 + 4x$ **17.** $-x^3y + xy^3$ **19.** $2x^4 - 3x^2 + 2x$ **21.** $2a^3 + 3a^2 + 2a$
23. $3x^6 - 3x^4 - 2x^2$ **25.** $-6y^4 - 12y^3 + 14y^2$ **27.** $-2a^3 - 6a^2 + 8a$ **29.** $6y^4 - 3y^3 + 6y^2$
31. $x^3y - 3x^2y^2 + xy^3$ **33.** $x^3 + 4x^2 + 5x + 2$ **35.** $a^3 - 6a^2 + 13a - 12$ **37.** $-2b^3 + 7b^2 + 19b - 20$
39. $-6x^3 + 31x^2 - 41x + 10$ **41.** $x^3 - 3x^2 + 5x - 15$ **43.** $x^4 - 4x^3 - 3x^2 + 14x - 8$
45. $15y^3 - 16y^2 - 70y + 16$ **47.** $5a^4 - 20a^3 - 5a^2 + 22a - 8$ **49.** $y^4 + 4y^3 + y^2 - 5y + 2$ **51.** $x^2 + 4x + 3$
53. $a^2 + a - 12$ **55.** $y^2 - 5y - 24$ **57.** $y^2 - 10y + 21$ **59.** $2x^2 + 15x + 7$ **61.** $3x^2 + 11x - 4$
63. $4x^2 - 31x + 21$ **65.** $3y^2 - 2y - 16$ **67.** $9x^2 + 54x + 77$ **69.** $21a^2 - 83a + 80$ **71.** $15b^2 + 47b - 78$
73. $2a^2 + 7ab + 3b^2$ **75.** $6a^2 + ab - 2b^2$ **77.** $2x^2 - 3xy - 2y^2$ **79.** $10x^2 + 29xy + 21y^2$
81. $6a^2 - 25ab + 14b^2$ **83.** $2a^2 - 11ab - 63b^2$ **85.** $100a^2 - 100ab + 21b^2$ **87.** $15x^2 + 56xy + 48y^2$
89. $14x^2 - 97xy - 60y^2$ **91.** $56x^2 - 61xy + 15y^2$ **93.** $y^2 - 25$ **95.** $4x^2 - 9$ **97.** $x^2 + 2x + 1$
99. $9a^2 - 30a + 25$ **101.** $9x^2 - 49$ **103.** $4a^2 + 4ab + b^2$ **105.** $x^2 - 4xy + 4y^2$ **107.** $16 - 9y^2$
109. $25x^2 + 20xy + 4y^2$ **111.** The area is $(2L^2 - 2L)$ ft^2.   **113.** The total area is $(90x + 2025)$ ft^2.
115. The area is $3.14x^2 + 18.84x + 28.26$.

SECTION 5.4   *pages 189–192*

1. $\frac{1}{5^2} = \frac{1}{25}$ **3.** $8^2 = 64$ **5.** $\frac{1}{3^3} = \frac{1}{27}$ **7.** $2^0 = 1$ **9.** $\frac{1}{x^2}$ **11.** a^6 **13.** $\frac{y}{x^3}$ **15.** $\frac{b}{a^2}$ **17.** 1

19. $\frac{y^4}{x^4}$ **21.** $-\frac{8x^3}{y^6}$ **23.** $\frac{9}{x^2y^4}$ **25.** $\frac{2}{x^4}$ **27.** $-\frac{5}{a^8}$ **29.** $-\frac{a^5}{8b^4}$ **31.** $\frac{10y^3}{x^4}$ **33.** $\frac{1}{2x^3}$ **35.** $\frac{3}{x^3}$

37. $\frac{1}{2x^2y^6}$ **39.** $\frac{1}{x^6y}$ **41.** $\frac{a^4}{y^{10}}$ **43.** $-\frac{1}{6x^3}$ **45.** $-\frac{a^2b}{6c^2}$ **47.** $-\frac{7b^6}{a^2}$ **49.** $\frac{t^4s^8}{4r^{12}}$ **51.** $\frac{125p^3}{27m^{15}n^6}$

53. $x + 1$ **55.** $2a - 5$ **57.** $3a + 2$ **59.** $4b^2 - 3$ **61.** $x - 2$ **63.** $-x + 2$ **65.** $x^2 + 3x - 5$

67. $x^4 - 3x^2 - 1$ **69.** $xy + 2$ **71.** $-3y^3 + 5$ **73.** $3x - 2 + \frac{1}{x}$ **75.** $-3x + 7 - \frac{6}{x}$ **77.** $4a - 5 + 6b$

79. $9x + 6 - 3y$ **81.** $x + 1$ **83.** $a - 3$ **85.** $x + 2$ **87.** $2x + 1$ **89.** $2x - 4$ **91.** $x + 1 + \frac{2}{x-1}$

93. $2x - 1 - \frac{2}{3x-2}$ **95.** $5x + 12 + \frac{12}{x-1}$ **97.** $a + 3 + \frac{4}{a+2}$ **99.** $y - 6 + \frac{26}{2y+3}$ **101.** $2x + 5 + \frac{8}{2x-1}$

103. $5a - 6 + \frac{4}{3a+2}$ **105.** $3x - 5$ **107.** $x^2 + 2x + 3$ **109.** $y^2 + 3y - 5 + \frac{10}{y+3}$

111. $2a^2 + 3a + 2 + \frac{9}{3a-2}$ **113.** $a^2 - 4a + 6 - \frac{16}{2a+3}$ **115.** $x^2 - 3$

CHAPTER REVIEW *pages 195–196*

1. $21y^2 + 4y - 1$ **2.** $2x^3 + 9x^2 - 3x - 12$ **3.** $10a^2 + 12a - 22$ **4.** $2x^2 + 3x - 8$ **5.** $13y^3 - 12y^2 - 5y - 1$
6. $-2y^2 + y - 5$ **7.** $-20x^3y^5$ **8.** $x^4y^8z^4$ **9.** $-54a^{13}b^5c^7$ **10.** 64 **11.** $9x^4y^6$
12. $100a^{15}b^{13}$ **13.** $-8x^3 - 14x^2 + 18x$ **14.** $8a^3b^3 - 4a^2b^4 + 6ab^5$ **15.** $6y^3 + 17y^2 - 2y - 21$
16. $12b^5 - 4b^4 - 6b^3 - 8b^2 + 5$ **17.** $10a^2 + 31a - 63$ **18.** $8b^2 - 2b - 15$ **19.** $a^2 - 49$

20. $25y^2 - 70y + 49$ **21.** The area is $(2w^2 - w)$ ft². **22.** The area is $(9x^2 - 12x + 4)$ in². **23.** $\frac{x^4y^6}{9}$

24. b^2 **25.** $-4y + 8$ **26.** $4b^4 + 12b^2 - 1$ **27.** $2y - 9$ **28.** $b^2 + 5b + 2 + \frac{7}{b-7}$

CHAPTER TEST *pages 197–198*

1. $3x^3 + 6x^2 - 8x + 3$ (5.1A) **2.** $-5a^3 + 3a^2 - 4a + 3$ (5.1B) **3.** a^4b^7 (5.2A) **4.** $-6x^3y^6$ (5.2A)
5. x^8y^{12} (5.2B) **6.** $-8a^6b^3$ (5.2B) **7.** $4x^3 - 6x^2$ (5.3A) **8.** $6y^4 - 9y^3 + 18y^2$ (5.3A)
9. $x^3 - 7x^2 + 17x - 15$ (5.3B) **10.** $-4x^4 + 8x^3 - 3x^2 - 14x + 21$ (5.3B) **11.** $a^2 + 3ab - 10b^2$ (5.3C)
12. $10x^2 - 43xy + 28y^2$ (5.3C) **13.** $16y^2 - 9$ (5.3D) **14.** $4x^2 - 20x + 25$ (5.3D) **15.** The area is

$3.14x^2 - 31.4x + 78.5.$ (5.3E) **16.** $-\frac{4}{x^6}$ (5.4A) **17.** $-2x^3$ (5.4A) **18.** $8ab^4$ (5.4A) **19.** $\frac{9y^{10}}{x^{10}}$ (5.4A)

20. $4a - 7$ (5.4B) **21.** $4x^4 - 2x^2 + 5$ (5.4B) **22.** $4x - 1 + \frac{3}{x^2}$ (5.4B) **23.** $x + 7$ (5.4C)

24. $x - 1 + \frac{2}{x+1}$ (5.4C) **25.** $2x + 3 + \frac{2}{2x-3}$ (5.4C)

CUMULATIVE REVIEW *pages 199–200*

1. $\frac{5}{144}$ (1.4B) **2.** $\frac{5}{3}$ (1.5A) **3.** $\frac{25}{11}$ (1.5B) **4.** $-\frac{22}{9}$ (2.1A) **5.** $5x - 3xy$ (2.2A)
6. $-9x$ (2.2B) **7.** $-18x + 12$ (2.2D) **8.** $x = -16$ (3.1C) **9.** $x = -16$ (3.3A)
10. $x = 15$ (3.3B) **11.** 22% (4.2A) **12.** $4b^3 - 4b^2 - 8b - 4$ (5.1A) **13.** $3y^3 + 2y^2 - 10y$ (5.1B)
14. a^9b^{15} (5.2B) **15.** $-8x^3y^6$ (5.2A) **16.** $6y^4 + 8y^3 - 16y^2$ (5.3A) **17.** $10a^3 - 39a^2 + 20a - 21$

(5.3B) **18.** $15b^2 - 31b + 14$ (5.3C) **19.** $9b^2 + 12b + 4$ (5.3D) **20.** $\frac{1}{2b^2}$ (5.4A) **21.** $6a - 4 + \frac{2}{a^2}$

(5.4B) **22.** $a - 7$ (5.4C) **23.** $\frac{6y^2}{x^6}$ (5.4A) **24.** $8x - 2x = 18; \ x = 3$ (3.4A) **25.** The selling price is
$43.20. (4.3A) **26.** The resulting mixture is 28% orange juice. (4.5B) **27.** The car overtakes the cyclist
25 mi from the starting point. (4.6A) **28.** The length is 15 m. The width is 6 m. (4.7A) **29.** Twenty
years ago, the gold coin was twice the age of the silver coin. (4.8C) **30.** The area is
$4x^2 + 12x + 9.$ (5.3E)

ANSWERS to Chapter 6 Odd-numbered Exercises

SECTION 6.1 *pages 207–208*

1. $5(a + 1)$ **3.** $8(2 - a^2)$ **5.** $4(2x + 3)$ **7.** $6(5a - 1)$ **9.** $x(7x - 3)$ **11.** $a^2(3 + 5a^3)$
13. $y(14y + 11)$ **15.** $2x(x^3 - 2)$ **17.** $2x^2(5x^2 - 6)$ **19.** $4a^5(2a^3 - 1)$ **21.** $xy(xy - 1)$
23. $3xy(xy^3 - 2)$ **25.** $xy(x - y^2)$ **27.** There is no common factor other than 1. **29.** $6b^2(a^2b - 2)$
31. $a(a^2 - 3a + 5)$ **33.** $5(x^2 - 3x + 7)$ **35.** $3x(x^2 + 2x + 3)$ **37.** $2x^2(x^2 - 2x + 3)$ **39.** $2x(x^2 + 3x - 7)$
41. $y^3(2y^2 - 3y + 7)$ **43.** $xy(x^2 - 3xy + 7y^2)$ **45.** $5y(y^2 + 2y - 5)$ **47.** $3b^2(a^2 - 3a + 5)$
49. $(b + 4)(x + 3)$ **51.** $(y - x)(a - b)$ **53.** $(x - 2)(x - y)$ **55.** $(2r - s)(4x + 3y)$ **57.** $(4a + 3b)(7p - 2r)$
59. $(4a - b)(2y + 1)$ **61.** $(x + 2)(x + 2y)$ **63.** $(p - 2)(p - 3r)$ **65.** $(a + 6)(b - 4)$ **67.** $(2z - 1)(z + y)$
69. $(2v - 3y)(4v + 7)$ **71.** $(2x - 5)(x - 3y)$ **73.** $(y - 2)(3y - a)$ **75.** $(3x - y)(y + 1)$ **77.** $(3s + t)(t - 2)$
79. $(2x + a)(3 - 2a)$ **81.** $(6ab - 5)(b + 2)$

Content and Format © 1991 HMCo.

SECTION 6.2 *pages 213–216*

1. $(x + 1)(x + 2)$ **3.** $(x + 1)(x - 2)$ **5.** $(a + 4)(a - 3)$ **7.** $(a - 2)(a - 1)$ **9.** $(a + 2)(a - 1)$
11. $(b - 3)(b - 3)$ **13.** $(b + 8)(b - 1)$ **15.** $(y + 11)(y - 5)$ **17.** $(y - 2)(y - 3)$ **19.** $(z - 5)(z - 9)$
21. $(z + 8)(z - 20)$ **23.** $(p + 3)(p + 9)$ **25.** $(x + 10)(x + 10)$ **27.** $(b + 4)(b + 5)$ **29.** $(x + 3)(x - 14)$
31. $(b + 4)(b - 5)$ **33.** $(y + 3)(y - 17)$ **35.** $(p + 3)(p - 7)$ **37.** Nonfactorable over the integers.
39. $(x - 5)(x - 15)$ **41.** $(x - 7)(x - 8)$ **43.** $(x + 8)(x - 7)$ **45.** $(a + 3)(a - 24)$ **47.** $(a - 3)(a - 12)$
49. $(z + 8)(z - 17)$ **51.** $(c + 9)(c - 10)$ **53.** $(z + 4)(z + 11)$ **55.** $(c + 2)(c + 17)$ **57.** $(x + 8)(x - 12)$
59. $(x - 8)(x - 14)$ **61.** $(b + 15)(b - 7)$ **63.** $(a + 3)(a - 12)$ **65.** $(b - 6)(b - 17)$ **67.** $(a + 3)(a + 24)$
69. $(x + 12)(x + 13)$ **71.** $(x + 6)(x - 16)$ **73.** $2(x + 1)(x + 2)$ **75.** $3(a + 3)(a - 2)$ **77.** $a(b + 5)(b - 3)$
79. $x(y - 2)(y - 3)$ **81.** $z(z - 3)(z - 4)$ **83.** $3y(y - 2)(y - 3)$ **85.** $3(x + 4)(x - 3)$ **87.** $5(z + 4)(z - 7)$
89. $2a(a + 8)(a - 4)$ **91.** $(x - 2y)(x - 3y)$ **93.** $(a - 4b)(a - 5b)$ **95.** $(x + 4y)(x - 7y)$
97. Nonfactorable over the integers.   **99.** $z^2(z - 5)(z - 7)$ **101.** $b^2(b - 10)(b - 12)$
103. $2y^2(y + 3)(y - 16)$ **105.** $x^2(x + 8)(x - 1)$ **107.** $4y(x + 7)(x - 2)$ **109.** $8(y - 1)(y - 3)$
111. $c(c + 3)(c + 10)$ **113.** $3x(x - 3)(x - 9)$ **115.** $(x - 3y)(x - 5y)$ **117.** $(a - 6b)(a - 7b)$
119. $(y + z)(y + 7z)$ **121.** $3y(x + 21)(x - 1)$ **123.** $3x(x - 3)(x + 4)$ **125.** $4z(z + 11)(z - 3)$
127. $4x(x + 3)(x - 1)$ **129.** $5(p + 12)(p - 7)$ **131.** $p^2(p + 12)(p - 3)$ **133.** $(t - 5s)(t - 7s)$
135. $(a + 3b)(a - 11b)$ **137.** $5x^2(x - 2)(x - 4)$ **139.** $15a(b + 4)(b - 1)$ **141.** $3y(x + 15)(x - 3)$

SECTION 6.3 *pages 221–224*

1. $(x + 1)(2x + 1)$ **3.** $(y + 3)(2y + 1)$ **5.** $(a - 1)(2a - 1)$ **7.** $(b - 5)(2b - 1)$ **9.** $(x + 1)(2x - 1)$
11. $(x - 3)(2x + 1)$ **13.** $(t + 2)(2t - 5)$ **15.** $(p - 5)(3p - 1)$ **17.** $(3y - 1)(4y - 1)$ **19.** Nonfactorable over the integers.   **21.** $(2t - 1)(3t - 4)$ **23.** $(x + 4)(8x + 1)$ **25.** Nonfactorable over the integers.
27. $(3y + 1)(4y + 5)$ **29.** $(a + 7)(7a - 2)$ **31.** $(b - 4)(3b - 4)$ **33.** $(z - 14)(2z + 1)$ **35.** $(p + 8)(3p - 2)$
37. $2(x + 1)(2x + 1)$ **39.** $5(y - 1)(3y - 7)$ **41.** $x(x - 5)(2x - 1)$ **43.** $b(a - 4)(3a - 4)$
45. Nonfactorable over the integers.   **47.** $(3 - x)(4 + x)$ **49.** $4(4y - 1)(5y - 1)$ **51.** $z(2z + 3)(4z + 1)$
53. $y(2x - 5)(3x + 2)$ **55.** $5(t + 2)(2t - 5)$ **57.** $p(p - 5)(3p - 1)$ **59.** $2(z + 4)(13z - 3)$
61. $2y(y - 4)(5y - 2)$ **63.** $yz(z + 2)(4z - 3)$ **65.** $3a(2a + 3)(7a - 3)$ **67.** $y(3x - 5y)(3x - 5y)$
69. $xy(3x - 4y)(3x - 4y)$ **71.** $(2x - 3)(3x - 4)$ **73.** $(b + 7)(5b - 2)$ **75.** $(2a - 3)(3a + 8)$
77. $(z + 2)(4z + 3)$ **79.** $(2p + 5)(11p - 2)$ **81.** $(y + 1)(8y + 9)$ **83.** $(3t + 1)(6t - 5)$
85. $(b + 12)(6b - 1)$ **87.** $(3x + 2)(3x + 2)$ **89.** $(2b - 3)(3b - 2)$ **91.** $(3b + 5)(11b - 7)$
93. $(3y - 4)(6y - 5)$ **95.** $(3a + 7)(5a - 3)$ **97.** $(2y - 5)(4y - 3)$ **99.** $(2z + 3)(4z - 5)$
101. Nonfactorable over the integers.   **103.** $(2z - 5)(5z - 2)$ **105.** $(6z + 5)(6z + 7)$ **107.** $(x + y)(3x - 2y)$
109. $(a + 2b)(3a - b)$ **111.** $(y - 2z)(4y - 3z)$ **113.** $(4 + z)(7 - z)$ **115.** $(1 - x)(8 + x)$
117. $3(x + 5)(3x - 4)$ **119.** $4(2x - 3)(3x - 2)$ **121.** $a^2(5a + 2)(7a - 1)$ **123.** $5(b - 7)(3b - 2)$
125. $(x - 7y)(3x - 5y)$ **127.** $3(8y - 1)(9y + 1)$ **129.** $(1 - x)(21 + x)$ **131.** $(3a - 2b)(5a + 7b)$
133. $z(3 - z)(11 + z)$ **135.** $2x(x + 1)(5x + 1)$ **137.** $yz(z - 8)(2z - 1)$ **139.** $b^2(2b - 3)(3b - 2)$
141. $ab(a - 5b)(2a - b)$

SECTION 6.4 *pages 229–232*

1. $(x + 2)(x - 2)$ **3.** $(a + 9)(a - 9)$ **5.** $(y + 1)^2$ **7.** $(a - 1)^2$ **9.** $(2x + 1)(2x - 1)$ **11.** $(x^3 + 3)(x^3 - 3)$
13. Nonfactorable over the integers.   **15.** $(x + y)^2$ **17.** $(2a + 1)^2$ **19.** $(3x + 1)(3x - 1)$
21. $(1 + 8x)(1 - 8x)$ **23.** Nonfactorable over the integers.   **25.** $(3a + 1)^2$ **27.** $(b^2 + 4a)(b^2 - 4a)$
29. $(2a - 5)^2$ **31.** $(3a - 7)^2$ **33.** $(5z + y)(5z - y)$ **35.** $(ab + 5)(ab - 5)$ **37.** $(5x + 1)(5x - 1)$
39. $(2a - 3b)^2$ **41.** $(2y - 9z)^2$ **43.** $(\frac{1}{x} - 2)(\frac{1}{x} + 2)$ **45.** $(3ab - 1)^2$ **47.** $2(2y - 1)(2y + 1)$
49. $3a(a + 1)^2$ **51.** $(m^2 + 16)(m - 4)(m + 4)$ **53.** $(9x + 4)(x + 1)$ **55.** $4y^2(2y + 3)^2$
57. $(y^4 + 9)(y^2 - 3)(y^2 + 3)$ **59.** $(5 - 2p)^2$ **61.** $(4x - 3 - y)(4x - 3 + y)$ **63.** $(x - 2 - y)(x - 2 + y)$
65. $5(x + 1)(x - 1)$ **67.** $x(x + 2)^2$ **69.** $x^2(x + 7)(x - 5)$ **71.** $5(b + 3)(b + 12)$ **73.** Nonfactorable over the integers.   **75.** $2y(x + 11)(x - 3)$ **77.** $x(x^2 - 6x - 5)$ **79.** $3(y^2 - 12)$ **81.** $(2a + 1)(10a + 1)$
83. $y^2(x + 1)(x - 8)$ **85.** $5(a + b)(2a - 3b)$ **87.** $2(5 + x)(5 - x)$ **89.** $b^2(a - 5)^2$ **91.** $ab(4a + b)(3a - b)$
93. $3a(2a - 1)^2$ **95.** $3(81 + a^2)$ **97.** $2a(2a - 5)(3a - 4)$ **99.** $a(2a + 5)^2$ **101.** $3b(3a - 1)^2$
103. $6(4 + x)(2 - x)$ **105.** $x^2(x + y)(x - y)$ **107.** $2a(3a + 2)^2$ **109.** $b(2 - 3a)(1 + 2a)$ **111.** $8x(3y + 1)^2$
113. $y^2(5 + x)(3 - x)$ **115.** $3(x + 3y)(x - 3y)$ **117.** $y(y + 3)(y - 3)$ **119.** $x^2y^2(5x + 4y)(3x - 5y)$
121. $2(x - 1)(a + b)$ **123.** $(x - 2)(x + 1)(x - 1)$ **125.** $(x + 2)(x - 2)(a + b)$ **127.** $(x - 5)(2 + x)(2 - x)$
129. $(x + 2)(x - 2)^2$

SECTION 6.5   *pages 237–240*

1. The solutions are −3 and −2.   **3.** The solutions are 7 and 3.   **5.** The solutions are 0 and 5.

7. The solutions are 0 and 9.   **9.** The solutions are 0 and $-\frac{3}{2}$.   **11.** The solutions are 0 and $\frac{2}{3}$.

13. The solutions are −2 and 5.   **15.** The solutions are 9 and −9.   **17.** The solutions are $\frac{7}{2}$ and $-\frac{7}{2}$.

19. The solutions are $\frac{1}{3}$ and $-\frac{1}{3}$.   **21.** The solutions are −2 and −4.   **23.** The solutions are −7 and 2.

25. The solutions are 2 and 3.   **27.** The solutions are −7 and 3.   **29.** The solutions are $-\frac{1}{2}$ and 5.

31. The solutions are $-\frac{1}{3}$ and $-\frac{1}{2}$.   **33.** The solutions are 0 and 3.   **35.** The solutions are 0 and 7.

37. The solutions are −1 and −4.   **39.** The solutions are 2 and 3.   **41.** The solutions are $\frac{1}{2}$ and −4.

43. The solutions are $\frac{1}{3}$ and 4.   **45.** The solutions are 3 and 9.   **47.** The solutions are 9 and −2.

49. The solutions are −1 and −2.   **51.** The solutions are −9 and 5.   **53.** The solutions are −7 and 4.
55. The solutions are −2 and −3.   **57.** The solutions are −5 and −8.   **59.** The solutions are 1 and 3.
61. The solutions are 5 and −12.   **63.** The solutions are $-\frac{3}{2}$ and −2.   **65.** The solutions are $-\frac{1}{2}$ and −4.

67. The number is 6.   **69.** The numbers are 2 and 4.   **71.** The numbers are 6 and 7.   **73.** The numbers are 3 and 7.   **75.** There will be 12 consecutive numbers.   **77.** There are 8 teams.   **79.** The object will hit 10 seconds later.   **81.** It will be 2 seconds later.   **83.** The length is 15 in., and the width is 5 in. **85.** The length of a side of the original square is 10 in.   **87.** The dimensions of the type area are 4 in. by 7 in. **89.** The radius of the original circle is 3.8078556 in.

CHAPTER REVIEW   *pages 243–244*

1. $5x(x^2 + 2x + 7)$ **2.** $3ab(4a + b)$ **3.** $7y^3(2y^6 − 7y^3 + 1)$ **4.** $(x − 3)(4x + 5)$ **5.** $(2x + 5)(5x + 2y)$
6. $(3a − 5b)(7x + 2y)$ **7.** $(b − 3)(b − 10)$ **8.** $(c + 6)(c + 2)$ **9.** $(y − 4)(y + 9)$ **10.** $3(a + 2)(a − 7)$
11. $4x(x − 6)(x + 1)$ **12.** $n^2(n + 1)(n − 3)$ **13.** $(2x − 7)(3x − 4)$ **14.** $(6y − 1)(2y + 3)$
15. Nonfactorable over the integers.   **16.** $(3x − 2)(x − 5)$ **17.** $(2a + 5)(a − 12)$ **18.** $(6a − 5)(3a + 2)$
19. $(a^3 + 10)(a^3 − 10)$ **20.** $(3y^2 + 5z)(3y^2 − 5z)$ **21.** $(ab + 1)(ab − 1)$ **22.** $5(x + 2)(x − 3)$ **23.** $3(x + 6)^2$
24. $2b(3b − 4)(2b − 7)$ **25.** The solutions are $\frac{1}{4}$ and −7.   **26.** The solutions are −3 and 7.   **27.** The length is 100 yd. The width is 60 yd.   **28.** The distance between the screen and the projector is 20 ft.
29. The width of the frame is $\frac{3}{2}$ in.   **30.** The length of a side is 20 ft.

CHAPTER TEST   *pages 245–246*

1. $2x(3x^2 − 4x + 5)$ (6.1A)   **2.** $(b + 6)(a − 3)$ (6.1B)   **3.** $(p + 2)(p + 3)$ (6.2A)   **4.** $(a − 3)(a − 16)$ (6.2A)
5. $(x + 5)(x − 3)$ (6.2A)   **6.** $(x + 3)(x − 12)$ (6.2A)   **7.** $5(x^2 − 9x − 3)$ (6.1A)   **8.** $2y^2(y + 1)(y − 8)$
(6.2B)   **9.** Nonfactorable over the integers. (6.3A)   **10.** $(2x + 1)(3x + 8)$ (6.3A)   **11.** $4(x + 4)(2x − 3)$
(6.3B)   **12.** $3y^2(2x + 1)(x + 1)$ (6.3B)   **13.** $(x − 2)(a + b)$ (6.3B)   **14.** $(p + 1)(x − 1)$ (6.3B)
15. $(b + 4)(b − 4)$ (6.4A)   **16.** $(2x + 7y)(2x − 7y)$ (6.4A)   **17.** $(p + 6)^2$ (6.4A)   **18.** $(2a − 3b)^2$ (6.4A)
19. $3(a + 5)(a − 5)$ (6.4B)   **20.** $3(x + 2y)^2$ (6.4B)   **21.** The solutions are $\frac{3}{2}$ and −7. (6.5A)   **22.** The
solutions are $\frac{1}{2}$ and $-\frac{1}{2}$ (6.5A)   **23.** The solutions are 3 and 5. (6.5A)   **24.** The two numbers are 3 and 7.
(6.5B)   **25.** The length is 15 cm. The width is 6 cm. (6.5B)

CUMULATIVE REVIEW   *pages 247–248*

1. 7 (1.2B)   **2.** 4 (1.5B)   **3.** −7 (2.1A)   **4.** $15x^2$ (2.2B)   **5.** 12 (2.2D)   **6.** $x = \frac{2}{3}$ (3.1C)

7. $x = \frac{7}{4}$ (3.3A)   **8.** $x = 3$ (3.3B)   **9.** 45 (4.2A)   **10.** $9a^6b^4$ (5.2B)   **11.** $x^3 − 3x^2 − 6x + 8$ (5.3B)

12. $4x + 8 + \frac{21}{2x − 3}$ (5.4C)   **13.** $\frac{y^6}{x^8}$ (5.4A)   **14.** $(a − b)(3 − x)$ (6.1B)   **15.** $5xy^2(3 − 4y^2)$ (6.1A)
16. $(x − 7y)(x + 2y)$ (6.2A)   **17.** $(p − 10)(p + 1)$ (6.2A)   **18.** $3a(2a + 5)(3a + 2)$ (6.3B)

19. $(6a - 7b)(6a + 7b)$ (6.4A)   **20.** $(2x + 7y)^2$ (6.4A)   **21.** $(3x - 2)(3x + 7)$ (6.3A)   **22.** $2(3x - 4y)^2$
(6.4B)   **23.** $(x - 3)(3y - 2)$ (6.1B)   **24.** The solutions are $\frac{2}{3}$ and -7. (6.5A)   **25.** The pieces measure
4 ft and 6 ft. (3.4B)   **26.** The discount rate is 40%. (4.3B)   **27.** \$6500 more must be invested. (4.4A)
28. The distance to the resort was 168 mi. (4.6A)   **29.** The integers are 10, 12, and 14. (4.8A)   **30.** The
length of the base is 12 in. (6.5B)

ANSWERS to Chapter 7 Odd-numbered Exercises

SECTION 7.1   *pages 255–258*

1. $\frac{3}{4x}$ **3.** $\frac{1}{x+3}$ **5.** -1 **7.** $\frac{2}{3y}$ **9.** $-\frac{3}{4x}$ **11.** $\frac{a}{b}$ **13.** $-\frac{2}{x}$ **15.** $\frac{y-2}{y-3}$ **17.** $\frac{x+5}{x+4}$ **19.** $\frac{x+4}{x-3}$

21. $-\frac{x+2}{x+5}$ **23.** $\frac{2(x+2)}{x+3}$ **25.** $\frac{2x-1}{2x+3}$ **27.** $-\frac{x+7}{x+6}$ **29.** $\frac{5ab^2}{12x^2y}$ **31.** $\frac{4x^3y^3}{3a^2}$ **33.** $\frac{3}{4}$ **35.** ab^2

37. $\frac{x^2(x-1)}{y(x+3)}$ **39.** $\frac{y(x-1)}{x^2(x+10)}$ **41.** $-ab^2$ **43.** $\frac{x+5}{x+4}$ **45.** 1 **47.** $-\frac{n-10}{n-7}$ **49.** $\frac{x(x+2)}{2(x-1)}$

51. $-\frac{x+2}{x-6}$ **53.** $\frac{x+5}{x-12}$ **55.** $\frac{3y+2}{3y+1}$ **57.** $-\frac{3x-5}{4x-5}$ **59.** $\frac{7a^3y^2}{40bx}$ **61.** $\frac{4}{3}$ **63.** $\frac{3a}{2}$ **65.** $\frac{x^2(x+4)}{y^2(x+2)}$

67. $\frac{x(x-2)}{y(x-6)}$ **69.** $-\frac{3by}{ax}$ **71.** $\frac{(x-3)(x+6)}{(x+7)(x-6)}$ **73.** 1 **75.** $-\frac{x+8}{x-4}$ **77.** $\frac{2n+1}{2n-3}$ **79.** $-\frac{3x+1}{2x-3}$

81. $\frac{(3x+2)(5-4x)(5x-4)}{(2x-3)(4x+5)(3x-5)}$

SECTION 7.2   *pages 261–262*

1. $24x^3y^2$ **3.** $30x^4y^2$ **5.** $8x^2(x+2)$ **7.** $6x^2y(x+4)$ **9.** $36x(x+2)^2$ **11.** $6(x+1)^2$
13. $(x-1)(x+2)(x+3)$ **15.** $(x-5)(2x+3)^2$ **17.** $(x-1)(x-2)$ **19.** $(x-3)(x+2)(x+4)$
21. $(x+4)(x+1)(x-7)$ **23.** $(x-6)(x+6)(x+4)$ **25.** $(x+3)(x-8)(x-10)$ **27.** $(x+2)(x-3)(3x-2)$
29. $(x+2)(x-3)$ **31.** $(x-5)(x+1)$ **33.** $(x-1)(x-2)(x-3)(x-6)$ **35.** $\frac{5}{ab^2}, \frac{6b}{ab^2}$ **37.** $\frac{15y^2}{18x^2y}, \frac{14x}{18x^2y}$

39. $\frac{ay+5a}{y^2(y+5)}, \frac{6y}{y^2(y+5)}$ **41.** $\frac{a^2y+7a^2}{y(y+7)^2}, \frac{ay}{y(y+7)^2}$ **43.** $\frac{b}{y(y-4)}, -\frac{b^2y}{y(y-4)}$ **45.** $-\frac{3y-21}{(y-7)^2}, \frac{2}{(y-7)^2}$

47. $\frac{2y^2}{y^2(y-3)}, \frac{3}{y^2(y-3)}$ **49.** $\frac{x^3+4x^2}{(2x-1)(x+4)}, \frac{2x^2+x-1}{(2x-1)(x+4)}$ **51.** $\frac{3x^2+15x}{(x-5)(x+5)}, \frac{4}{(x-5)(x+5)}$

53. $\frac{x-3}{(3x-2)(x+2)}, \frac{6x-4}{(3x-2)(x+2)}$ **55.** $\frac{x^2-1}{(x+5)(x-3)(x+1)}, \frac{x^2-3x}{(x+5)(x-3)(x+1)}$ **57.** $-\frac{2x^2-6x}{(x-3)(x-5)(x+2)}, \frac{x^2+4x+4}{(x-3)(x-5)(x+2)}$

59. $\frac{x^2-6x-7}{(x+5)(x-7)}, \frac{x^2+7x+10}{(x+5)(x-7)}, -\frac{3}{(x+5)(x-7)}$

SECTION 7.3   *pages 267–270*

1. $\frac{11}{y^2}$ **3.** $-\frac{7}{x+4}$ **5.** $\frac{8x}{2x+3}$ **7.** $\frac{5x+7}{x-3}$ **9.** $\frac{2x-5}{x+9}$ **11.** $\frac{-3x-4}{2x+7}$ **13.** $\frac{1}{x+5}$ **15.** $\frac{1}{x-6}$

17. $\frac{3}{2y-1}$ **19.** $\frac{1}{x-5}$ **21.** $\frac{4y+5x}{xy}$ **23.** $\frac{19}{2x}$ **25.** $\frac{5}{12x}$ **27.** $\frac{19x-12}{6x^2}$ **29.** $\frac{52y-35x}{20xy}$ **31.** $\frac{13x+2}{15x}$

33. $\frac{7}{24}$ **35.** $\frac{x+90}{45x}$ **37.** $\frac{x^2+2x+2}{2x^2}$ **39.** $\frac{2x^2+3x-10}{4x^2}$ **41.** $\frac{-3x^2+16x+2}{12x^2}$ **43.** $\frac{x^2-x+2}{x^2y}$

45. $\frac{16xy-12y+6x^2+3x}{12x^2y^2}$ **47.** $\frac{3xy-6y-2x^2-14x}{24x^2y}$ **49.** $\frac{9x+2}{(x-2)(x+3)}$ **51.** $\frac{2(x+23)}{(x-7)(x+3)}$ **53.** $\frac{2x^2-5x+1}{(x+1)(x-3)}$

55. $\frac{4x^2-34x+5}{(2x-1)(x-6)}$ **57.** $\frac{2a-5}{a-7}$ **59.** $\frac{4x+9}{(x-3)(x+3)}$ **61.** $\frac{-x+9}{(x+2)(x-3)}$ **63.** $\frac{14}{(x-5)(x-5)}$ **65.** $-\frac{2(x+7)}{(x+6)(x-7)}$

67. $\frac{x-4}{x-6}$ **69.** $\frac{2x+1}{x-1}$ **71.** $-\frac{3(x^2+8x+25)}{(x-3)(x+7)}$

SECTION 7.4   *pages 273–274*

1. $\frac{x}{x-3}$ **3.** $\frac{2}{3}$ **5.** $\frac{y+3}{y-4}$ **7.** $\frac{2(2x+13)}{5x+36}$ **9.** $\frac{x+2}{x+3}$ **11.** $\frac{x-6}{x+5}$ **13.** $-\frac{x-2}{x+1}$ **15.** $x-1$ **17.** $\frac{1}{2x-1}$

19. $\frac{x-3}{x+5}$ **21.** $\frac{x-7}{x-8}$ **23.** $\frac{2y-1}{2y+1}$ **25.** $\frac{x-2}{2x-5}$ **27.** $-\frac{x+1}{4x-3}$ **29.** $\frac{x+1}{2(5x-2)}$ **31.** $\frac{b+11}{4b-21}$

SECTION 7.5   *pages 277–278*

1. The solution is 3.   **3.** The solution is 1.   **5.** The solution is 9.   **7.** The solution is 1.   **9.** The solution is $\frac{1}{4}$.   **11.** The solution is 1.   **13.** The solution is -3.   **15.** The solution is $\frac{1}{2}$.   **17.** The solution is 8.   **19.** The solution is 5.   **21.** The solution is -1.   **23.** The solution is 5.   **25.** The equation has no solution.   **27.** The solutions are 2 and 4.   **29.** The solutions are $-\frac{3}{2}$ and 4.   **31.** The solution is 3.   **33.** The solution is 4.   **35.** The solution is -1.

SECTION 7.6   *pages 281–282*

1. The solution is 9.   **3.** The solution is 12.   **5.** The solution is 7.   **7.** The solution is 6.   **9.** The solution is 1.   **11.** The solution is -6.   **13.** The solution is 4.   **15.** The solution is $-\frac{2}{3}$.   **17.** There will be 20,000 voters voting in favor.   **19.** The building will need 140 air vents.   **21.** Yes, the shipment will be accepted.   **23.** It will take 18 min to print the document.   **25.** There are an estimated 75 elk in the preserve.   **27.** The health department will need 6 additional vials.   **29.** The area for a window should be 40 ft^2.

SECTION 7.7   *pages 285–286*

1. $y = -3x + 10$ **3.** $y = 4x - 3$ **5.** $y = -\frac{3}{2}x + 3$ **7.** $y = \frac{2}{5}x - 2$ **9.** $y = -\frac{2}{7}x + 2$ **11.** $y = -\frac{1}{3}x + 2$ **13.** $y = \frac{1}{4}x - 3$ **15.** $y = \frac{7}{2}x - 7$ **17.** $y = 3x + 7$ **19.** $x = -3y + 6$ **21.** $x = \frac{1}{3}y + 1$ **23.** $x = -\frac{5}{2}y + 5$ **25.** $x = 2y - 1$ **27.** $x = -\frac{4}{5}y - 4$ **29.** $x = \frac{2}{3}y + 5$ **31.** $h = \frac{2A}{b}$ **33.** $t = \frac{d}{r}$ **35.** $T = \frac{PV}{nR}$ **37.** $l = \frac{P - 2w}{2}$ **39.** $b_1 = \frac{2A - hb_2}{h}$ **41.** $h = \frac{3V}{A}$ **43.** $S = C - Rt$ **45.** $P = \frac{A}{1 + rt}$ **47.** $w = \frac{A}{S + 1}$

SECTION 7.8   *pages 291–294*

1. It would take the experienced painter 6 h working alone.   **3.** It would take both skiploaders 3 h working together.   **5.** It would take 30 h.   **7.** It would take 6 min with both air conditioners working.   **9.** It would take the second welder 15 h working alone.   **11.** It would take the second harvester 3 h working alone.   **13.** It would take the second mason 6 h.   **15.** It would take the second technician 3 h.   **17.** It would take one welder 40 h working alone.   **19.** It would take one machine 28 h working alone.   **21.** The camper hiked at a rate of 5 mph.   **23.** The rate of the jogger is 8 mph, and the rate of the cyclist is 20 mph.   **25.** The rate of the jet is 360 mph.   **27.** The rate of the second plane is 150 mph.   **29.** The rate of the car is 48 mph.   **31.** The rate for the first 9 mi was 3 mph.   **33.** The rate of the current is 2 mph.   **35.** The rate of the gulf current is 6 mph.   **37.** The rate of the trucker for the first 330 mi is 55 mph.   **39.** The rate of the current is 5 mph.

CHAPTER REVIEW   *pages 297–298*

1. $\frac{2x^4}{3y^7}$ **2.** $-\frac{x + 6}{x + 3}$ **3.** $\frac{by^3}{6ax^2}$ **4.** $\frac{1}{x}$ **5.** $\frac{8x + 5}{3x - 4}$ **6.** $\frac{b^3y}{10ax}$ **7.** $\frac{1}{x^2}$ **8.** $\frac{(3y - 2)^2}{(y - 1)(y - 2)}$ **9.** $(5x - 3)(2x - 1)(4x - 1)$ **10.** $\frac{3x^2 - x}{(2x + 3)(6x - 1)(3x - 1)}, \frac{24x^3 - 4x^2}{(2x + 3)(6x - 1)(3x - 1)}$ **11.** $\frac{1}{x + 3}$ **12.** $\frac{7x + 22}{60x}$ **13.** $\frac{2y - 3}{5y - 7}$ **14.** $\frac{3x - 1}{x - 5}$ **15.** $\frac{x}{x - 7}$ **16.** $x - 2$ **17.** $\frac{x - 2}{3x - 10}$ **18.** The solution is 2.   **19.** The equation has no solution.   **20.** The solution is 5.   **21.** The solution is 12.   **22.** The solution is 62.   **23.** The solution is 10.   **24.** The pitcher's ERA is 1.35.   **25.** $y = -\frac{4}{9}x + 2$ **26.** $c = \frac{100m}{i}$ **27.** $a = \frac{T - 2bc}{2b + 2c}$ **28.** It would take 6 h to fill the pool using both hoses.   **29.** The rate of the car is 45 mph.   **30.** The rate of the wind is 20 mph.

CHAPTER TEST   *pages 299–300*

1. $\frac{2x^3}{3y^3}$ (7.1A)   **2.** $-\frac{x+5}{x+1}$ (7.1A)   **3.** $\frac{x+1}{x^3(x-2)}$ (7.1B)   **4.** $\frac{(x-5)(2x-1)}{(x+3)(2x+5)}$ (7.1B)   **5.** $\frac{x+5}{x+4}$ (7.1C)

6. $3(2x-1)(x+1)$ (7.2A)   **7.** $\frac{3x+6}{x(x-2)(x+2)}, \frac{x^2}{x(x-2)(x+2)}$ (7.2B)   **8.** $\frac{2}{x+5}$ (7.3A)   **9.** $\frac{5}{(2x-1)(3x+1)}$ (7.3B)

10. $\frac{x^2-4x+5}{(x-2)(x+3)}$ (7.3B)   **11.** $\frac{x-3}{x-2}$ (7.4A)   **12.** The solution is 2. (7.5A)   **13.** The equation has no

solution. (7.5A)   **14.** The solution is -1. (7.6A)   **15.** Two additional pounds of salt are required. (7.6B)

16. Fifty-four sprinklers are needed. (7.6B)   **17.** $y = \frac{3}{8}x - 2$ (7.7A)   **18.** $t = \frac{d-s}{r}$ (7.7A)   **19.** It would

take 4 h to fill the pool with both pipes turned on. (7.8A)   **20.** The rate of the wind is 20 mph. (7.8B)

CUMULATIVE REVIEW   *pages 301–302*

1. $\frac{31}{30}$ (1.5B)   **2.** 21 (2.1A)   **3.** $5x - 2y$ (2.2A)   **4.** $-8x + 26$ (2.2D)   **5.** $x = -\frac{9}{2}$ (3.2A)

6. $x = -12$ (3.3B)   **7.** 10 (4.2A)   **8.** a^3b^7 (5.2A)   **9.** $a^2 + ab - 12b^2$ (5.3C)   **10.** $3b^3 - b + 2$

(5.4B)   **11.** $x^2 + 2x + 4$ (5.4C)   **12.** $(4x+1)(3x-1)$ (6.3A)   **13.** $(y-6)(y-1)$ (6.2A)

14. $a(2a-3)(a+5)$ (6.3B)   **15.** $4(b+5)(b-5)$ (6.4B)   **16.** The solutions are -3 and $\frac{5}{2}$. (6.5A)

17. $\frac{2x^3}{3y^5}$ (7.1A)   **18.** $-\frac{x-2}{x+5}$ (7.1A)   **19.** 1 (7.1C)   **20.** $\frac{3}{(2x-1)(x+1)}$ (7.3B)   **21.** $\frac{x+3}{x+5}$ (7.4A)

22. The solution is 4. (7.5A)   **23.** The solution is 3. (7.6A)   **24.** $t = \frac{f-v}{a}$ (7.7A)   **25.** $5x - 13 = -8$;

$x = 1$ (3.4A)   **26.** The 120 gram alloy is 70% silver. (4.5B)   **27.** The base is 10 in. and the height is 6

in. (6.5B)   **28.** A policy of $5000 would cost $80. (7.6B)   **29.** Working together, it would take the pipes 6

min to fill the tank. (7.8A)   **30.** The rate of the current is 2 mph. (7.8B)

ANSWERS to Chapter 8 Odd-numbered Exercises

SECTION 8.1   *pages 309–312*

1.   **3.**   **5.**

7. A is $(2, 3)$, B is $(4, 0)$, C is $(-4, 1)$ and D is $(-2, -2)$.   **9.** A is $(-2, 5)$, B is $(3, 4)$, C is $(0, 0)$ and D is $(-3, -2)$.
11. a) The abscissa of point A is 2. The abscissa of point C is -4.   **b)** The ordinate of point B is 1. The
ordinate of point D is -3.   **13.** Yes, $(3, 4)$ is a solution of $y = -x + 7$.   **15.** No, $(-1, 2)$ is not a solution of
$y = \frac{1}{2}x - 1$.   **17.** No, $(4, 1)$ is not a solution of $y = \frac{1}{4}x + 1$.   **19.** Yes, $(0, 4)$ is a solution of $y = \frac{3}{4}x + 4$.
21. No, $(0, 0)$ is not a solution of $y = 3x + 2$.   **23.** The ordered pair solution is $(3, 7)$.   **25.** The ordered pair
solution is $(6, 3)$.   **27.** The ordered pair solution is $(0, 1)$.   **29.** The ordered pair solution is $(-5, 0)$.

31. 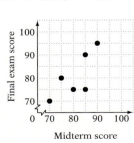   **33.** 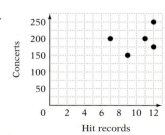   **35.**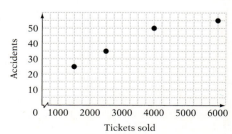

SECTION 8.2   *pages 317–320*

1.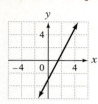

3.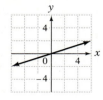

5.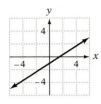

7.

9.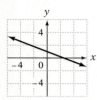

11.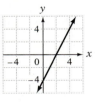

13.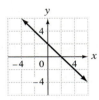

15.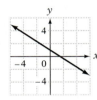

17.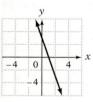

19.

21.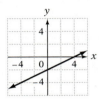

23.

25.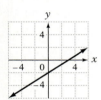

27.

29.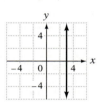

31.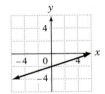

SECTION 8.3   *pages 327–330*

1. The *x*-intercept is (3, 0), and the *y*-intercept is (0, −3).   **3.** The *x*-intercept is (2, 0), and the *y*-intercept is (0, −6).   **5.** The *x*-intercept is (10, 0), and the *y*-intercept is (0, −2).   **7.** The *x*-intercept is (−4, 0) and the *y*-intercept is (0, 12).   **9.** The *x*-intercept and the *y*-intercept are both (0, 0).
11. The *x*-intercept is (6, 0), and the *y*-intercept is (0, 3).

13. *x*-intercept: (2, 0)
    y-intercept: (0, 5)
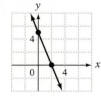

15. *x*-intercept: (4, 0)
    y-intercept: (0, −3)
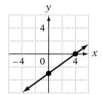

17. The slope is −2.   **19.** The slope is $\frac{1}{3}$.   **21.** The slope is $-\frac{5}{2}$.   **23.** The slope is $-\frac{1}{2}$.   **25.** The slope is −1.   **27.** The slope is undefined.   **29.** The line has zero slope.   **31.** The slope is $-\frac{1}{3}$.   **33.** The line has zero slope.   **35.** The slope is −5.   **37.** The slope is undefined.   **39.** The slope is $-\frac{2}{3}$.

41.   **43.** 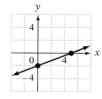   **45.** 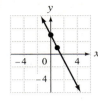   **47.**

49.   **51.**   **53.** 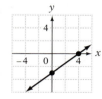   **55.**

SECTION 8.4   *pages 333–334*

1. The equation of the line is $y = 2x + 2$.   **3.** The equation of the line is $y = -3x - 1$.   **5.** The equation of the line is $y = \frac{1}{3}x$.   **7.** The equation of the line is $y = \frac{3}{4}x - 5$.   **9.** The equation of the line is $y = -\frac{3}{5}x$.

11. The equation of the line is $y = \frac{1}{4}x + \frac{5}{2}$.   **13.** The equation of the line is $y = 2x - 3$.   **15.** The equation of the line is $y = -2x - 3$.   **17.** The equation of the line is $y = \frac{2}{3}x$.   **19.** The equation of the line is $y = \frac{1}{2}x + 2$.

21. The equation of the line is $y = -\frac{3}{4}x - 2$.   **23.** The equation of the line is $y = \frac{3}{4}x + \frac{5}{2}$.

CHAPTER REVIEW   *pages 337–338*

1. 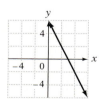   **2.** **a)** The abscissa of point A is 3.   **b)** The ordinate of point B is 0.

3. No, (6, 3) is not a solution of $y = \frac{2}{3}x + 1$.   **4.** The ordered pair solution is (9, −13).

5. 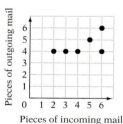   **6.** 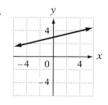   **7.** 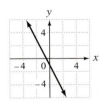   **8.**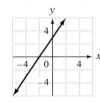

9. 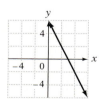   **10.** 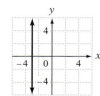   **11.** The x-intercept is (8, 0), and the y-intercept is (0, −12).

12. The slope is $\frac{7}{11}$.   **13.** The line has zero slope.

14.
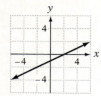

15.

16.
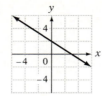

17. The equation of the line is $y = 4x - 7$.   **18.** The equation of the line is $y = -\frac{2}{5}x + 7$.

19. The equation of the line is $y = \frac{3}{5}x - 3$.   **20.** The equation of the line is $y = -\frac{1}{3}x$.

CHAPTER TEST   *pages 339–340*

1.   (8.1A)

2. The ordered pair solution is (3, 0).   (8.1B)

3. The ordered pair solution is (3, 1).   (8.1B)   **4.**
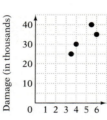
(8.1C)

5.   (8.2A)

6.   (8.2A)

7.   (8.2B)
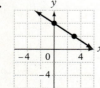

8.   (8.2B)

9. The *x*-intercept is (2, 0), and the *y*-intercept is (0, −3).   (8.3A)   **10.** The *x*-intercept is (−2, 0), and the *y*-intercept is (0, 1).   (8.3A)   **11.** The slope is 2.   (8.3B)   **12.** The line has zero slope.   (8.3B)   **13.** The slope is undefined.   (8.3B)   **14.** The slope is $-\frac{2}{3}$.   (8.3C)

15.   (8.3C)

16.   (8.3C)
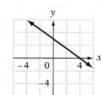

17. The equation of the line is $y = 3x - 1$.   (8.4A)   **18.** The equation of the line is $y = \frac{2}{3}x + 3$.   (8.4A)

19. The equation of the line is $y = \frac{1}{2}x + 2$.   (8.4B)   **20.** The equation of the line is $y = -\frac{2}{3}x + \frac{4}{3}$.   (8.4B)

CUMULATIVE REVIEW *pages 341–342*

1. -12 (1.5B)   **2.** $-\frac{5}{8}$ (2.1A)   **3.** $-17x + 28$ (2.2D)   **4.** $x = \frac{3}{2}$ (3.2A)   **5.** $x = \frac{19}{18}$ (3.3B)

6. $\frac{1}{15}$ (4.1A)   **7.** $-32x^8y^7$ (5.2A)   **8.** $-3x^2$ (5.4A)   **9.** $x + 3$ (5.4C)   **10.** $5(x + 2)(x + 1)$ (6.2B)

11. $(a + 2)(x + y)$ (6.1A)   **12.** The solutions are 4 and -2. (6.5A)   **13.** $\frac{x^3(x + 3)}{y(x + 2)}$ (7.1B)   **14.** $\frac{3}{x + 8}$

(7.3A)   **15.** $x = 2$ (7.5A)   **16.** $y = \frac{4}{5}x - 3$ (7.7A)   **17.** The ordered pair solution is $(-2, -5)$. (8.1B)

18. The line has zero slope. (8.3B)   **19.** The equation of the line is $y = \frac{1}{2}x - 2$. (8.4A)   **20.** The equation

of the line is $y = -3x + 2$. (8.4A)   **21.** The equation of the line is $y = 2x + 2$. (8.4B)   **22.** The equation of

the line is $y = \frac{2}{3}x - 3$. (8.4B)   **23.** The sale price is \$62.30. (4.3B)   **24.** The present age of the gold coin is

135 years and the present age of the silver coin is 75 years. (4.8C)   **25.** The value of the home is

\$110,000. (7.6B)   **26.** It would take $3\frac{3}{4}$ h with both the electrician and the apprentice working. (7.8A)

27. 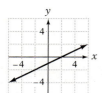 (8.2A)   **28.** 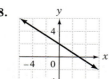 (8.3C)

ANSWERS to Chapter 9 Odd-numbered Examples

SECTION 9.1 *pages 349–352*

1. Yes, (2, 3) is a solution of the system of equations.   **3.** Yes, (1, -2) is a solution of the system of equations.
5. No, (4, 3) is not a solution of the system of equations.   **7.** No, (-1, 3) is not a solution of the system of
equations.   **9.** No, (0, 0) is not a solution of the system of equations.   **11.** Yes, (2, -3) is a solution of the
system of equations.   **13.** Yes, (5, 2) is a solution of the system of equations.   **15.** Yes, (-2, -3) is a
solution of the system of equations.   **17.** No, (0, -3) is not a solution of the system of equations.

19.   **21.**   **23.** 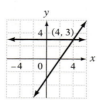   **25.**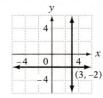

27. 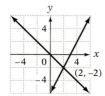   **29.** 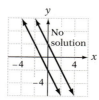   **31.** 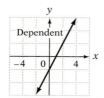   **33.**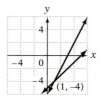

35. 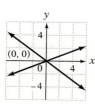   **37.** 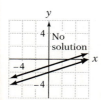   **39.** 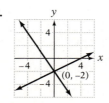   **41.**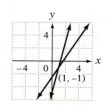

SECTION 9.2   *pages 355–356*

1.   The solution is (2, 1).   **3.**   The solution is (4, 1).   **5.**   The solution is (−1, 1).   **7.**   The solution is (3, 1).
9.   The solution is (1, 1).   **11.**   The solution is (−1, 1).   **13.**   The system of equations is inconsistent and has
no solution.   **15.**   The system of equations is inconsistent and has no solution.   **17.**   The solution is $\left(-\frac{3}{4}, -\frac{3}{4}\right)$.
19.   The solution is (5, 7).   **21.**   The solution is (1, 7).   **23.**   The solution is $\left(\frac{17}{5}, -\frac{7}{5}\right)$.   **25.**   The solution is
$\left(-\frac{6}{11}, \frac{31}{11}\right)$.   **27.**   The solution is (2, 3).   **29.**   The solution is (0, 0).   **31.**   The system of equations is
dependent. The solutions are the ordered pairs that satisfy the equation $3x + y = 4$.   **33.**   The solution is
$\left(\frac{20}{17}, -\frac{15}{17}\right)$.   **35.**   The solution is (5, 2).   **37.**   The solution is (−17, −8).   **39.**   The solution is $\left(-\frac{5}{7}, \frac{13}{7}\right)$.
41.   The solution is (3, −2).

SECTION 9.3   *pages 361–362*

1.   The solution is (5, −1).   **3.**   The solution is (1, 3).   **5.**   The solution is (1, 1).   **7.**   The solution is
(3, −2).   **9.**   The system is dependent. The solutions are the ordered pairs that satisfy the equation $2x - y = 1$.
11.   The solution is (3, 1).   **13.**   The system is dependent. The solutions are the ordered pairs that satisfy the
equation $2x - 3y = 1$.   **15.**   The solution is $\left(-\frac{13}{17}, -\frac{24}{17}\right)$.   **17.**   The solution is (2, 0).   **19.**   The solution is
(0, 0).   **21.**   The solution is (5, −2).   **23.**   The solution is $\left(\frac{32}{19}, -\frac{9}{19}\right)$.   **25.**   The solution is (3, 4).   **27.**   The
solution is (1, −1).   **29.**   The system is dependent. The solutions are the ordered pairs that satisfy the equation
$5x + 15y = 20$.   **31.**   The solution is (3, 1).   **33.**   The solution is (−1, 2).   **35.**   The solution is (1, 1).
37.   The solution is $\left(\frac{1}{2}, -\frac{1}{2}\right)$.   **39.**   The solution is $\left(\frac{2}{3}, \frac{1}{9}\right)$.   **41.**   The solution is $\left(\frac{7}{25}, -\frac{1}{25}\right)$.

SECTION 9.4   *pages 367–368*

1.   The rate of the plane is 400 mph. The rate of the wind is 50 mph.   **3.**   The rate of the boat is 7 mph. The
rate of the current is 3 mph.   **5.**   The rate of the plane is 125 mph. The rate of the wind is 25 mph.   **7.**   The
rate of the plane is 105 mph. The rate of the wind is 15 mph.   **9.**   The rate of the football pass is 55 ft/s. The
rate of the wind is 15 ft/s.   **11.**   The word processing program costs $245. The spreadsheet program costs $325.
13.   The dividend per share of oil company stock is $0.25. The dividend per share of the movie company stock is
$0.45.   **15.**   There would have been 3 touchdowns and 4 field goals.   **17.**   There are 15 dimes and 25
quarters.   **19.**   The present age of the coin is 75 and the age of the stamp is 50.

CHAPTER REVIEW   *pages 371–372*

1.   Yes, (−1, −3) is a solution of the system.   **2.**   No, (−2, 0) is not a solution of the system.

3. 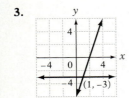   **4.** 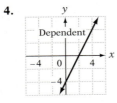   **5.**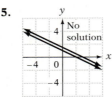

6.   The solution is (−1, 1).   **7.**   The solution is (1, 6).   **8.**   The solution is (1, 6).   **9.**   The solution is
(−1, −3).   **10.**   The system of equations is inconsistent and has no solution.   **11.**   The system of equations is
dependent. The solutions are the ordered pairs that satisfy the equation $4x + 3y = 12$.   **12.**   The solution is
(−3, 1).   **13.**   The solution is (1, −5).   **14.**   The solution is $\left(-\frac{5}{6}, \frac{1}{2}\right)$.   **15.**   The system of equations is

inconsistent and has no solution.   **16.**   The system of equations is dependent. The solutions are the ordered pairs that satisfy the equation $3x + y = -2$.   **17.**   The solution is $(-1, -2)$.   **18.**   The solution is $\left(\frac{2}{3}, -\frac{1}{6}\right)$.
19.   The rate of the canoeist in still water is 8 mph. The rate of the current is 2 mph.   **20.**   The rate of the flight crew in calm air is 125 km/h. The rate of the wind is 15 km/h.   **21.**   The rate of the plane in calm air is 105 mph. The rate of the wind is 15 mph.   **22.**   The rate of the sculling team in calm water is 9 mph. The rate of the current is 3 mph.   **23.**   There are 130 advertisements requiring 25¢ postage and 60 advertisements requiring 45¢ postage.   **24.**   There are 350 bushels of lentils and 200 bushels of corn in the mixture.
25.   There were 1300 shares at $6 per share and 200 shares at $25 per share purchased.

CHAPTER TEST   *pages 373–374*

1.   Yes, $(-2, 3)$ is a solution of the system.   (9.1A)   **2.**   Yes, $(1, -3)$ is a solution of the system.   (9.1A)

3.   (9.1A)   **4.**   The solution is $(3, 1)$.   (9.2A)   **5.**   The solution is $(1, -1)$.   (9.2A)

6.   The solution is $(2, -1)$.   (9.2A)   **7.**   The solution is $\left(\frac{22}{7}, -\frac{5}{7}\right)$.   (9.2A)   **8.**   The system is inconsistent and has no solution.   (9.2A)   **9.**   The solution is $(2, 1)$.   (9.3A)   **10.**   The solution is $\left(\frac{1}{2}, -1\right)$.   (9.3A)   **11.**   The system of equations is dependent. The solutions are the ordered pairs that satisfy the equation $x + 2y = 8$.   (9.3A)
12.   The solution is $(2, -1)$.   (9.3A)   **13.**   The solution is $(1, -2)$.   (9.3A)   **14.**   The rate of the plane in calm air is 100 mph. The rate of the wind is 20 mph.   (9.4A)   **15.**   The price of a reserved-seat ticket is $10. The price of a general-admission ticket is $6.   (9.4B)

CUMULATIVE REVIEW   *pages 375–376*

1. $\frac{3}{2}$ (2.1A)   **2.** $x = -\frac{3}{2}$ (3.1C)   **3.** $x = -\frac{7}{2}$ (3.3B)   **4.** $-6a^3 + 13a^2 - 9a + 2$ (5.3B)

5. $-2x^5y^2$ (5.4A)   **6.** $2b - 1 + \frac{1}{2b - 3}$ (5.4C)   **7.** $-\frac{4y}{x^3}$ (5.5A)   **8.** $4y^2(xy - 4)(xy + 4)$ (6.4B)

9.   The solutions are 4 and -1.   (6.5A)   **10.** $x - 2$ (7.1C)   **11.** $\frac{x^2 + 2}{(x - 1)(x + 2)}$ (7.3B)   **12.** $\frac{x - 3}{x + 1}$ (7.4A)

13. $x = -\frac{1}{5}$ (7.5A)   **14.** $r = \frac{A - P}{Pt}$ (7.7A)   **15.**   The x-intercept is $(6, 0)$, and the y-intercept is $(0, -4)$.

(8.3A)   **16.**   The slope is $-\frac{7}{5}$.   (8.3B)   **17.**   The equation of the line is $y = -\frac{3}{2}x$.   (8.4A)

18.   Yes, $(2, 0)$ is a solution of the system of equations.   (9.1A)   **19.**   The solution is $(-6, 1)$.   (9.2A)
20.   The solution is $(4, -3)$.   (9.3A)   **21.**   The amount invested at 9.6% is $3750. The amount invested at 7.2% is $5000.   (4.4A)   **22.**   The rate of the freight train is 48 mph. The rate of the passenger train is 56 mph.
(4.6A)   **23.**   A side of the original square measures 8 in.   (6.5B)   **24.**   The rate of the wind is 30 mph.
(7.8B)

25.   (8.2B)   **26.**   (9.1A)

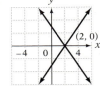
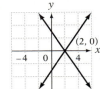

27.   The rate of the boat in calm water is 14 mph.   (9.4A)   **28.**   There are 40 dimes in the first bank.   (9.4B)

ANSWERS to Chapter 10 Odd-numbered Exercises

SECTION 10.1   *pages 383–384*

1. $A = \{16, 17, 18, 19, 20, 21\}$ **3.** $A = \{9, 11, 13, 15, 17\}$ **5.** $A = \{b, c\}$ **7.** $A = \{1, 4, 9, 16, 25, 36, 49\}$
9. $A \cup B = \{3, 4, 5, 6\}$ **11.** $A \cup B = \{-10, -9, -8, 8, 9, 10\}$ **13.** $A \cup B = \{a, b, c, d, e, f\}$
15. $A \cup B = \{1, 3, 7, 9, 11, 13\}$ **17.** $A \cap B = \{4, 5\}$ **19.** $A \cap B = \varnothing$ **21.** $A \cap B = \{c, d, e\}$
23. $A \cap B = \{7, 11\}$ **25.** $\{x \mid x > -5, x \text{ is a negative integer}\}$ **27.** $\{x \mid x > 30, x \text{ is an integer}\}$
29. $\{x \mid x > 5, x \text{ is an even integer}\}$ **31.** $\{x \mid x > 8, x \in \text{real numbers}\}$ **33.** $\{x \mid x > -5, x \in \text{real numbers}\}$

35.   **37.**   **39.**

41.   **43.**

SECTION 10.2   *pages 389–392*

1. $x < 2$ **3.** $x > 3$ **5.** $n \geq 3$

7. $x \leq -4$ **9.** $x \geq -1$ **11.** $y \geq -9$ **13.** $x < 12$

15. $x \geq 5$ **17.** $x < -11$ **19.** $x \leq 10$ **21.** $x \geq -6$ **23.** $x > 2$ **25.** $d < -\frac{1}{6}$ **27.** $x \geq -\frac{31}{24}$

29. $x < \frac{5}{8}$ **31.** $x < \frac{5}{4}$ **33.** $x > \frac{5}{24}$ **35.** $x < -3.8$ **37.** $x \leq -1.2$ **39.** $x < 5.6$ **41.** $x > -1.48$

43. $x < 4$ **45.** $y \geq 3$ **47.** $x \leq 1$

49. $x < -1$ **51.** $b < -4$ **53.** $y \leq -4$ **55.** $x > \frac{2}{7}$

57. $x \leq -\frac{5}{2}$ **59.** $x < 16$ **61.** $x \geq 16$ **63.** $x \geq -14$ **65.** $x \leq 21$ **67.** $x < \frac{7}{6}$ **69.** $x \leq -\frac{12}{7}$
71. $x > \frac{2}{3}$ **73.** $x \leq \frac{2}{3}$ **75.** $x \leq 2.3$ **77.** $x < -3.2$ **79.** $x \leq 5$ **81.** $x < -5.4$ **83.** $y \geq -3.15$
85. $x > -8.22$ **87.** $x \leq 7$ **89.** The possible diameters of a major league baseball will be in the range
2.87 in. $\leq d \leq$ 2.95 in.   **91.** A person, with an annual income tax liability of $3500, must pay $\geq$ $3150.
93. A patient, with a cholesterol level of 275 units, should reduce his/her cholesterol level by $\geq$ 55 units.
95. The student must receive $\geq$ 78 on the fifth test to earn a B grade.   **97.** A dollar amount > $3125 will
make the commission offer more attractive.

SECTION 10.3   *pages 395–396*

1. $x < 4$ **3.** $x < -4$ **5.** $x \geq 1$ **7.** $x \leq 5$ **9.** $x < 0$ **11.** $x < 20$ **13.** $x > 500$ **15.** $x > 2$
17. $x \leq -5$ **19.** $y \leq \frac{5}{2}$ **21.** $x < \frac{25}{11}$ **23.** $x > 11$ **25.** $n \leq \frac{11}{18}$ **27.** $x \geq 6$ **29.** The agent expects
to sell at most $20,000.   **31.** To exceed the $10 fee, a person must use the service more than 60 min.
33. The maximum amount that can be added is 8 oz.   **35.** The distance to the ski area is greater than 76 mi.
37. The crew can prepare between 8 and 12 aircraft.

SECTION 10.4   *pages 399–400*

1. 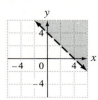   **3.** 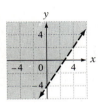   **5.** 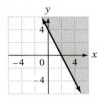   **7.**

9.   **11.** 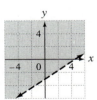   **13.**   **15.**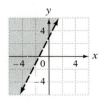

CHAPTER REVIEW   *pages 403–404*

1. $A = \{1, 3, 5, 7\}$ **2.** $A \cup B = \{2, 4, 6, 8, 10\}$ **3.** $A \cap B = \varnothing$ **4.** $A \cap B = \{1, 5, 9\}$
5. $\{x | x > -8, x \text{ is an odd integer}\}$ **6.** $\{x | x > 3, x \in \text{real numbers}\}$

7. 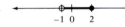   **8.**    **9.**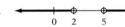

10. $x > 2$ 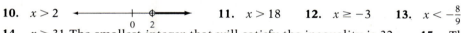   **11.** $x > 18$ **12.** $x \geq -3$ **13.** $x < -\frac{8}{9}$
14. $x > 31$ The smallest integer that will satisfy the inequality is 32.   **15.**   The lowest score the student can
receive is 72.   **16.** $x \geq -4$ **17.** $x \geq 4$ **18.** $x \geq 4$ **19.** $x > -18$ **20.** $x < \frac{1}{2}$
21.   The minimum length is 24 ft.   **22.**   Five or more residents make Florist B the more economical florist.

23. 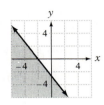   **24.** 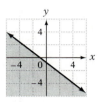   **25.**

CHAPTER TEST   *pages 405–406*

1. $A = \{4, 6, 8\}$ (10.1A)   **2.** $A \cap B = \{12\}$ (10.1A)   **3.** $\{x | x < 50, x \text{ is a positive integer}\}$ (10.1B)
4. $\{x | x > -23, x \in \text{real numbers}\}$ (10.1B)   **5.**  (10.1C)

6. 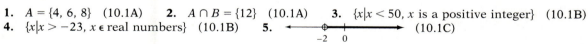 (10.1C)   **7.** $x < -3$ 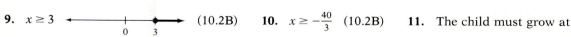 (10.2A)   **8.** $x > \frac{1}{8}$ (10.2A)

9. $x \geq 3$ (10.2B)   **10.** $x \geq -\frac{40}{3}$ (10.2B)   **11.**   The child must grow at

least 5 in. (10.2C)   **12.**   The allowable diameters are between 0.0389 in. and 0.0395 in. (10.2C)
13. $x < -1$ (10.3A)   **14.** $x \geq -4$ (10.3A)   **15.** $x \leq -3$ (10.3A)   **16.** $x < -\frac{22}{7}$ (10.3A)   **17.** The
maximum width is 11 ft. (10.3B)   **18.**   The broker processed less than \$75,000. (10.3B)

19.   (10.4A)   **20.**   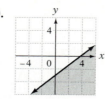   (10.4A)

CUMULATIVE REVIEW   *pages 407–408*

1. $40a - 28$ (2.2D)   **2.** $x = \frac{1}{8}$ (3.2A)   **3.** $x = 4$ (3.3B)   **4.** $-12a^7b^4$ (5.2B)   **5.** $-\frac{1}{b^4}$ (5.4A)

6. $4x - 2 - \frac{4}{4x - 1}$ (5.4C)   **7.** $(4x - 1)(x - 5)$ (6.3A)   **8.** $3a^2(3x - 1)(3x + 1)$ (6.4D)   **9.** $\frac{1}{x + 2}$ (7.1C)

10. $\frac{18a}{(2a - 3)(a + 3)}$ (7.3B)   **11.** $y = -\frac{5}{9}$ (7.5A)   **12.** $C = S + Rt$ (7.7A)   **13.**   The slope is $-\frac{7}{3}$.   (8.3B)

14.   The equation of the line is $y = -\frac{3}{2}x - \frac{3}{2}$.   (8.4B)   **15.**   The solution is (4, 1).   (9.2A)   **16.**   The solution is (1, −4).   (9.3A)   **17.** $A \cup B = \{-10, -2, 0, 1, 2\}$ (10.1A)   **18.** $\{x | x < 48, x \in$ real numbers$\}$ (10.1B)

19.   (10.1C)   **20.** $x > -2$ (10.2B)   **21.** $x < -15$ (10.2B)

22. $x > 2$ (10.3A)   **23.**   All integers less than or equal to −26 will satisfy the inequality.   (10.2C)
24.   The maximum number of miles is 359.   (10.3B)   **25.**   There are 5000 fish in the lake.   (7.6B)
26.   There are seven 13¢ stamps in the drawer.   (4.8B)

27.   (8.2B)   **28.**   (10.4A)

ANSWERS to Chapter 11 Odd-numbered Exercises

SECTION 11.1   *pages 415–416*

1.   4   **3.**   7   **5.** $4\sqrt{2}$ **7.** $2\sqrt{2}$ **9.** $18\sqrt{2}$ **11.** $10\sqrt{10}$ **13.** $\sqrt{15}$ **15.** $\sqrt{29}$ **17.** $-54\sqrt{2}$
19. $3\sqrt{5}$ **21.**   0   **23.** $48\sqrt{2}$ **25.** $\sqrt{105}$ **27.**   30   **29.** $30\sqrt{5}$ **31.** $5\sqrt{10}$ **33.** $4\sqrt{6}$
35.   18   **37.**   15.492   **39.**   16.968   **41.**   16   **43.**   16.585   **45.**   15.652   **47.**   18.76   **49.** x^3
51. $y^7\sqrt{y}$ **53.** a^{10} **55.** x^2y^2 **57.** $2x^2$ **59.** $2x\sqrt{6}$ **61.** $xy^3\sqrt{xy}$ **63.** $ab^5\sqrt{ab}$ **65.** $2x^2\sqrt{15x}$
67. $7a^2b^4$ **69.** $3x^2y^3\sqrt{2xy}$ **71.** $2x^5y^3\sqrt{10xy}$ **73.** $4a^4b^5\sqrt{5a}$ **75.** $8ab\sqrt{b}$ **77.** x^3y
79. $8a^2b^3\sqrt{5b}$ **81.** $6x^2y^3\sqrt{3y}$ **83.** $4x^3y\sqrt{2y}$ **85.** $5a + 20$ **87.** $2x^2 + 8x + 8$ **89.** $x + 2$
91. $y + 1$ **93.** $x + 4$

SECTION 11.2   *pages 419–420*

1. $3\sqrt{2}$ **3.** $-\sqrt{7}$ **5.** $-11\sqrt{11}$ **7.** $10\sqrt{x}$ **9.** $-2\sqrt{y}$ **11.** $-11\sqrt{3b}$ **13.** $2x\sqrt{2}$
15. $-3a\sqrt{3a}$ **17.** $-5\sqrt{xy}$ **19.** $8\sqrt{5}$ **21.** $8\sqrt{2}$ **23.** $15\sqrt{2} - 10\sqrt{3}$ **25.** $\sqrt{x}$ **27.** $-12x\sqrt{3}$
29. $2xy\sqrt{x} - 3xy\sqrt{y}$ **31.** $-9x\sqrt{3x}$ **33.** $-13y^2\sqrt{2y}$ **35.** $4a^2b^2\sqrt{ab}$ **37.** $7\sqrt{2}$ **39.** $6\sqrt{x}$
41. $-3\sqrt{y}$ **43.** $-45\sqrt{2}$ **45.** $13\sqrt{3} - 12\sqrt{5}$ **47.** $32\sqrt{3} - 3\sqrt{11}$ **49.** $6\sqrt{x}$ **51.** $-34\sqrt{3x}$
53. $10a\sqrt{3b} + 10a\sqrt{5b}$ **55.** $-2xy\sqrt{3}$ **57.** $(-7b + 4a)\sqrt{ab}$ **59.** $3ab\sqrt{2a} - ab + 4ab\sqrt{3b}$

SECTION 11.3   *pages 425–426*

1.   5   **3.**   6   **5.** x **7.** x^3y^2 **9.** $3ab^6\sqrt{2a}$ **11.** $12a^4b\sqrt{b}$ **13.** $2 - \sqrt{6}$ **15.** $x - \sqrt{xy}$
17. $5\sqrt{2} - \sqrt{5x}$ **19.** $4 - 2\sqrt{10}$ **21.** $x - 6\sqrt{x} + 9$ **23.** $3a - 3\sqrt{ab}$ **25.** $10abc$

27. $15x - 22y\sqrt{x} + 8y^2$ **29.** $x - y$ **31.** $10x + 13\sqrt{xy} + 4y$ **33.** 4   **35.** 7   **37.** 3   **39.** $x\sqrt{5}$
41. $\frac{a^2}{7}$ **43.** $\frac{\sqrt{3}}{3}$ **45.** $\frac{3\sqrt{x}}{x}$ **47.** $\frac{2\sqrt{y}}{xy}$ **49.** $\frac{2\sqrt{3x}}{3y}$ **51.** $-\frac{\sqrt{2}+3}{7}$ **53.** $\frac{15-3\sqrt{5}}{20}$ **55.** $3 + \sqrt{6}$
57. $-\frac{20+7\sqrt{3}}{23}$ **59.** $\frac{6+5\sqrt{x}-x}{4-x}$ **61.** $\frac{x\sqrt{y}+y\sqrt{x}}{x-y}$

SECTION 11.4   *pages 431–432*

1. The solution is 25.   **3.** The solution is 144.   **5.** The solution is 5.   **7.** The solution is 16.   **9.** The solution is 8.   **11.** The equation has no solution.   **13.** The solution is 6.   **15.** The solution is 24.
17. The solution is -1.   **19.** The solution is $-\frac{2}{5}$.   **21.** The solution is $\frac{4}{3}$.   **23.** The solution is 15.
25. The solution is 1.   **27.** The solution is 2.   **29.** The equation has no solution.   **31.** The pitcher's mound is less than half-way between home plate and second base.   **33.** The height of the periscope must be 12.76 ft above the water.   **35.** The height of the bridge is 36 ft.   **37.** The height is 8.7 in.   **39.** The length of the pendulum is 3.25 ft.

CHAPTER REVIEW   *pages 435–436*

1. 12   **2.** $20\sqrt{3}$ **3.** $-6\sqrt{30}$ **4.** $20\sqrt{10}$ **5.** $9a^2\sqrt{2ab}$ **6.** $2y^4\sqrt{6}$ **7.** $36x^8y^5\sqrt{3xy}$
8. $26\sqrt{3x}$ **9.** $7x^2y\sqrt{15xy}$ **10.** $18a\sqrt{5b} + 5a\sqrt{b}$ **11.** $-6x^3y^2\sqrt{2y}$ **12.** 3   **13.** $3a\sqrt{2} + 2a\sqrt{3}$
14. $8y + 10\sqrt{5y} - 15$ **15.** $7x^2y^4$ **16.** $\frac{16\sqrt{a}}{a}$ **17.** $\frac{8\sqrt{x}+24}{x-9}$ **18.** $-x\sqrt{3} - x\sqrt{5}$ **19.** The solution is 20.
20. The equation has no solution.   **21.** The solution is 3.   **22.** The solution is $\frac{3}{5}$.   **23.** The explorer would weigh 144 lb on the surface of the earth.   **24.** The depth of the water is 100 ft.   **25.** The radius of the sharpest corner is 25 ft.

CHAPTER TEST   *pages 437–438*

1. $3\sqrt{5}$ (11.1A)   **2.** $5\sqrt{3}$ (11.1A)   **3.** $11x^4y$ (11.1B)   **4.** $6x^3y\sqrt{2x}$ (11.1B)
5. $4a^2b^5\sqrt{2ab}$ (11.1B)   **6.** $-5\sqrt{2}$ (11.2A)   **7.** $21\sqrt{2y} - 12\sqrt{2x}$ (11.2A)   **8.** $-2xy\sqrt{3xy} - 3xy\sqrt{xy}$ (11.2A)   **9.** $4x^2y^2\sqrt{5y}$ (11.3A)   **10.** $6x^2y\sqrt{y}$ (11.3A)   **11.** $a - \sqrt{ab}$ (11.3A)   **12.** $y + 2\sqrt{y} - 15$
(11.3A)   **13.** 9 (11.3B)   **14.** $7ab\sqrt{a}$ (11.3B)   **15.** $\sqrt{3} + 1$ (11.3B)   **16.** $\frac{17-8\sqrt{5}}{31}$ (11.3B)
17. The solution is 25. (11.4A)   **18.** The equation has no solution. (11.4A)   **19.** The larger integer is 51. (11.4B)   **20.** The length of the pendulum is 7.30 ft. (11.4B)

CUMULATIVE REVIEW   *pages 439–440*

1. $-\frac{1}{12}$ (1.5B)   **2.** $2x + 18$ (2.2D)   **3.** $x = \frac{1}{13}$ (3.3B)   **4.** $6x^5y^5$ (5.2A)   **5.** $-2b^2 + 1 - \frac{1}{3b^2}$ (5.4B)
6. $3x^2y^2(4x - 3y)$ (6.1A)   **7.** $2a(a - 5)(a - 3)$ (6.3B)   **8.** $\frac{1}{4(x+1)}$ (7.1B)   **9.** $\frac{x+3}{x-3}$ (7.3B)
10. $x = \frac{5}{3}$ (7.5A)   **11.** The equation of the line is $y = \frac{1}{2}x - 2$. (8.4A)   **12.** The solution is (1, 1). (9.2A)
13. The solution is $(3, -2)$. (9.3A)   **14.** $x \le -\frac{9}{2}$ (10.3A)   **15.** $6\sqrt{3}$ (11.1A)   **16.** $-4\sqrt{2}$ (11.2A)
17. $4ab\sqrt{2ab} - 5ab\sqrt{ab}$ (11.2A)   **18.** $14a^5b^2\sqrt{2a}$ (11.3A)   **19.** $3\sqrt{2} - x\sqrt{3}$ (11.3A)   **20.** 8 (11.3B)
21. $-6 - 3\sqrt{5}$ (11.3B)   **22.** The solution is 6. (11.4A)   **23.** The book costs $24.50. (4.3A)   **24.** 56 oz of pure water must be added. (4.5B)   **25.** The numbers are 13 and 8. (6.5B)   **26.** Working alone, it would take the small pipe 48 h to fill the tank. (7.8A)

27. The solution is $(2, -1)$.   (9.1B)   **28.**   (10.4A)

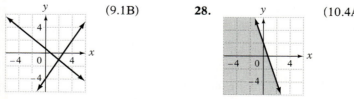

29. The smaller integer is 40. (11.4B)   **30.** The building is 400 ft high. (11.4B)

ANSWERS to Chapter 12 Odd-numbered Exercises

SECTION 12.1   *pages 447–448*

1. The solutions are −5 and 3.   **3.** The solutions are 1 and 3.   **5.** The solutions are −1 and −2.

7. The solution is 3.   **9.** The solutions are 0 and $-\frac{2}{3}$.   **11.** The solutions are −2 and 5.   **13.** The solutions are $\frac{2}{3}$ and 1.   **15.** The solutions are −3 and $\frac{1}{3}$.   **17.** The solution is $\frac{2}{3}$.   **19.** The solutions are $-\frac{1}{2}$ and $\frac{3}{2}$.   **21.** The solution is $\frac{1}{2}$.   **23.** The solutions are −3 and 3.   **25.** The solutions are $-\frac{1}{2}$ and $\frac{1}{2}$.

27. The solutions are −3 and 5.   **29.** The solutions are 1 and 5.   **31.** The solutions are −1 and $\frac{13}{2}$.

33. The solutions are 7 and −7.   **35.** The solutions are 8 and −8.   **37.** The solutions are $\frac{8}{3}$ and $-\frac{8}{3}$.

39. The solutions are $\frac{5}{2}$ and $-\frac{5}{2}$.   **41.** The solutions are $\frac{8}{5}$ and $-\frac{8}{5}$.   **43.** The equation has no real number solution.   **45.** The solutions are $4\sqrt{3}$ and $-4\sqrt{3}$.   **47.** The solutions are 5 and −9.   **49.** The solutions are 8 and −2.   **51.** The solutions are $\frac{3}{2}$ and $-\frac{15}{2}$.   **53.** The solutions are $\frac{26}{9}$ and $\frac{10}{9}$.   **55.** The solutions are $-5 + 5\sqrt{2}$ and $-5 - 5\sqrt{2}$.   **57.** The equation has no real number solution.   **59.** The solutions are $-\frac{3}{4} + 2\sqrt{3}$ and $-\frac{3}{4} - 2\sqrt{3}$.

SECTION 12.2   *pages 453–454*

1. The solutions are 1 and −3.   **3.** The solutions are 8 and −2.   **5.** The solution is 2.   **7.** The quadratic equation has no real number solution.   **9.** The solutions are −1 and −4.   **11.** The solutions are −8 and 1.   **13.** The solutions are $-2 + \sqrt{3}$ and $-2 - \sqrt{3}$.   **15.** The solutions are $-3 + \sqrt{14}$ and $-3 - \sqrt{14}$.   **17.** The solutions are $1 + \sqrt{2}$ and $1 - \sqrt{2}$.   **19.** The solutions are $\frac{-3 + \sqrt{13}}{2}$ and $\frac{-3 - \sqrt{13}}{2}$.   **21.** The solutions are 2 and 1.   **23.** The solutions are $\frac{-1 + \sqrt{13}}{2}$ and $\frac{-1 - \sqrt{13}}{2}$.   **25.** The solutions are $-5 + 4\sqrt{2}$ and $-5 - 4\sqrt{2}$.   **27.** The solutions are $\frac{3 + \sqrt{29}}{2}$ and $\frac{3 - \sqrt{29}}{2}$.   **29.** The solutions are $\frac{1 + \sqrt{17}}{2}$ and $\frac{1 - \sqrt{17}}{2}$.   **31.** The quadratic equation has no real number solution.   **33.** The solutions are 1 and $\frac{1}{2}$.   **35.** The solutions are −3 and $\frac{1}{2}$.

37. The solutions are 2 and $\frac{3}{2}$.   **39.** The solutions are 1 and $-\frac{1}{2}$.   **41.** The solutions are −2 and $\frac{1}{3}$.

43. The solutions are −2 and $-\frac{2}{3}$.   **45.** The solutions are $\frac{1}{2}$ and $-\frac{3}{2}$.   **47.** The solutions are $\frac{1}{3}$ and $-\frac{3}{2}$.

49. The solutions are $-\frac{1}{2}$ and $\frac{4}{3}$.   **51.** The solutions are $\frac{1 + \sqrt{2}}{2}$ and $\frac{1 - \sqrt{2}}{2}$.   **53.** The solutions are $\frac{2 + \sqrt{5}}{2}$ and $\frac{2 - \sqrt{5}}{2}$.   **55.** The solutions are 2 and 4.   **57.** The solutions are $1 + \sqrt{6}$ and $1 - \sqrt{6}$.   **59.** The solutions are approximately −4.193 and 1.193.   **61.** The solutions are approximately 2.766 and −1.266.   **63.** The solutions are approximately −1.652 and 0.152.

SECTION 12.3   *pages 457–458*

1. The solutions are 5 and −1.   **3.** The solutions are −3 and 5.   **5.** The solutions are −7 and 1.   **7.** The solutions are 2 and −3.   **9.** The solutions are 3 and −1.   **11.** The solutions are −5 and 1.

13. The solutions are $-\frac{1}{2}$ and 1.   **15.** The quadratic equation has no real number solution.   **17.** The solutions are 0 and 1.   **19.** The solutions are $\frac{3}{2}$ and $-\frac{3}{2}$.   **21.** The solutions are $\frac{3}{2}$ and $-\frac{5}{2}$.   **23.** The solutions are 3 and $-\frac{2}{3}$.   **25.** The solutions are −3 and $\frac{4}{5}$.   **27.** The solutions are $-\frac{1}{2}$ and $\frac{2}{3}$.   **29.** The quadratic equation has no real number solution.   **31.** The solutions are $1 + \sqrt{6}$ and $1 - \sqrt{6}$.   **33.** The solutions are $-3 + \sqrt{10}$ and $-3 - \sqrt{10}$.   **35.** The solutions are $2 + \sqrt{13}$ and $2 - \sqrt{13}$.   **37.** The solutions are $\frac{1+\sqrt{2}}{2}$ and $\frac{1-\sqrt{2}}{2}$.   **39.** The solutions are $-3+2\sqrt{2}$ and $-3-2\sqrt{2}$.   **41.** The solutions are $\frac{3+2\sqrt{6}}{2}$ and $\frac{3-2\sqrt{6}}{2}$.   **43.** The solutions are $\frac{-1+\sqrt{2}}{3}$ and $\frac{-1-\sqrt{2}}{3}$.   **45.** The solution is $-\frac{1}{2}$.   **47.** The quadratic equation has no real number solution.   **49.** The solutions are $\frac{-4+\sqrt{5}}{2}$ and $\frac{-4-\sqrt{5}}{2}$.   **51.** The solutions are $\frac{1+2\sqrt{3}}{2}$ and $\frac{1-2\sqrt{3}}{2}$.   **53.** The solutions are $\frac{-5+\sqrt{2}}{3}$ and $\frac{-5-\sqrt{2}}{3}$.   **55.** The solutions are approximately 5.690 and -3.690.

57. The solutions are approximately 7.690 and -1.690.   **59.** The solutions are approximately 4.590 and -1.090.   **61.** The solutions are approximately -2.118 and 0.118.   **63.** The solutions are approximately 1.105 and -0.905.

SECTION 12.4   *pages 461–462*

1. 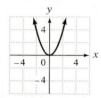   **3.** 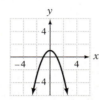   **5.** 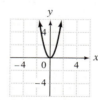   **7.**

9. 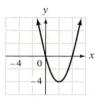   **11.** 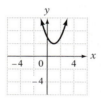   **13.**   **15.**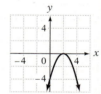

SECTION 12.5   *pages 465–466*

1. The length is 6 ft, and the width is 4 ft.   **3.** The length of the pool is 100 ft, and the width is 50 ft.   **5.** The integers are 7 and 9.   **7.** The two integers are 5 and 7.   **9.** The integer is either 0 or 2.   **11.** The silver coin is 10 years old, and the gold coin is 5 years old.   **13.** The first coin is 8 years old, and the second coin is 6 years old.   **15.** The first computer alone would take 35 min. The second computer working alone would take 14 min.   **17.** The first engine alone would take 12 h. The second engine working alone would take 6 h.   **19.** The rate of the plane is 100 mph.   **21.** The rate of the ship during the last 75 mi was 25 mph.

CHAPTER REVIEW   *pages 469–470*

1. The solutions are $\frac{4}{3}$ and $-\frac{7}{2}$.   **2.** The solutions are 2 and $\frac{5}{12}$.   **3.** The solutions are -7 and -10.   **4.** The solutions are $-\frac{1}{3}$ and $-\frac{1}{2}$.   **5.** The solutions are $-\frac{5}{7}$ and $\frac{5}{7}$.   **6.** The equation has no real number solution.   **7.** The solutions are -9 and 1.   **8.** The solutions are $-2-2\sqrt{6}$ and $-2+2\sqrt{6}$.   **9.** The solutions are -1 and 2.   **10.** The solutions are -6 and 4.   **11.** The solutions are -4 and $\frac{3}{2}$.   **12.** The solutions are $2-\sqrt{3}$ and $2+\sqrt{3}$.   **13.** The equation has no real number solution.   **14.** The solutions are $\frac{-4-\sqrt{23}}{2}$ and $\frac{-4+\sqrt{23}}{2}$.   **15.** The solutions are -6 and 1.   **16.** The solutions are 1 and $\frac{3}{2}$.   **17.** The equation has no real number solution.   **18.** The solutions are $\frac{3-\sqrt{29}}{2}$ and $\frac{3+\sqrt{29}}{2}$.   **19.** The solutions are -2 and $-\frac{1}{2}$.

20. 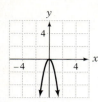 **21.** 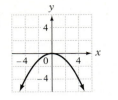 **22.** 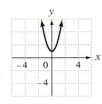 **23.**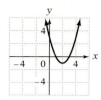

24. 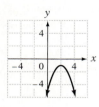   **25.**   The rate of the hawk in calm air is 75 mph.

CHAPTER TEST   *pages 471–472*

1.   The solutions are 6 and -1.   (12.1A)   **2.**   The solutions are -4 and $\frac{5}{3}$.   (12.1A)   **3.**   The solutions are 0 and 10.   (12.1B)   **4.**   The solutions are $-4 + 2\sqrt{5}$ and $-4 - 2\sqrt{5}$.   (12.1B)   **5.**   The solutions are $-2 + 2\sqrt{5}$ and $-2 - 2\sqrt{5}$.   (12.2A)   **6.**   The solutions are $\frac{-3 + \sqrt{41}}{2}$ and $\frac{-3 - \sqrt{41}}{2}$.   (12.2A)   **7.**   The solutions are $\frac{3 + \sqrt{7}}{2}$ and $\frac{3 - \sqrt{7}}{2}$.   (12.2A)   **8.**   The solutions are $\frac{-4 + \sqrt{22}}{2}$ and $\frac{-4 - \sqrt{22}}{2}$.   (12.2A)   **9.**   The solutions are $-2 + \sqrt{2}$ and $-2 - \sqrt{2}$.   (12.3A)   **10.**   The solutions are $\frac{3 + \sqrt{33}}{2}$ and $\frac{3 - \sqrt{33}}{2}$.   (12.3A)   **11.**   The solutions are $-\frac{1}{2}$ and 3.   (12.3A)   **12.**   The solutions are $\frac{1 + \sqrt{13}}{6}$ and $\frac{1 - \sqrt{13}}{6}$.   (12.3A)

13. 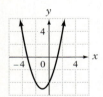   (12.4A)

14.   The length is 8 ft. The width is 5 ft.   (12.5A)   **15.**   The rate of the boat in calm water is 11 mph.   (12.5A)

CUMULATIVE REVIEW   *pages 473–474*

1. $-28x + 27$ (2.2D)   **2.** $x = \frac{3}{2}$ (3.1C)   **3.** $x = 3$ (3.3B)   **4.** $-12a^8b^4$ (5.2B)   **5.** $x + 2 - \frac{4}{x - 2}$ (5.4C)   **6.** $x(3x - 4)(x + 2)$ (6.3B)   **7.** $\frac{9x^2(x - 2)^2}{(2x - 3)^2}$ (7.1C)   **8.** $\frac{x + 2}{2(x + 1)}$ (7.3B)   **9.** $\frac{x - 4}{2x + 5}$ (7.4A)

10.   The x-intercept is $(3,0)$, and the y-intercept is $(0,-4)$.   (8.3A)   **11.**   The equation of the line is $y = -\frac{4}{3}x - 2$.   (8.4B)   **12.**   The solution is $(2,1)$.   (9.2A)   **13.**   The solution is $(2,-2)$.   (9.3A)

14. $x > \frac{1}{9}$ (10.3A)   **15.** $a - 2$ (11.3A)   **16.** $6ab\sqrt{a}$ (11.3B)   **17.** $\frac{-6 + 5\sqrt{3}}{13}$ (11.3B)

18.   The solution is 5.   (11.4A)   **19.**   The solutions are $\frac{5}{2}$ and $\frac{1}{3}$.   (12.1A)   **20.**   The solutions are $5 + 3\sqrt{2}$ and $5 - 3\sqrt{2}$.   (12.1B)   **21.**   The solutions are $\frac{-7 + \sqrt{13}}{6}$ and $\frac{-7 - \sqrt{13}}{6}$.   (12.2A)   **22.**   The solutions are 2 and $-\frac{1}{2}$.   (12.3A)   **23.**   The selling price of the mixture is $2.25/lb.   (4.5A)   **24.**   250 additional shares are required.   (7.6B)   **25.**   The rate of the plane in still air is 200 mph. The rate of the wind is 40 mph.   (9.4A)

26. The student must receive a score of 77 or above.   (10.2C)   **27.** The integer is -5 or 5.   (12.5A)
28. The rate for the last 8 miles was 4 mph.   (12.5A)

29. 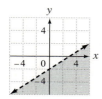   (10.4A)   **30.**   (12.4A)

FINAL EXAM   *pages 475–478*

1. -3 (1.1B)   **2.** -6 (1.2B)   **3.** -256 (1.5A)   **4.** -11 (1.5B)   **5.** $-\dfrac{15}{2}$ (2.1A)

6. $9x + 6y$ (2.2A)   **7.** $6z$ (2.2B)   **8.** $16x - 52$ (2.2D)   **9.** $x = -50$ (3.1C)   **10.** $x = -3$ (3.3B)
11. 12.5% (4.1B)   **12.** 15.2 (4.2A)   **13.** $-3x^2 - 3x + 8$ (5.1B)   **14.** $81x^4y^{12}$ (5.2B)

15. $6x^3 + 7x^2 - 7x - 6$ (5.3B)   **16.** $-\dfrac{x^4y}{2}$ (5.4A)   **17.** $\dfrac{3x}{y} - 4x^2 - \dfrac{5}{x}$ (5.4B)   **18.** $5x - 12 + \dfrac{23}{x + 2}$ (5.4C)

19. $\dfrac{4y^6}{x^6}$ (5.4A)   **20.** $(2x - 3)(x + 1)$ (6.3A)   **21.** $(x - 6)(x + 1)$ (6.2A)   **22.** $(3x + 2)(2x - 3)$ (6.3A)

23. $4x(2x - 1)(x - 3)$ (6.3B)   **24.** $(5x - 4)(5x + 4)$ (6.4A)   **25.** $2(a + 3)(4 - x)$ (6.1B)

26. $3y(5 - 2x)(5 + 2x)$ (6.4B)   **27.** The solutions are $\dfrac{1}{2}$ and 3. (6.5A)   **28.** $\dfrac{2(x + 1)}{x - 1}$ (7.1B)

29. $\dfrac{-3x^2 + x - 25}{(2x - 5)(x + 3)}$ (7.3B)   **30.** $\dfrac{x^2 - 2x}{x - 1}$ (7.4A)   **31.** The solution is 2. (7.5A)   **32.** The solution is b.

(7.7A)   **33.** The slope is $\dfrac{2}{3}$. (8.3B)   **34.** The equation of the line is $y = -\dfrac{2}{3}x - 2$. (8.4A)   **35.** The solution

is (6, 17). (9.2A)   **36.** The solution is (2, -1). (9.3A)   **37.** $x \le -3$ (10.2A)   **38.** $y \ge \dfrac{5}{2}$ (10.3A)

39. $7x^3$ (11.1B)   **40.** $38\sqrt{3a}$ (11.2A)   **41.** $\sqrt{15} + 2\sqrt{3}$ (11.3B)   **42.** The solution is 5. (11.4A)

43. The solutions are -1 and $\dfrac{4}{3}$. (12.1A)   **44.** The solutions are $\dfrac{1 + \sqrt{5}}{4}$ and $\dfrac{1 - \sqrt{5}}{4}$. (12.3A)

45. $2x + 3(x - 2); 5x - 6$ (2.3C)   **46.** The original value is \$3000. (4.2B)   **47.** The markup rate is
65% (4.3A)   **48.** An additional \$6000 must be invested. (4.4A)   **49.** The mixture costs \$3 per
pound. (4.5A)   **50.** The percent concentration of the acid in the mixture is 36%. (4.5B)   **51.** In the first
hour, the plane flew 215 km. (4.6A)   **52.** The measures of the angles are $50°$, $60°$, and $70°$. (4.7B)
53. There are 60 dimes in the bank. (4.8B)   **54.** The length is 10 m. The width is 5 m. (6.5B)
55. Sixteen oz of dye are required for 120 oz of base paint. (7.6B)   **56.** Working together, it would take 0.6
h to prepare the dinner. (7.8A)   **57.** The rate of the boat in calm water is 15 mph. The rate of the current is 5
mph. (9.4A)   **58.** The rate of the wind is 25 mph. (12.5A)

59. 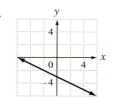   (8.3C)   **60.** 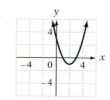   (12.4A)

Index